Mathematical Methods
in
Chemical Engineering

Mathematical Methods in Chemical Engineering

Volume 3

Process Modeling, Estimation, and Identification

JOHN H. SEINFELD

Department of Chemical Engineering
California Institute of Technology

LEON LAPIDUS

Department of Chemical Engineering
Princeton University

PRENTICE-HALL, INC.
Englewood Cliffs, New Jersey

Library of Congress Cataloging in Publication Data

Main entry under title:
Mathematical methods in chemical engineering.
(Prentice-Hall international series in the physical and chemical sciences)

Includes bibliographies.
CONTENTS: [1] Amundson, N. R. Matrices and their application.—v. 2. Aris R. and Amundson, N. R. First-order partial differential equations with applications.—v. 3. Seinfeld, J. H. and Lapidus, L. Process modeling, estimation, and identification.
1. Chemical engineering—Mathematics—Collected works. I. Amundson, Neal Russell. II. Title.
TP149.M33 660.2'01'51 66–11768
ISBN 0–13–561118–0

10 9 8 7 6 5 4 3 2 1

Printed in the United States of America

PRENTICE-HALL, INC.
PRENTICE-HALL INTERNATIONAL, INC., UNITED KINGDOM AND EIRE
PRENTICE-HALL OF CANADA, LTD., CANADA

To

Connie

and

Mary and Jay

Contents

Preface

Over the last several years chemical engineering, as a discipline, has witnessed a tremendous diversification in the types of systems being studied. One need only glance at the contributions to any recent issue of a chemical engineering journal for evidence of this diversity.

In spite of the variety of systems studied, there is, more often than not, a unifying theme of modeling of the system, that is, the description of the phenomenon in terms of a mathematical model which can simulate the observed output from the system. In the present volume we have attempted to present a self-contained, comprehensive treatment of the mathematical techniques of process modeling. Since this is primarily a book on mathematical methods, we attempt to stress the generality of the methods without tying them necessarily to specific applications. Nevertheless, applications to real systems provide the ultimate motivation for the methods, and therefore appear frequently as illustrations of the theory.

We have chosen to emphasize two major areas of mathematical methods fundamental to process modeling: (1) probability theory and stochastic processes, and (2) parameter estimation and system identification. These represent two subject areas of great importance that have not generally been part of a chemical engineer's training. We hope to fill this gap with the present volume. Each of these areas could clearly occupy a book of its own. Nevertheless, we have eschewed presentations of great detail in either area in favor of a rather broad treatment suitable for a one year course. The level of the material presented is, we feel, sufficient to prepare the student for reading the literature in these areas.

The contents of the book together with the motivation underlying what material is covered is summarized in Chapter 1. The material is intended for the introductory graduate level, or possibly the senior level. We assume that the reader has had the traditional undergraduate courses in transport phenomena and chemical reactor analysis. Although the volume is intended for a one year course, it is entirely suitable for a one semester, or two quarter, course, particularly at the senior level, with the omission of Sections 5.5, 6.8 and Chapters 9 and 10. Chapters 4 and 5 can serve as reference material for a general course in chemical engineering analysis, while Chapters 7-10 can supplement courses in process dynamics or chemical reactor analysis. The material in Chapters 9 and 10 is somewhat more advanced than that in the remainder of the book, but it is important fundamental material for those students with special interests in control theory. The material is based in part on courses taught by the authors at the California Institute of Technology and Princeton University.

We would like to express special appreciation to Dr. Robert Rossen who helped with the preparation of Chapter 9. Also, we are indebted to the many inspiring students at Caltech and Princeton, our experiences with whom were clearly responsible for our undertaking this task.

JOHN H. SEINFELD

LEON LAPIDUS

How often have I seen you sitting at your desk
lost
You always seem so tranquil, so at peace with
yourself
You are writing, working, creating and you are so
happy
I have frequently interrupted you
I was unaware
You never once refused me your time, only your
thoughts
You appear to be a different man when you are working
alone
Your mind is alive, industrious, diligent with
knowledge
Knowledge concerning subjects I can barely pronounce
I am a stranger
To see you so captivated with your work is to see your
individuality
How nice this is for
me
I rarely see you as being anything but my
father
The next time I see you sitting at your desk
lost
I will watch

MARY LAPIDUS

Introduction 1

The rational, intelligent analysis of observations has long been one of the key problems facing the scientist and engineer. There has developed over the years a body of knowledge and techniques that has almost universal applicability to problems of the analysis of observational data from physical systems. This book presents the elements of this body of knowledge in the particular framework of chemical engineering problems. The basic question underlying the book is, How do I extract as much information as possible from data from a given system, so that I might formulate a mathematical model of the system?

For sure, this is not a new question; it has always been an integral part of the process of understanding a real system. It usually involves the tentative selection of a mathematical model, the matching of the model to data, and subsequent reevaluation of the model, and possibly additional experiments, until the analyst is satisfied that he understands the system.

Process modeling has not generally been the subject of a separate course in chemical engineering curricula but rather has been treated as it arose in the traditional disciplines of transport phenomena, chemical kinetics and reactor design, and thermodynamics. It is our hope, with this textbook, to provide a means for the systematic instruction, either in the classroom or out, of the fundamentals of process modeling and analysis. We shall draw on a number of examples of chemical engineering systems to illustrate the techniques about which the book is centered.

The first part of the book (Chapters 2–5) develops the mathematical tools of process modeling. Certainly one could argue that all of applied mathematics is relevant to mathematical modeling. We grant this, however, the choice of material must be limited to that which is most directly applicable to the majority of problems encountered. Thus the first half of the book is concerned with *process modeling*. We do not dwell extensively on the physical considerations in selecting a model but rather on the mathematical ones. The physical considerations belong more appropriately in the traditional disciplines of chemical engineering.

Chapter 2 defines the problem of process modeling and introduces the common mathematical forms of models: differential equations, difference equations, and integral equations (and certain combinations thereof). We shall present several typical systems in Chapter 2 to illustrate the form of common models. In certain cases, explicit solutions are presented. The purpose of Chapter 2 is to define the types of models with which we shall most often deal. The material will be basically a review for those students who have had a course in engineering mathematics.

Because the Laplace transform is so central to the analysis of linear systems, we have chosen to devote Chapter 3 to its properties and use. While several other transform techniques exist for the solution of the classic boundary-value problems of mathematical physics, the Laplace transform has proved the most useful. Of primary interest is the concept of the system transfer function. The Laplace transform is generally used to analyze dynamic systems wherein time varies continuously. However, many important problems involve discrete changes with mathematical descriptions in the form of difference equations. The analog of the Laplace transform for discrete systems, the z-transform, is developed and illustrated. The purpose of Chapter 3 is to present to the student a brief but reasonably complete treatment of the Laplace transform, primarily so that the book will be self-contained.

Whereas the material in Chapters 2 and 3 will be generally familiar to most senior and graduate engineers, that in Chapters 4 and 5 may not be. Chapters 4 and 5, which complete the first half of the book, are devoted to the analysis of nondeterministic systems. So many of the systems of practical interest involve limited information and uncertainty that the theory of probability and stochastic processes is rapidly becoming a necessary component of the graduate education of an engineer. Chapter 4 presents an introduction to the concepts of random variables and probability distributions. We shall present several probability distributions that arise frequently in engineering applications. The characterization of probability distributions by moments is also stressed, as this aspect will be important in the second half of the book.

Chapter 5 is an introduction to the theory of stochastic processes. This is a subject which is often not included in traditional engineering curricula but which is of ever-increasing importance in the analysis of complex engineering systems. Of special importance is the concept of the Markov property. Stochastic processes deal, in general, with the evolution of random variables with time (either continuously or discretely), and the Markov property allows us to make an important simplification in describing the time behavior of a random (stochastic) process. Whereas Chapter 4 deals with the properties of random variables and their distributions, Chapter 5 considers the properties of random variables that are functions of time. Two examples, developed in some detail, which illustrate certain aspects of stochastic mathematical models, are the axial dispersion of a tracer in a packed bed and a continuous stirred tank reactor subject to random fluctuations in inlet conditions and rate of heating. Chapter 5 completes much of the mathematical background necessary for the formulation of a deterministic or stochastic mathematical model of a complex engineering system.

The next five chapters are devoted to *estimation* and *identification*. A fundamental consideration in the mathematical modeling of a real system is the gathering of observations from the system. Carefully designed experiments can spell the difference between success and failure in modeling. This theme will pervade each of the last five chapters. We shall distinguish estimation and identification in the following way. Estimation refers to the process of fitting a model of known form to experimental data by determination of the values of unknown coefficients, or parameters, in the model. This is perhaps the most common situation encountered in the analysis of experimental data, that is, in which the form of the model can be specified from conservation principles, chemical kinetic mechanisms, etc., with the freedom residing in adjustable parameters. Chapters 6, 7, and 8 are devoted to this case.

Many systems of chemical engineering interest, such as fixed and fluidized beds, separation units, and turbulent flow reactors, involve the passage of fluid through them for the purpose of heat or mass transfer or chemical reaction (or combinations of these). To predict the rate of transfer or the extent of reaction it is necessary to know the way in which the fluid passes through the system. The passage of an element of fluid through such a system can be viewed as a stochastic process in which the residence time of the element in the system is a continuous random variable. In dealing with the types of systems mentioned, there has grown a variety of techniques for analyzing tracer experiments called *residence time distribution theory*. Chapter 6 presents the elements of residence time distribution theory. The essential problem of residence time distribution theory is to match the predictions of a model to the experimental tracer concentrations obtained from a well-defined input, such as a pulse or step function of tracer. The matching of the model to the residence time data lies in selecting certain model parameters that characterize the degree and type of mixing that takes place in the system. Since the residence time distribution is simply a probability distribution, classic techniques used in characterizing probability distributions are of great utility. The most widely used procedure is to match the moments of the model and the data. In Chapter 6 we shall discuss the types of mixing models and residence time experiments. The actual mechanics of parameter estimation is the subject of Chapter 7.

Chapter 7 is a comprehensive treatment of *parameter estimation.* The main question faced is, Now that I have a model with unknown parameters and experimental data, what criterion should I use in matching model to data and *how* do I actually find the "best" parameter values according to this criterion? Various criteria based on the form of the model, and whether the model is linear or not, are outlined, and detailed algorithms are given for the determination of the parameter values using each criterion. Chapter 7 is developed around two problems, the first being the analysis of a packed bed reactor with adsorption of material on the packing and the second being the determination of kinetic parameters from experimental reactor data. Other systems, such as a trickle-flow reactor and a fluidized bed, are also considered.

As we noted, running through Chapters 6 and 7 is the supposition that experimental data have been obtained. As we have also noted, the proper design of an experiment can spell the difference between modeling success and failure. What do we mean by success and failure? It depends on our objective. If our aim is to estimate parameters accurately, we would want to perform those experiments which lead to the most accurate estimates possible. If, on the other hand, we want to find the "best" model from among several contenders, we want our data to amplify the differences among the models as much as possible. Chapter 8 considers the problem of designing experiments with these two objectives in mind. It is by no means an exhaustive treatment of experimental design. Rather, Chapter 8 stands as a complement to Chapter 7, since the estimates of parameters in themselves are of little value until they have been analyzed and used.

In Chapters 6–8 it is supposed that enough is known beforehand about the system so that the form of a mathematical model can be deduced from basic principles. This is usually a safe supposition for rather well-understood systems such as packed beds, distillation columns, etc. However, engineering is now concerned with systems so large and/or so complex that we cannot even write down a tentative mathematical model. Such situations often arise in the attempt to model biological systems such as the dynamics of nerve cells, the process of vision, or the response of an organism to an applied stimulus. In cases such as this we must essentially "throw up our hands" and treat the system as a black box. What we really have to do in such cases is gather input/output (stimulus/response) data and from that data formulate some empirical model of the system; that is, we must *identify* the system. Chapters 9 and 10 present an introduction to system identification.

Chapter 9 considers the case in which the system is assumed to be linear. Thus we seek a representation of the system that obeys the conditions of linearity (Chapter 2). In fact, we seek a representation in terms of fewest possible variables, called a *minimum realization.* Algorithms are given for developing a minimum realization from input/output data. The "black box" is thus developed as a linear system.

Chapter 10 considers the most complex case, that in which the system is nonlinear. Based on the original work of Volterra and Wiener, techniques are developed for the construction of an empirical, nonlinear model from input/output data. Concepts of expansions of orthogonal polynomials and the use of a special type of random input are important.

Types of Process Models 2

With the advent of high-speed large-memory computers, the mathematical modeling of processes has become probably the single most important aspect of the overall design and analysis of engineering systems. While avoiding the costly and time-consuming necessity of actually constructing a small-scale replica of the process for observing its behavior under a variety of conditions, mathematical modeling enables the rapid and thorough exploration of such questions as extrapolation of the process to operating ranges not being employed, the controllability and stability of the process, the sensitivity of the process to changes in operating variables, and the optimal economic operating conditions of the process. Mathematical modeling is taken to mean the formulation of mathematical relationships which describe the behavior of the actual system such that the dependent and independent variables and parameters of the model are directly related to the physical and chemical quantities in the real system.

The basic problem in model building is to deduce a structure or an equation for a physical system on the basis of observations of the real system. Hopefully this is not done by intuition but rather by a rigorous and systematic mathematical procedure. The model which results is then a quantitative summary of these observations, which, in some sense, can then be used to replace the system itself. This replacement of the original physical system by a mathematical representation may then be used to explore a number of aspects of the system, thereby perhaps suggesting additional experiments which would confirm or contradict features of the model.

Only in very special cases will a mathematical model be able to represent all the details of a process. In general, the most we can aim for is a description of the most important and relevant observable features of the system, those which include the effects of changes in key operating variables and parameters. Also, the modeling procedure is an adaptive procedure in which information gained from each attempt is used to improve the next try. Since it is reasonable to suspect that many different mathematical interpretations of physical processes can exist, it may be too much ever to expect a unique model. If this is true, we may be led into a whole spectrum of possible models, some of which are "better" than others.

The general approach to process modeling consists of a number of steps:

1. *Definition of the problem.* We should initially face the questions

a. What is the physical system we wish to represent mathematically?
b. To what uses will the model be put?
c. How complex a description do we seek?
d. What kind of data is available against which to test a model?

Clearly, these questions are not independent. For example, the complexity of description is related to the ultimate uses of the model and the quality and quantity of data available.

2. *Collection of data from the process.* The data can be that from a special experimental test or can be the normal operating data from the system. If a special test is to be conducted, two questions must be faced:

a. Which variables are to be perturbed?
b. What type of perturbation should be used?

The form of the mathematical model may have an important bearing on the type of experimental test performed. For example, pulse, step, and sinusoidal inputs are often associated with particular types of models.

3. *Theoretical analysis of the nature of the system to generate a mathematical model.* We can categorize the origin of mathematical models as follows:

a. The model can be specified directly from basic principles of physics and chemistry; that is, we know the underlying physical and chemical principles of the system and can write down the model equations on this basis. At most, certain parameters or coefficients of the model might be unknown and these might have to be estimated from the system's data. Models of this type, relevant to chemical engineering systems, are often called transport phenomena models. Details on the actual derivation of the model of an enormous number of systems of chemical engineering interest may be found in Bird et al. [9], Carslaw and Jaeger [12], Crank [14], Himmelblau and Bischoff [25], and Richtmyer and Morton [38].

b. At the other extreme is the so-called "black box" model where essentially no a priori information about the system is known. The model must be determined solely on the basis of data available from the actual system.

c. Most real situations lie somewhat intermediate of these two extremes. We often have some information on the basic nature of the system but not enough to set down precisely the relations among all variables. It is in this case that the modeling procedure often includes extensive comparisons of alternative models to determine a "best" model. It is also in this case that stochastic models find extensive usefulness.

4. *Solution of the model.* Once the model has been formulated, the next step is its solution. In general, linear models will be amenable to analytical techniques, whereas nonlinear models must be handled by numerical methods on a computer. We shall not discuss the solution of nonlinear models since numerical techniques are covered elsewhere, for example, [31] and [38]. We shall, however, cover two of the most general and powerful analytical methods available for the solution of linear process models. These techniques will be the subject of Chapter 3.

5. *Evaluation of the model.* This step refers to the comparison of the model and the data. Depending on the type of model and data, the matching of model and data can follow a number of paths:

a. If the form of the model is specified from basic principles, the matching of model to data is usually accomplished by estimation of model parameters to produce an optimum fit. This is referred to as *parameter estimation* and is the subject of Chapters 6–8.

b. If no prior information on the system exists other than input/output data, we have to construct an empirical model of the system. This is referred to as the *identification problem* and is the subject of Chapters 9 and 10.

Once a model is constructed it may be used for predicting the future behavior of a system, reconstructing the past behavior of a system, outlining a form of control of the system, or indicating a better and improved set of operating conditions. One of the most exciting and active areas of science and engineering at this time is the modeling of large, complex systems. For example, such systems include the frontal motion of air streams [2], mountain winds [26], ocean circulation [34], general meteorology [28], traffic signal allocation in high automobile densities [13], induced aeration of polluted streams [15], transport of air pollutants globally [29] and in urban atmospheres [44], reaction systems [4, 5, 21, 33], structural design [36], socio-economic problems [27] including economic theory [16], and many biological problems including organisms [20], pulmonary circulation [39], and arterial blood flow [46]. Further, in order to handle the analysis of these many diverse models, specialized computer programs have become available [10, 11, 22, 42] which can be used with a minimum of experience.

An example of the type of reasoning that is inherent in mathematical modeling of complex systems has been summarized in the case of air pollution modeling as follows [45]:

> A critical element in developing a viable simulation model of an urban airshed is the achievement of a balance between accuracy of prediction and computational feasibility. Three primary component models will comprise the overall airshed model: (1) the meteorological model, describing transport and dispersion in the atmosphere, (2) the reaction kinetics model, describing the rates of atmospheric chemical reactions, and (3) the source model, describing the spatial distribution and mass of emissions from motor vehicles and major fixed sources.
>
> The meteorological model includes the kinetics model and the source model, however the major descriptive aspect of the meteorological model is the representation of the transport and dispersion processes. The primary purpose of the reaction kinetics model is to describe the rates of atmospheric chemical reactions as a function of concentration, intensity of radiation, and temperature. Since there are many contaminant species that are reactants, intermediates, and products of atmospheric reactions, a complete description of rates of reactions would be an enormous undertaking. The objective is to develop a simplified yet acceptably accurate model, one that is comprised of a manageable number of reactions species, or groups of species, lumped together, capable of predicting contaminant concentrations as a function of time and location.
>
> As a sound mathematical description of all ingredients outlined constitutes a substantial undertaking, it becomes important to consider carefully the objectives of a mathematical simulation of an urban atmosphere. Analysis of uses for the model, the most important of which will usually be evaluation of the effect of emission control measures on air quality, and consideration of the atmospheric and source data available for validation will dictate the degree of complexity which can be justified. In general, for a particular air quality region, the model would accept an input, emissions data (or estimates of reduced emissions associated with a particular control strategy) for motor vehicles, power plants, refineries, area sources, etc., and, for a particular set of meteorological conditions, predict concentrations of primary (those emitted directly from sources) and secondary (those formed in the atmosphere by chemical reaction) contaminants as a function of time and location.

It is our objective in this book to develop the fundamental aspects underlying the steps of mathematical modeling. We cannot hope to cover all the possible types of systems for which modeling is necessary. Therefore we shall illustrate the fundamental principles with examples drawn largely from chemical engineering applications. In fact, we shall consider these chemical engineering applications in significant detail, enough to illustrate the full power, as well as the shortcomings, of mathematical modeling techniques.

2.1 Basic Classifications of Mathematical Models

The classification of mathematical models is a difficult and somewhat arbitrary task, much like that faced by Lewis Carroll's walrus.† Nevertheless, there are a number of general features which can be used to group the types of models in common use. We shall choose as the most fundamental division, deterministic vs. stochastic models. Deterministic models are those in which there are no elements of randomness in the mathematical description—that is, the values of variables and parameters are definite fixed numbers—and in which the solution of the model yields the exact value of the response. Stochastic models, on the other hand, admit elements of randomness in the mathematical description. In fact, in a stochastic model we *never* know the exact value of a quantity which is considered as random but only the probabilities associated with observing certain values of the variable. Thus the output of a stochastic mathematical model is a probability rather than a definite fixed number. We shall delay discussion of stochastic variables and models until Chapter 4.

In this chapter we shall briefly outline the classifications of models that we shall consider in this book. The models are summarized in Table 2.1 using continuous and discrete variables as distinct entities. Before we discuss each category we shall introduce some general aspects of models for dynamic systems.

2.2 Fundamental Features of Models

There are certain basic, although rather abstract, concepts which apply to almost all mathematical models. Certain of these concepts will be introduced

†" 'The time has come,' the Walrus said,
'To talk of many things:
Of shoes—and ships—and sealing wax—
Of cabbages and kings—' "

Lewis Carroll, *Through the Looking Glass and What Alice Found There*, in *The Annotated Alice* (Cleveland: World Publishing Co., 1960).

TABLE 2.1
BASIC CLASSIFICATIONS OF MATHEMATICAL MODELS

	Continuous variables	Discrete variables
One independent variable	Ordinary differential equations	Ordinary difference equations
	Integral equations	Steady-state staged system
	Unsteady lumped[1] system	
	Steady-state distributed[2] system	
Two or more independent variables	Partial differential equations	Differential-difference equations
	Integral-differential equations	Unsteady staged system
	Unsteady distributed system	
	Steady-state distributed system	

[1]Lumped parameter systems are those in which spatial variations are ignored and the dependent variables (concentrations, pressures, temperatures, etc.) are uniform throughout the entire system. The only variations are those with time leading to ordinary differential equations.

[2]Distributed parameter systems are those in which spatial variations are taken into account. The dependent variables can vary from point to point within the system as well as with time, leading to description in terms of partial differential equations.

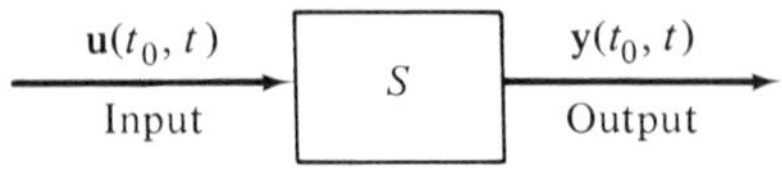

Figure 2.1. Evolution of abstract input/output system S.

here as well as in later sections of this book. A more complete treatment of these ideas can be found in the book by Zadeh and Desoer [48].

Consider the abstract system S in Figure 2.1 which is evolving in time. At time t_0 an *input* is applied to S and an *output* is observed from $t = t_0$ until $t = t$. We shall denote the input by the vector $\mathbf{u}(t_0, t)$,† where the argument (t_0, t) indicates that the input is applied over that interval and depends, in general, on the time t_0 at which the input is applied to S.‡ The output is represented by the vector $\mathbf{y}(t_0, t)$. The system is assumed to be deterministic. $\mathbf{y}(t_0, t)$ will obviously depend on $\mathbf{u}(t_0, t)$ and on the condition of S at time t_0. We now must introduce the concept of the *state* of the system S. Interpreted loosely, the state of a system is taken as that information needed at any time t_0 to determine the behavior of the system from that time on, for $t \geq t_0$. Thus the initial condition of S is the initial state of S. If we call this initial state the vector $\mathbf{x}(t_0)$, $\mathbf{y}(t_0, t)$ depends on $\mathbf{x}(t_0)$ and $\mathbf{u}(t_0, t)$. More explicit terminology will be given shortly. Note, however, that we do not consider what are termed anticipative systems, i.e., those in which the present value of the output depends on future values of the input.

Figure 2.1 shows a schematic of the arrangement under discussion. The pair $(\mathbf{u}, \mathbf{y})$ is termed an *input/output pair*. For example, if S is characterized by a scalar, linear ordinary differential equation (an input/output relation)

$$\frac{dy}{dt} + a_0 y = u$$

the solution of this equation will constitute the input/output pair. We note that in this case the state of S is identical to the output of S. In more complex cases this may not necessarily be so. Sedlar and Bekey [43] show alternative ways to describe Figure 2.1 by finite state and signal flow diagrams.

†For conciseness of presentation, we shall often use vector notation in this book. If not otherwise stated, we shall always take a vector, such as $\mathbf{u}(t_0, t)$, to be a *column vector*, that is,

$$\mathbf{u}(t_0, t) = \begin{bmatrix} u_1(t_0, t) \\ u_2(t_0, t) \\ \cdot \\ \cdot \\ \cdot \\ u_r(t_0, t) \end{bmatrix}$$

The material in Chapters 2–8 will require a knowledge only of very basic matrix algebra in instances when vector-matrix notation is used. Chapters 9 and 10 require some additional concepts, which will be explained as they appear.

‡For the purposes of illustration, in this section we shall consider variables which depend only on time. These concepts readily follow for multiple independent variables.

We also need to assume a set T for values of the independent variable t. If T is the real line, then S is said to be *continuous;* if T is only the integer reals, then S is said to be *discrete*. Each of these two frameworks usually occurs naturally in a model; e.g., homogeneous systems lead to continuous models and stagewise systems lead to discrete models (see Table 2.1).

Two other concepts of importance are *time-invariant* or *constant* systems and *linear* systems. Those systems which do not fall into these categories are termed *time-varying* and *nonlinear*, respectively. In simple terms time invariance is related to the invariance of input/output pairs under translation in time. Alternatively, the input/output pair is independent of the time interval in which the action occurs. This usually means that parameters or coefficients in a system model are constant.

A system S is linear if and only if (iff) any linear combination of any two input/output pairs in S is also an input/output pair in S. Thus, if $\tau[(\mathbf{u}, \mathbf{y})]$ is a transformation which converts $\mathbf{u}$ into $\mathbf{y}$ and c is any scalar,

$$\tau[(\mathbf{u}_0, \mathbf{y}_0) + (\mathbf{u}_1, \mathbf{y}_1)] = \tau[(\mathbf{u}_0, \mathbf{y}_0)] + \tau[(\mathbf{u}_1, \mathbf{y}_1)]$$

$$\tau[c(\mathbf{u}, \mathbf{y})] = c\tau[(\mathbf{u}, \mathbf{y})]$$

that is, the system obeys the *principle of superposition.*

For example, if we have the two input/output relationships

$$\frac{dy_0}{dt} + a_0 y_0 = u_0$$

$$\frac{dy_1}{dt} + a_0 y_1 = u_1$$

then a linear combination yields

$$\frac{d}{dt}[c_1 y_0 + c_2 y_1] + c_1 a_0 y_0 + c_2 a_0 y_1 = c_1 u_0 + c_2 u_1$$

where c_1 and c_2 are real. Thus the original input/output pairs (u_0, y_0) and (u_1, y_1) are also an input/output pair when linearly combined.

Linearity is the single most important property that a model can possess, since a wealth of analytical tools becomes available in that case. Unfortunately, an overwhelming number of physical systems must be represented by nonlinear models, greatly limiting the analytical techniques applicable.

Another important point is the question of the nonuniqueness of models or, in other terms, *indistinguishability* or *equivalence*. In these cases one cannot determine a single input/output pair from an experiment. Thus, consider two systems S_0 and S_1 and try to specify from an input/output pair $(\mathbf{u}, \mathbf{y})$ which system is being investigated. If this cannot be done, i.e., either S_0 or S_1 is indistinguishable from a single or multiple set of experiments, we term S_0 and S_1 equivalent and $S_0 = S_1$.

The last indefinite case does not, however, rule out the obvious statement that for every system there is a model. The question is how to find at least one model, called a *realization*, and then to find a *minimal model*. This point will be considered in detail in Chapter 9.

2.3 Ordinary Differential Equations

We have defined the state of a system to be that information at time t_0, together with a specification of the input $\mathbf{u}(t)$ for $t \geq t_0$, which enables the computation of the output for all time subsequent to t_0, $\mathbf{y}(t)$. Concisely stated,

$$\mathbf{y}(t) = \mathbf{y}[\mathbf{x}(t_0), \mathbf{u}(t_0, t)], \qquad t > t_0$$

In addition, the future state depends only on $\mathbf{x}(t_0)$ and $\mathbf{u}(t_0, t)$:

$$\mathbf{x}(t) = \mathbf{x}[\mathbf{x}(t_0), \mathbf{u}(t_0, t)], \qquad t > t_0$$

A large class of physical systems may be described by an n-dimensional state vector $\mathbf{x}(t)$ which satisfies a vector ordinary differential equation of the form

$$\dot{\mathbf{x}} = \frac{d\mathbf{x}}{dt} = \mathbf{f}(t, \mathbf{x}(t), \mathbf{u}(t)) \tag{2.3.1}$$

Since we are interested in the modeling of real systems, we shall always assume that a solution of (2.3.1) exists and is unique. The conditions for existence and uniqueness of (2.3.1) can be found in any standard text on ordinary differential equations. Written in terms of the elements of $\mathbf{x}$ and $\mathbf{u}$, (2.3.1) is

$$\begin{aligned}
\dot{x}_1(t) &= f_1(t; x_1(t), x_2(t), \ldots, x_n(t); u_1(t), \ldots, u_r(t)) \\
&\vdots \\
\dot{x}_n(t) &= f_n(t; x_1(t), x_2(t), \ldots, x_n(t); u_1(t), \ldots, u_r(t))
\end{aligned}$$

In this section we shall present the solution of the general linear form of (2.3.1). First, however, we shall consider the reduction of an nth-order ordinary differential equation (O.D.E.) to n-coupled first-order O.D.E. of the form (2.3.1).

EXAMPLE 2.1. Show that the general O.D.E. of the nth order

$$a_n \frac{d^n y}{dt^n} + a_{n-1} \frac{d^{n-1} y}{dt^{n-1}} + \cdots + a_0 y = u \tag{A}$$

where $a_n, a_{n-1}, \ldots, a_0$ are constants and $a_n \neq 0$, can be written in the general form

$$\dot{\mathbf{x}} = \mathbf{F}\mathbf{x} + \mathbf{G}\mathbf{u} \tag{B}$$

$$\mathbf{y} = \mathbf{H}\mathbf{x} \tag{C}$$

We define a new set of variables $\mathbf{x}(t) = [x_1, x_2, \ldots, x_n]^T$ (the superscript T indicates transpose) in terms of the derivatives of $y(t)$, namely

$$\begin{aligned} x_1 &= y \\ x_2 &= \frac{dy}{dt} \\ &\ \vdots \\ x_n &= \frac{d^{n-1}y}{dt^{n-1}} \end{aligned} \tag{D}$$

It then follows directly from (A) and (D) that

$$\begin{aligned} \frac{dx_1}{dt} &= x_2 \\ \frac{dx_2}{dt} &= x_3 \\ &\ \vdots \\ \frac{dx_n}{dt} &= \frac{1}{a_n}[u - a_0 x_1 - \cdots - a_{n-1}x_n] \end{aligned} \tag{E}$$

Then if we define

$$\mathbf{F} = \begin{bmatrix} 0 & 1 & 0 & & \cdots & 0 \\ 0 & 0 & 1 & 0 & \cdots & 0 \\ & & & & & 0 \\ & & & & & 1 \\ -\dfrac{a_0}{a_n} & & & & \cdots & -\dfrac{a_{n-1}}{a_n} \end{bmatrix}, \quad \mathbf{G} = \begin{bmatrix} 0 \\ 0 \\ \cdot \\ \cdot \\ \cdot \\ \dfrac{1}{a_n} \end{bmatrix}, \quad \mathbf{H} = [1 \quad 0 \quad \cdots \quad 0]$$

there results the set of equations (B). However, the solution of (B), $y(t)$, is then determined by an initial condition on $\mathbf{x}(t)$ and $u(t_0, t)$. Calling this initial condition $\mathbf{x}(t_0)$ and recalling that for (A) the initial conditions would be $y(t_0), \ldots, d^{n-1}y(t_0)/dt^{n-1}$, we can let

$$\mathbf{x}(t_0) = \left[y(t_0), \ldots, \frac{d^{n-1}y(t_0)}{dt^{n-1}}\right]^T$$

Thus $\mathbf{x}(t_0)$ is an initial condition for $y(t)$, and since only it, along with $u(t_0, t)$, is needed to determine $y(t)$ at any t, $\mathbf{x}(t)$ serves as a state vector for S.

The general linear form of (2.3.1),

$$\dot{\mathbf{x}} = \mathbf{F}\mathbf{x} + \mathbf{G}\mathbf{u} \tag{2.3.2}$$

is an extremely important equation in process modeling. We shall now discuss the solution of (2.3.2) subject to $\mathbf{x}(t_0) = \mathbf{x}_0$. Note particularly that the matrices **F** and **G** do not depend on time; that is, they are time-invariant or constant. We assume that **F** has dimension $n \times n$ and that **G** has dimension $n \times r$, where $\mathbf{u}(t)$ is an r-dimensional vector. The solution of the homogeneous form of (2.3.2),

$$\dot{\mathbf{x}} = \mathbf{F}\mathbf{x} \tag{2.3.3}$$

is

$$\mathbf{x}(t) = e^{\mathbf{F}(t-t_0)}\mathbf{x}_0 = \exp[\mathbf{F}(t - t_0)]\mathbf{x}_0 \tag{2.3.4}$$

The exponential matrix is defined by the infinite series

$$e^{\mathbf{F}t} = \mathbf{I} + \mathbf{F}t + \mathbf{F}^2\frac{t^2}{2!} + \mathbf{F}^3\frac{t^3}{3!} + \cdots \tag{2.3.5}$$

We define the state transition matrix as

$$\mathbf{\Phi}(t - t_0) = e^{\mathbf{F}(t-t_0)} \tag{2.3.6}$$

Therefore (2.3.4) can be written as

$$\mathbf{x}(t) = \mathbf{\Phi}(t - t_0)\mathbf{x}_0 \tag{2.3.7}$$

We shall not discuss the calculation of $\mathbf{\Phi}(t - t_0)$ at this point.

Now let us solve the nonhomogeneous case, (2.3.2). We let

$$\mathbf{x}(t) = \mathbf{\Phi}(t - t_0)\,\mathbf{z}(t) \tag{2.3.8}$$

Differentiating,

$$\dot{\mathbf{x}} = \dot{\mathbf{\Phi}}\mathbf{z}(t) + \mathbf{\Phi}\dot{\mathbf{z}} \tag{2.3.9}$$

We note from the definition of $\mathbf{\Phi}(t - t_0)$ that

$$\dot{\mathbf{\Phi}} = \mathbf{F}\mathbf{\Phi} \tag{2.3.10}$$

$$\mathbf{\Phi}(0) = \mathbf{I} \tag{2.3.11}$$

Thus (2.3.9) is

$$\dot{\mathbf{x}} = \mathbf{F}\mathbf{x} + \mathbf{\Phi}\dot{\mathbf{z}} \tag{2.3.12}$$

Comparing (2.3.2) and (2.3.12), we find that

$$\mathbf{\Phi}(t - t_0)\dot{\mathbf{z}} = \mathbf{G}\mathbf{u}(t) \tag{2.3.13}$$

Solving for $\dot{\mathbf{z}}$ yields

$$\dot{\mathbf{z}} = \mathbf{\Phi}^{-1}(t - t_0)\,\mathbf{G}\mathbf{u}(t) \tag{2.3.14}$$

where we assume that the inverse exists. Integrating (2.3.14),

$$\mathbf{z}(t) = \mathbf{z}(t_0) + \int_{t_0}^{t} \mathbf{\Phi}^{-1}(\tau - t_0)\,\mathbf{G}\mathbf{u}(\tau)d\tau \tag{2.3.15}$$

Using (2.3.15) in (2.3.8) and solving for $\mathbf{x}(t)$ gives (using $\mathbf{z}(t_0) = \mathbf{x}_0$)

$$\mathbf{x}(t) = \mathbf{\Phi}(t - t_0)\mathbf{x}_0 + \mathbf{\Phi}(t - t_0)\int_{t_0}^{t} \mathbf{\Phi}^{-1}(\tau - t_0)\,\mathbf{G}\mathbf{u}(\tau)d\tau \tag{2.3.16}$$

The integral is the contribution to the solution resulting from the input $\mathbf{u}(t)$.

A more general form of the vector linear O.D.E. is that in which the matrices **F** and **G** are functions of t. Such a system is called nonstationary, nonautonomous, nonconstant, or, simply, time-varying. The system is written as

$$\dot{\mathbf{x}}(t) = \mathbf{F}(t)\mathbf{x}(t) + \mathbf{G}(t)\mathbf{u}(t)$$
$$\mathbf{x}(t_0) = \mathbf{x}_0 \tag{2.3.17}$$

As in the time-invariant case, the solution can be obtained by means of the state transition matrix. Consider the homogeneous system

$$\dot{\mathbf{x}} = \mathbf{F}(t)\mathbf{x} \tag{2.3.18}$$

If the $n \times n$ matrix $\mathbf{\Phi}(t; t_0)$ satisfies

$$\dot{\mathbf{\Phi}}(t; t_0) = \frac{\partial \mathbf{\Phi}(t; t_0)}{\partial t} = \mathbf{F}(t)\mathbf{\Phi}(t; t_0) \tag{2.3.19}$$

$$\mathbf{\Phi}(t_0; t_0) = \mathbf{I} \tag{2.3.20}$$

then by direct substitution a solution of (2.3.18) is

$$\mathbf{x}(t) = \mathbf{\Phi}(t; t_0)\mathbf{x}_0 \tag{2.3.21}$$

The analogy to (2.3.7) is obvious. However, in the time-varying case, the transition matrix depends in general on the value of t_0.

The state transition matrix $\mathbf{\Phi}(t; t_0)$ obeys the following properties:

$$\mathbf{\Phi}(t_2; t_0) = \mathbf{\Phi}(t_2; t_1)\mathbf{\Phi}(t_1; t_0) \tag{2.3.22}$$

$$\mathbf{\Phi}(t; t_0) = \mathbf{\Phi}^{-1}(t_0; t) \tag{2.3.23}$$

A proof of (2.3.23) is given by Athans and Falb [6]. For the stationary case $\mathbf{\Phi}(t; t_0) = \mathbf{\Phi}(t - t_0)$, and

$$\mathbf{\Phi}(t - t_0) = \mathbf{\Phi}(t - \tau)\mathbf{\Phi}(\tau - t_0) \tag{2.3.24}$$

$$\mathbf{\Phi}^{-1}(t - t_0) = \mathbf{\Phi}(t_0 - t) \tag{2.3.25}$$

Using (2.3.24) and (2.3.25) in (2.3.16), we can rewrite (2.3.16) in the equivalent form

$$\mathbf{x}(t) = \mathbf{\Phi}(t - t_0)\mathbf{x}_0 + \int_{t_0}^{t} \mathbf{\Phi}(t - \tau)\mathbf{G}\mathbf{u}(\tau)d\tau \tag{2.3.26}$$

Of fundamental importance also is that if $\mathbf{F}(t)$ is continuous over any finite time interval, $\mathbf{\Phi}(t - t_0)$ is nonsingular (the column vectors of $\mathbf{\Phi}$ are linearly independent) for all finite t.

Finally, consider the general time-varying, forced case of (2.3.17). If we let

$$\mathbf{x}(t) = \mathbf{\Phi}(t; t_0)\mathbf{z}(t)$$

we can parallel the procedure in (2.3.8)–(2.3.16) to obtain

$$\mathbf{x}(t) = \mathbf{\Phi}(t; t_0)\mathbf{x}_0 + \mathbf{\Phi}(t; t_0) \int_{t_0}^{t} \mathbf{\Phi}^{-1}(\tau; t_0)\mathbf{G}(\tau)\mathbf{u}(\tau)d\tau \tag{2.3.27}$$

Using (2.3.23) we can rewrite (2.3.27) as

$$\mathbf{x}(t) = \mathbf{\Phi}(t; t_0)\mathbf{x}_0 + \int_{t_0}^{t} \mathbf{\Phi}(t; \tau)\mathbf{G}(\tau)\mathbf{u}(\tau)d\tau \tag{2.3.28}$$

A full discussion of techniques for explicitly computing $\mathbf{\Phi}(t; t_0)$ is given by Athans and Falb [6] and Lapidus and Luus [30]. The most obvious procedure is to use the infinite series expansion of (2.3.5) but truncated at some finite set of terms. The accuracy of this series expansion method depends on the magnitude of the elements of $\mathbf{F}$, the value of t, and the number of terms retained. Note also that

$$e^{\mathbf{F}t} = (e^{\mathbf{F}\tau})^{2^r}$$

where $t = 2^r\tau$ and r is an integer. One can choose τ to control quick convergence and obtain $e^{\mathbf{F}t}$ by squaring r times. Francis [19] has analyzed this procedure in depth.

It should be pointed out that the series method for evaluating the transition matrix is simple to use, is easy to program for digital computation, and does not require the eigenvalues of $\mathbf{F}$. All other methods, among which are the inverse Laplace transform, the Hamilton-Cayley theorem, and similarity transformations of $\mathbf{F}$, require the eigenvalues. As such they can only be used with ease for low-order systems. At least one of these methods will, however, be mentioned in Chapter 3.

Of further interest is the *adjoint* system associated with the time-varying or invariant form of (2.3.2). Thus for

$$\dot{\mathbf{x}} = \mathbf{F}\mathbf{x} + \mathbf{G}\mathbf{u}$$

we shall define the adjoint system as

$$\dot{\mathbf{z}} = -\mathbf{F}^T\mathbf{z} \tag{2.3.29}$$

For $\mathbf{F}$ and $\mathbf{G}$ as scalars the adjoint systems are

$$\dot{x} = fx + gu$$
$$\dot{z} = -fx$$

Note that there is no effect of the input in the adjoint equation and that the adjoint equation is equivalent to running the original homogeneous equation backwards in time. Obviously, when the homogeneous solution of the original system decreases as time increases, the adjoint solution increases in the same direction. This means that computational problems may arise when t is large since instability (nonbounding of the numbers) may occur if both sets of equations are evaluated in an increasing time direction. In many applications the original system and the adjoint system are actually evaluated in opposite time directions to bypass this numerical problem.

If the state transition matrix of the adjoint equations (2.3.29) is defined as $\mathbf{\Psi}(t; t_0)$, then it follows that

$$\dot{\mathbf{\Psi}}(t; t_0) = -\mathbf{F}^T\mathbf{\Psi}(t; t_0) \tag{2.3.30}$$

with

$$\mathbf{\Psi}(t_0; t_0) = \mathbf{I} \tag{2.3.31}$$

Also of interest is that

$$\boldsymbol{\Psi}^T(t; t_0) = \boldsymbol{\Phi}(t; t_0)^{-1} = \boldsymbol{\Phi}(t_0; t) \tag{2.3.32}$$

In general the nonlinear form (2.3.1) cannot be solved analytically. The most common procedure is a numerical solution on a digital computer [31].

EXAMPLE 2.2. For the matrix

$$\mathbf{F} = \begin{bmatrix} -\frac{3}{2} & \frac{3}{2} \\ \frac{1}{6} & -\frac{3}{2} \end{bmatrix}$$

evaluate the transition matrix $\boldsymbol{\Phi}(t)$ using the exponential expansion of (2.3.5). Note that

$$\mathbf{F}^2 = \begin{bmatrix} \frac{5}{2} & -\frac{9}{2} \\ -\frac{1}{2} & \frac{5}{2} \end{bmatrix}, \qquad \mathbf{F}^3 = \begin{bmatrix} -\frac{9}{2} & \frac{21}{2} \\ \frac{7}{6} & -\frac{9}{2} \end{bmatrix}, \ldots$$

and thus

$$\boldsymbol{\Phi}(t) = \exp[\mathbf{F}t] = \begin{bmatrix} 1 & 0 \\ 0 & 1 \end{bmatrix} + \begin{bmatrix} -\frac{3}{2} & \frac{3}{2} \\ \frac{1}{6} & -\frac{3}{2} \end{bmatrix} t + \begin{bmatrix} \frac{5}{2} & -\frac{9}{2} \\ -\frac{1}{2} & \frac{5}{2} \end{bmatrix} \frac{t^2}{2} + \cdots \tag{A}$$

Combining the matrices in (A) yields

$$\boldsymbol{\Phi}(t) = \begin{bmatrix} 1 - \frac{3}{2}t + \frac{5}{2}\frac{t^2}{2} - \cdots & 0 + \frac{3}{2}t - \frac{9}{2}\frac{t^2}{2} + \cdots \\ 0 + \frac{1}{6}t - \frac{1}{2}\frac{t^2}{2} + \cdots & 1 - \frac{3}{2}t + \frac{5}{2}\frac{t^2}{2} - \cdots \end{bmatrix}$$

But each element in the matrix can be related to the series for $e^{\alpha t}$ and thus

$$\boldsymbol{\Phi}(t) = \begin{bmatrix} \frac{1}{2}(e^{-t} + e^{-2t}) & \frac{3}{2}(e^{-t} - e^{-2t}) \\ \frac{1}{6}(e^{-t} - e^{-2t}) & \frac{1}{2}(e^{-t} + e^{-2t}) \end{bmatrix} \tag{B}$$

The reader can readily determine that for a given value of t the determinant of $\boldsymbol{\Phi}(t)$ is positive and thus that $\boldsymbol{\Phi}(t)$ is nonsingular.

Obviously the very simple form of $\mathbf{F}(t)$ has allowed us to carry out the calculation for $\boldsymbol{\Phi}(t)$. In [30] many examples are given for more complicated Fs with relevance to more realistic systems.

EXAMPLE 2.3. Consider the free or homogeneous form of (2.3.2), namely

$$\dot{\mathbf{x}} = \mathbf{F}\mathbf{x} \tag{A}$$

with $\mathbf{x}_0$ as the initial condition. For the special form for $\mathbf{F}$,

$$\mathbf{F} + \mathbf{F}^T = \mathbf{0} \tag{B}$$

show that the Euclidean norm of $\mathbf{x}(t)$,

$$||\mathbf{x}(t)|| = \sqrt{x_1^2 + \cdots + x_n^2} = \text{constant} \tag{C}$$

From (2.3.29) we can see that (B) implies that the adjoint system is the same as the original system (A). However, this means that

$$\boldsymbol{\Phi}(t)^T\boldsymbol{\Phi}(t) = \mathbf{I} \tag{D}$$

and for

$$\mathbf{x}(t) = \mathbf{\Phi}(t)\mathbf{x}_0$$

then

$$\mathbf{x}^T(t)\mathbf{x}(t) = [\mathbf{\Phi}(t)\mathbf{x}_0]^T[\mathbf{\Phi}(t)\mathbf{x}_0]$$

Using (D),

$$||\mathbf{x}(t)|| = ||\mathbf{x}_0||$$

and (C) is satisfied.

2.4 Difference Equations

If the independent variable t assumes only integer values, then the physical system can be governed by a vector difference equation of the form

$$\mathbf{x}(t_{k+1}) = \mathbf{f}[t_k, \mathbf{x}(t_k), \mathbf{u}(t_k)] \tag{2.4.1}$$

where the state $\mathbf{x}$ is defined only at discrete values of t, denoted by $t_0, t_1, \ldots, t_k, \ldots$. The input vector $\mathbf{u}$ also assumes values only at the discrete values of t. The state concepts at the beginning of Section 2.1 carry over directly to the discrete case with t assuming only discrete values. The output vector of the discrete system whose state vector is $\mathbf{x}(t_k)$ can be expressed in the general form

$$\begin{aligned}\mathbf{y}(t_k) &= \mathbf{y}[\mathbf{x}(t_k)] \\ &= \mathbf{y}[\mathbf{x}(t_0), \mathbf{u}(t_0, t_k)]\end{aligned} \tag{2.4.2}$$

where $\mathbf{u}(t_0, t_k)$ is the set of inputs $\mathbf{u}(t_0), \mathbf{u}(t_1), \ldots, \mathbf{u}(t_k)$. For simplicity, we shall denote $\mathbf{x}(t_k)$ and $\mathbf{u}(t_k)$ by $\mathbf{x}(k)$ and $\mathbf{u}(k)$, respectively.

Consider the linear form of (2.4.1) in which the coefficients do not depend on k.

$$\mathbf{x}(k+1) = \mathbf{F}\mathbf{x}(k) + \mathbf{G}\mathbf{u}(k) \tag{2.4.3}$$

As in the O.D.E. case, let us first solve the homogeneous form of (2.4.3),

$$\mathbf{x}(k+1) = \mathbf{F}\mathbf{x}(k) \tag{2.4.4}$$

Since $\mathbf{x}(1) = \mathbf{F}\mathbf{x}(0)$ and $\mathbf{x}(2) = \mathbf{F}\mathbf{x}(1) = \mathbf{F}^2\mathbf{x}(0)$, the general solution is obviously

$$\mathbf{x}(k) = \mathbf{F}^k\mathbf{x}(0) \tag{2.4.5}$$

If we define the $n \times n$ transition matrix

$$\mathbf{\Phi}(k) = \mathbf{F}^k \tag{2.4.6}$$

then (2.4.5) can be written as

$$\mathbf{x}(k) = \mathbf{\Phi}(k)\mathbf{x}(0) \tag{2.4.7}$$

The nonhomogeneous case (2.4.3) can be solved similarly. We note that

$$\begin{aligned}\mathbf{x}(1) &= \mathbf{F}\mathbf{x}(0) + \mathbf{G}\mathbf{u}(0) \\ \mathbf{x}(2) &= \mathbf{F}\mathbf{x}(1) + \mathbf{G}\mathbf{u}(1) \\ &= \mathbf{F}^2\mathbf{x}(0) + \mathbf{F}\mathbf{G}\mathbf{u}(0) + \mathbf{G}\mathbf{u}(1)\end{aligned} \tag{2.4.8}$$

Continuing, we can show that

$$\mathbf{x}(k) = \mathbf{F}^k\mathbf{x}(0) + \sum_{i=0}^{k-1} \mathbf{F}^{k-i-1}\mathbf{G}\mathbf{u}(i), \qquad k \geq 1 \tag{2.4.9}$$

An equivalent way of writing (2.4.9) is

$$\mathbf{x}(k) = \mathbf{\Phi}(k)\mathbf{x}(0) + \sum_{i=0}^{k-1} \mathbf{\Phi}(k - i - 1)\mathbf{G}\mathbf{u}(i), \qquad k \geq 1 \tag{2.4.10}$$

Next we shall consider the class of linear difference equations with nonconstant coefficients governed by

$$\mathbf{x}(k + 1) = \mathbf{F}(k)\mathbf{x}(k) + \mathbf{G}(k)\mathbf{u}(k) \tag{2.4.11}$$

As usual, we shall consider first the homogeneous case, which now is

$$\mathbf{x}(k + 1) = \mathbf{F}(k)\mathbf{x}(k) \tag{2.4.12}$$

If the $n \times n$ matrix $\mathbf{\Phi}(k; 0)$ satisfies

$$\mathbf{\Phi}(k + 1; 0) = \mathbf{F}(k)\mathbf{\Phi}(k; 0) \tag{2.4.13}$$

$$\mathbf{\Phi}(0; 0) = \mathbf{I} \tag{2.4.14}$$

then by direct substitution a solution of (2.4.12) is

$$\mathbf{x}(k) = \mathbf{\Phi}(k; 0)\mathbf{x}(0) \tag{2.4.15}$$

The discrete transition matrix $\mathbf{\Phi}(k; 0)$ is given directly by

$$\begin{aligned} \mathbf{\Phi}(1; 0) &= \mathbf{F}(0) \\ \mathbf{\Phi}(2; 0) &= \mathbf{F}(1)\mathbf{F}(0) \\ &\vdots \\ \mathbf{\Phi}(k; 0) &= \mathbf{F}(k - 1)\mathbf{F}(k - 2) \cdots \mathbf{F}(0) \end{aligned} \tag{2.4.16}$$

In addition, the discrete transition matrix obeys the following properties:

$$\mathbf{\Phi}(k; 0) = \mathbf{\Phi}(k; j)\mathbf{\Phi}(j; 0) \tag{2.4.17}$$

$$\mathbf{\Phi}^{-1}(k; 0) = \mathbf{\Phi}(0; k) \tag{2.4.18}$$

The requirement that the inverse exists is that all $\mathbf{F}(k)$ must be nonsingular.

Finally, we shall consider the nonhomogeneous case (2.4.11), the solution of which is

$$\mathbf{x}(k) = \mathbf{\Phi}(k; 0)\mathbf{x}(0) + \sum_{i=0}^{k-1} \mathbf{\Phi}(k; i + 1)\mathbf{G}(i)\mathbf{u}(i), \qquad k \geq 1 \tag{2.4.19}$$

As in the O.D.E. case, the general nonlinear form (2.4.1) cannot be solved analytically and is usually solved sequentially on a digital computer. Typical systems which are described by difference equations are sequences of staged separation units, such as extractors and the individual plates of a distillation column.

From the results of this section it is apparent that there exists a theory of difference equations paralleling the theory of O.D.E. We have already presented the relevant aspects of the theory of vector, first order linear difference equations in this section. However we will find that for single second-order difference equations it may indeed be easier to solve the equation directly using the so-called characteristic equation rather than to convert the system to a two-dimensional vector, first-order difference equation. Because the reader may not be familiar with the general techniques involved for solving a second-order, nonhomogeneous difference equation, we present these here.

We consider the equation

$$a_0 y(k+2) + a_1 y(k+1) + a_2 y(k) = b(k) \tag{2.4.20}$$

where a_0 and a_2 are nonzero. If $y_1(k)$ and $y_2(k)$ are any two solutions of the homogeneous form of (2.4.20), then $c_1 y_1(k) + c_2 y_2(k)$, where c_1 and c_2 are arbitrary constants, is also a solution. In addition, if the determinant of

$$\begin{bmatrix} y_1(k) & y_2(k) \\ y_1(k+1) & y_2(k+1) \end{bmatrix}$$

is nonzero, then any solution $y_3(k)$ of the homogeneous equation can be written as $c_1 y_1(k) + c_2 y_2(k)$. The similarity of this determinant to the Wronskian in O.D.E. is clearly apparent. If this condition is satisfied, $c_1 y_1(k) + c_2 y_2(k)$ is called a *complete solution* to the homogeneous equation. If $w(k)$ is any solution of the nonhomogeneous equation (2.4.20), and if $c_1 y_1(k) + c_2 y_2(k)$ is the complete solution of the homogeneous equation, the $c_1 y_1(k) + c_2 y_2(k) + w(k)$ is the complete solution of the nonhomogeneous equation.

To solve second order O.D.E. we usually assume a solution of the form $y(t) = e^{mt}$. In the case of homogeneous difference equations we assume the solution to be of the form

$$y(k) = r^k \tag{2.4.21}$$

which is equivalent to e^{mt} if $r = e^m$. Substituting (2.4.21) into the homogeneous form of (2.4.20) gives

$$a_0 r^{k+2} + a_1 r^{k+1} + a_2 r^k = 0$$

or, dividing by r^k,

$$a_0 r^2 + a_1 r + a_2 = 0 \tag{2.4.22}$$

This is called the *characteristic equation* of (2.4.20). If the roots are distinct, then $y_1(k) = r_1^k$ and $y_2(k) = r_2^k$, and

$$y(k) = c_1 r_1^k + c_2 r_2^k \tag{2.4.23}$$

If r_1 and r_2 are complex, it is customary to reduce (2.4.23) to a more convenient form. Let the complex roots be

$$r_{1,2} = p \pm iq = \alpha e^{\pm i\theta}$$

$$\alpha = \sqrt{p^2 + q^2}; \qquad \tan\theta = \frac{q}{p}$$

Then

$$\begin{aligned} y(k) &= c_1(\alpha e^{i\theta})^k + c_2(\alpha e^{-i\theta})^k \\ &= \alpha^k[c_1 e^{i\theta k} + c_2 e^{-i\theta k}] \\ &= \alpha^k[d_1 \cos \theta k + d_2 \sin \theta k] \end{aligned}$$

where $d_1 = c_1 + c_2$ and $d_2 = i(c_1 - c_2)$.

If $r_1 = r_2$, we need to find another independent solution. We try kr_1^k and by substitution find that this is indeed a solution when the characteristic equation has equal roots. Thus,

$$y(k) = c_1 r_1^k + c_2 k r_1^k \tag{2.4.24}$$

Now, we proceed to the nonhomogeneous equation, where we must add a particular solution to the complete solution of the homogeneous equation. The procedure is analogous to that for O.D.E. Table 2.2 summarizes particular solutions $w(k)$ for several common $b(k)$.

TABLE 2.2
PARTICULAR SOLUTIONS $w(k)$ OF SECOND ORDER DIFFERENCE EQUATION $a_0 y(k+2) + a_1 y(k+1) + a_2 y(k) = b(k)$ [47]

$b(k)$	$w(k)$
1. b	B
2. bk^m, m a positive integer	$B_0 k^m + B_1 k^{m-1} + \cdots + B_{m-1}k + B_m$
3. bm^k	Bm^k
4. $b \cos mk$ }	$A \cos mk + B \sin mk$
5. $b \sin mk$ }	

Note: When $b(k)$ consists of a sum of several terms, $w(k)$ is chosen as the sum of the appropriate individual terms. If a term in $w(k)$ duplicates one in the homogeneous solution, all terms in $w(k)$ must be multiplied by k (or the lowest power needed to eliminate the duplication).

EXAMPLE 2.4. The *Fibonacci numbers* form the following sequence

$$0, 1, 1, 2, 3, 5, 8, 13, 21, \ldots$$

each number of which is the sum of the previous two numbers. Show that the kth number in the sequence can be obtained as the solution of a linear difference equation.

From the definition of the numbers it follows that

$$y(k+2) - y(k+1) - y(k) = 0 \tag{A}$$

where $y(k)$ is the kth number in the sequence. This is a second order linear difference equation. Let us solve it by the general theory just outlined. We first want to convert (A) to the vector form of (2.4.3). We do this by defining (by a procedure completely analogous to that for O.D.E. in Example 2.1)

$$x_1(k) = y(k)$$

$$x_2(k) = y(k+1)$$

so that (A) can be written as

$$\begin{aligned} x_1(k+1) &= x_2(k) \\ x_2(k+1) &= x_2(k) + x_1(k) \end{aligned} \tag{B}$$

which is in the form of (2.4.3) with $\mathbf{G} = \mathbf{0}$ and

$$\mathbf{F} = \begin{bmatrix} 0 & 1 \\ 1 & 1 \end{bmatrix} \tag{C}$$

The initial condition is

$$\mathbf{x}(0) = \begin{bmatrix} x_1(0) \\ x_2(0) \end{bmatrix} = \begin{bmatrix} 0 \\ 1 \end{bmatrix} \tag{D}$$

Thus, the solution of (B) is given by (2.4.5), so that for any k it is only necessary to compute $\mathbf{F}^k$.

As we have seen, an nth order difference equation can be solved either by the procedure just used, namely reduction to a system of n coupled first-order difference equations, or by a direct procedure based on finding the roots of the characteristic equation of the difference equation. We now illustrate this second procedure.

The characteristic equation of (A) is

$$r^2 - r - 1 = 0 \tag{E}$$

of which the roots are $r_{1,2} = \frac{1}{2}[1 \pm \sqrt{5}]$. The general solution of (A) can then be written

$$y(k) = c_1 r_1^k + c_2 r_2^k \tag{F}$$

as long as the roots are distinct. The arbitrary constants c_1 and c_2 are evaluated from the initial conditions, $y(0) = 0, y(1) = 1$, and are found to be $c_{1,2} = \pm 1/\sqrt{5}$. Therefore,

$$y(k) = \frac{1}{\sqrt{5}} \left\{ \left[\frac{1+\sqrt{5}}{2}\right]^k - \left[\frac{1-\sqrt{5}}{2}\right]^k \right\} \tag{G}$$

We now present a second example in which the general technique of determining the roots of the characteristic equation of the difference equation will be used. The advantage of using this technique as opposed to (2.4.5) for simple systems is that $\mathbf{F}^k$ need not be calculated.

EXAMPLE 2.5. A laboratory experiment has shown that a new type of porous material is selective in allowing only component 1 of a two-component mixture to pass through it. The ratio of component 1 passing through the porous material to the total amount of components 1 and 2 initially contacting the porous material is a constant α. It is now desired to devise a continuous system which will produce a product of component 1.

Figure 2.2 shows a suggested steady-state arrangement of stages with L the feed rate to the system, $c(k) = c_k$ the rate of component-1 (lean) flow from the kth stage, and $q(k) = q_k$ the rate of component-1 (rich) flow from the kth stage. Derive an equation for q_k in terms of the inlet and outlet rates.

A material balance around the $(k + 1)$th stage yields

$$q_k + c_{k+2} = q_{k+1} + c_{k+1} \tag{A}$$

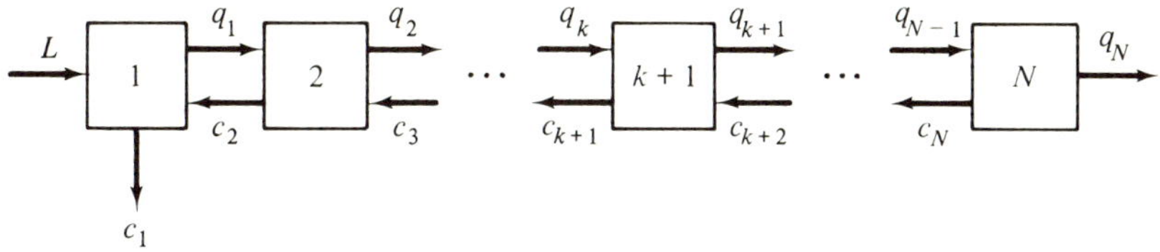

Figure 2.2. Countercurrent staged separation system for Example 2.5.

but since

$$\alpha = \frac{q_{k+1}}{q_{k+1} + c_{k+1}}$$

it follows that (A) can be written as

$$q_{k+2} - \frac{1}{1-\alpha} q_{k+1} + \frac{\alpha}{1-\alpha} q_k = 0 \tag{B}$$

This is a homogeneous, second-order, linear difference equation the roots of the characteristic equation of which are

$$r_{1,2} = \frac{1}{2}\left[\left(\frac{1}{1-\alpha}\right) \pm \sqrt{\left(\frac{1}{1-\alpha}\right)^2 - \frac{4\alpha}{1-\alpha}}\right]$$

However, simplification of the square root shows that

$$r_1 = 1 \quad \text{and} \quad r_2 = \frac{\alpha}{1-\alpha}$$

Thus

$$q_k = K_1(1)^k + K_2\left(\frac{\alpha}{1-\alpha}\right)^k \tag{C}$$

where K_1 and K_2 are arbitrary constants. To evaluate these constants we note that

$$q_k = q_N, \qquad k = N$$

and that

$$L + c_2 = q_1 + c_1, \qquad k = 1$$

or

$$q_2 = \frac{q_1}{1-\alpha} - \frac{\alpha}{1-\alpha} L, \qquad k = 1$$

From these conditions (C) may be used to explicitly find K_1 and K_2 and this yields

$$q_k = \frac{q_N - L[\alpha/(1-\alpha)]^N}{1 - [\alpha/(1-\alpha)]^N} + \frac{L - q_N}{1 - [\alpha/(1-\alpha)]^N}\left(\frac{\alpha}{1-\alpha}\right)^k$$

2.5 Differential-Difference Equations

Inherent in the use of an O.D.E. to represent a physical process is the assumption that the current rate of change of the state, $\dot{\mathbf{x}}(t)$, depends only on t and its current value $\mathbf{x}(t)$, leading to $\dot{\mathbf{x}} = \mathbf{f}(t, \mathbf{x})$. However, a number of real systems force us to take into account the fact that the current rate of change of the state may depend not only on the present state but also on the past history. In such a case it is necessary to write

$$\dot{\mathbf{x}}(t) = \mathbf{f}(t, \mathbf{x}(t), \mathbf{x}(s)) \tag{2.5.1}$$

where s ranges over a set of values less than t. Perhaps the most common equation of this type is a differential-difference equation (D.D.E.)

$$\dot{\mathbf{x}}(t) = \mathbf{f}(t, \mathbf{x}(t), \mathbf{x}(t - t_1), \ldots, \mathbf{x}(t - t_l)) \tag{2.5.2}$$

where $0 < t_1 < \ldots < t_l$. Instead of an initial condition prescribed at a *point* $t = t_0$ as in the O.D.E., it is now necessary to specify an initial function over an initial *interval*. In the case of (2.5.2) the general initial condition would be

$$\mathbf{x}(t) = \mathbf{g}(t), \qquad 0 \leq t \leq t_l \tag{2.5.3}$$

We shall not discuss techniques of solution of D.D.E here. However, we will find in Chapter 3 that the Laplace transform may be used to solve linear D.D.E. The standard treatment of D.D.E. is the book by Bellman and Cooke [8].

Differential difference equations arise naturally in the description of the transient behavior of staged processes, such as a distillation column. A number of examples of this type are given by Gould [23]. Processes in which there is a delay in the transfer of "information" also lead to D.D.E. For example, the feedback control of a reactor based on measurements of the effluent taken down-stream of the reactor, will result in a time delay t_1 equal to the time it takes the fluid to flow from the reactor to the point of measurement. Other situations which lead to D.D.E models are the dynamics of drug distribution in the human body, control mechanisms in the heart-lung system, and the dynamics of rocket or aircraft engines that have valve delays in piping systems from fuel tank to engine. Finally, partial differential equations are sometimes converted to D.D.E. for the purpose of numerical solution [24].

2.6 Partial Differential Equations

We present in this section some common classes of partial differential equations (P.D.E.) that arise often in engineering problems.

The following scalar P.D.E. is termed a *first-order* linear P.D.E. in two independent variables:

$$L(z) = a(x, y)\frac{\partial z}{\partial x} + b(x, y)\frac{\partial z}{\partial y} = c(x, y) \tag{2.6.1}$$

This equation is first-order because no derivatives of order higher than the first appear, and it is linear because of the following identity, which is valid for arbitrary constants α_1 and α_2 and arbitrary (differentiable) functions z_1

and z_2:

$$L(\alpha_1 z_1 + \alpha_2 z_2) = \alpha_1 L(z_1) + \alpha_2 L(z_2) \tag{2.6.2}$$

The general nonlinear form of (2.6.1) can be written as

$$f\left(x, y, z, \frac{\partial z}{\partial x}, \frac{\partial z}{\partial y}\right) = 0 \tag{2.6.3}$$

We note that when dealing with systems described by P.D.E.s we may use the notation z for the system state (which may be a vector) and x, y, and t for independent variables. No confusion should result with the O.D.E. and D.E. cases in which x denotes the state variable and t the single independent variable.

The second category we shall consider is the P.D.E. of the form

$$a(x, y)\frac{\partial^2 z}{\partial x^2} + 2b(x, y)\frac{\partial^2 z}{\partial x\,\partial y} + c(x, y)\frac{\partial^2 z}{\partial y^2} - f\left(x, y, z, \frac{\partial z}{\partial x}, \frac{\partial z}{\partial y}\right) = 0 \tag{2.6.4}$$

which is called a second-order P.D.E. with linear principal part, *second-order* referring to the fact that (2.6.4) contains no derivatives of higher order than second, while the linear principal part is the combination of terms containing second derivatives,

$$L(z) = a(x, y)\frac{\partial^2 z}{\partial x^2} + 2b(x, y)\frac{\partial^2 z}{\partial x \partial y} + c(x, y)\frac{\partial^2 z}{\partial y^2} \tag{2.6.5}$$

$L(z)$ is linear since (2.6.2) holds for (2.6.5). Many of the P.D.E.s of mathematical physics are of the general form (2.6.4), for example,

1. Wave equation:

$$\frac{\partial^2 z}{\partial x^2} - \frac{\partial^2 z}{\partial y^2} = 0$$

2. Heat equation:

$$\frac{\partial^2 z}{\partial x^2} - \frac{\partial z}{\partial y} = 0$$

3. Laplace's equation:

$$\frac{\partial^2 z}{\partial x^2} + \frac{\partial^2 z}{\partial y^2} = 0$$

It is convenient to classify the P.D.E. (2.6.4) in one of three ways. Let us assume that we can define the two new independent variables

$$\xi = \xi(x, y), \qquad \eta = \eta(x, y) \tag{2.6.6}$$

The principal part of (2.6.4), namely (2.6.5), may be expressed in terms of the new variables as follows (Problem 2.19):

$$a\frac{\partial^2 z}{\partial x^2} + 2b\frac{\partial^2 z}{\partial x\,\partial y} + c\frac{\partial^2 z}{\partial y^2} = \alpha\frac{\partial^2 z}{\partial \xi^2} + 2\beta\frac{\partial^2 z}{\partial \xi\,\partial \eta} + \gamma\frac{\partial^2 z}{\partial \eta^2} + \cdots \tag{2.6.7}$$

where

$$\alpha = a\xi_x^2 + 2b\xi_x\xi_y + c\xi_y^2 \qquad \left(\xi_x = \frac{\partial \xi}{\partial x}, \text{ etc.}\right)$$
$$\beta = a\xi_x\eta_x + b(\xi_x\eta_y + \xi_y\eta_x) + c\xi_y\eta_y \tag{2.6.8}$$
$$\gamma = a\eta_x^2 + 2b\eta_x\eta_y + c\eta_y^2$$

and the quantities indicated in (2.6.7) by $\cdots$ involve only z and its first derivatives. The key point is that (2.6.4) is transformed into another second-order equation the principal part of which is $\alpha(\partial^2 z/\partial\xi^2) + 2\beta(\partial^2 z/\partial\xi\,\partial\eta)$ $\gamma(\partial^2 z/\partial\eta^2)$. It can then be shown that (Problem 2.20)

$$\beta^2 - \alpha\gamma = (b^2 - ac)(\xi_x\eta_y - \xi_y\eta_x)^2 \tag{2.6.9}$$

Therefore the quantities $b^2 - ac$ and $\beta^2 - \alpha\gamma$ are both positive, both zero, or both negative at any given pair of corresponding points (note that $\xi_x\eta_y - \xi_y\eta_x \neq 0$). If (2.6.4) is classified according to the sign of the quantity $b^2 - ac$, its classification will not change under change of variables. Therefore we classify (2.6.4) according to

1. $b^2 - ac > 0$: hyperbolic.
2. $b^2 - ac = 0$: parabolic.
3. $b^2 - ac < 0$: elliptic.

In this book we shall confine our attention to P.D.E.s of either the first- or second-order type, usually with a linear principal part. Some of the problems at the end of this chapter deal with the proper specification of boundary conditions for some common P.D.E.s. In general terms, the boundary conditions for hyperbolic systems are determined for a field within a restricted semiinfinite range, the parabolic systems to a field within a semi-infinite range, and the elliptic to a finite closed field. In many of the coming chapters of this book the complete derivations of certain P.D.E.s will be developed.

EXAMPLE 2.6. To illustrate the development of a P.D.E., we shall take the case of steady-state gaseous diffusion of component A in a long tube with chemical reaction, i.e., $A \xrightarrow{k} B$, at the tube wall. The gas is assumed to move in rodlike, frictionless, and nonturbulent flow with v_0 the average velocity. Using c_A as the concentration of A, D as the axial diffusion coefficient, r and x the radial and axial independent variables, and d the diameter of the tube, a balance is taken on a control volume in the tube (see Figure 2.3). This control volume is taken as $2\pi r\,\Delta r\,\Delta x$ where Δr and Δx are infinitesimal values. The balance is

$$2\pi r\,\Delta r v_0 c_A\Big|_x + \left(-2\pi r\,\Delta r D\frac{\partial c_A}{\partial x}\right)\Big|_x + \left(-2\pi r\,\Delta x D\frac{\partial c_A}{\partial r}\right)\Big|_r = 2\pi r\,\Delta r v_0 c_A\Big|_{x+\Delta x}$$
$$+ \left(-2\pi r\,\Delta r D\frac{\partial c_A}{\partial x}\right)\Big|_{x+\Delta x} + \left(-2\pi r\,\Delta x D\frac{\partial c_A}{\partial r}\right)\Big|_{r+\Delta r}$$

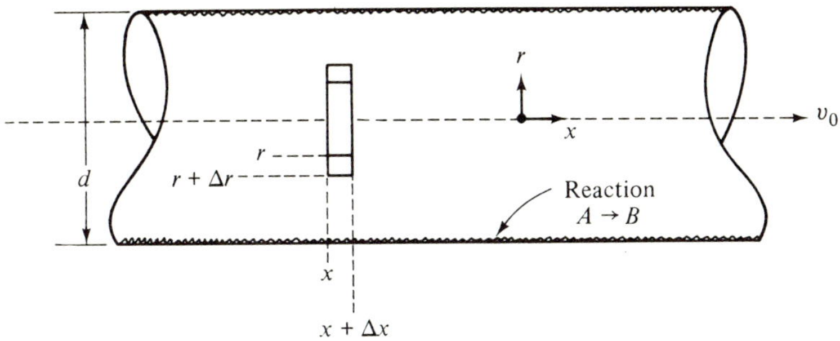

Figure 2.3. Steady-state diffusion of A in a long tube. Chemical reaction at wall. The system of Example 2.6.

which equates input to output for the control volume. The terminology $|_x$, $|_r$, etc., means that that term is evaluated at the position x, r, etc., in the control volume. Dividing each term by $2\pi r\,\Delta r\,\Delta x$ and taking the limit as $\Delta x, \Delta r \longrightarrow 0$ results in

$$D\left[\frac{1}{r}\frac{\partial}{\partial r}\left(r\frac{\partial c_A}{\partial r}\right) + \frac{\partial^2 c_A}{\partial x^2}\right] - v_0\frac{\partial c_A}{\partial x} = 0 \qquad \text{(A)}$$

The reader may wish to check that (A) is parabolic.

Typical boundary conditions would include

$$c_A = \text{finite} \quad \text{or} \quad \frac{\partial c_A}{\partial r} = 0 \qquad \text{at } r = 0 \qquad \text{(B)}$$

$$-D\frac{\partial c_A}{\partial r} = -kc_A \qquad \text{at } r = \frac{d}{2} \qquad \text{(C)}$$

indicating symmetry at the tube centerline (B) and that the rate of chemical reaction at the wall is equal to the flux of A to the wall (C). In addition, a condition on c_A at $x = 0$ and at the tube outlet, $x = L$ or $x = +\infty$, needs to be specified.

2.7 Integral Equations

In previous sections of this chapter we have considered three major types of model equations for systems: O.D.E.s, D.D.E.s, and P.D.E.s. Now we shall turn to a fourth type, namely integral equations (I.E.), which while not as frequently studied in elementary courses as the previous types are quite important. We shall only cover the highlights of this type of equation, but adequate references are available for the interested reader [37, 40].

We shall follow the conventional practice of referring to integral equations with fixed limits in the integral as equations of the *Fredholm type* and to those with a variable limit as equations of *Volterra type*. The equations of

Fredholm type of the first and second kinds, respectively, are given by

$$g(t) = \int_a^b K(t, \zeta)y(\zeta)\, d\zeta \tag{2.7.1}$$

(Fredholm)

$$y(t) = G(t) + \int_a^b K(t, \zeta)y(\zeta)\, d\zeta \tag{2.7.2}$$

The corresponding equations of Volterra type can be written as

$$G(t) = \int_a^t K(t, \zeta)y(\zeta)\, d\zeta \tag{2.7.3}$$

(Volterra)

$$y(t) = G(t) + \int_a^t K(t, \zeta)y(\zeta)\, d\zeta \tag{2.7.4}$$

In (2.7.1) to (2.7.4), $y(t)$ is the desired function, all other functions being known either analytically, graphically, or numerically. If the limits of the integral are infinite the result is termed a *singular integral equation;* $K(t, \zeta)$ is termed the *kernel* of the equation.

2.7.1 Connection to O.D.E.

Many O.D.E.s can be reformulated as I.E.s. In certain cases it may be easier to solve the I.E. than the original O.D.E. Thus it is of interest to show how this conversion occurs. To start we shall select the nonlinear form of (2.3.1) as

$$\ddot{y}(t) = f(t, y) \tag{2.7.5}$$

Integration of (2.7.5) in two steps yields

$$\dot{y}(t) = \int_0^t f\{\zeta, y(\zeta)\}\, d\zeta + c_1 \tag{2.7.6}$$

and

$$y(t) = \int_0^t \left[\int_0^t f\{(\zeta, y(\zeta)\}\, d\zeta\right] dt + c_1 t + c_2 \tag{2.7.7}$$

where c_1 and c_2 are arbitrary constants. An alternative form for $y(t)$ in the last expression is

$$y(t) = \int_0^t (t - \zeta) f\{\zeta, y(\zeta)\}\, d\zeta + G(t) \tag{2.7.8}$$

where $G(t) = c_1 t + c_2$. Note that (2.7.8) is a Volterra equation of the second kind. Equation (2.7.8) required the use of

$$\underbrace{\int_a^t \cdots \int_a^t}_{n \text{ times}} y(t)\, \underbrace{dt \cdots dt}_{n \text{ times}} = \frac{1}{(n-1)!} \int_a^t (t - \zeta)^{n-1} y(\zeta)\, d\zeta$$

To determine the constants c_1 and c_2, boundary conditions are needed. These may be, as examples, (1) $y(0) = a$ and $\dot{y}(0) = b$ or (2) $y(0) = a$ and

$y(1) = b$. In the first case, (2.7.8) becomes

$$y(t) = \int_0^t (t - \zeta)f\{\zeta, y(\zeta)\}\, d\zeta + bt + a \tag{2.7.9}$$

Note that (2.7.9), a Volterra equation, results from an initial-value O.D.E. with $c_1 = b$ and $c_2 = a$. In the second case,

$$c_1 = (b - a) - \int_0^1 (1 - \zeta)f\{\zeta, y(\zeta)\}\, d\zeta$$

$$c_2 = a$$

Substituting c_1 and c_2 into (2.7.8) there results

$$y(t) = a + (b - u)t + \int_0^t \zeta(t - 1)f\, d\zeta + \int_t^1 t(\zeta - 1)f\, d\zeta \tag{2.7.10}$$

with $f = f\{\zeta, y(\zeta)\}$. Alternatively, we may write (2.7.10), which results from a boundary-value O.D.E., as

$$y(t) = G(t) + \int_0^1 K(t, \zeta)f\, d\zeta \tag{2.7.11}$$

with the kernel defined as

$$K(t, \zeta) = \begin{cases} \zeta(t - 1), & t \geq \zeta \\ t(\zeta - 1), & t \leq \zeta \end{cases} \tag{2.7.12}$$

and $G(t) = a + (b - a)t$. In this case we have a Fredholm-type integral equation.

The kernel $K(t, \zeta)$ in (2.7.12) has different expressions in $\zeta < t$ and $\zeta > t$, but the expressions are equivalent when $\zeta = t$. Thus in terms of a fixed value of ζ but K a function of t, the K is continuous at $t = \zeta$. However, the derivative $\partial K/\partial t$ is discontinuous at $t = \zeta$, but $\partial^2 K/\partial t^2 = 0$ in each section. Finally, we shall note that the kernel $K(t, \zeta)$ is symmetric in the sense that $K(t, \zeta) = K(\zeta, t)$; i.e., the kernel is unchanged if t and ζ are interchanged.

Special (linear) cases of (2.7.5) and the boundary conditions might be

$$\begin{aligned} \ddot{y} + \lambda y &= f(t) \\ y(0) = 1, \qquad \dot{y}(0) &= 0 \end{aligned} \tag{2.7.13}$$

and

$$\begin{aligned} \ddot{y} + \lambda y &= 0 \\ y(0) = 0 \qquad y(a) &= 0 \end{aligned} \tag{2.7.14}$$

In the same way as before, (2.7.13) yields

$$y(t) = \lambda \int_0^t (\zeta - t)y(\zeta)\, d\zeta + 1 - \int_0^t (\zeta - t)f(t)\, dt \tag{2.7.15}$$

analogous to (2.7.9), while (2.7.14) yields

$$y(t) = \lambda \int_0^t \frac{\zeta}{a}(a - t)y(\zeta)\, d\zeta + \lambda \int_t^a \frac{t}{a}(a - \zeta)y(\zeta)\, d\zeta \tag{2.7.16}$$

If we define

$$K(t, \zeta) = \begin{cases} \dfrac{\zeta}{a}(a - t), & \zeta \leq t \\ \dfrac{t}{a}(a - \zeta), & \zeta \geq t \end{cases}$$

then (2.7.16) becomes

$$y(t) = \lambda \int_0^a K(t, \zeta) y(\zeta)\, d\zeta \tag{2.7.17}$$

analogous to (2.7.11)

EXAMPLE 2.7. Consider the famous Blasius equation (see Ames [3])

$$\dddot{y} + y\ddot{y} = 0 \tag{A}$$

with

$$y(0) = \dot{y}(0) = \dot{y}(\infty) - 2 = 0$$

$y(t)$ is the stream function and defines a problem of steady laminar boundary-layer flow over a semiinfinite flat plate. Rewriting (A) as

$$\frac{d\ddot{y}}{\ddot{y}} = -y\, dt$$

and integrating from zero to t yields

$$\ddot{y} = c_1 \exp\left[-\int_0^t y(\zeta)\, d\zeta\right] \tag{B}$$

Integrating twice more yields

$$y(t) = c_1 \int_0^t \int_0^t \exp\left[-\int_0^t y(\zeta)\, d\zeta\right] d\zeta\, d\zeta + c_2 t + c_3 \tag{C}$$

Substituting $y(0) = \dot{y}(0) = 0$ yields $c_2 = c_3 = 0$. Thus (C) becomes

$$y(t) = c_1 \int_0^t (t - \zeta) \exp\left[-\int_0^\zeta y(\zeta)\, d\zeta\right] d\zeta \tag{D}$$

The use of $\dot{y}(\infty) - 2 = 0$ then evaluates c_1. We shall not detail this point here but wait until Example 2.9.

Thus we see that O.D.E.s may be converted to equivalent I.E.s. This conversion incorporates, in an easy manner, the boundary conditions, and the result possesses the *smoothing* properly inherent in all integrals as contrasted to differentials or derivatives.

2.7.2 Singular Integral Equations

The singular I.E. has already been defined as an integral equation in which at least one limit is infinite. One case is of importance here since it forms the basis for the transform techniques discussed in Chapter 3.

If we have the double integral

$$y(t) = \int_a^b \int_a^b \Phi(t, \zeta_1) K(\zeta, \zeta_1) y(\zeta)\, d\zeta\, d\zeta_1$$

we may consider the replacement by the two integrals

$$\bar{y}(s) = \int_a^b K(s, \zeta) y(\zeta)\, d\zeta \tag{2.7.18}$$

$$y(t) = \int_a^b \Phi(t, \zeta) \bar{y}(\zeta)\, d\zeta \tag{2.7.19}$$

Here we may visualize solving (2.7.18) for $\bar{y}(s)$ and then solving (2.7.19) for the variable $y(t)$. Convention calls (2.7.18) the *transform of* $y(t)$ and (2.7.19) the *inverse transform of* $\bar{y}(s)$. Depending on the specific values used for a and b and the functions $\Phi(t, \zeta)$ and $K(s, \zeta)$, many *transform pairs* result. Frequently, however, in the actual use of such transforms only (2.7.18) is required because of certain unique properties of the transforms.

Typical examples are

1. $a = 0$, $b = \infty$, $K(s, \zeta) = e^{-s\zeta}$: Laplace transform.
2. $a = 0$, $b = \infty$, $K(s, \zeta) = \sin(s\zeta)$: sine transform.

EXAMPLE 2.8. For the special case $y(t) = e^{-\alpha t}$, find the Laplace transform of $y(t)$. Using (2.7.18) with the conditions 1 above,

$$\bar{y}(s) = \int_0^\infty e^{-s\zeta} y(\zeta)\, d\zeta$$

Thus for $y(t) = e^{-\alpha t}$

$$\bar{y}(s) = \int_0^\infty e^{-s\zeta} e^{-\alpha\zeta}\, d\zeta = \int_0^\infty e^{-(s+\alpha)\zeta}\, d\zeta$$

Carrying out the indicated integration and inserting the appropriate limits, there results

$$\bar{y}(s) = \frac{1}{s + \alpha}$$

as the Laplace transform of $y(t)$.

2.7.3 Approximate Solution Methods

Since many integral equations are nonlinear, a number of methods including numerical techniques have been devised for their solution. The work of Linz [32] and the material in Section 3.5 of Chapter 3 are illustrative of those methods which rely on quadrature methods to convert the integral to a more easily handled form.

Here we wish to comment briefly on an iteration technique and on the methods using variational techniques; the latter are frequently labeled as weighted residual, collocation, Ritz, and Galerkin methods depending on the specific procedure used.

The iteration method can be visualized using the Fredholm equation

$$y(t) = G(t) + \int_a^b K(t, \zeta) y(\zeta)\, d\zeta \tag{2.7.20}$$

We shall make an initial approximation for the unknown $y(t)$ which we shall call $y^{(0)}$. This approximation is inserted inside the integral and the right-hand side of the integral equation, (2.7.20), is evaluated to yield $y^{(1)}$. Assuming that this iteration converges, $y^{(1)}$ is closer in some sense to the true $y(t)$ than $y^{(0)}$, and we thus insert $y^{(1)}$ inside the integral to yield $y^{(2)}$, etc. This is continued in the form

$$y^{(n+1)}(t) = G(t) + \int_a^b K(t, \zeta) y^{(n)}(\zeta)\, d\zeta, \qquad n = 0, 1, 2, \ldots$$

Conditions for convergence are, of course, difficult to specify except in special cases.

EXAMPLE 2.9. Let us return to Example 2.7 which discussed the solution of the Blasius equation but in which we left the arbitrary constant c_1 unevaluated. In (D) of that example $y(t)$ occurs in both sides of the equation, and to apply the iteration technique we need to estimate $y^{(0)}(t)$. To do this we use (B) but with t small. As a result

$$\ddot{y}(t) = c_1$$

and it follows that

$$\dot{y}(t) = c_1 t$$

$$y(t) = \frac{c_1 t^2}{2} \tag{A}$$

using the boundary conditions on the problem. Equation (A) serves then as the first approximation since it was obtained on the basis that t was small.

From (B) in Example 2.7 we may write

$$\dot{y}(t) = c_1 \int_0^t \exp\left[-\int_0^t y(\zeta)\, d\zeta\right] d\zeta$$

Now using the approximation $y^{(0)}(t) = c_1 t^2/2$ inside the integral,

$$\dot{y}^{(1)}(t) = c_1 \int_0^t \exp\left[-\frac{c_1 \zeta^3}{3!}\right] d\zeta$$

$$= c_1^{2/3}(3!)^{1/3} \int_0^\alpha e^{-\alpha^3}\, d\alpha$$

where $\alpha = (c_1/3!)^{1/3}\zeta$. The resulting integral no longer contains $y(t)$ and may thus be evaluated from appropriate tables. However, since the boundary condition $\dot{y}(\infty) - 2 = 0$ has not been used previously, we employ it now and then evaluate the only remaining unknown c_1. As shown by Ames [3], this yields $c_1 = 1.37$ as compared to the exact value of $c_1 = 1.3282$. Thus one iteration has already yielded $y(t)$ or c_1 which is close to the correct final answer.

Next we shall turn to a brief discussion of the variational methods for solving integral equations. As we have indicated, such methods are frequently used with the titles of weighted residuals, collocation, Ritz, and Galerkin methods. The books of Ames [3], Rektorys [37], and Schechter [41] and papers by Finlayson and Scriven [18] and Paterson and Cresswell [35] are indicative of the work in this area. In the past 3–4 years this area of approximation has been the subject of numerous publications.

The basic concepts can be ascertained by means of the Fredholm integral equation (2.7.20). An approximate solution to $y(t)$ is proposed as the linear combination of functions $\phi_1(t), \phi_2(t), \ldots, \phi_n(t)$ in the form

$$y(t) = \sum_{i=1}^{n} c_i \phi_i(t) \tag{2.7.21}$$

The c_i are weighting factors to be determined in such a manner as to satisfy (2.7.20) as closely as possible in some manner to be defined. We may now write (2.7.20) in the form

$$G(t) = \sum_{i=1}^{n} c_i[\phi_i(t) - \theta_i(t)] \tag{2.7.22}$$

or

$$G(t) = \sum_{i=1}^{n} c_i f_i(t) \tag{2.7.23}$$

where we have used

$$\theta_i(t) = \int_a^b K(t, \zeta)\phi_i(\zeta)\, d\zeta \tag{2.7.24}$$

$$f_i(t) = \phi_i(t) - \theta_i(t) \tag{2.7.25}$$

In the collocation method the n constants $c_1, c_2, \ldots, c_n$ are evaluated by requiring that (2.7.23) be a true equality at n distinct t points in $\{a, b\}$. Since the unknowns are the c_i, these equalities yield a set of n linear algebraic equations for the c_i. As an example the choice

$$y(t) = c_1 + c_2 t + c_3 t^2 + c_4 t^3 \tag{2.7.26}$$

might be made for the weighted functions ϕ_i; i.e., $\phi_1 = 1, \phi_2 = t, \phi_3 = t^2$, and $\phi_4 = t^3$. On the basis of a given a and b and a specified $K(t, \zeta)$, (2.7.24) may be used to generate $\theta_1, \ldots, \theta_4$. In turn, $f_1, \ldots, f_4$ can be specified from (2.7.25) and (2.7.23) used to generate an equation with t as the variable. Now by specifying four values of t in $\{a, b\}$, four equations in the four unknowns $c_1, \ldots, c_4$ are generated and the c_i determined.

Obviously the choice of a cubic in (2.7.26) is only for convenience since any degree polynomial can be arbitrarily specified.

In the Galerkin method the difference between the two sides of (2.7.23) is specified as orthogonal to n linearly independent functions $\psi_j(t), j = 1, 2, \ldots, n$. This implies that

$$\int_a^b \psi_j G\, d\zeta = \sum_{i=1}^{n} c_i \int_a^b \psi_j f_i\, d\zeta, \qquad j = 1, 2, \ldots, n \tag{2.7.27}$$

However, once the linearly independent set ψ_j is specified the integrals $\int_a^b \psi_j G\, d\zeta$ and $\int_a^b \psi_j f_i\, d\zeta$ can be evaluated leading once again to a set of n

linear algebraic equations in the c_i. A convenient choice of the ψ_i are $1, t, t^2, t^3, \ldots$ as in the collocation method.

For more specific details on these and other allied methods the reader is referred to the excellent review by Finlayson and Scriven [18].

2.8 Integro-Differential Equations

The final class of models we delineate are integro-differential equations. Examples of equations of this type are

$$\dot{y}(t) = ay(t) + \int_0^t K(t - \zeta)y(\zeta)\,d\zeta$$

and

$$\frac{\partial y(x, t)}{\partial t} = -y(x, t) + \int_0^\infty K(t - \zeta)y(\zeta)\,d\zeta$$

Illustrations of this class of equations are given in the books of Bellman [7] and Saaty [40].

EXAMPLE 2.10. Aris and Gavalas [5] have developed a theory of chemical reactions in mixtures with an infinite number of components. In a conventional mixture of discrete chemical species, the species are denoted by A_s, where s is an integer $1 \le s \le S$. If one now considers a mixture with an infinite number of components, the "species" present in such a mixture may be denoted by $A(x)$, where x is a real variable $a \le x \le b$. The concept of a continuous mixture of species has application to extremely complex mixtures (e.g. of hydrocarbons) of which no precise analysis is possible. In such a case one may refer not to individual species but to the fraction boiling between temperatures x and $x + dx$, or having a retention time on a chromatographic column of x to $x + dx$. One lets $c(x, t)$ be the molar concentration distribution at time t.

Consider a monomolecular reaction system, that is, one in which the coupling between each pair of species can be represented by

$$A(x) \underset{k(x', x)}{\overset{k(x, x')}{\rightleftarrows}} A(x')$$

Then the rate of change of the concentration distribution $c(x, t)$ can be written as

$$\begin{aligned}\frac{\partial c(x, t)}{\partial t} &= -\int_a^b k(x, x')c(x, t)\,dx' + \int_a^b k(x', x)c(x', t)\,dx' \\ &= -\kappa(x)c(x, t) + \int_a^b k(x, x')c(x', t)\,dx'\end{aligned}$$

where

$$\kappa(x) = \int_a^b k(x, x')\,dx'$$

Thus, we see that integro-differential equations arise naturally in the theory of reactions in continuous mixtures.

2.9 Summary

In this chapter we have introduced the concept of mathematical modeling and some common types of mathematical models. We have not attempted to face the question of where the models really come from, for example, why in some cases we obtain ordinary differential equation models, whereas in others we obtain partial differential or integral equation models. The selection of the type of mathematical model we use must come before everything else. The type of mathematical model we obtain will depend on the assumptions made in initially defining the system. In general, the more assumptions made, the simpler the structure of the mathematical model.

To illustrate this point, consider the description of the transient cooling of a steel sphere of radius R, initially at temperature T_0, in an air stream of temperature T_1. The equation governing the sphere temperature $T(r, t)$ is

$$\frac{\partial T}{\partial t} = \alpha\left(\frac{\partial^2 T}{\partial r^2} + \frac{2}{r}\frac{\partial T}{\partial r}\right)$$

$$T(r, 0) = T_0$$

$$\frac{\partial T}{\partial r} = 0, \qquad r = 0$$

$$-k\frac{\partial T}{\partial r} = h(T(R, t) - T_1), \qquad r = R$$

where $\alpha = k/\rho C_p$ is the thermal diffusivity of the sphere. Now, if α is very large (large thermal conductivity k), then, for all practical purposes, variations of temperature within the sphere will be negligible. In this case, $T(r, t)$ can be simplified to $T(t)$ only. The problem of describing the transient cooling of the sphere reduces to the solution of the overall energy balance,

$$\rho C_p\left(\frac{4}{3}\pi R^3\right)\frac{dT(t)}{dt} = h(4\pi R^2)(T(t) - T_1)$$

$$T(0) = T_0$$

a lumped as opposed to a distributed parameter model.

As noted in Table 2.1, a distributed parameter model (P.D.E.) takes into account variations in the state of the system (temperature, concentration, velocity) from point to point in the system. On the other hand, a lumped parameter model (O.D.E.) ignores spatial variation of the system states. In reality, all systems are distributed. However, it is often possible, as in the case of the steel sphere with large thermal conductivity, to neglect spatial variation of properties. In general, if the response time of the state profile within the system is fast compared to the time over which external changes are taking place, a lumped parameter approximation will suffice.

Some examples of systems which are usually modeled by lumped and distributed representations are

Lumped	*Distributed*
Vigorously mixed vessel	Packed bed reactor
Plate in a distillation column	Heat exchanger
Electro-mechanical systems	Tubular flow reactor

We cannot give general rules for whether a particular system should be modeled by ordinary or partial differential equations. In the final analysis it depends on how much time and money one is willing to spend for a description of the system.

We shall conclude this chapter with a set of problems. These problems, as in the following chapters, are intended to allow the student to test concepts as well as to introduce certain new direct extensions of the material already covered. We urge the student to work each of the problems.

REFERENCES

1. Acrivos, A., and N. R. Amundson, "Applications of Matrix Mathematics to Chemical Engineering Problems," *Ind. Eng. Chem. 47*, 1533 (1955).
2. Alterman, Z. S., and E. Isaacson, "A Method for Calculating Frontal Motion," *J. Comp. Phys. 4*, 67 (1969).
3. Ames, W. F., *Nonlinear Ordinary Differential Equations in Transport Processes*, Academic Press, New York, 1968.
4. Aris, R., *Elementary Chemical Reactor Analysis*, Prentice-Hall, Inc., Englewood Cliffs, N.J., 1969.
5. Aris, R., and G. R. Gavalas, "On the Theory of Reactions in Continuous Mixtures," *Phil. Trans. Roy. Soc. London A260*, 35 (1966).
6. Athans, M., and P. L. Falb, *Optimal Control*, McGraw-Hill, New York, 1966.
7. Bellman, R., *Methods of Nonlinear Analysis*, vol. 1, Academic Press, New York, 1970.
8. Bellman, R., and K. Cooke, *Differential-Difference Equations*, Academic Press, New York, 1963.
9. Bird, R. B., W. E. Stewart, and E. N. Lightfoot, *Transport Phenomena*, Wiley, New York, 1960.
10. Brennan, R. D., and M. Y. Silberberg, "Two Continuous System Modeling Programs," *IBM System J. 6*, 242 (1967).
11. Cardenas, A., and W. J. Karplus, "PDEL—A Language for Partial Differential Equations," *Commun. ACM 13*, 184 (1970).
12. Carslaw, H. S., and J. C. Jaeger, *Conduction of Heat in Solids*, 2nd ed., Oxford University Press, Inc., New York, 1959.
13. Chang, A., "Synchronization of Traffic Signals in Grid Networks," *IBM J. Res. Develop. 11*, 436 (1967).

14. Crank, J., *The Mathematics of Diffusion*, Oxford University Press, Inc., New York, 1956.

15. Davidson, B., and R. W. Bradshaw, "A Steady State Optimal Design of Artificial Induced Aeration in Polluted Streams," *Water Resources Res. 6*, 383 (1970).

16. Dobell, A. R., and Y. C. Ho, "Optimal Investment Policy: An Example of a Control Problem in Economic Theory," *IEEE Trans. Auto. Cont. 12*, 4 (1967).

17. Edeskuty, F. J., and N. R. Amundson, "Mathematics of Adsorption. IV. Effect of Intraparticle Diffusion in Agitated Static Systems," *J. Phys. Chem. 56*, 148 (1952).

18. Finlayson, B. A., and L. E. Scriven, "The Method of Weighted Residuals—A Review," *Appl. Mech. Rev. 19*, 735 (1966).

19. Francis, N. D., "On the Computation of the Transition Matrix," *Proc. IEEE 57*, 2079 (1969).

20. Fredrickson, A. G., and H. M. Tsuchiya, "Continuous Propagation of Microorganisms," *AIChE J. 9*, 459 (1963).

21. Gall, C. E., and R. Aris, "The Dynamics of Reactors of Mixed Type," *Can. J. Chem. Eng. 43*, 16 (1965).

22. Garfinkel, D., "A Machine-Independent Language for the Simulation of Complex Chemical and Biochemical Systems," *Comp. Biomed. Res. 2*, 31 (1968).

23. Gould, L. A., *Chemical Process Control: Theory and Applications*, Addison-Wesley, Reading, Mass. 1969.

24. Hicks, J. S., and J. Wei, "Numerical Solution of Parabolic Partial Differential Equations with Two Point Boundary Conditions by Use of the Method of Lines," *J. ACM 14*, 549 (1967).

25. Himmelblau, D. M., and K. B. Bischoff, *Process Analysis and Simulation*, Wiley, New York, 1968.

26. Houghton, D. D., and E. Isaacson, "Mountain Winds," *Studies Numer. Anal. 2*, 21 (1968).

27. Koenig, H. E., "Mathematical Models of Socio-Economic Systems," *IEEE Trans. Systems Sci. Cybernetics 1*, 11 (1965).

28. Kolsky, H. G., "Some Computer Aspects of Meteorology," *IBM J. Res. Develop. 11*, 584 (1967).

29. Kwok, H. C. W., W. E. Langlois, and R. A. Ellefsen, "Digital Simulation of the Global Transport of Carbon Monoxide," *IBM J. Res. Develop. 15*, 3 (1971).

30. Lapidus, L., and R. Luus, *Optimal Control of Engineering Processes*, Ginn/Blaisdell, Waltham, Mass. 1967.

31. Lapidus, L., and J. H. Seinfeld, *Numerical Solution of Ordinary Differential Equations*, Academic Press, New York, 1971.

32. Linz, P., "Numerical Methods for Volterra Integral Equations of the First Kind," *Comp. J. 12*, 393 (1969).

33. McGuire, M., and L. Lapidus, "On the Stability of a Detailed Packed-Bed Reactor," *AIChE J. 11*, 819 (1965).

34. O'Brien, J. J., "The Numerical Simulation of Large-Scale Ocean Circulation," in *Proceedings of the 1970 Summer Computer Simulation Conference*, Denver, Col., June 1970.

35. Paterson, W. R., and D. L. Cresswell, "A Simple Method for the Calculation of Effectiveness Factors," *Chem. Eng. Sci. 26*, 605 (1971).

36. Prager, W., "Optimization of Structural Design," *J. Optimization Theory Appl. 6*, 1 (1970).

37. Rektorys, K., *Survey of Applicable Mathematics*, M.I.T. Press, Cambridge, Mass. 1969.

38. Richtmyer, R. D., and K. W. Morton, *Difference Methods for Initial-Value Problems*, Wiley-Interscience, New York, 1967.

39. Rideout, V. C., and J. A. Katra, "Computer Simulation of the Pulmonary Circulation," *Simulation 14*, 239 (1969).

40. Saaty, T. L., *Modern Nonlinear Equations*, McGraw-Hill, New York, 1967.

41. Schechter, R. S., *The Variational Method in Engineering*, McGraw-Hill, New York, 1967.

42. Schiesser, W. E., "A Digital Simulation System for Mixed Ordinary/Partial Differential Equation Models," in *Proceedings of the IFAC Symposium*, Budapest, Hungary, Sept. 1971.

43. Sedlar, M., and G. A. Bekey, "Signal Flow Graphs of Sampled-Data Systems," *IEEE Trans. Auto. Cont. 12*, 154 (1967).

44. Seinfeld, J. H., S. D. Reynolds, and P. M. Roth, "Simulation of Urban Air Pollution," in *Photochemical Smog and Ozone Reactions, Advances in Chemistry 113*, 58 (1972).

45. Seinfeld, J. H., "Modeling Problems in Air Pollution," *Simulation Councils Proc. 1*, 33 (1971).

46. Snyder, M. F., and V. C. Rideout, "Computer Modeling of the Human Systemic Arterial Tree," *J. Biomech. 1*, 341 (1968).

47. Wylie, C. R., *Advanced Engineering Mathematics*, McGraw-Hill, New York, 1960.

48. Zadeh, L. A., and C. A. Desoer, *Linear System Theory*, McGraw-Hill, New York, 1963.

PROBLEMS

2.1 Solve the differential equation

$$\frac{d^2y}{dt^2} - a^2y = 0$$

by conversion to state variable representation and use of the state transition matrix.

2.2 Find the general solution of

$$\frac{d^2y}{dt^2} - a^2y = u(t)$$

2.3 Solve the differential equation

$$\dot{\mathbf{x}} = \mathbf{F}(t)\mathbf{x}$$

for

$$\mathbf{F}(t) = \begin{bmatrix} 0 & 1 \\ -\frac{2}{t^2} & -\frac{4}{t} \end{bmatrix}$$

by use of the state transition matrix.

2.4 Set up the state equations for the system characterized by the equation

$$\frac{d^2y}{dt^2} + 3\frac{dy}{dt} + 2y = u$$

For the matrices **F** and **G** which result, show that the vectors $\mathbf{F}^{n-1}\mathbf{G}, \ldots,$ $\mathbf{FG}, \mathbf{G}$ are linearly independent.

2.5 Consider the case of

$$\mathbf{F} = \begin{bmatrix} -\frac{1}{2} & -\frac{1}{2} \\ \frac{1}{2} & -\frac{1}{2} \end{bmatrix}; \qquad \mathbf{x}_0 = \begin{bmatrix} 1 \\ 0 \end{bmatrix}$$

with an overall time interval of (0, 2). Using the time increment $\Delta t = 0.05$, calculate $\mathbf{x}(t)$ at $t = 2$ via the transition matrix by stepping along in increments of 0.05. Show the influence of error by single and double precision calculations.

2.6 Consider the O. D. E. solution of the form of (2.3.26), namely

$$\mathbf{x}(t) = \mathbf{\Phi}(t)\mathbf{x}_0 + \int_0^t \mathbf{\Phi}(t - \tau)\mathbf{G}\mathbf{u}(\tau)\,d\tau$$

For **G** a constant matrix and $\mathbf{u}(\tau)$ a constant vector, show that the calculation of the integral can then be handled by the formula

$$\int_0^t e^{\mathbf{F}\lambda}\,d\lambda = \mathbf{F}^{-1}[e^{\mathbf{F}t} - \mathbf{I}]$$

Further show that

$$\begin{aligned}\mathbf{F}^{-1}[e^{\mathbf{F}t} - \mathbf{I}]\mathbf{G} &= [(e^{\mathbf{F}t} - \mathbf{I})(e^{\mathbf{F}\tau} - \mathbf{I})^{-1}][\mathbf{F}^{-1}(e^{\mathbf{F}\tau} - \mathbf{I})\mathbf{G}] \\ &= (e^{\mathbf{F}t/2} + \mathbf{I})(e^{\mathbf{F}t/4} + \mathbf{I}) \cdots (e^{\mathbf{F}\tau} + \mathbf{I})[e^{\mathbf{F}\tau} - \mathbf{I}]\mathbf{G}\end{aligned}$$

where $t = 2^r\tau$.

2.7 Show for

$$\mathbf{F} = \begin{bmatrix} 2 & 1 & 1 \\ 1 & 2 & 1 \\ 1 & 1 & 2 \end{bmatrix}, \qquad \mathbf{A} = \begin{bmatrix} 1 & 1 & 1 \\ 1 & 1 & 1 \\ 1 & 1 & 1 \end{bmatrix}$$

that

$$\exp[\mathbf{F}t] = e^t\mathbf{I} + \tfrac{1}{3}[e^{4t} - e^t]\mathbf{A}$$

2.8 Show that

$$\exp[\mathbf{F}t_1]\exp[\mathbf{F}t_2] = \exp[\mathbf{F}(t_1 + t_2)]$$

and that

$$\exp[-\mathbf{F}t] = \{\exp[\mathbf{F}t]\}^{-1}$$

2.9 Stability of linear time-invariant systems. The system (2.3.3) is *stable* if the Euclidean norm

$$||\mathbf{x}(t)|| = \sqrt{x_1^2(t) + x_2^2(t) + \cdots + x_n^2(t)}$$

remains bounded as $t \longrightarrow \infty$ for every solution $\mathbf{x}(t)$. We call the system *strictly* stable if it is stable and if

$$\lim_{t\to\infty} ||\mathbf{x}(t)|| \longrightarrow 0$$

If we denote the eigenvalues of the matrix $\mathbf{F}$ by $\lambda_1, \lambda_2, \ldots, \lambda_n$, the system is stable iff

1. The real part of all λs $[\text{Re}(\lambda_i)] \leq 0$ for all i.
2. If λ_k is a multiple root of the characteristic polynomial of $\mathbf{F}$, then $\text{Re}(\lambda_k) < 0$.

The system is strictly stable iff $\text{Re}(\lambda_i) < 0$ for all i. Investigate the stability of the system (2.3.3) with

$$\mathbf{F} = \begin{bmatrix} a_{11} & a_{12} \\ a_{21} & a_{22} \end{bmatrix}$$

2.10 Show that if λ is an eigenvalue of $\mathbf{F}$ and the system is strictly stable, then the adjoint is unstable. *Hint:* You need to determine that $-\lambda$ is an eigenvalue of $-\mathbf{F}^T$.

2.11 Solve the difference equation

$$y(t_{k+2}) - a^2 y(t_k) = 0$$

by conversion to state variable representation and use of the state transition matrix.

2.12 Find the general solution of

$$y(t_{k+2}) - a^2 y(t_k) = u(t_k)$$

2.13 Solve the difference equation

$$\mathbf{x}(k+1) = \mathbf{F}(k)\mathbf{x}(k) \qquad \text{for } \mathbf{F}(k) = \begin{bmatrix} 0 & 1 \\ -\dfrac{2}{k^2} & -\dfrac{4}{k} \end{bmatrix}$$

2.14 If $y(k+1)$ satisfies the difference equation

$$y(k+1) - 2y(k)\cos\alpha + y(k-1) = 0, \qquad k = 1, 2, \ldots$$

and the initial values $y(0) = 0$, $y(1) = 1$, determine $y(2)$, $y(3)$, and $y(4)$ in terms of α. Show that these results agree with

$$y(k) = \frac{\sin k\alpha}{\sin\alpha}$$

2.15 The three-term difference equation

$$T(k+1) - xT(k) + \tfrac{1}{4}T(k-1) = 0, \qquad k = 1, 2, \ldots$$

may be used to generate successive Chebyshev polynomials

$$T(k) = \frac{1}{2^{k-1}} \cos(k \cos^{-1} x), \qquad |x| \leq 1$$
$$k = 0, 1, 2, \ldots$$

starting with $T(0) = 2$ and $T(1) = x$. Show that this is true and generate $T(2)$, $T(3)$, and $T(4)$.

2.16 Given the one-dimensional heat equation

$$\frac{\partial^2 z}{\partial x^2} = \frac{\partial z}{\partial y}$$

without specifying the boundary conditions, show that the approximations

$$\frac{\partial^2 z}{\partial x^2} \simeq \frac{z_{r+1,s} - 2z_{r,s} + z_{r-1,s}}{\Delta x^2}$$

$$\frac{\partial z}{\partial y} \simeq \frac{z_{r,s+1} - z_{r,s}}{\Delta y}$$

lead to a difference equation in $z_{r,s+1}$. In these differences r and s refer to discrete points in the x- and y-spaces, respectively.

2.17 Consider the scalar differential-difference equation

$$\dot{x}(t) = x(t-1)$$
$$x(t) = 1, \qquad 0 \leq t \leq 1$$

Show by induction that the solution of this equation subject to the given initial conditions is

$$x(t) = \sum_{j=0}^{N} \frac{(t-j)^j}{j}, \qquad N \leq t \leq N+1$$
$$N = 0, 1, 2, \ldots$$

2.18 Acrivos and Amundson [1] have considered the chemical reaction

$$A_1 \underset{k_1'}{\overset{k_1}{\rightleftharpoons}} A_2 \underset{k_2'}{\overset{k_2}{\rightleftharpoons}} A_3$$

taking place in a sequence of continuous stirred tank reactors (CSTRs). If $x_i(k)$ is the concentration of the ith species ($i = 1, 2, 3$) in the kth CSTR and θ is a constant average residence time in each CSTR, balances of the form

$$\dot{x}_1(k) = -\left(k_1 + \frac{1}{\theta}\right)x_1(k) + k_1' x_2(k) + \frac{1}{\theta}x_1(k-1)$$

$$\dot{x}_2(k) = k_1 x_1(k) - \left(k_1' + k_2 + \frac{1}{\theta}\right)x_2(k) + k_2' x_3(k) + \frac{1}{\theta}x_2(k-1)$$

$$\dot{x}_3(k) = k_2 x_2(k) - \left(k_2' + \frac{1}{\theta}\right)x_3(k) + \frac{1}{\theta}x_3(k-1)$$

result. Place these equations in the form

$$\dot{\mathbf{x}}(k) = \mathbf{F}\mathbf{x}(k) + \mathbf{G}\mathbf{x}(k-1)$$

2.19 Derive the complete form of (2.6.7).

2.20 Derive (2.6.9).

2.21 Typical boundary conditions for the heat equation

$$\frac{\partial z}{\partial y} - \alpha \frac{\partial^2 z}{\partial x^2} = 0$$

depend on the nature of the physical boundaries of the region. Assume that this equation governs the temperature distribution in a region bounded by $x = 0$ and $x = l$. Specify the proper boundary conditions if

a. There exist constant heat fluxes Q_0 and Q_l (calories per area-time) at $x = 0$ and $x = l$, respectively.

b. The heat fluxes at $x = 0$ and $x = l$ are proportional to the difference between the temperature of the region and the ambient temperatures z_0 and z_l.

2.22 Write any consistent set of general boundary conditions for (2.6.1).

2.23 Consider the case of an infinitely rapid chemical reaction which takes place on the surface of a solid slab such that the solid surface is continuously removed, causing it to move inward as the reaction progresses and leaving a porous inert layer. The situation is depicted in Fig. P2.23 at time $t = 0$

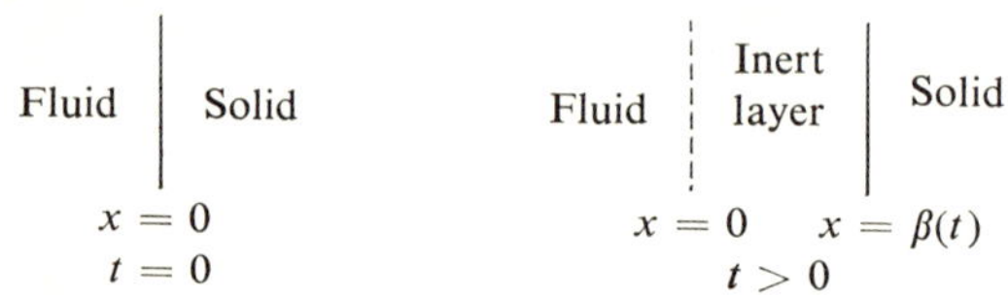

Figure P2.23

and some time $t > 0$ after which the solid interface has progressed back to $x = \beta(t)$. The concentration of fluid in the inert layer is described by the diffusion equation (same as the heat equation)

$$\frac{\partial z}{\partial t} - \alpha \frac{\partial^2 z}{\partial x^2} = 0$$

If $\beta(t)$ is the position of the moving solid interface and the chemical reaction is of the type fluid + solid $\longrightarrow$ products with no change in volume, write a consistent set of boundary and initial conditions for this problem.

2.24 Consider an agitated vessel containing W grams of spherical porous adsorbent and V liters of adsorbate-free solvent [17]. At time zero, adsorbate is admitted to the vessel so that the concentration in the solution is C_0. The agitation is sufficient to prevent gradients in the solution but a gradient remains around the porous adsorbent. Adsorbate must go from the solution, C, through the gradient; diffuse into the adsorbent, c; and then adsorb on the surface, n. Show that the appropriate balances are

$$D\left[c_{rr} + \frac{2}{r} c_r\right] = c_t + \frac{1}{\alpha} n_t \qquad \text{(inside the adsorbent)}$$

$$n = f(c) \qquad \text{(adsorption)}$$

$$-Dc_r = k(c - C), \quad r = R \qquad \text{(sphere surface)}$$

$$-VC_t = \frac{3D\alpha W}{R\rho} C_r, \quad r = R \qquad \text{(in the solution)}$$

and that boundary conditions are

$$c = c_0 \qquad t = 0$$
$$C = C_0$$
$$c, C \text{ remain finite for all } r \text{ and } t$$

D, α, k, and ρ are all real constants.

2.25 Show that the O.D.E. and boundary conditions

$$\ddot{y} + y = 0, \qquad y(0) = \ddot{y}(0) = 0, \qquad \dot{y}(0) = 1$$

are equivalent to the integral equation

$$y(t) = t + \int_0^t (\zeta - t)y(\zeta)\, d\zeta$$

2.26 Show that (2.7.15) and (2.7.16) follow from (2.7.13) and (2.7.14), respectively.

2.27 Show that for the O.D.E.

$$\ddot{y} + ty = 1$$

and for the conditions $y(0) = \dot{y}(0) = 0$, the Volterra equation

$$y(t) = \int_0^t \zeta(\zeta - t)y(\zeta)\, d\zeta + \frac{1}{2}t^2$$

satisfies $y(t)$.

2.28 For the O.D.E. and boundary conditions

$$\ddot{y} + a\dot{y} + by = 0$$
$$y(0) = y(1) = 0$$

show that the integral equation

$$y(t) = \int_0^1 K(t, \zeta)y(\zeta)\, d\zeta$$

with

$$K(t, \zeta) = \begin{cases} b\zeta(1 - t) + at - a, & \zeta \leq t \\ bt(1 - \zeta) + at, & \zeta \geq t \end{cases}$$

results. Is this kernel symmetric? Is it discontinuous? What influence does $a = 0$ have on the kernel?

2.29 Show that the Laplace transform of $y(t) = \alpha$ = constant is given by $\bar{y}(s) = \alpha/s$.

2.30 For the integral equation

$$y(t) = \lambda \int_0^1 t\zeta y(\zeta)\, d\zeta + 1$$

show that an iterative procedure yields

$$y(t) = 1 + t\left[\frac{\lambda}{2} + \frac{\lambda^2}{6} + \frac{\lambda^3}{18} + \cdots\right]$$

The Laplace Transform 3

In this chapter we shall consider in detail the properties and uses of the Laplace transform, which we mentioned previously in Section 2.7.2. While this is only one of many integral transforms, the Laplace transform is perhaps the most useful, and thus special care is given to its development.

3.1 General Development

In general, integral transforms, and particularly the Laplace transform (denoted by L.T. from this point on), are most useful for the solution of linear ordinary and partial differential equations.

We might compare the L.T. operation to the evaluation of D in $D = AB/C$ by taking logarithms of both sides. We need to use only addition and subtraction rather than multiplication and division and then finally take the antilog to recover the value of D. Figure 3.1 shows a schematic diagram of the process of solving linear O.D.E. and P.D.E. by the L.T. First, the transform converts the O.D.E. to an algebraic equation or the P.D.E. to an O.D.E. Next, these are solved in the transform domain, finally being converted back to the original space (often called the time domain space if time is the independent variable).

In more precise terms the integral L.T. is a continuous linear transformation which projects a function in a given space onto one or more continuous

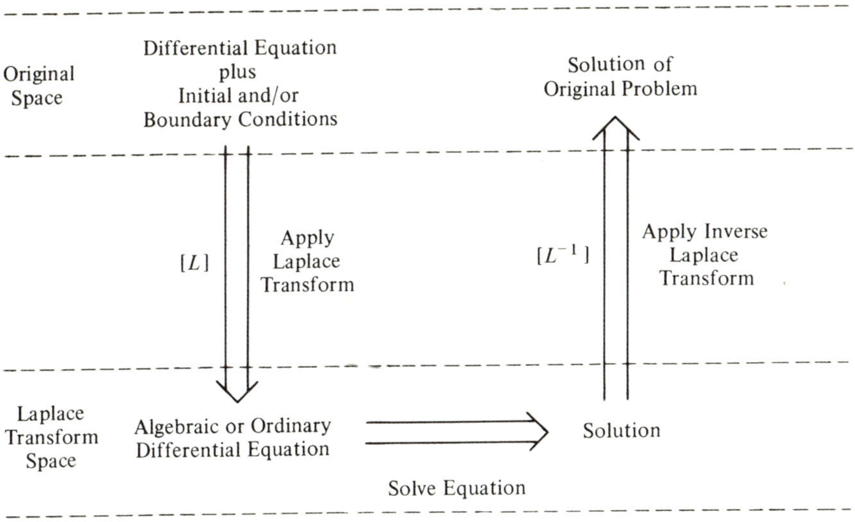

Figure 3.1. Schematic for application of Laplace transform.

bases which span the space. Furthermore, this linear transformation has a unique inverse of the same form; that is, we can recover the originally transformed function by means of a similar continuous linear transformation.

3.1.1 Definition and nomenclature

Consider the operation of multiplying a function $F(t)$, defined for positive t, by e^{-st}, integrating the result from zero to infinity and thereby generating a new function denoted by $f(s)$ of the real or complex variable s. In effect the function $F(t)$ has been mapped from the t-domain to the s-domain and in doing so the L.T. of $F(t)$ has been formed. Symbolically this procedure is given by

$$f(s) = \int_0^\infty e^{-st}F(t)\,dt \tag{3.1.1}$$

where $f(s)$ is called the Laplace transform of $F(t)$. $f(s)$ may also be denoted by

$$f(s) = L[F(t)] = \bar{F}(s) \tag{3.1.2}$$

where lowercase letters are taken for the transformed variables and uppercase letters for the original function. In some instances it may also prove advantageous to use the uppercase letter with a bar over it to indicate the transformed function.

Since $F(t)$ may be transformed into $f(s)$, a table of transform pairs can be developed in the same way that a table of integrals is connected to defined equations of integration. In other words, an image table can be developed between the functions $F(t)$ and $f(s)$. Once such a table is available it is possible to move back and forth for specific functions between the t- and s-domains. Since the Laplace transform of $F(t)$ is denoted by $f(s)$ (moving from the t-domain to the s-domain), the reverse process may be denoted (moving from the s-domain to the t-domain) by

$$L^{-1}[f(s)] = F(t) \tag{3.1.3}$$

where $L^{-1}[f(s)]$ is called the inverse transform of $f(s)$. This inverse process may be looked upon as solving the integral equation

$$f(s) = \int_0^\infty e^{-st}F(t)\,dt$$

where $F(t)$ is to be determined with $f(s)$ given.

Thus we now have the two symbolic paths

$$\begin{aligned} &F(t) \xrightarrow{L} f(s) \qquad \text{(direct L.T.)} \\ &F(t) \xleftarrow{L^{-1}} f(s) \qquad \text{(inverse L.T.)} \end{aligned} \tag{3.1.4}$$

as indicated in Figure 3.1.

It is, of course, possible that for certain $F(t)$ the integral of (3.1.1) will fail to converge. In actuality, almost all functions representing real physical systems will lead to convergence. However, it is well at this point to describe the *sufficient* conditions on $F(t)$ which will allow the integral to exist:

1. If $F(t)$ is continuous or sectionally continuous in any finite interval $a \leq t \leq b$, where $a > 0$,
2. And if $e^{-\alpha t}|F(t)|$ is bounded for large values of t, with $\alpha =$ constant,
3. And if $t^n|F(t)|$ is bounded near $t = 0$, where n is some number less than 1.0,

then the Laplace transform of $F(t)$ exists.

A function $F(t)$ is said to be sectionally continuous over a finite range if the interval can be divided into a finite number of intervals such that $F(t)$ is continuous within each interval and also approaches finite values as t tends toward either end of the interval proceeding from the interior out. Thus a function which has *finite* discontinuities within a finite range is sectionally continuous. At a point within this interval, say $t = t_1$, $f(t)$ may have one limit as approached from the left and another limit as t_1 is approached from the right. For such a sectionally continuous or, of course, a continuous function, the integral $\int_{t_1}^{T} e^{-st}F(t)\,dt$ converges.

A function $F(t)$ is termed of exponential order if as $t \longrightarrow \infty$ the product $e^{-\alpha t}|F(t)|$ is bounded for $t > T$, where α and T are finite constants. In effect any increase in $F(t)$ is damped by the exponential product. Thus when $s > \alpha$ the integral $\int_T^\infty e^{-st}F(t)\,dt$ converges.

In addition, if $F(t)$ approaches a finite limit as $t \longrightarrow 0^+$ or if $|F(t)| \longrightarrow \infty$ as $t \longrightarrow 0^+$ in such a way that for $n < 1.0$, $t^n|F(t)|$ remains bounded near $t = 0$, then the integral $\int_0^{t_1} e^{-st}F(t)\,dt$ converges. Note that this does not rule out an infinite discontinuity at $t = 0$.

Since the sum of the three convergent integrals represents $\int_0^\infty e^{-st}F(t)\,dt$, the Laplace transform of $F(t)$ exists if statements 1–3 are fulfilled.

The reader may wish to show that $F(t) = \exp[t^2]$ has no L.T. This function is not of exponential order since $|\exp[t^2]|$ grows more rapidly as t tends toward infinity than $e^{\alpha t}$, where α is constant.

It is important to note that from the conditions imposed upon the convergence near $t = 0$ the Laplace transform integral should really be written

$$\int_{0^+}^{\infty} e^{-st}F(t)\,dt \tag{3.1.5}$$

indicating that t approaches zero from the plus side. This point should be kept in mind.

Further, it must be realized that a L.T. can only be applied to functions which are not restricted to a finite interval of t; i.e., if $c(t)$ represents the concentration output of a continuous stirred reactor as a function of time, then the L.T. of $c(t)$ exists since $c(t)$ is defined as $t \longrightarrow \infty$. However, if $T(x)$ represents the temperature distribution of a finite slab with dimension $0 \leq x \leq 1$, we cannot take the L.T. of $T(x)$ since the function is defined only over a finite region.

The question of uniqueness of the transform is also of interest. Stated simply, $f(s)$ is unique for each function $F(t)$ that has such a transform. However, the inverse process may not be unique (although this is not of importance in real physical systems). Different discontinuous functions may have the same transform; i.e., for a step function (Section 3.2) the Laplace transform is the same irrespective of the value of $F(t)$ at $t = 0$. However, when two or more functions are continuous (Lerch's theorem) they must be the same function to yield the same transform. Thus a specific $f(s)$ cannot have more than one inverse transform continuous for all $t > 0$. It then follows that for almost all $t > 0$ a unique $F(t)$ corresponds to an $f(s)$ [8].

3.1.2 The relationship of the Fourier transform to the Laplace transform

We can express a *periodic* function $F(t)$ in a Fourier series

$$F(t) = A_0 + \sum_{k=1}^{\infty} (A_k \cos k\omega t + B_k \sin k\omega t) \tag{3.1.6}$$

where, if the period of $F(t)$ is τ, the coefficients in the Fourier series are

$$A_0 = \frac{1}{\tau}\int_{-\tau/2}^{\tau/2} F(t)\,dt \tag{3.1.7}$$

$$A_k = \frac{2}{\tau}\int_{-\tau/2}^{\tau/2} F(t)\cos k\omega t\,dt \tag{3.1.8}$$

$$B_k = \frac{2}{\tau}\int_{-\tau/2}^{\tau/2} F(t)\sin k\omega t\,dt \tag{3.1.9}$$

and $\omega = 2\pi/\tau$. We can write these relations in the compact form

$$F(t) = \sum_{k=-\infty}^{\infty} f(i\omega_k)e^{i\omega_k t} \tag{3.1.10}$$

$$f(i\omega_k) = \frac{1}{\tau}\int_{-\tau/2}^{\tau/2} F(t)e^{-i\omega_k t}\,dt \tag{3.1.11}$$

where

$$\omega_k = k\omega, \qquad k = 0, \pm 1, \pm 2, \ldots \tag{3.1.12}$$

Equations (3.1.10) and (3.1.11) are the Fourier transform pair for a periodic function $F(t)$.

The more important case from our standpoint is that of an *aperiodic* function. We may consider this case as a limit case of the periodic transform with $\tau \longrightarrow \infty$. We then allow the separation between frequencies to approach zero:

$$\lim_{\tau\to\infty} \Delta\omega_k = d\omega$$

Thus the discrete frequency $\omega_k = 2\pi k/\tau$ becomes a continuous frequency ω, and we replace summations by integrations.

We therefore obtain the continuous Fourier transform pair:

$$F(t) = \frac{1}{2\pi}\int_{-\infty}^{\infty} f(i\omega)e^{i\omega t}\,d\omega \tag{3.1.13}$$

$$f(i\omega) = \int_{-\infty}^{\infty} F(t)e^{-i\omega t}\,dt \tag{3.1.14}$$

$f(i\omega)$ is called the Fourier transform of $F(t)$. It is a continuous function of the frequency ω and is, in general, complex. The Fourier transform of a function $F(t)$ is not guaranteed to exist unless the function is absolutely integrable, namely, unless

$$\int_{-\infty}^{\infty} |F(t)|\,dt < \infty \tag{3.1.15}$$

Many of the functions in which we shall be interested unfortunately do not meet this requirement, typical examples being the unit step function and the sine function. It is convenient to modify (3.1.15) in the form (consider now only functions that are nonzero for $t \geq 0$)

$$\int_0^t e^{-at} |F(t)| \, dt < \infty \tag{3.1.16}$$

so that there exists a constant $a \geq 0$ such that (3.1.16) holds. The smallest a for which (3.1.16) holds is called the abscissa for convergence and is denoted a_0.

If (3.1.16) holds, we then consider the Fourier transform of $e^{-at}F(t)$, which for $a > a_0$ is guaranteed to exist,

$$f(a + i\omega) = \int_0^\infty F(t)e^{-(a+i\omega)t} \, dt \tag{3.1.17}$$

If we define the complex frequency s by

$$s = a + i\omega \tag{3.1.18}$$

then the Fourier transform of $e^{-at}F(t)$ becomes the L.T. of $F(t)$,

$$f(s) = \int_0^\infty F(t)e^{-st} \, dt$$

Therefore the L.T. is a natural result of providing the Fourier transform with a built-in convergence factor.

3.1.3 Some simple transforms

At this point it will be instructive to illustrate how the corresponding $f(s)$ may be obtained for some elementary $F(t)$.

1. $F(t) = e^{-at}$, $a =$ constant:

$$f(s) = \int_{0^+}^\infty e^{-st}e^{-at} \, dt = \int_{0^+}^\infty e^{-(s+a)t} \, dt$$

$$= -\frac{e^{-(s+a)t}}{s + a}\bigg|_{0^+}^\infty = \frac{1}{s + a}$$

In the nomenclature of (3.1.4)

$$e^{-at} \xrightarrow{L} \frac{1}{s + a}, \qquad e^{-at} \xleftarrow{L^{-1}} \frac{1}{s + a}$$

2. $F(t) = \sin at$, $a =$ constant:

$$L[\sin at] = f(s) = \int_{0^+}^\infty e^{-st} \sin at \, dt$$

$$= \frac{e^{-st}}{s^2 + a^2}[-a \sin at + a \cos at]\bigg|_{0^+}^\infty$$

$$= \frac{a}{s^2 + a^2}$$

3. $F(t) = \begin{cases} 0, & t < 0 \\ a, & t > 0 \end{cases}$, $F(t) =$ step function of height a at $t = 0$ or $F(t) =$ the

constant a:

$$L[F(t)] = f(s) = \int_{0^+}^{\infty} e^{-st}a\,dt$$

$$= -\frac{a}{s}e^{-st}\Big|_{0^+}^{\infty} = \frac{a}{s}$$

If $a = 1.0$, then $f(s) = 1/s$.

4. $F(t) = t^n$:

$$L[t^n] = f(s) = \int_{0^+}^{\infty} e^{-st}t^n\,dt$$

By calling $st = x$, $s\,dt = dx$ there results

$$f(s) = \int_{0^+}^{\infty} \left(\frac{x}{s}\right)^n e^{-x}\frac{dx}{s} = \frac{1}{s^{n+1}}\int_{0^+}^{\infty} x^n e^{-x}\,dx$$

and recognizing the integral as the definition of the *gamma function*, $\Gamma(n + 1)$,

$$f(s) = \frac{\Gamma(n+1)}{s^{n+1}}$$

5. $F(t) = e^{-bt}\sin at$, a, b = constants:

$$L[e^{-bt}\sin at] = f(s) = \int_{0^+}^{\infty} e^{-st}e^{-bt}\sin at\,dt$$

$$= \int_{0^+}^{\infty} e^{-(s+b)t}\sin at\,dt$$

From 2,

$$f(s) = \frac{a}{(s+b)^2 + a^2}$$

Based on this simple material we could begin to construct a table of Laplace transform pairs, i.e.,

$F(t)$	$f(s)$
e^{-at}	$\frac{1}{s+a}$
$\sin at$	$\frac{a}{s^2 + a^2}$
	$\vdots$

which allows travel back and forth between the t- and s-domains. The analogy to log and antilog tables and to differential and integral tables is obvious. However, to go beyond the exercise of evaluating specific integrals, we now need to turn to more general properties of the L.T.

TABLE 3.1

$$L[F(t)] = \int_0^\infty e^{-st}F(t)\,dt = f(s)$$

$$F(t) \xrightarrow{L} f(s) \qquad F(t) \xleftarrow{L^{-1}} f(s)$$

A. Laplace Transform Operations

$F(t)$	$f(s)$	Comments
$aF(t)$	$af(s)$	Linearity, $a =$ constant
$\dot{F}(t)$	$sf(s) - F(0^+)$	Differentiation of $F(t)$
$\overset{[n]}{F}(t)$	$s^n f(s) - s^{n-1}F(0^+) - s^{n-2}\dot{F}(0^+)\cdots\overset{[n-1]}{F}(0^+)$	Differentiation of $F(t)$
$\int_0^t F(t)\,dt$	$\dfrac{f(s)}{s}$	Integration of $F(t)$
$F(t-a)$ $F(t+a)$	$e^{-as}f(s)$ $e^{+as}f(s)$	Shifting or translation, $a =$ constant
$(-t)^n F(t)$	$\overset{[n]}{f}(s)$	Differentiation of $f(s)$
$\dfrac{F(t)}{t}$	$\int_s^\infty f(s)\,ds$	Integration of $f(s)$
$F(t) = F(t+\tau) = F(t+n\tau)$	$\dfrac{\int_0^\tau e^{-st}F(t)\,dt}{1-e^{-s\tau}}$	$F(t)$ is periodic with period τ
$\int_0^t F_1(t-\lambda)F_2(\lambda)\,d\lambda = \int_0^t F_2(t-\lambda)F_1(\lambda)\,d\lambda$	$f_1(s)f_2(s)$	Convolution
$\lim_{t\to 0} F(t) = \lim_{s\to\infty} sf(s)$		Initial value theorem
$\lim_{t\to\infty} F(t) = \lim_{s\to 0} sf(s)$		Final value theorem
$F(t) = \sum_1^i \dfrac{p(s_n)}{\dot{q}(s_n)} e^{s_n t}$	$f(s) = \dfrac{p(s)}{q(s)}$	Heaviside expansion

i roots of $q(s) = 0$ are real and nonrepeated.
For cases where roots are repeated or complex, see the text.

B. Laplace Transform Pairs

$F(t)$	$f(s)$	Comments
a	$\dfrac{a}{s}$	$a =$ constant
$\dfrac{t^n}{\Gamma(n+1)}$	$\dfrac{1}{s^{n+1}}$	$\Gamma(\cdot) =$ gamma function
t^n	$\dfrac{\Gamma(n+1)}{s^{n+1}}$	
e^{-at}	$\dfrac{1}{s+a}$	

TABLE 3.1 (Continued)

$F(t)$	$f(s)$	Comments
$\dfrac{e^{-at}-e^{-bt}}{b-a}$	$\dfrac{1}{(s+a)(s+b)}$	
$\dfrac{(c-a)e^{-at}-(c-b)e^{-bt}}{b-a}$	$\dfrac{s+c}{(s+a)(s+b)}$	
$\dfrac{1}{ab}+\dfrac{be^{-at}-ae^{-bt}}{ab(a-b)}$	$\dfrac{1}{s(s+a)(s+b)}$	
$\dfrac{e^{-at}}{(b-a)(c-a)}+\dfrac{e^{-bt}}{(a-b)(c-b)}+\dfrac{e^{-ct}}{(a-c)(b-c)}$	$\dfrac{1}{(s+a)(s+b)(s+c)}$	
$\dfrac{(d-a)e^{-at}}{(b-a)(c-a)}+\dfrac{(d-b)e^{-bt}}{(a-b)(c-b)}+\dfrac{(d-c)e^{-ct}}{(a-c)(b-c)}$	$\dfrac{s+d}{(s+a)(s+b)(s+c)}$	
$\dfrac{1}{a}\sin at$	$\dfrac{1}{s^2+a^2}$	
$\dfrac{1}{a}\sinh at$	$\dfrac{1}{s^2-a^2}$	
$\cos at$	$\dfrac{s}{s^2+a^2}$	
$\cosh at$	$\dfrac{s}{s^2-a^2}$	
te^{-at}	$\dfrac{1}{(s+a)^2}$	
$[(b-a)t+1]e^{-at}$	$\dfrac{s+b}{(s+a)^2}$	
$\dfrac{1-(1+at)e^{-at}}{a^2}$	$\dfrac{1}{s(s+a)^2}$	
$\dfrac{e^{-at}+at-1}{a^2}$	$\dfrac{1}{s^2(s+a)^2}$	
$\dfrac{(b-a)}{a^2}e^{-at}+\dfrac{bt}{a}+\dfrac{a-b}{a^2}$	$\dfrac{s+b}{s^2(s+a)}$	
$\dfrac{a^2-ab+c}{a^2}e^{-at}+\dfrac{ct}{a}+\dfrac{ba-c}{a^2}$	$\dfrac{s^2+bs+c}{s^2(s+a)}$	
$\dfrac{1}{a^2}t-\dfrac{1}{a^3}\sin at$	$\dfrac{1}{s(s^2+a^2)}$	
$\dfrac{1}{b^2-a^2}\left[\dfrac{1}{a}\sin at+\dfrac{1}{b}\sin bt\right]$	$\dfrac{1}{(s^2+a^2)(s^2+b^2)}$	
$\dfrac{1}{2a}t\sin at$	$\dfrac{s}{(s^2+a^2)^2}$	
$t\cos at$	$\dfrac{s^2-a^2}{(s^2+a^2)^2}$	
$\sin at \sinh at$	$\dfrac{s}{s^4+a^4}$	
$\dfrac{1}{2a^3}[\sinh at-\sin at]$	$\dfrac{1}{s^4-a^4}$	

TABLE 3.1 (Continued)

$F(t)$	$f(s)$	Comments
$e^{-at}\sin bt$	$\dfrac{b}{(s+a)^2+b^2}$	
$e^{-at}\cos bt$	$\dfrac{s+a}{(s+a)^2+b^2}$	
$\dfrac{1}{a}\operatorname{erf}(a\sqrt{t})$	$\dfrac{1}{s\sqrt{s+a^2}}$	erf(·) = error function
$\operatorname{erf}\left(\dfrac{at}{2}\right)$	$\dfrac{1}{s}e^{(s/a)^2}\left[1-\operatorname{erf}\dfrac{s}{a}\right]$	
$e^t[\operatorname{erf}\sqrt{t}-1]+\dfrac{1}{\sqrt{\pi t}}$	$\dfrac{1}{1+\sqrt{s}}$	
$e^t[1-\operatorname{erf}\sqrt{t}]$	$\dfrac{1}{s+\sqrt{s}}$	
$\dfrac{a}{2\sqrt{\pi t^3}}\exp\left(-\dfrac{a^2}{4t}\right)$	$e^{-a\sqrt{s}}$	
$\dfrac{1}{\sqrt{\pi t}}\exp\left(-\dfrac{a^2}{4t}\right)$	$\dfrac{e^{-a\sqrt{s}}}{\sqrt{s}}$	
$\operatorname{erfc}\left(\dfrac{a}{2\sqrt{t}}\right)$	$\dfrac{e^{-a\sqrt{s}}}{s}$	erfc(·)=complimentary error function
$ab\left[1-\operatorname{erf}\left(b\sqrt{t}+\dfrac{a}{2\sqrt{t}}\right)\right]+e^{b^2t}$	$\dfrac{e^{-a\sqrt{s}}}{s+b\sqrt{s}}$	
$2\sqrt{\dfrac{t}{\pi}}\exp\left(-\dfrac{a^2}{4t}\right)-a\operatorname{erfc}\left(\dfrac{a}{2\sqrt{t}}\right)$	$\dfrac{e^{-a\sqrt{s}}}{s^{3/2}}$	
$\dfrac{1}{a\sqrt{\pi}}\exp\left(-\dfrac{t^2}{4a^2}\right)$	$e^{a^2s^2}\operatorname{erfc}(as)$	
$\operatorname{erf}\left(\dfrac{t}{2a}\right)$	$\dfrac{1}{s}e^{a^2s^2}\operatorname{erfc}(as)$	
$J_0(at)$	$\dfrac{1}{\sqrt{s^2+a^2}}$	$J_0(\cdot)$ = Bessel function
$I_0(at)$	$\dfrac{1}{\sqrt{s^2-a^2}}$	
$\dfrac{1}{a^n}J_n(at)$	$\dfrac{1}{\sqrt{s^2+a^2}(s+\sqrt{s^2+a^2})^n}$	
$\dfrac{1}{a^n}I_n(at)$	$\dfrac{1}{\sqrt{s^2-a^2}(s+\sqrt{s^2+a^2})^n}$	$I_n(\cdot)$ = Bessel function
$I_0(2\sqrt{at})$	$\dfrac{1}{s}e^{+a/s}$	
$J_0(a\sqrt{t^2+2bt})$	$\dfrac{e^{b(s-\sqrt{s^2+a^2})}}{\sqrt{s^2+a^2}}$	

TABLE 3.1 (Continued)

C. Input Functions

$U(t)$	$\frac{1}{s}$	Unit step function	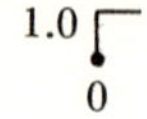
$U(t-t_1)$	$\frac{e^{-st_1}}{s}$	Displaced unit step function	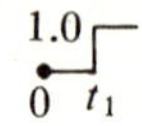
$U(t-t_1)-U(t-t_2)$	$\frac{e^{-st_1}-e^{-st_2}}{s}$	Displaced unit pulse function	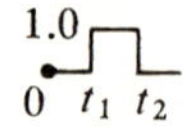
$\delta(t)$	1.0	Dirac or unit impulse function	
$\delta(t-t_1)$	e^{-st_1}	Displaced Dirac function	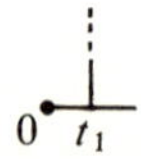
$\frac{1}{t_1}[tU(t)-(t-t_1)U(t-t_1)]$	$\frac{1}{t_1 s^2}[1-e^{-st_1}]$	Ramp function	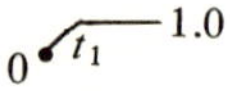
$tU(t)$	$\frac{1}{s^2}$	Linear scale	
$tU(t)-(t-t_1)U(t-t_1)$ $-t_1U(t-t_1)$	$\frac{1-(1+st_1)e^{-st_1}}{s^2}$	Sawtooth pulse	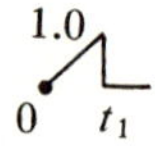
$(t-\tau)U(t-\tau)$ $-2(t-2\tau)U(t-2\tau)$ $+(t-3\tau)U(t-3\tau)$	$\frac{e^{-s}[1-e^{-s}]^2}{s}$	Displaced triangular pulse	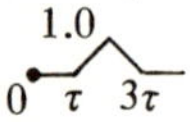
	$\frac{\tau}{s[1-e^{-s\tau}]}$	Staircase function	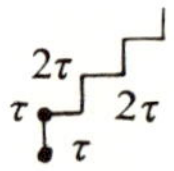
	$\frac{1}{1-e^{-s\tau}}$	Impulse train	
	$\frac{1}{s}\frac{1-e^{-st_1}}{1-e^{-st_2}}$	Square-wave pulse train	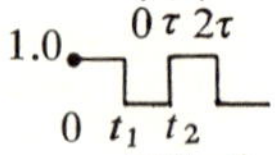
	$\frac{1-e^{-s\tau}-s\tau e^{-s\tau}}{s^2(1-e^{-s\tau})}$	Periodic sawtooth	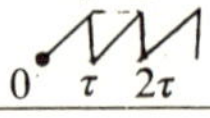

D. z-Transforms (τ = sampling period)

$F(t)$	Laplace transform	z-Transform	Comments
$U(t)$	$\frac{1}{s}$	$\frac{z}{z-1}$	Step function
$\delta(t)$	1	1	Impulse at $t=0$
$\delta(t-kt)$	$e^{-k\tau s}$	$\frac{1}{z^k}$	Impulse at $t=k\tau$
t	$\frac{1}{s^2}$	$\frac{\tau z}{(z-1)^2}$	Ramp
e^{-at}	$\frac{1}{s+a}$	$\frac{z}{z-e^{-a\tau}}$	Exponential

TABLE 3.1 (Continued)

$F(t)$	Laplace transform	z-Transform	Comments
$\frac{t^2}{2}$	$\frac{1}{s^3}$	$\frac{\tau^2 z(z+1)}{z(z-1)^3}$	
te^{-at}	$\frac{1}{(s+a)^2}$	$\frac{\tau z e^{-a\tau}}{(z-e^{-a\tau})^2}$	
$1-e^{-at}$	$\frac{a}{s(s+a)}$	$\frac{z(1-e^{-a\tau})}{(z-1)(z-e^{-a\tau})}$	
$\sin at$	$\frac{a}{s^2+a^2}$	$\frac{z \sin a\tau}{z^2 - 2z\cos a\tau + 1}$	Sine
$\cos at$	$\frac{s}{s^2+a^2}$	$\frac{z(z-\cos a\tau)}{z^2 - 2z\cos a\tau + 1}$	Cosine
	$\frac{1}{1-e^{-\tau s}}$	$\frac{z}{z-1}$	Impulse train
$\frac{d^k}{k!}\int_0^b e^{-ab}b^k\,db$		$\frac{\exp\{[(d/z)-a]b\}-1}{(d/z)-a}$	Modified incomplete gamma integral
	$\lim_{t\to 0} F(t) = \lim_{z\to\infty} f(z)$		Initial value theorem
	$\lim_{t\to\infty} F(t) = \lim_{z\to 1} \frac{z-1}{z} f(z)$		Final value theorem
$F(t-\tau)$		$\frac{f(z)}{z}$	τ translation
$F(t+\tau)$		$zf(z) - zF(0)$	Shifting theorem

3.1.4 General properties

Here we shall develop a number of important properties of the L.T. They are tabulated in Table 3.1 along with a specific series of transforms. For a more detailed table, see [27].

1. *The L.T. is a linear operation.* Thus, for a = constant,

$$\begin{aligned} F_1(t) + F_2(t) &\xrightarrow{L} f_1(s) + f_2(s) \\ F(at) = aF(t) &\xrightarrow{L} af(s) \\ F_1(t) + F_2(t) &\xleftarrow{L^{-1}} f_1(s) + f_2(s) \\ F(at) = aF(t) &\xleftarrow{L^{-1}} af(s) \end{aligned} \tag{3.1.19}$$

2. *The L.T. of a derivative.* Assuming that the transforms exist and that $L[F(t)] = f(s)$, we write

$$L\left[\frac{dF}{dt} = \dot{F}(t)\right] = \int_{0^+}^{\infty} e^{-st}\dot{F}(t)\,dt$$

Integrating by parts,

$$L[\dot{F}(t)] = e^{-st}F(t)\Big|_{0^+}^{\infty} + s\int_{0^+}^{\infty} e^{-st}F(t)\,dt$$

The second term on the right is just $sf(s)$ and if $F(t)$ is of exponential order, the first term vanishes as $t \longrightarrow \infty$. Thus

$$L[\dot{F}(t)] = sf(s) - F(0^+)$$

or

$$\dot{F}(t) \xrightarrow{L} sf(s) - F(0^+) \tag{3.1.20}$$

where $F(0^+)$ designates the limit of $F(t)$ as $t \longrightarrow 0$ from the positive side. By induction, it also follows that

$$\ddot{F}(t) \xrightarrow{L} s^2 f(s) - sF(0^+) - \dot{F}(0^+) \tag{3.1.21}$$

$$\overset{[n]}{F}(t) \xrightarrow{L} s^n f(s) - s^{n-1}F(0^+) - \cdots - \overset{[n-1]}{F}(0^+) \tag{3.1.22}$$

This property of the L.T. may be one of the most important noted.

The reverse operation is also of interest since if $F(0^+) = 0$,

$$\dot{F}(t) = G(t) \xleftarrow{L^{-1}} sf(s) \tag{3.1.23}$$

In other words, the inverse of $sf(s)$ is obtained by differentiating $F(t)$. In a similar manner if the inverse of $s^n f(s)$ is required, $F(t)$ is differentiated n times assuming that $F(0^+), \ldots, \overset{[n-1]}{F}(0^+)$ are zero.

This property of the L.T. of a differentiated function has dealt with a function of only one variable t. The extension to a two-or-more variable function is also of definite interest. Thus if $F(x, t)$ is transformable with respect to t and has the transform $f(x, s)$,

$$F(x, t) \xrightarrow{L} f(x, s)$$

then

$$\frac{\partial F(x, t)}{\partial x} \xrightarrow{L} \frac{\partial f(x, s)}{\partial x} \tag{3.1.24}$$

and

$$\frac{\partial F(x, t)}{\partial t} \xrightarrow{L} sf(x, s) - F(x, 0^+)$$

EXAMPLE 3.1. To illustrate the explicit feature of (3.1.24) we shall consider the simple one-dimensional transient heat equation for a body of length 1:

$$\frac{\partial T(x, t)}{\partial t} = k\frac{\partial^2 T(x, t)}{\partial x^2}$$

$$T(x, 0) = T_0$$

$$T(0, t) = T_1$$

$$T(1, t) = T_2$$

Defining

$$T(x, t) \xrightarrow{L} \bar{T}(x, s) = \bar{T}(s)$$

(3.1.24) yields

$$s\bar{T}(s) - T(0^+) = k\frac{\partial^2 \bar{T}(s)}{\partial x^2}$$

The reader can check that this follows by an interchange of the order of integration with respect to t and differentiation with respect to x. Since s is only a parameter and not a variable and using $T(0^+) = T_0$, we may now write

$$s\bar{T}(s) - T_0 = k\frac{d^2\bar{T}(s)}{dx^2}$$

or

$$\frac{d^2\bar{T}(s)}{dx^2} - \frac{s}{k}\bar{T}(s) = -\frac{T_0}{k} \tag{A}$$

Observe that the application of the L.T. to the original partial differential equation has transformed the equation to an ordinary differential equation and at the same time has already directly incorporated the initial condition. Equation (A) may now be solved by standard methods and the general solution will involve two arbitrary constants. To evaluate these constants, use is made of the boundary conditions at $x = 0$ and $x = 1$ after taking the L.T. of each. Thus, using the L.T. of a constant,

$$T_1 \xrightarrow{L} \frac{T_1}{s}, \qquad x = 0$$

$$T_2 \xrightarrow{L} \frac{T_2}{s}, \qquad x = 1$$

The problem of inverting the solution $\bar{T}(s)$ back to the original function $T(x, t)$ will be discussed later.

3. *The initial and final value theorems.* Assuming the appropriate limits exist, it is possible to show that

$$\textit{Initial value theorem:} \quad \lim_{s\to\infty} sf(s) = \lim_{t\to 0} F(t) = F(0^+) \tag{3.1.25}$$

and

$$\textit{Final value theorem:} \quad \lim_{s\to 0} sf(s) = \lim_{t\to\infty} F(t) \tag{3.1.26}$$

EXAMPLE 3.2. Show, as an example, the derivation of the initial value theorem:

$$\lim_{s\to\infty} L[\dot{F}(t)] = \lim_{s\to\infty} \int_{0^+}^{\infty} e^{-st}\dot{F}(t)\,dt = \lim_{s\to\infty} [sf(s) - F(0^+)]$$

from (3.1.20). However, it immediately follows that

$$0 = \lim_{s\to\infty} [sf(s) - F(0^+)]$$

or

$$\lim_{s\to\infty} sf(s) = F(0^+)$$

4. *The L.T. of an integral.* For an $F(t)$ which is integrated over $(0, t)$ it is possible to show that

$$\int_{0^+}^{t} F(t)\,dt \xrightarrow{L} \frac{f(s)}{s} \tag{3.1.27a}$$

and also that

$$\int_a^t F(t)\,dt \xrightarrow{L} \frac{f(s)}{s} + \frac{1}{s}\int_a^0 F(t)\,dt \tag{3.1.27b}$$

The equivalent inverse operation also holds. This has application since if $g(s) = f(s)/s$, the inverse of $g(s)$, denoted as $G(t)$, can be obtained by inverting $f(s)$ and integrating the result between $(0, t)$. Extension to $f(s)/s^n$, $n = 1, 2, 3, \ldots$, follows in an analogous way.

5. *Differentiation and integration of* $f(s)$. By suitable differentiation and integration it is possible to show that

$$\frac{df(s)}{ds} = \dot{f}(s) = L[-tF(t)] \tag{3.1.28}$$

and that

$$\int_s^\infty f(s)\,ds = L\left[\frac{F(t)}{t}\right] \tag{3.1.29}$$

6. *The shifting or translation theorem.* If $F(t)$ is multiplied by e^{at}, $a =$ constant, then the L.T. of the product is immediately available if the transform of $F(t)$ is known:

$$L[e^{at}F(t)] = \int_{0^+}^\infty e^{-st}e^{+at}F(t)\,dt$$
$$= \int_{0^+}^\infty e^{-(s-a)t}F(t)\,dt$$

Since s is only a parameter in the integration, a new parameter is defined by

$$s - a = r$$

and

$$e^{at}F(t) \xrightarrow{L} f(r) = f(s - a) \tag{3.1.30}$$

Thus the transform of the product of e^{at} and any $F(t)$ will be the same as $f(s)$ except that wherever there is an s in the transform there will now be $s - a$.

The inverse of (3.1.30) is

$$e^{at}F(t) \xleftarrow{L^{-1}} f(s - a) \tag{3.1.31}$$

and if $s - a$ is replaced by s, then the inverse of this new function must be multiplied by e^{at} in order to obtain the inverse of the original transform function. The application of the shifting or translation theorem is very important and the reader will find that it is used extensively in many problems.

7. *Periodic functions.* Consider a periodic function as shown, for example, in Figure 3.2, where

$$F(t) = \begin{cases} t, & 0 < t < \tau/2 \\ \tau - t, & \tau/2 < t < \tau \end{cases} \tag{3.1.32}$$

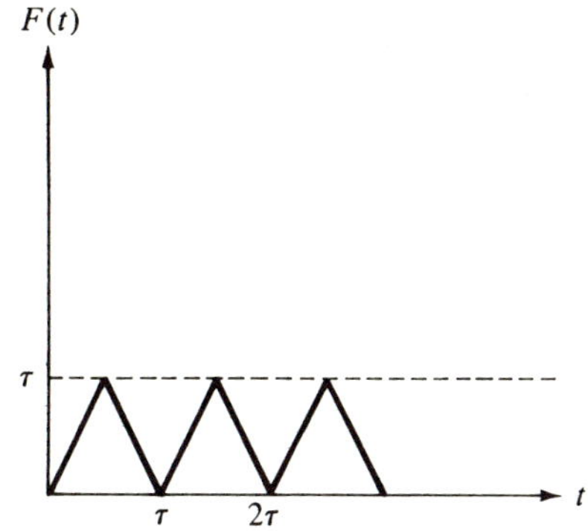

Figure 3.2. Periodic triangular wave.

and

$$F(t) = F(t + \tau), \qquad t > 0$$

In this case

$$L[F(t)] = \int_{0^+}^{\tau} e^{-st}F(t)\,dt + \int_{\tau}^{2\tau} e^{-st}F(t)\,dt + \cdots$$
$$= \sum_{k=0}^{\infty} \int_{k\tau}^{(k+1)\tau} e^{-st}F(t)\,dt$$

Substituting $t = z + k\tau$, $dt = dz$, and noting that because $F(t)$ is periodic, $F(z) = F(z + k\tau)$, there results

$$L[F(t)] = \int_{0^+}^{\tau} e^{-sz}F(z)\,dz + \int_{0^+}^{\tau} e^{-s(z+\tau)}F(z + \tau)\,dz + \cdots$$
$$= \sum_{k=0}^{\infty} e^{-ks\tau} \int_{0^+}^{\tau} e^{-sz}F(z)\,dz$$

The summation portion of this equation is a geometric progression such that

$$\sum_{k=0}^{\infty} e^{-ks\tau} = \frac{1}{1 - e^{-s\tau}}$$

and there finally results

$$L[F(t)] = \frac{\int_{0^+}^{\tau} e^{-st}F(t)\,dt}{1 - e^{-s\tau}} \tag{3.1.33}$$

This equation holds for any periodic function $F(t)$ with period τ.

8. *The convolution integral.* Perhaps one of the most general and important properties of the L.T. relates to the convolution integral (or the Faltung formula) for two different time functions $F_1(t)$ and $F_2(t)$. Without showing the derivation (lengthy; see [30]), we merely note that if

$$F_1(t) \xrightarrow{L} f_1(s)$$
$$F_2(t) \xrightarrow{L} f_2(s)$$

then the product of these two transforms is

$$L[F_1(t)]L[F_2(t)] = f_1(s)f_2(s) = L\left[\int_0^t F_1(\lambda)F_2(t-\lambda)\,d\lambda\right] \tag{3.1.34}$$

where λ is the variable of integration. From simple symmetry $F_1(t)$ and $F_2(t)$ commute in the convolution so that we may also write

$$f_1(s)f_2(s) = L\left[\int_0^t F_2(\lambda)F_1(t-\lambda)\,d\lambda\right] \tag{3.1.35}$$

The utility of this formula lies in the inverse process (s-domain to t-domain), which may be obtained by applying L^{-1} to (3.1.34) and (3.1.35):

$$\int_0^t F_1(\lambda)F_2(t-\lambda)\,d\lambda = \int_0^t F_2(\lambda)F_1(t-\lambda)\,d\lambda \xleftarrow{L^{-1}} f_1(s)f_2(s) \tag{3.1.36}$$

Note that if the inverse L.T.s of $f_1(s)$ and $f_2(s)$ are known, the inverse L.T. of the product can be obtained by an integration.

EXAMPLE 3.3. To illustrate certain features of the previous rules and properties of the L.T. we shall here consider the purging of a continuous stirred tank reactor (CSTR) (see Figure 3.3). Assume that the reactor has a volume V and that for $t < 0$ it is fed a salt solution ($c_i(t) = c_0$) at a rate q. The reactor is assumed completely mixed; i.e., the tank has a single uniform mixed salt concentration equal to the concentration out of the CSTR, $c(t)$. The rate of flow out of the tank is equal to the flow in. Assuming no reaction, the exit concentration is c_0. Consider three cases:

1. At time $t = 0$, the feed stream is changed to pure solvent (no salt), $c_i(t) = 0$, and we desire to predict the outgoing salt concentration from the CSTR as a function of time, $t > 0$. Obviously, we can predict a priori that since there is no longer any salt being fed to the CSTR the output will decrease from the initial concentration c_0 to zero; the rate of decrease will depend on the speed of purging of the salt held in the volume of the CSTR at time, $t = 0$.

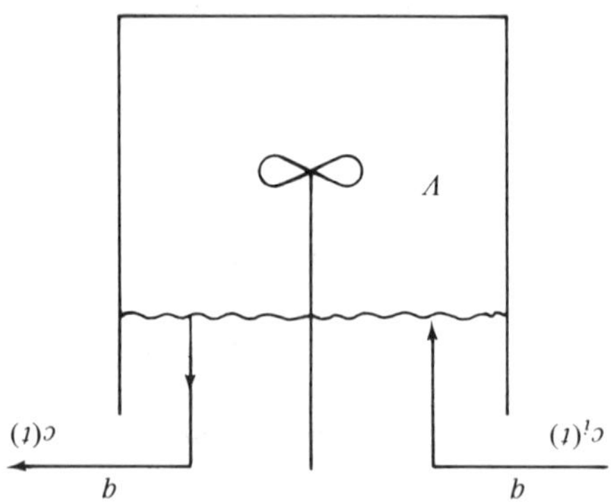

Figure 3.3. Schematic for CSTR system of Example 3.3.

A material balance around the CSTR yields

$$qc_i = qc(t) + V\frac{dc(t)}{dt}, \qquad t > 0 \tag{A}$$

$$\text{input} = \text{output} + \text{accumulation}$$

where c_i is the input salt concentration (zero in this particular arrangement) and $c(t)$ is the output. Rearranging and letting $q/V = \alpha$,

$$\frac{dc(t)}{dt} + \alpha c(t) = 0 \tag{B}$$

In addition, the following initial condition is known:

$$c = c_0, \qquad t = 0 \tag{C}$$

To solve (B) and (C) we let

$$c(t) \xrightarrow{L} \bar{c}(s) = \bar{c}$$

and with (3.1.20) we obtain

$$s\bar{c} - c_0 + \alpha\bar{c} = 0$$

or

$$\bar{c} = \frac{c_0}{s + \alpha} \tag{D}$$

Note that the initial condition, (C), is already included.

As an aside the reader will note that we here use $\bar{c}(s)$ to denote the transformed variable of $c(t)$. Alternatively, we could have used $C(t)$ and $c(s)$ but the present notation seems more convenient here.

To pass back to the t-domain we recall that (see Table 3.1)

$$e^{-at} \xleftarrow{L^{-1}} \frac{1}{s + a}$$

and

$$c(t) \xleftarrow{L^{-1}} \bar{c}$$

Thus

$$c(t) = c_0 e^{-\alpha t} \tag{E}$$

and the salt concentration decreases exponentially with time.

2. If, instead of feeding pure solvent, the feed had contained solute of constant concentration $c_i = c_1$, $t > 0$, then the answer can be obtained in an analogous manner. The material balance becomes

$$\frac{dc}{dt} + \alpha c = \alpha c_1 \tag{F}$$

Applying the L.T. there results (with αc_1 = constant)

$$s\bar{c} - c_0 + \alpha\bar{c} = \frac{\alpha c_1}{s}$$

or

$$\bar{c}(s) = \frac{\alpha c_1}{s(s + \alpha)} + \frac{c_0}{s + \alpha} \tag{G}$$

The inverse of the second term is identical to that considered in part 1 (D). To evaluate the first term on the right-hand side, at least two procedures are possible. First, from Table 3.1 we note that

$$\frac{1}{b-a}\left[e^{-at}-e^{-bt}\right] \xleftarrow{L^{-1}} \frac{1}{(s+a)(s+b)}$$

so that with $b = 0$ and $a = \alpha$ there results

$$c_1\left[1-e^{-\alpha t}\right] \xleftarrow{L^{-1}} \frac{c_1}{s(s+\alpha)}$$

Second, we may use a partial fraction expansion (see 9 following this example)

$$\frac{1}{s(s+\alpha)} = \frac{K_1}{s} + \frac{K_2}{s+\alpha} = \frac{1}{\alpha s} - \frac{1}{\alpha(s+\alpha)}$$

The inverse of this expression term by term and the inclusion of αc_1 leads to the same result as previously. Thus

$$c(t) = c_1[1 - e^{-\alpha t}] + c_0 e^{-\alpha t} \tag{H}$$

which becomes (E) if $c_1 = 0$.

3. Consider now that any arbitrary time-varying concentration $c_i(t)$ is used as feed to the reactor. The material balance becomes

$$\frac{dc}{dt} + \alpha c = \alpha c_i(t) \tag{I}$$

The L.T. of (I) is

$$s\bar{c}(s) - c_0 + \alpha\bar{c}(s) = \alpha\bar{c}_i(s)$$

or

$$\bar{c}(s) = \frac{\alpha\bar{c}_i(s)}{s+\alpha} + \frac{c_0}{s+\alpha} \tag{J}$$

where $c_i(t) \xrightarrow{L} \bar{c}_i(s)$. For the first term in the form

$$\frac{\alpha}{s+\alpha}\cdot\bar{c}_i(s) = f_1(s)f_2(s)$$

we note that

$$\alpha e^{-\alpha t} \xleftarrow{L^{-1}} \frac{\alpha}{s+\alpha} = f_1(s)$$

$$c_i(t) \xleftarrow{L^{-1}} \bar{c}_i(s) = f_2(s)$$

Thus the convolution integral in the form of (3.1.36) may be used for the first term of (J). The total inverse is merely

$$c(t) = \alpha\int_0^t c_i(t-\lambda)e^{-\alpha\lambda}\,d\lambda + c_0 e^{-\alpha t} \tag{K}$$

For $c_i(t) = c_1 =$ constant, (J) becomes (H), as expected.

It is important to note that the solution of the most complicated problem, part 3, is almost as simple as that for the easiest problem, part 1. It is this ability to easily handle time-varying inputs or boundary conditions which distinguishes the L.T. as a method of solution from most other methods.

9. *Expansion theorems*. As the reader begins to use the L.T. he will find that in a large number of cases the transform function which must be inverted is made up of algebraic functions of s. It would, therefore, seem advisable to have available certain standard techniques for inverting these particular functions.

Consider the following function, which is to be inverted:

$$f(s) = \frac{p(s)}{q(s)} = \frac{A_j s^j + A_{j-1}s^{j-1} + \cdots + A_1 s + A_0}{s^i + B_{i-1}s^{i-1} + \cdots + B_1 s + B_0} \tag{3.1.37}$$

where the order of the polynomial $q(s)$ is greater than that of the polynomial $p(s)$; i.e., $i > j$ and the As and Bs are real constants. If the order of these polynomials is not too large, then it is possible to use a partial fraction expansion as was done in the previous simple example. Assuming that $q(s)$ has roots which are all distinct and real, i.e., $q(s) = (s - s_1)(s - s_2) \cdots (s - s_i)$, (3.1.37) becomes

$$\frac{p(s)}{q(s)} = \frac{p(s)}{(s - s_1)(s - s_2)(s - s_3)\cdots(s - s_i)} \tag{3.1.38}$$

Expanding this ratio of polynomials in s by partial fractions,

$$\frac{p(s)}{q(s)} = \frac{K_1}{s - s_1} + \frac{K_2}{s - s_2} + \frac{K_3}{s - s_3} + \cdots + \frac{K_n}{s - s_n} + \cdots + \frac{K_i}{s - s_i} \tag{3.1.39}$$

where the Ks are coefficients to be determined. Now multiply both sides of (3.1.39) by $s - s_n$

$$\frac{(s - s_n)p(s)}{q(s)} = \frac{(s - s_n)K_1}{s - s_1} + \frac{(s - s_n)K_2}{s - s_2} + \cdots + K_n + \cdots + \frac{(s - s_n)K_i}{s - s_i} \tag{3.1.40}$$

and let $s = s_n$. The left-hand side of (3.1.40) becomes just a number since $s - s_n$ cancels out of $q(s)$ and the right hand side becomes K_n since all other terms become zero. Thus

$$K_n = \left[\frac{(s - s_n)p(s)}{q(s)}\right]_{s = s_n} \tag{3.1.41}$$

and all the coefficients in (3.1.39) may be evaluated in the same manner. Further, note that

$$\left.\frac{q(s)}{s - s_n}\right|_{s = s_n} = \frac{d}{ds}[q(s)]_{s = s_n} = \dot{q}(s)\Big|_{s = s_n}$$

and that (3.1.39) and (3.1.41) become

$$\frac{p(s)}{q(s)} = \sum_{n=1}^{i} \frac{p(s_n)}{\dot{q}(s_n)} \frac{1}{s - s_n}$$

$$K_n = \left.\frac{p(s)}{\dot{q}(s)}\right|_{s = s_n}$$

Thus

$$\sum_{n=1}^{i} \frac{p(s_n)}{\dot{q}(s_n)} e^{s_n t} \xleftarrow{L^{-1}} \frac{p(s)}{q(s)} = \sum_{n=1}^{i} \frac{p(s_n)}{\dot{q}(s_n)} \frac{1}{s - s_n} \tag{3.1.42}$$

Equation (3.1.42) is one form of Heaviside's expansion formula for $F(t)$ assuming that the denominator of $F(t)$ has only real and nonrepeated roots.

An important extension of (3.1.42) occurs when one of the roots of $q(s)$ is zero; i.e., $s_1 = 0$. In this case

$$\frac{p(s)}{q(s)} = \frac{p(s)}{sq_1(s)}$$

where $q_1(s) = q(s)/s$. Using the essentially identical approach we now obtain

$$\frac{p(0)}{q_1(0)} + \sum_{n=2}^{i} \frac{p(s_n)}{\dot{q}(s_n)} e^{s_n t} \xleftarrow{L^{-1}} \frac{p(s)}{q(s)} \qquad (s_1 = 0) \tag{3.1.43}$$

The extension of these ideas to repeated and complex roots can be carried out, but we shall not do so here, since the details are excessive (see [30]). Equation (3.1.42) has been used to develop many of the transform pairs in Table 3.1.

If $f(s)$ has the fractional form $p(s)/q(s)$, where $p(s)$ and $q(s)$ are differentiable at $s = s_n$, $q(s_n) = 0$, and $p(s_n) \neq 0$, and if no multiple roots occur, (3.1.42) may be extended to cover an infinite number of roots. Thus

$$\sum_{n=1}^{\infty} \frac{p(s_n)}{\dot{q}(s_n)} e^{s_n t} \xleftarrow{L^{-1}} \frac{p(s)}{q(s)} \tag{3.1.44}$$

where the summation is taken over all the roots of $q(s) = 0$. The utility of this equation will be seen in one of the examples to follow.

EXAMPLE 3.4. Find the inverse L.T. of the function

$$f(s) = \frac{1}{(s-1)(s-2)(s-3)}$$

$p(s) = 1$, and the roots of $q(s)$ are real and distinct, $s_1 = 1$, $s_2 = 2$, $s_3 = 3$. Thus

$$p(s) = 1$$

$$q(s) = (s-1)(s-2)(s-3)$$

$$\dot{q}(s) = (s-2)(s-3) + (s-1)(s-3) + (s-1)(s-2)$$

$$\dot{q}(s_1) = 2, \qquad \dot{q}(s_2) = -1, \qquad \dot{q}(s_3) = 3$$

$$i = 3$$

Thus (3.1.42) yields

$$\frac{1}{2}e^{t} + e^{2t} + \frac{1}{2}e^{3t} \xleftarrow{L^{-1}} \frac{1}{(s-1)(s-2)(s-3)}$$

EXAMPLE 3.5. Consider n CSTRs each with fluid volume V_n and initial solute concentration A_n (see Figure 3.4). At time $t = 0$ solvent is added to reactor 1 with a time-varying solute concentration $c_0(t)$ at a flow rate q. Equal amounts of fluid overflow each reactor and enter the next one. We desire an equation for the solute effluent concentration, $c_n(t)$, at any time from any of the reactors.

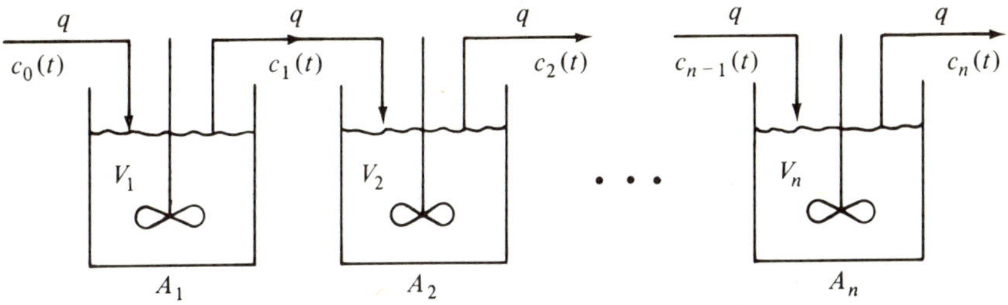

Figure 3.4. Schematic of n-CSTR system of Example 3.5.

A material balance around the nth reactor may be written as

$$qc_{n-1}(t) = qc_n(t) + V_n\frac{dc_n(t)}{dt} \tag{A}$$

$$\text{input} = \text{output} + \text{accumulation}$$

In addition, the following initial and boundary conditions apply:

$$c_n(t) = A_n, \qquad t = 0 \tag{B}$$

$$c_n(t) = c_0(t), \qquad n = 0 \tag{C}$$

Note that this is a fairly complex system consisting of a differential-difference equation with arbitrary initial and boundary conditions.

Defining

$$c_n(t) \xrightarrow{L} \bar{c}_n(s)$$

and letting $\theta_n = V_n/q$, (3.1.20) when applied to (A)–(C) yields

$$\bar{c}_{n-1}(s) = \bar{c}_n(s) + \theta_n[s\bar{c}_n(s) - A_n]$$

or

$$\bar{c}_n(s) = \frac{\bar{c}_{n-1}(s) + \theta_n A_n}{1 + \theta_n s} \tag{D}$$

and

$$\bar{c}_n(s) = \bar{c}_0(s), \qquad n = 0 \tag{E}$$

Equation (D) is a nonhomogeneous difference equation which can be solved by iteration for $\bar{c}_n$ in terms of only $\bar{c}_0$ by successively considering $n = 1, 2, 3, \ldots, n$:

$$n = 1: \quad \bar{c}_1 = \frac{\bar{c}_0}{1 + \theta_1 s} + \frac{\theta_1 A_1}{1 + \theta_1 s}$$

$$n = 2: \quad \bar{c}_2 = \frac{\bar{c}_1}{1 + \theta_2 s} + \frac{\theta_2 A_2}{1 + \theta_2 s}$$

$$= \frac{\bar{c}_0}{(1 + \theta_1 s)(1 + \theta_2 s)} + \frac{\theta_1 A_1}{(1 + \theta_1 s)(1 + \theta_2 s)} + \frac{\theta_2 A_2}{(1 + \theta_2 s)}$$

$$\vdots \qquad \vdots$$

$$n = n: \quad \bar{c}_n = \frac{\bar{c}_0}{(1 + \theta_1 s)(1 + \theta_2 s)\cdots(1 + \theta_n s)} + \frac{\theta_1 A_1}{(1 + \theta_1 s)(1 + \theta_2 s)\cdots(1 + \theta_n s)}$$

$$+ \frac{\theta_2 A_2}{(1 + \theta_2 s)(1 + \theta_3 s)\cdots(1 + \theta_n s)} + \cdots + \frac{\theta_n A_n}{1 + \theta_n s}$$

or

$$\bar{c}_n = \frac{\bar{c}_0}{\prod_{i=1}^{n}(1+\theta_i s)} + \frac{\theta_1 A_1}{\prod_{i=1}^{n}(1+\theta_i s)} + \frac{\theta_2 A_2}{\prod_{i=2}^{n}(1+\theta_i s)} + \cdots + \frac{\theta_n A_n}{\prod_{i=n}^{n}(1+\theta_i s)} \tag{F}$$

To take the inverse of (F) it is noted that each term is of the form $p(s)/q(s)$ with the order of $q(s) > p(s)$. Thus (3.1.42) may be used. Consider the term

$$\frac{1}{\prod_{i=1}^{n}(1+\theta_i s)} = \frac{1}{q(s)}$$

for which

$$\dot{q}(s) = \sum_{i=1}^{n} \theta_i \prod_{\substack{k=1\\k\neq i}}^{n} (1+\theta_k s)$$

with the roots $s_i = -(1/\theta_i)$. Thus

$$\sum_{1}^{n} \frac{p(s_i)}{\dot{q}(s_i)} e^{s_i t} = \sum_{i=1}^{n} \frac{e^{-t/\theta_i}}{\theta_i \prod_{\substack{k=1\\k\neq i}}^{n} (1 - \theta_k/\theta_i)}$$

Now the inverse transform of (F) may be taken term by term:

$$L^{-1}\left[\frac{\bar{c}_0}{\prod_{i=1}^{n}(1+\theta_i s)}\right] = \int_0^t c_0(t-\lambda) \sum_{i=1}^{n} \frac{e^{-\tau/\theta_i}}{\theta_i \prod_{\substack{k=1\\k\neq i}}^{n}(1-\theta_k/\theta_i)}\, d\lambda \tag{G}$$

$$L^{-1}\left[\frac{\theta_m A_m}{\prod_{i=m}^{n}(1+\theta_i s)}\right] = \theta_m A_m \sum_{i=m}^{n} \frac{e^{-t/\theta_i}}{\theta_i \prod_{\substack{k=m\\k\neq i}}^{n}\theta_i(1-\theta_k/\theta_i)}, \qquad m = 1, 2, 3, \ldots \tag{H}$$

Note that in (G) the convolution integral is used. Finally,

$$c_n(t) = L^{-1}\left[\frac{\bar{c}_0}{\prod_{i=1}^{n}(1+\theta_i s)}\right] + \sum_{m=1}^{n} L^{-1}\left[\frac{\theta_m A_m}{\prod_{i=m}^{n}(1+\theta_i s)}\right]$$

10. *The complex inversion integral.* The L.T. pair indicated schematically in (3.1.4) is

$$f(s) = \int_0^\infty F(t)e^{-st}\, dt \tag{3.1.45}$$

(L.T. pair)

$$F(t) = \frac{1}{2\pi i}\int_{a-i\infty}^{a+i\infty} f(s)e^{st}\, ds \tag{3.1.46}$$

The second equation is termed the inversion integral, which is a line integral in the complex plane, taken along a vertical line to the right of all the singularities of $f(s)$ or along any other path into which this can legitimately be deformed [30]. Thus, in theory, we can recover the $F(t)$ corresponding to any $f(s)$ if (3.1.46) can be evaluated.

A powerful theorem for the evaluation of line integrals in the complex plane is the *residue theorem:*

RESIDUE THEOREM: If C is a closed curve and $\phi(s)$ is *analytic* within and on C except at a finite number of singular points in the interior of C, then

$$\int_C \phi(s)\, ds = 2\pi i(r_1 + r_2 + \cdots + r_m)$$

where $r_1, r_2, \ldots, r_m$ are the *residues* of $\phi(s)$ at its singular points within C.

By *analytic* at a point s_0 we mean that $\phi(s)$ possesses a derivative at $s = s_0$ and at every point in the neighborhood of s_0. The *residues* of $\phi(s)$ are found as follows. We expand $\phi(s)$ in partial fractions (a Laurent expansion). The coefficient of $(s - s_k)^{-1}$ in the expansion of $\phi(s)$ in the neighborhood of an isolated singular point is called the residue of $\phi(s)$ at the point $s = s_k$.

Consider as an example the inversion of

$$f(s) = \frac{s}{s^2 + k^2}$$

by the residue theorem [30]. The singular points (poles) of $f(s)$ are at $s = \pm ik$ on the imaginary axis. As a closed curve enclosing $\pm ik$ we choose the semicircular contour consisting of the vertical line at $s = a$ extending from $-ib$ to $+ib$ (C_1) and the semicircle connecting the ends of the line and crossing the imaginary axis (C_2).

Applying the residue theorem,

$$\begin{aligned} F(t) &= \frac{1}{2\pi i}\int_{a-ib}^{a+ib} \frac{se^{st}}{s^2 + k^2}ds + \frac{1}{2\pi i}\int_{C_2} \frac{se^{st}}{s^2 + k^2}ds \\ &= 2\pi i \sum \text{residues of } \frac{se^{st}}{2\pi i(s^2 + k^2)} \qquad \text{at } s = \pm ik \\ &= \sum \text{residues of } \frac{se^{st}}{s^2 + k^2} \qquad \text{at } s = \pm ik \end{aligned}$$

At $s = \pm ik$ the residues are

$$\lim_{s \to ik}\left[\frac{(s - ik)se^{st}}{(s - ik)(s + ik)}\right] = \frac{ike^{ikt}}{2ik} = \frac{1}{2}e^{ikt}$$

$$\lim_{s \to -ik}\left[\frac{(s + ik)se^{st}}{(s - ik)(s + ik)}\right] = \frac{-ike^{-ikt}}{-2ik} = \frac{1}{2}e^{-ikt}$$

Thus

$$\begin{aligned} F(t) &= \frac{1}{2}(e^{ikt} + e^{-ikt}) \\ &= \cos kt \end{aligned}$$

We shall not deal further with the complex inversion integral in this chapter. However, in many cases the use of the inversion integral and the residue theorem may be the only way to invert a complicated L.T.

3.2 Input and Transfer Functions

We shall now consider the L.T. of certain inputs and the concept of a system transfer function. These items will be of great importance for the identification problem. Table 3.1 tabulates many of the inputs we shall discuss.

3.2.1 The unit step function

Consider the time function shown in Figure 3.5(a), which has the form

$$F(t) = \begin{cases} 0, & t < t_1 \\ 1, & t > t_1 \end{cases} \tag{3.2.1}$$

In symbolic terminology we call $F(t) = U(t - t_1)$, indicating that this step-function is zero for all $t < t_1$.

The L.T. can be obtained in a simple manner:

$$\begin{aligned} L[U(t - t_1)] &= L[\_\ulcorner] = \int_{0^+}^{\infty} e^{-st}U(t - t_1)dt \\ &= \int_{t_1}^{\infty} e^{-st}\,dt = \frac{e^{-st_1}}{s} \end{aligned} \tag{3.2.2}$$

If $t_1 = t_0 = 0$, the function begins at the origin $t = t_0$ [Figure 3.5(b)] and is called the Heaviside unit step function with the L.T.

$$L[U(t - 0)] = L[\ulcorner] = \frac{1}{s} \tag{3.2.3}$$

From the linearity of the L.T. it follows immediately that if the step is a units high, the transform function is multiplied by a. This was previously shown in Section 3.14.

Having the definition of the unit step function, compounding can be used to obtain the L.T. of many other inputs. Thus consider the square wave of Figure 3.5(c) in which $F(t)$ is zero for $t < t_1$ and then for $t > t_2$. This may be represented as

$$F(t) = U(t - t_1) - U(t - t_2) \tag{3.2.4}$$

resulting from the compounding of the two step functions. It follows immediately from (3.2.2) that

$$L[U(t - t_1) - U(t - t_2)] = L[\_\sqcap\_] = \frac{e^{-st_1} - e^{-st_2}}{s} \tag{3.2.5}$$

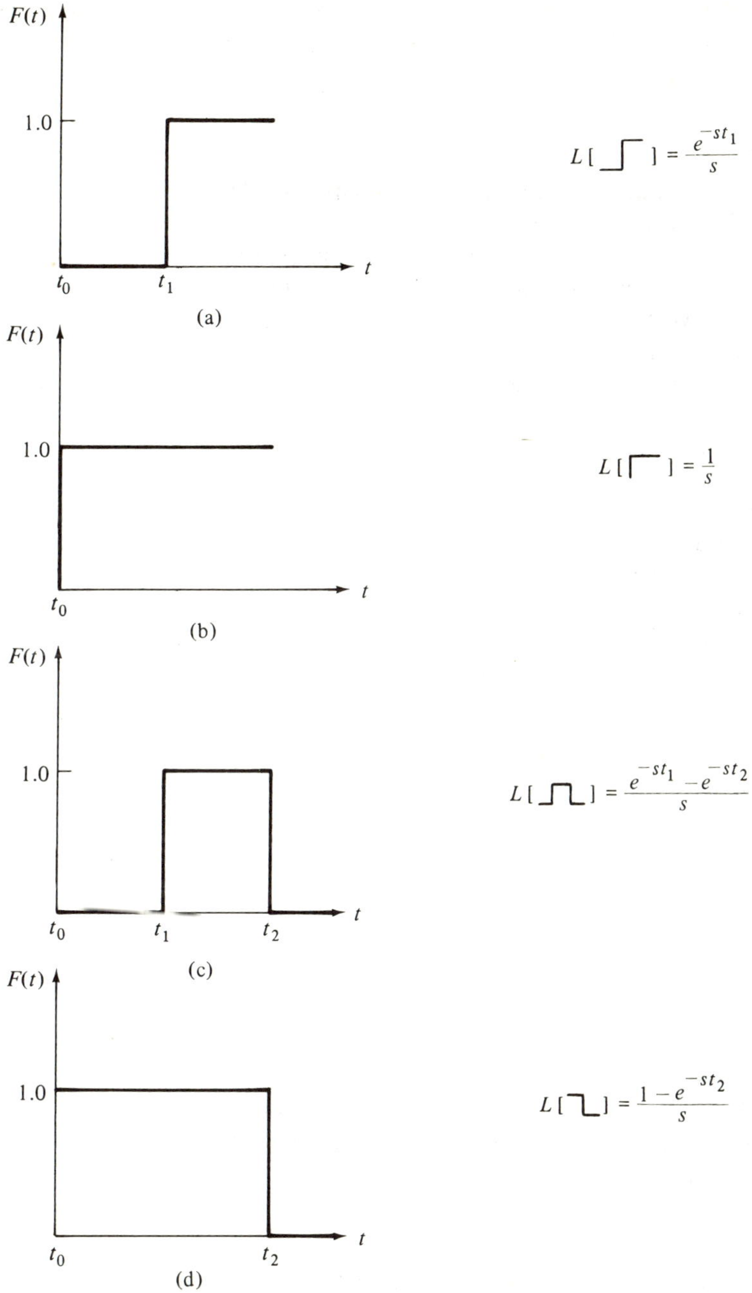

Figure 3.5. Unit step and allied inputs.

This square wave is frequently termed a rectangular pulse of unit intensity. By letting $t_1 \longrightarrow t_0 = 0$ the transform of Figure 3.5(d) follows immediately:

$$L[U(t) - U(t - t_2)] = L[\,\urcorner\!\llcorner\,] = \frac{1 - e^{-st_2}}{s} \tag{3.2.6}$$

Obviously such functions are idealized ones since it is essentially impossible in practice to obtain the true discontinuity. Thus various authors have investigated allied functions with continuous first-order derivatives for all t. While interesting, we shall not discuss this point here.

The step function is also referred to as a filter function since any other function which is multiplied by the step is either annihilated completely or passed without any change.

Considering the product $\sin(t)U(t - t_1)$, it can be quickly observed that because the step function is zero for $t < t_1$, the product of the two functions must also be zero for $t < t_1$. Thus the step function can be thought of as a filtering function, in this case suppressing the sine function in $t < t_1$. From the L.T. of the unit step function starting at the origin, (3.2.3), as compared to the transform for the displaced unit step function, (3.2.2), it can be seen that the term e^{-st_1} is a delay factor. Thus if the transform function has an exponential, e^{-st_1}, it may be suppressed and the remaining function inverted. The delay adjustment may then be made by changing t to $t - t_1$. This represents another form of the shift theorem. This may be expressed as

$$L[F(t)] = e^{-st_1}L[G(t)] \tag{3.2.7}$$

where $G(t)$ represents the particular function originating at $t = 0$ and $F(t)$ the same function originating at or delayed to $t = t_1$.

3.2.2 The unit impulse function

As a second class of input functions we shall consider a limiting case of the special square wave in Figure 3.6(a), which is

$$F(t) = \frac{1}{\tau}[U(t) - U(t - \tau)] \tag{3.2.8}$$

The limiting process is to let $\tau \longrightarrow 0$ so that the width goes to zero and the height goes to infinity at $t = 0$ but the area remains constant at 1.0. The result is called the *unit impulse* or *Dirac delta function*, which we write symbolically as

$$\delta(t) = \lim_{\tau \to 0} \frac{1}{\tau}[U(t) - U(t - \tau)] \tag{3.2.9}$$

The unit impulse function is the derivative of the unit step function [the right-hand side of (3.2.9) is merely equivalent to the usual derivative formula].

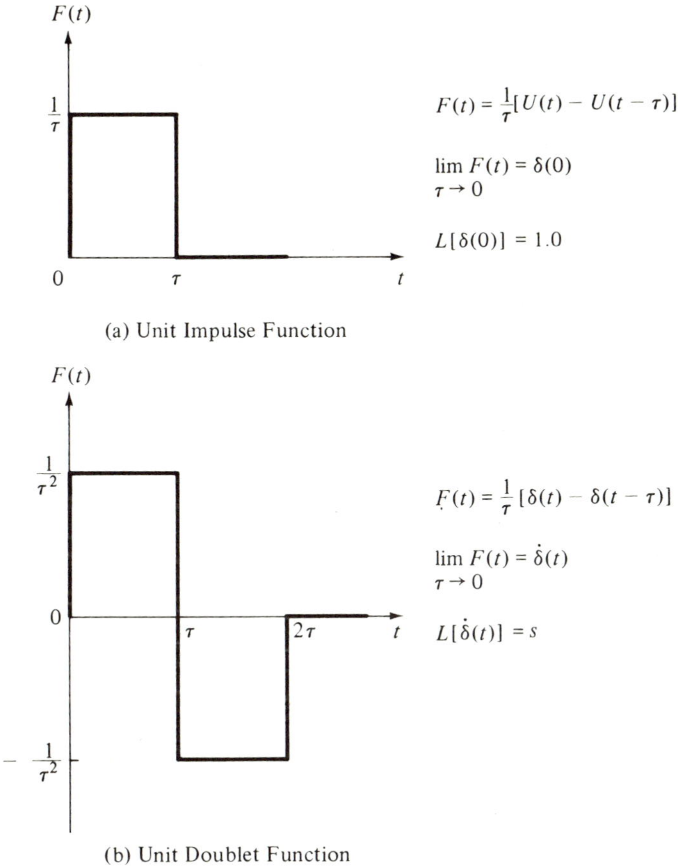

Figure 3.6. Special rectangular functions.

Since the area is unity, it follows that

$$\int_0^\infty \delta(t)\, dt = U(t) = 1 \tag{3.2.10}$$

A unit impulse may occur at a point other than $t_0 = 0$, i.e., at t_1 such that

$$\delta(t - t_1) = \lim_{\tau \to 0} \frac{1}{\tau} [U(t - t_1) - U(t - t_1 - \tau)] \tag{3.2.11}$$

In terms of (3.2.11)

$$L[\delta(t - t_1)] = \lim_{\tau = 0} \frac{1}{\tau} \int_{t_1}^{t_1 + \tau} e^{-st}\, dt$$

since $\delta(t - t_1) = 1/\tau$ between t_1 and $t_1 + \tau$ and zero everywhere else. Carrying out the integration,

$$L[\delta(t-t_1)] = \lim_{\tau \to 0} \frac{1}{\tau s}[e^{-t_1 s} - e^{-s(t_1+\tau)}] \tag{3.2.12}$$

This is indeterminate in the limit, but the application of L'Hospital's rule gives

$$L[\delta(t-t_1)] = \lim_{\tau \to 0} e^{-s(t_1+\tau)}$$

$$L[\perp] = e^{-st_1} \tag{3.2.13}$$

and if $t_1 = t_0 = 0$,

$$L[\llcorner] = L[\delta(t)] = 1.0 \tag{3.2.14}$$

Thus the L.T. of a unit impulse at $t = 0$ is 1.0, while at any other t, $t = t_1$, the delay term applies. Obviously, if the area were not 1.0 but a finite value A, then all these results would be multiplied by A.

The unit impulse function also exhibits a so-called sifting property. If a unit impulse function at t_1, $\delta(t - t_1)$, is multiplied by an arbitrary time function $F(t)$ and integrated over the entire t-axis, the value of $F(t)$ at t_1 is the result. Thus

$$\int_0^\infty \delta(t-t_1)F(t)\,dt = F(t_1) \tag{3.2.15}$$

This follows from the fact that the integrand is zero for all t except $t = t_1$. As a result $F(t) = F(t_1)$ can be removed from within the integral leading to (3.2.15).

Having shown that the first derivative of the unit step function is the unit impulse with a single singularity, higher-order derivatives can be obtained. As an illustration the limiting process in the case of Figure 3.6(b) can be written as

$$\dot{\delta}(t) = \lim_{\tau \to 0} \frac{1}{\tau}[\delta(t) - \delta(t-\tau)] \tag{3.2.16}$$

This is termed the *unit doublet function* with the properties

$$\int_0^\infty \dot{\delta}(t)\,dt = \delta(t)$$
$$L[\dot{\delta}(t)] = s \tag{3.2.17}$$

Other higher derivatives can be obtained, but we shall not pursue this.

3.2.3 Ramp and sawtooth functions

From this information we can now cover many other input functions. To illustrate this point, consider in Figure 3.7(a) the sawtooth function; also shown are a number of related functions and their analytical representations. Thus Figure 3.7(b) is a ramp function which is represented as $F(t) = tU(t)$;

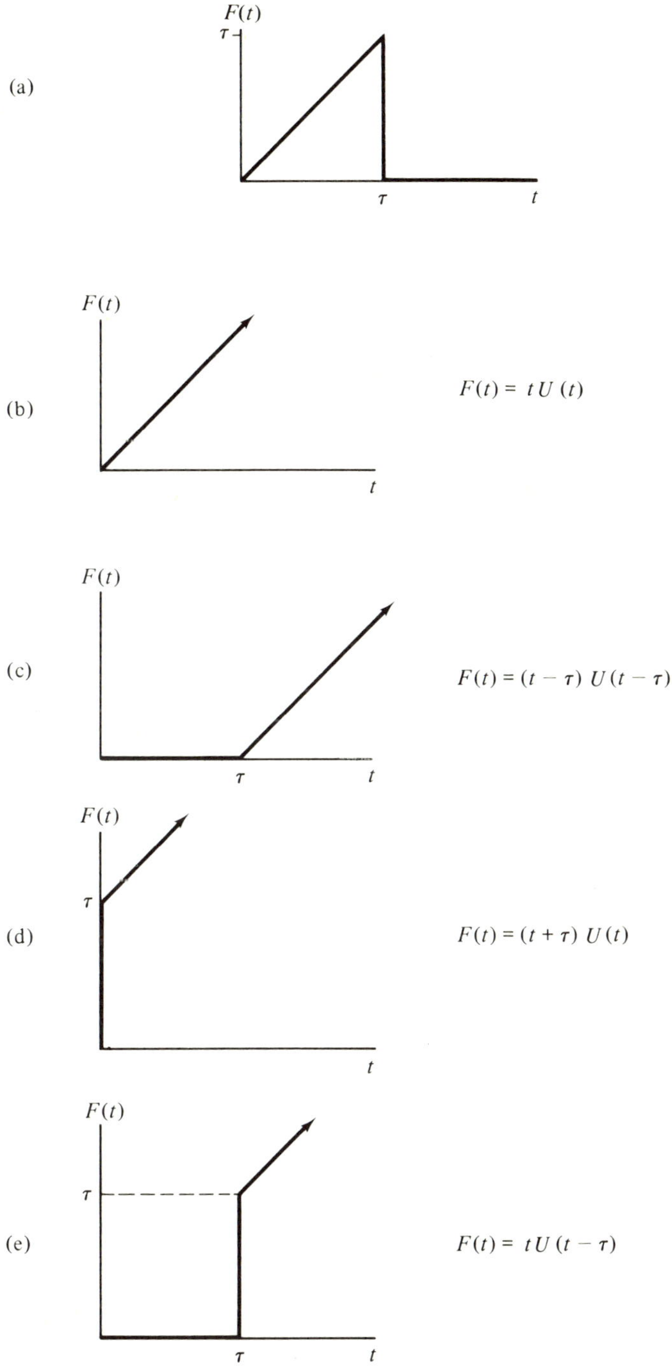

Figure 3.7. Various ramp functions.

note that the step function multiplication does not change $F(t)$. In Figure 3.7(c), however, the multiplication by $U(t - \tau)$ cuts off the straight line. The reader should prove that these forms are correct, including parts (d) and (e) of Figure 3.7. The sawtooth function may be represented by combinations of these curves, namely

$$F(t) = tU(t) - (t - \tau)U(t - \tau) - \tau U(t - \tau) \qquad (3.2.18)$$

The first term represents a 45-degree line going to infinity. The second term corresponds to a negative 45-degree line originating at $t = \tau$, which, when subtracted from the first term, will leave a constant with magnitude τ from $t > \tau$ extending to $t \longrightarrow \infty$. Since the sawtooth function is zero from τ on, a further subtraction of a step function originating at τ and with a magnitude τ yields the desired result. Thus by a proper choice of unit step functions a functional relation for the sawtooth pulse has been obtained. The L.T. follows immediately by using the previously developed relations for step functions and $L[t] = 1/s^2$. Thus

$$L[\diagup\!\!\llcorner] = \frac{1}{s^2} - \frac{1}{s^2}e^{-s\tau} - \frac{\tau e^{-s\tau}}{s} = \frac{1 - (1 + s\tau)e^{-s\tau}}{s^2} \qquad (3.2.19)$$

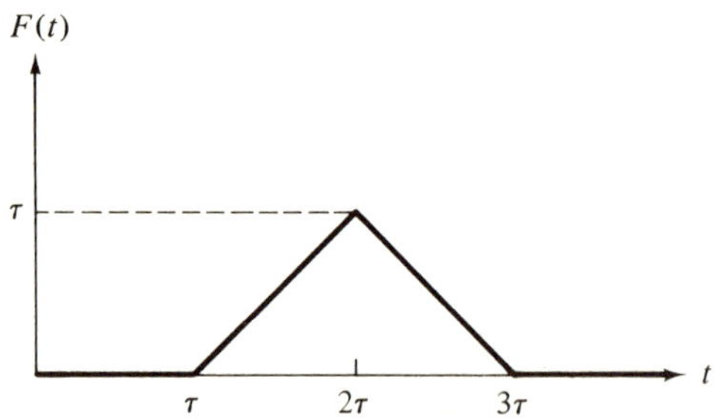

Figure 3.8. Triangular pulse.

An extension of the sawtooth pulse to form a triangular pulse is shown in Figure 3.8. In a manner analogous to the development of (3.2.18) and (3.2.19)

$$F(t) = (t - \tau)U(t - \tau) - 2(t - 2\tau)U(t - 2\tau) + (t - 3\tau)U(t - 3\tau) \qquad (3.2.20)$$

and

$$L[\_\wedge\_] = \frac{e^{-s}[1 - e^{-s}]^2}{s} \qquad (3.2.21)$$

3.2.4 Periodic or repeated input functions

The L.T. of periodic or repeated input functions can also be developed in a straightforward manner. In Figure 3.9(a) we show a repeated sawtooth

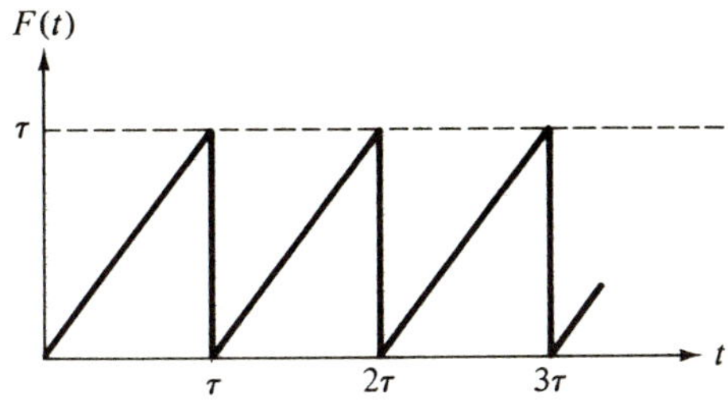

(a) Repeated Sawtooth Function

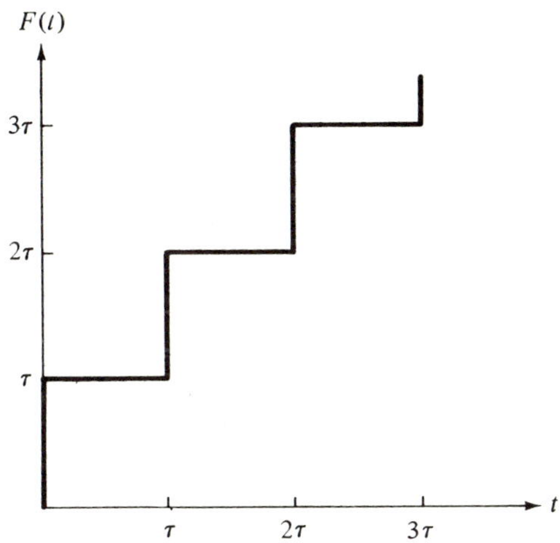

(b) Staircase Function

Figure 3.6. Periodic functions.

pulse. Using (3.1.33) with $F(t) = t$,

$$L[F(t)] = \frac{\int_0^\tau F(t)e^{-st}\,dt}{1 - e^{-s\tau}}$$

$$= \frac{\int_0^\tau te^{-st}\,dt}{1 - e^{-s\tau}}$$

$$= \frac{1}{1 - e^{-s\tau}}\left[\frac{1}{s^2}(1 - e^{-s\tau}) - \frac{\tau}{s}e^{-s\tau}\right]$$

$$L[\diagup\!\!\mid\!\diagup\!\!\mid\!\diagup] = \frac{1 - e^{-s\tau} - s\tau e^{-s\tau}}{s^2[1 - e^{-s\tau}]} \qquad (3.2.22)$$

Actually (3.2.22) could have been derived without using the periodic transform formula. Once an analytical expression for the function during its first period, i.e., $F_1(t) = t$, is available, the time is merely delayed by τ for the second period. This is repeated for n periods. Thus

$$
\begin{array}{ll}
F_1(t) = t, & 0 < t < \tau \\
F_2(t) = t - \tau, & \tau < t < 2\tau \\
F_3(t) = t - 2\tau, & 2\tau < t < 3\tau \\
\vdots & \vdots \\
F_n(t) = t - (n-1)\tau, & (n-1)\tau < t < n\tau \\
\vdots & \vdots
\end{array}
$$

Further,

$$F(t) = F_1(t) + F_2(t) + F_3(t) + \cdots + F_n(t) + \cdots \tag{3.2.23}$$

since each term is defined only in a single period. By taking the L.T. of both sides of (3.2.23) and recalling the shifting theorem, there results

$$L[F(t)] = L[F_1(t)]\{1 + e^{-s\tau} + e^{-2s\tau} + e^{-3s\tau} + \cdots + e^{-ns\tau} + \cdots\}$$

$L[$  $] = \dfrac{L[F_1(t)]}{1 - e^{-s\tau}}$ (3.2.24)

For the sawtooth function, $F_1(t)$ is given by (3.2.18), and the L.T. is given by (3.2.19). As a result

$$L[F_1(t)] = \frac{1}{s^2} - \frac{1}{s^2}e^{-s\tau} - \frac{\tau}{s}e^{-s\tau}$$

and

$L[$  $] = \dfrac{1 - e^{-s\tau} - s\tau e^{-s\tau}}{s^2[1 - e^{-s\tau}]}$

which is identical to the previous transform.

To illustrate further how repetitive source functions can be handled, consider the staircase function of Figure 3.9(b). In the first period

$$F_1(t) = \tau$$

for which

$$L[F_1(t)] = \frac{\tau}{s}$$

Thus the L.T. of the staircase function is

$L[$  $] = \dfrac{L[F_1(t)]}{1 - e^{-s\tau}} = \dfrac{\tau}{s[1 - e^{-s\tau}]}$ (3.2.25)

The unit step function representation of Figure 3.9(b) is

$$F(t) = U(t) + \tau U(t - \tau) + \tau U(t - 2\tau) + \cdots$$

and the L.T. is identical to (3.2.25).

The L.T. of repetitive unit impulses, referred to as a *pulse train*, may be obtained in a very simple way since $L[F_1(t)] = 1.0$. Thus

$$L[\,|\,|\,|\,|\,] = 1 + e^{-s\tau} + e^{-2s\tau} + e^{-3s\tau} + \cdots \tag{3.2.26}$$

$$= \frac{1}{1 - e^{-s\tau}} \tag{3.2.27}$$

3.2.5 Transfer functions

Let us return for a moment to the transform that we obtained for the single CSTR system, Eq. (J) in Example 3.3:

$$\bar{c}(s) = \frac{\alpha}{s + \alpha}\bar{c}_i(s) + \frac{c_0}{s + \alpha} \tag{3.2.28}$$

The first term on the right-hand side includes the input function, while the second term relates to the initial condition of the CSTR. If we had started with the system *at rest* or with *zero initial conditions*, $c_0 = 0$, (3.2.28) could be written as

$$\bar{c}(s) = g(s)f(s) \tag{3.2.29}$$

where $g(s) = \alpha/(s + \alpha)$ and $f(s) = \bar{c}_i(s)$. $g(s)$, termed the *transfer* or the *system function*, relates the input/output pair in the transform domain. It depends solely on the inner structure of the model for the system. As we shall show later (Section 3.3), there are analogous terms in the state or time domain where the terms *response* function, *weighting* function, and *Green's* function apply.

As a second simple example consider the model equation

$$a_0\ddot{y} + a_1\dot{y} + a_2 = b_0 u$$

The L.T. of this equation is

$$\bar{y}(s) = \frac{b_0}{a_0 s^2 + a_1 s + a_2}\bar{u}(s) \tag{3.2.30}$$

when all the initial conditions are zero. Thus the term $b_0/(a_0 s^2 + a_1 s + a_2)$ is now the transfer function for this model.

The transfer function is seen to be a linear operator (in the s-domain) which operates on the input to yield the output. As such it simplifies much of the calculation on linear models of finite and even infinite dimensional systems since the complex t-domain equations (the solutions) are not required; only image space information is needed. To illustrate this point, consider Figure 3.10(a) where two transfer functions $g_1(s)$ and $g_2(s)$ are cascaded in series (two CSTRs in series). Using (3.2.29), it follows that

$$\bar{y}_1(s) = g_1(s)f(s)$$

$$\bar{y}_2(s) = g_2(s)\bar{y}_1(s)$$

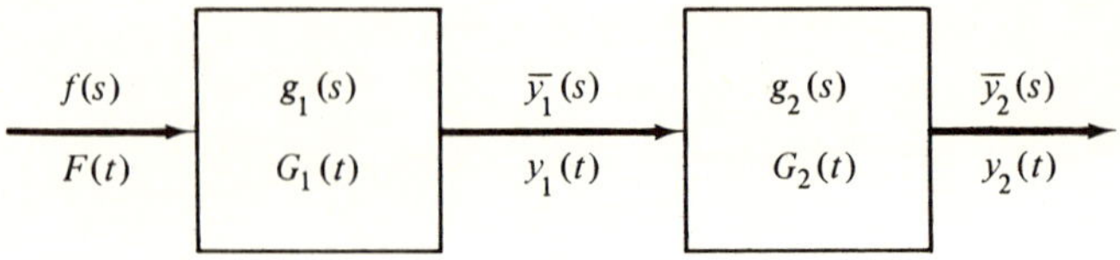

(a) Series Transfer Functions

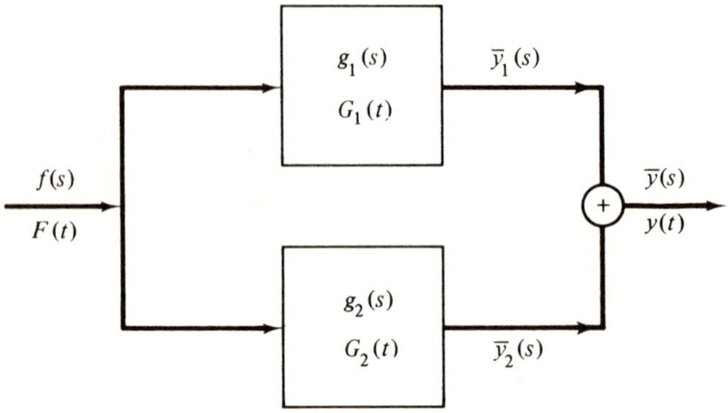

(b) Parallel Transfer Functions

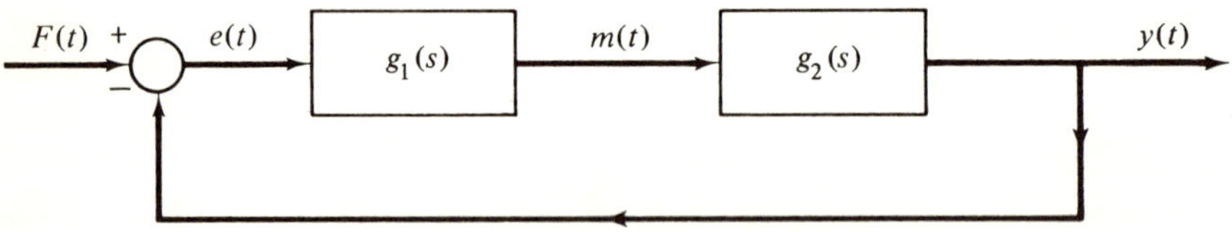

(c) Combined Feedback Transfer Functions

Figure 3.10. Serial, parallel, and combined transfer functions.

and thus

$$\bar{y}_2(s) = g_1(s)g_2(s)f(s) = g(s)f(s)$$

where

$$g(s) = g_1(s)g_2(s) \tag{3.2.31}$$

Thus when transfer functions are arranged in series they merely multiply together to form an overall transfer function between the input and output. By contrast, Figure 3.10(b) shows a parallel arrangement for which

$$\bar{y}_1(s) = g_1(s)f(s)$$
$$\bar{y}_2(s) = g_2(s)f(s)$$

and since

$$\bar{y}(s) = \bar{y}_1(s) + \bar{y}_2(s)$$

it follows that

$$\begin{aligned}\bar{y}(s) &= [g_1(s) + g_2(s)]f(s) \\ &= g(s)f(s) \qquad (3.2.32)\end{aligned}$$

Thus parallel transfer functions add to form an overall transfer function.

Finally in Figure 3.10(c) we show a more complicated feedback system. In obvious notation

$$\begin{aligned}\bar{y}(s) &= g_2(s)\bar{m}(s) \\ \bar{m}(s) &= g_1(s)\bar{e}(s) \\ \bar{e}(s) &= f(s) - \bar{y}(s)\end{aligned}$$

and thus

$$\bar{y}(s) = \frac{g_1(s)g_2(s)}{1 + g_1(s)g_2(s)} f(s) \qquad (3.2.33)$$

Once again the transfer function relating the input/output pair is obtained in a simple algebraic manner.

The utility of the unit impulse function becomes immediately apparent now since, using (3.2.29), $f(s) = 1.0$ and the output is equal to the transfer function, $\bar{c}(s) = g(s)$. This is, if the input to a system is an impulse, the output in the t-domain (or the s-domain) is connected directly to the system model parameters. This feature will be elucidated in detail in Chapter 9.

We shall now derive *Duhamel's formula* or *the superposition formula*. This formula permits the determination of the response of any linear system to any forcing function when its response, $A(t)$, to a unit step function is known. Consider an input $F(t) \xrightarrow{L} f(s)$ to a system which is characterized by an output $c(t) \xrightarrow{L} \bar{c}(s)$. It then follows that

$$c(s) = g(s)f(s)$$

where $g(s)$ is the transfer function for the system. However, if $F(t)$ is a unit step function, $f(s) = 1/s$ and

$$\bar{c}(s) = \frac{g(s)}{s} = a(s) = L[A(t)]$$

As a result the response function to the unit step function is

$$c(t) = A(t)$$

Also,

$$\bar{c}(s) = \frac{g(s)}{s} \cdot sf(s) = a(s)sf(s)$$

or, using the differentiation property,

$$\bar{c}(s) = a(s)L[\dot{F}(t)] + a(s)F(0^+)$$

Convoluting each term now leads to [with $F(0^+)$ a constant]

$$c(t) = A(t)F(0^+) + \int_0^t \dot{F}(\lambda)A(t-\lambda)d\lambda \tag{3.2.34}$$

which is one form of Duhamel's formula. Other equivalent forms are

$$\begin{aligned} c(t) &= \int_0^t F(\lambda)\dot{A}(t-\lambda)\,d\lambda \\ &= \int_0^t F(t-\lambda)\dot{A}(\lambda)\,d\lambda \end{aligned} \tag{3.2.35}$$

EXAMPLE 3.6. It is apparent from the above material that for a single CSTR (or well-mixed stage) the transfer function is

$$\text{1 CSTR:}\quad g(s) = \frac{1}{1+\theta s} \tag{A}$$

Here $\theta = 1/\alpha$ in our previous notation with units of time in physical situations. For two or more CSTRs in series, all of equal volume,

$$n\ \text{CSTRs:}\quad g(s) = \frac{1}{(1+\theta s)^n} \tag{B}$$

In the same manner, for two CSTRs in series, each with different volumes, we can write

$$g(s) = \frac{1}{(1+\theta_1 s)(1+\theta_2 s)} \tag{C}$$

The inverse L.T. of (C) is

$$\frac{1}{(\theta_1-\theta_2)}[e^{-t/\theta_1} - e^{-t/\theta_2}] \xleftarrow{L^{-1}} \frac{1}{(1+\theta_1 s)(1+\theta_2 s)}$$

Next consider n CSTRs with total volume V and each CSTR with volume V/n. The transfer function is then

$$g(s) = \frac{1}{[1+(\theta/n)s]^n}$$

and in the limit as $n \longrightarrow \infty$,

$$g(s) = e^{-\theta s}$$

which is a simple plug flow or pure time delay case. In fact, then, it is seen that a pure time delay can be approximated by a large set of CSTRs in series.

Looking further at n equal-volume CSTRs, the transfer functions and the corresponding time functions can be written as

$$n=1:\quad g(s) = \frac{1}{1+\theta s}, \qquad G(t) = \frac{1}{\theta}e^{-t/\theta}$$

$$n=2:\quad g(s) = \frac{1}{(1+\theta s)(1+\theta s)}, \qquad G(t) = \frac{t}{\theta^2}e^{-t/\theta}$$

$$n=n:\quad g(s) = \frac{1}{(1+\theta s)^n}, \qquad G(t) = \frac{1}{(n-1)!\,\theta}\left(\frac{t}{\theta}\right)^{n-1} e^{-t/\theta}$$

If a plot is made of $G(t)$ vs. t/θ, the case $n > 1$ yields a family of curves, each curve of which starts at zero, goes up to a maximum value, and then decreases to zero.

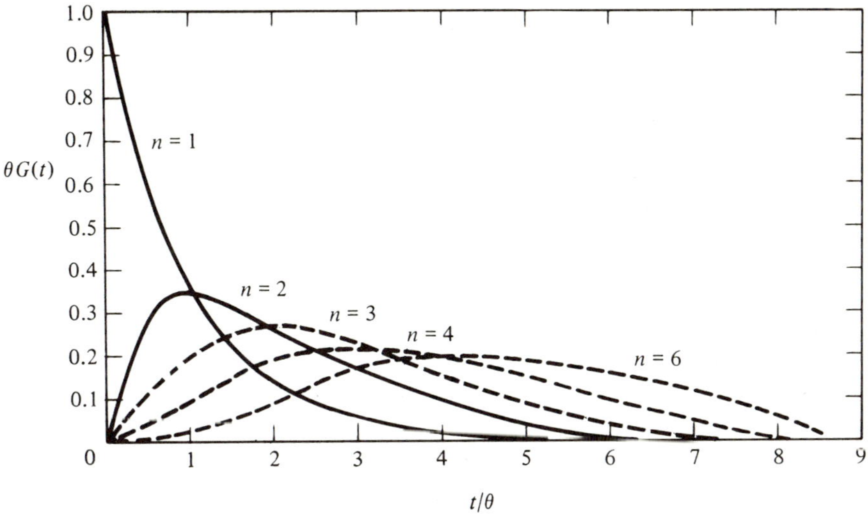

Figure 3.11. Impulse response of n-CSTRs of Example 3.6.

As n increases, the location of the maximum point moves to the right (Figure 3.11). Analogous results can be obtained for systems governed by P.D.E.s. Thus for the first-order hyperbolic system representing simple plug flow

$$\frac{\partial c}{\partial x} + \frac{\partial c}{\partial t} = 0$$

with initial and boundary conditions

$$c(x, 0) = 0$$
$$c(0, t) = F(t)$$

the response $c(x, t)$ to an arbitrary input $F(t)$ is

$$c(x, t) = F(t - x)$$

Thus the output is identical to the input but delayed by x. When a specific value of $x = \bar{x}$ is chosen it also follows directly that

$$\bar{c}(\bar{x}, s) = \exp[-\theta s] f(s) \tag{D}$$

This yields the transfer function for a pure plug flow system. In other words, any input $F(t)$ is transmitted unchanged to the point $x = \bar{x}$ except that it requires a time period $t = \theta$ to reach $\bar{x}$.

Further, we note that for a plug flow system of time θ (any of the types above) the transfer function is

$$\text{plug flow:} \quad g(s) = \exp[-\theta s] \tag{E}$$

corresponding to the delay in the input/output pair. A serial combination of a CSTR and a plug flow unit then has a transfer function

$$\text{CSTR + plug flow:} \quad g(s) = \frac{\exp[-\theta s]}{1 + \theta s} \tag{F}$$

Here the times θ have been assumed equal, but this is obviously unnecessary. In fact, a 2-CSTR system, with unequal volumes, and a plug flow unit have a transfer function

$$\text{2 CSTR + plug flow:} \quad g(s) = \frac{\exp[-\theta_1 s]}{(1+\theta_2 s)(1+\theta_3 s)} \tag{G}$$

3.2.6 Frequency response

There remains one further forcing function of importance, the sine or cosine function. In fact, one of the most important techniques in process identification consists of subjecting a system to a sinusoidal input and measuring the output as a function of the frequency of the input function. Let us consider the motivation for using such a method, which is called *frequency response testing.*

Consider a linear system with input $u(t)$ and output $y(t)$ and transfer function $g(s)$. Let the input be given by

$$u(t) = u_0 \cos \omega t$$

Then

$$\bar{u}(s) = \frac{u_0 s}{s^2 + \omega^2}$$

The resulting output is found from

$$y(t) = L^{-1}\left\{g(s)\frac{u_0 s}{s^2+\omega^2}\right\}$$

Let us assume that the poles (roots of the denominator) of the transfer function are $\lambda_1, \lambda_2, \ldots$, each of which has nonpositive real parts (the system is stable). In terms of partial fractions

$$y(t) = L^{-1}\left\{\frac{A_1}{s+i\omega} + \frac{A_2}{s-i\omega} + \frac{A_3}{s-\lambda_1} + \frac{A_4}{s-\lambda_2} + \cdots\right\}$$

Inverting,

$$y(t) = A_1 e^{-i\omega t} + A_2 e^{i\omega t} + A_3 e^{\lambda_1 t} + A_4 e^{\lambda_2 t} + \cdots$$

Since all the λs are in the left half-plane, as $t \to \infty$, all the exponential terms vanish, so that the ultimate output $y(t)$ corresponding to the sustained periodic input is

$$y(t) = A_1 e^{-i\omega t} + A_2 e^{i\omega t}$$

The amplitudes A_1 and A_2 are found from

$$A_1 = \left.\frac{u_0 s}{s - i\omega} g(s)\right|_{s=-i\omega} = \frac{u_0}{2} g(-i\omega)$$

$$A_2 = \left.\frac{u_0 s}{s + i\omega} g(s)\right|_{s=i\omega} = \frac{u_0}{2} g(i\omega)$$

Therefore

$$y(t) = \frac{u_0}{2} g(-i\omega)e^{-i\omega t} + \frac{u_0}{2} g(i\omega)e^{i\omega t}$$
$$= u_0 |g(i\omega)| \cos[\omega t + \angle g(i\omega)]$$

where $|g(i\omega)|$ is the magnitude of $g(i\omega)$ and $\angle g(i\omega)$ is the phase angle of $g(i\omega)$.

Thus we obtain the significant result that a linear, time-invariant system when subjected to a harmonic input signal will, after attaining a steady state, have a harmonic output signal of the same frequency as the input but with the amplitude altered by a factor of $|g(i\omega)|$ and with the phase shifted by $\angle g(i\omega)$. Even though we demonstrated this for a system whose transfer function $g(s)$ has poles $\lambda_1, \lambda_2, \ldots$, this relation will hold for any stable, linear, time-invariant system.

It may be difficult to interrupt an operating system to perform pulse or step testing. It may be possible, however, to test the system while in operation by superimposing on the normal input a sinusoidal signal and measuring the resulting output. By varying the frequency of the sinusoidal test input, one can obtain the system response as a function of frequency. In conducting experimental testing it is customary to vary ω and plot the resulting amplitude change $|g(i\omega)|$, and phase shift $\angle g(i\omega)$, as a function of ω. A plot of this type is often called a Bode plot.

3.2.7 Response of a CSTR to test inputs

Let us now return to the single CSTR of Example 3.3. For $c_0 = 0$ and an input $c_i(t)$ we have the convolution integral

$$c(t) = \alpha \int_0^t c_i(t-\lambda)e^{-\alpha\lambda}\, d\lambda = \alpha \int_0^t c_i(\lambda)e^{-\alpha(t-\lambda)}\, d\lambda \tag{3.2.36}$$

and

$$\bar{c}(s) = \frac{\alpha}{s+\alpha}\bar{c}_1(s) = g(s)\bar{c}_1(s) \tag{3.2.37}$$

The transfer function is $g(s) = \alpha/(s+\alpha)$ and the corresponding time response to an impulse is $g(t) = \alpha e^{-\alpha t}$. We may now consider some more complicated inputs.

1. $c_i(t)$ is a square wave of intensity c_1 and duration t_1:

$$c_i(t) = \begin{cases} c_1, & 0 < t < t_1 \\ 0, & t_1 < t \end{cases}$$

or

$$c_i(t) = c_1[U(t) - U(t-t_1)] \tag{3.2.38}$$

First let us use (3.2.37). Since

$$L[\,\urcorner\!\_\!\lrcorner\,] = \frac{c_1[1 - e^{-st_1}]}{s}$$

$$\bar{c}(s) = c_1\alpha\left[\frac{1}{s(s+\alpha)} - \frac{e^{-st_1}}{s(s+\alpha)}\right]$$

However, the inverse of the first term has already been evaluated, Eq. (H) in Example 3.3, and the second term is merely delayed by t_1 units. Thus

$$c_1[1 - e^{-\alpha t}] \xleftarrow{L^{-1}} \frac{c_1\alpha}{s(s+\alpha)}$$

$$c_1[1 - e^{-\alpha(t-t_1)}] \xleftarrow{L^{-1}} \frac{c_1\alpha e^{-st_1}}{s(s+\alpha)}$$

and

$$c(t) = \begin{cases} c_1[1 - e^{-\alpha t}], & t < t_1 \\ c_1[e^{-\alpha(t-t_1)} - e^{-\alpha t}], & t > t_1 \end{cases} \tag{3.2.39}$$

Using the final value theorem (3.1.26), we also note that

$$\lim_{s\to 0} s\bar{c}(s) = \lim_{s\to 0} c_1\alpha\left[\frac{1}{s+\alpha} - \frac{e^{-st_1}}{s+\alpha}\right]$$
$$= 0$$

which is the correct concentration as $t \to \infty$.

To illustrate the use of the convolution integral we note that

$$c(t) = \alpha\int_0^t c_1 e^{-\alpha(t-\lambda)}\,d\lambda = \alpha c_1 e^{-\alpha t}\int_0^t e^{\alpha\lambda}\,d\lambda, \qquad \text{for } t < t_1$$

$$= \alpha\int_0^{t_1} c_1 e^{-\alpha(t-\lambda)}\,d\lambda \qquad \text{for } t > t_1$$

yielding results identical to (3.2.39).

2. $c_i(t)$ is a sinusoidal input $= A \sin \omega t$. The steady-state response can be obtained by replacing s in the transfer function with $i\omega$ and then calculating the modulus of this complex number. Thus

$$\begin{aligned} g(s) &= \frac{\alpha}{s+\alpha} \\ g(i\omega) &= \frac{\alpha}{i\omega+\alpha} \\ |g(i\omega)| &= \frac{\alpha}{\sqrt{\omega^2+\alpha^2}} \\ c_{ss}(t) &= \frac{\alpha A}{\sqrt{\omega^2+\alpha^2}}\sin(\omega t + \phi) \end{aligned} \tag{3.2.40}$$

where $\phi = \tan^{-1}(-\omega/\alpha) = \angle g(i\omega)$.

3. $c_i(t)$ is a repeated sawtooth function with period t_1 and height c_1. From (3.2.19) the transformed function for a single sawtooth is

$$L[\diagup\!\_\!\_] = \bar{c}_i(s) = \frac{c_1}{t_1}\left[\frac{1}{s^2} - \frac{1}{s^2}e^{-st_1} - \frac{t_1}{s}e^{-st_1}\right]$$

The response to this first sawtooth is given by

$$\begin{aligned}\bar{c}(s) &= g(s)\bar{c}_i(s)\\ &= \frac{c_1\alpha}{t_1(s+\alpha)}\left[\frac{1}{s^2} - \frac{e^{-st_1}}{s^2} - \frac{t_1}{s}e^{-st_1}\right]\\ &= \frac{c_1\alpha}{t_1}\left[\left(\frac{1}{s^2(s+\alpha)} - \frac{e^{-st_1}}{s^2(s+\alpha)}\right) - \frac{t_1e^{-st_1}}{s(s+\alpha)}\right] \qquad (3.2.41)\end{aligned}$$

The inverse of the terms in brackets is

$$\frac{c_1}{\alpha t_1}\left[e^{-\alpha t} + \alpha t - 1 - e^{-\alpha(t-t_1)} - \alpha(t-t_1) + 1\right], \qquad t > t_1$$

and the inverse of the term not in brackets is

$$c_1[1 - e^{-\alpha(t-t_1)}], \qquad t > t_1$$

Adding the two inverses, rearranging, and inserting a step function representation,

$$c(t) = \frac{c_1}{\alpha t_1}[e^{-\alpha t}U(t-t_1) - (1-\alpha t_1)e^{-\alpha(t-t_1)}U(t-t_1)] \qquad (3.2.42)$$

$c(t)$ represents the response of the system to the first sawtooth function. The response to the nth sawtooth function will be identical to (3.2.42) except delayed by nt_1. Since the system is linear, all the components can be added to yield

$$c(t) = \frac{c_1}{\alpha t_1}\sum_{k=0}^{n-1}[e^{-\alpha(t-kt_1)}U(t-kt_1) - (1-\alpha t_1)e^{-\alpha(t-(k+1)t_1)}U(t-(k+1)t_1)]$$

where $c(t)$ is the response function for $(n-1)t_1 < t < nt_1$.

4. We may also develop a few interesting points using the first-order system in the form

$$\frac{dc}{dt} + \beta c = u(t) \qquad (3.2.43)$$

[compare this to Eq. (I) in Example 3.3]. The response to a unit impulse is then

$$c(t) = e^{-\beta t}$$

for the system originally at zero. However, suppose that we use a nonunit impulse, i.e., one whose area is not 1.0 but rather A. In this case it follows that

$$c(t) = Ae^{-\beta t}$$

However, this is the same response that one would get if the system had a zero input but started at an initial condition $c_0 = A$.

Thus when a first-order system is at rest the action of an impulse is equivalent to jarring (an instantaneous motion) the system to a new condition (which is a function of A); from this condition the system follows an exponential decay back to the zero state. Obviously, the action of $A\delta(t - \tau)$ merely delays the start of this trajectory.

In actual operation the impulse is really a square wave of finite height and short duration in which the system follows a path upwards (the response to an equivalent step input) until the input is removed and then decreases exponentially.

Note also that if $\beta = 0$ in (3.2.43), the response to an impulse is a step function. Thus if we integrate the area under a pulse response curve we obtain the step response.

Finally, we shall look at the response of (3.2.43) to a general input $u(t)$. If at time τ an impulse of width $d\tau$ is applied, then

$$A = u(\tau)\,d\tau$$

It follows then that

$$u(t) = A\delta(t - \tau) = [u(\tau)\,d\tau]\delta(t - \tau)$$

and the response is

$$c(t) = [u(\tau)\,d\tau]e^{-\beta(t-\tau)}U(t - \tau)$$

or

$$c(t) = [u(\tau)\,d\tau]H(t - \tau)$$

where

$$H(t - \tau) = e^{-\beta(t-\tau)}U(t - \tau)$$

If we represent $u(t)$ by a series of such impulses occurring at different times, then the total response is obtained by an integration over $0 \longrightarrow t$. Thus

$$c(t) = \int_0^t H(t - \tau)u(\tau)\,d\tau \tag{3.2.44}$$

which is nothing more than the convolution integral for the first-order system (3.2.43).

3.3 Relationship of the Laplace Transform to the State Space Representation

Recall that in Chapter 1 we defined the n-dimensional linear time-invariant system

$$\dot{\mathbf{x}}(t) = \mathbf{F}\mathbf{x}(t) + \mathbf{G}\mathbf{u}(t) \tag{3.3.1}$$

with solution

$$\mathbf{x}(t) = \mathbf{\Phi}(t)\mathbf{x}(0) + \int_0^t \mathbf{\Phi}(t - \lambda)\mathbf{G}\mathbf{u}(\lambda)\,d\lambda \tag{3.3.2}$$

where $\mathbf{\Phi}(t)$ is the $(n \times n)$-dimensional state transition matrix. The forced motion $\mathbf{x}(t)$ due to $\mathbf{u}(t)$ is the superposition of the free motion, the homogeneous solution, and the effect of the forcing $\mathbf{u}(t)$. In this section we shall show how the fundamental matrix of a time-invariant linear system can be computed by the L.T.

Consider first the homogeneous system $\dot{\mathbf{x}} = \mathbf{F}\mathbf{x}$, the solution of which has already been shown to be

$$\mathbf{x}(t) = e^{\mathbf{F}t}\mathbf{x}(0) = \mathbf{\Phi}(t)\mathbf{x}(0) \tag{3.3.3}$$

The L.T. of $\mathbf{x}(t)$ is the n vector whose elements are the L.T. of the elements of $\mathbf{x}(t)$. Taking the L.T. of $\dot{\mathbf{x}} = \mathbf{F}\mathbf{x}$, we obtain [using (3.1.20)]

$$s\bar{\mathbf{x}}(s) - \mathbf{x}(0) = \mathbf{F}\bar{\mathbf{x}}(s) \tag{3.3.4}$$

or

$$(s\mathbf{I} - \mathbf{F})\bar{\mathbf{x}}(s) = \mathbf{x}(0) \tag{3.3.5}$$

We note that $\det(s\mathbf{I} - \mathbf{F}) = 0$ if the n roots of this algebraic equation are the n eigenvalues of $\mathbf{F}$, $\lambda_1, \lambda_2, \ldots, \lambda_n$. Therefore the matrix $s\mathbf{I} - \mathbf{F}$ is nonsingular for all $s \neq \lambda_i$, $i = 1, 2, \ldots, n$, and so we have

$$\bar{\mathbf{x}}(s) = (s\mathbf{I} - \mathbf{F})^{-1}\mathbf{x}(0) \tag{3.3.6}$$

Taking the inverse L.T.,

$$\mathbf{x}(t) = L^{-1}\{(s\mathbf{I} - \mathbf{F})^{-1}\}\mathbf{x}(0) \tag{3.3.7}$$

Therefore

$$\mathbf{\Phi}(t) \xleftarrow{L^{-1}} (s\mathbf{I} - \mathbf{F})^{-1} \tag{3.3.8}$$

that is, we must invert the matrix $s\mathbf{I} - \mathbf{F}$ and then find its inverse L.T. to evaluate $\mathbf{\Phi}(t) = e^{\mathbf{F}t}$.

EXAMPLE 3.7 Determine $\mathbf{\Phi}(t)$ for the system $\dot{\mathbf{x}} = \mathbf{F}\mathbf{x}$, having

$$\mathbf{F} = \begin{bmatrix} 0 & 1 \\ 0 & -2 \end{bmatrix}$$

by means of the L.T. First,

$$s\mathbf{I} - \mathbf{F} = \begin{bmatrix} s & -1 \\ 0 & s+2 \end{bmatrix}$$

and

$$(s\mathbf{I} - \mathbf{F})^{-1} = \begin{bmatrix} \dfrac{1}{s} & \dfrac{1}{s(s+2)} \\ 0 & \dfrac{1}{s+2} \end{bmatrix}$$

Inverting each element,

$$\mathbf{\Phi}(t) = e^{\mathbf{F}t} = \begin{bmatrix} 1 & \frac{1}{2}(1 - e^{-2t}) \\ 0 & e^{-2t} \end{bmatrix}$$

EXAMPLE 3.8. Determine $\mathbf{\Phi}(t)$ for the system with

$$\mathbf{F} = \begin{bmatrix} -2 & 3 \\ -3 & -2 \end{bmatrix}$$

Then,

$$s\mathbf{I} - \mathbf{F} = \begin{bmatrix} s+2 & -3 \\ 3 & s+2 \end{bmatrix}$$

and

$$(s\mathbf{I} - \mathbf{F})^{-1} = \begin{bmatrix} \dfrac{s+2}{(s+2)^2+9} & \dfrac{3}{(s+2)^2+9} \\ \dfrac{-3}{(s+2)^2+9} & \dfrac{s+2}{(s+2)^2+9} \end{bmatrix}$$

Inverting,

$$\mathbf{\Phi}(t) = e^{\mathbf{F}t} = \begin{bmatrix} e^{-2t}\cos 3t & e^{-2t}\sin 3t \\ -e^{-2t}\sin 3t & e^{-2t}\cos 3t \end{bmatrix}$$

Consider now the forced system (3.3.1), with the output transformation

$$\mathbf{y}(t) = \mathbf{H}\mathbf{x}(t) \tag{3.3.9}$$

Taking the L.T. of (3.3.1) and (3.3.9) yields

$$s\bar{\mathbf{x}}(s) - \mathbf{x}(0) = \mathbf{F}\bar{\mathbf{x}}(s) + \mathbf{G}\bar{\mathbf{u}}(s)$$
$$\bar{\mathbf{y}}(s) = \mathbf{H}\bar{\mathbf{x}}(s)$$

or

$$\bar{\mathbf{x}}(s) = (s\mathbf{I} - \mathbf{F})^{-1}[\mathbf{x}(0) + \mathbf{G}\bar{\mathbf{u}}(s)] \tag{3.3.10}$$
$$\bar{\mathbf{y}}(s) = \mathbf{H}(s\mathbf{I} - \mathbf{F})^{-1}[\mathbf{x}(0) + \mathbf{G}\bar{\mathbf{u}}(s)] \tag{3.3.11}$$

Taking the inverse L.T.,

$$\mathbf{x}(t) = L^{-1}\{(s\mathbf{I} - \mathbf{F})^{-1}\}\mathbf{x}(0) + L^{-1}\{(s\mathbf{I} - \mathbf{F})^{-1}\mathbf{G}\bar{\mathbf{u}}(s)\} \tag{3.3.12}$$

Let us provide an interpretation of the second term on the right-hand side (R.H.S.) of (3.3.12). From the convolution integral (3.1.36), we note that

$$L^{-1}\{(s\mathbf{I} - \mathbf{F})^{-1}\mathbf{G}\bar{\mathbf{u}}(s)\} = \int_0^t e^{\mathbf{F}(t-\lambda)}\mathbf{G}\mathbf{u}(\lambda)\,d\lambda \tag{3.3.13}$$

so that the contribution from $\mathbf{u}(t)$ to $\mathbf{x}(t)$ is simply a convolution of $e^{\mathbf{F}t}$ with $\mathbf{G}\mathbf{u}(t)$. Actually, we can interpret $(s\mathbf{I} - \mathbf{F})^{-1}\mathbf{G}$ and $\mathbf{H}(s\mathbf{I} - \mathbf{F})^{-1}\mathbf{G}$ as the transfer functions between the input $\bar{\mathbf{u}}(s)$ and the state $\bar{\mathbf{x}}(s)$ and the output $\bar{\mathbf{y}}(s)$, respectively.

3.3.1 The state equation from the transfer function

While it is relatively simple to determine the transfer function relating the input and output of a linear system, the reverse procedure is not necessarily so easy. For example, for the system

$$\overset{[n]}{y} + a_{n-1}\overset{[n-1]}{y} + \cdots + a_1\dot{y} + a_0 y = b_0 u(t) \tag{3.3.14}$$

the transfer function is

$$\frac{\bar{y}(s)}{\bar{u}(s)} = \frac{b_0}{s^n + a_{n-1}s^{n-1} + \cdots + a_1 s + a_0} \tag{3.3.15}$$

The roots of the denominator of (3.3.15), that is, of

$$s^n + a_{n-1}s^{n-1} + \cdots + a_1 s + a_0 = 0 \tag{3.3.16}$$

are called the *poles* of the transfer function. Also, for the system

$$\overset{[n]}{y} + a_{n-1}\overset{[n-1]}{y} + \cdots + a_1\dot{y} + a_0 = b_m\overset{[m]}{u} + b_{m-1}\overset{[m-1]}{u} + \cdots + b_1\dot{u} + b_0 u \tag{3.3.17}$$

the transfer function is

$$\frac{\bar{y}(s)}{\bar{u}(s)} = \frac{b_m s^m + b_{m-1}s^{m-1} + \cdots + b_1 s + b_0}{s^n + a_{n-1}s^{n-1} + \cdots + a_1 s + a_0}$$

The roots of the numerator polynomial are called the *zeros* of the transfer function.

Now let us consider the reverse problem, namely the determination of a state representation of a system given its transfer function. There are some inherent problems in this task, the most severe of which is that there may not be a unique differential equation corresponding to the transfer function. To illustrate this point, take the second-order equation

$$\ddot{x} + (a + b)\dot{x} + abx = u \tag{3.3.19}$$

with $\mathbf{H} = \mathbf{I}$. By the standard L.T. of this equation the transfer function is

$$g(s) = \frac{1}{s^2 + (a + b)s + ab} = \frac{1}{b - a}\left[\frac{1}{s + a} - \frac{1}{s + b}\right] \tag{3.3.20}$$

The time response to an impulse is then

$$x(t) = \frac{1}{a - b}[e^{-bt} - e^{-at}]$$

To obtain the state variable form of (3.3.19) we can (with $a = 1$ and $b = 2$)

1. Let $x = x_1$, yielding

$$\begin{bmatrix}\dot{x}_1\\ \dot{x}_2\end{bmatrix} = \begin{bmatrix}0 & 1\\ -2 & -3\end{bmatrix}\begin{bmatrix}x_1\\ x_2\end{bmatrix} + \begin{bmatrix}0\\ 1\end{bmatrix}u$$

2. Let $x = z_1 - z_2$, yielding

$$\begin{bmatrix}\dot{z}_1\\ \dot{z}_2\end{bmatrix} = \begin{bmatrix}-1 & 0\\ 0 & -2\end{bmatrix}\begin{bmatrix}z_1\\ z_2\end{bmatrix} + \begin{bmatrix}1\\ 1\end{bmatrix}u \tag{3.3.21}$$

3. Let $x = x_2$, yielding

$$\begin{bmatrix} \dot{x}_1 \\ \dot{x}_2 \end{bmatrix} = \begin{bmatrix} -1 & 0 \\ 1 & -2 \end{bmatrix} \begin{bmatrix} x_1 \\ x_2 \end{bmatrix} + \begin{bmatrix} 1 \\ 0 \end{bmatrix} u$$

In each case we may then calculate $(s\mathbf{I} - \mathbf{F})$, $(s\mathbf{I} - \mathbf{F})^{-1}$, and $(s\mathbf{I} - \mathbf{F})^{-1}\mathbf{G}$, with the last being the transfer function. Thus, in order,

$$\begin{aligned}
&1. \quad \begin{bmatrix} s & -1 \\ 2 & s+3 \end{bmatrix}; \quad \frac{1}{(s+1)(s+2)} \begin{bmatrix} s+3 & 1 \\ -2 & s \end{bmatrix}; \\
&\qquad \frac{1}{(s+1)(s+2)} \begin{bmatrix} 1 \\ s \end{bmatrix} \\
&2. \quad \begin{bmatrix} s+1 & 0 \\ 0 & s+2 \end{bmatrix}; \quad \frac{1}{(s+1)(s+2)} \begin{bmatrix} s+2 & 0 \\ 0 & s+1 \end{bmatrix}; \\
&\qquad \frac{1}{(s+1)(s+2)} \begin{bmatrix} s+2 \\ s+1 \end{bmatrix} \\
&3. \quad \begin{bmatrix} s+1 & 0 \\ -1 & s+2 \end{bmatrix}; \quad \frac{1}{(s+1)(s+2)} \begin{bmatrix} s+2 & 0 \\ 1 & s+1 \end{bmatrix}; \\
&\qquad \frac{1}{(s+1)(s+2)} \begin{bmatrix} s+2 \\ 1 \end{bmatrix}
\end{aligned} \tag{3.3.22}$$

While these three transfer functions are explicitly different, they really yield the same information. This can be seen by noting how x is related to u, i.e., in case 1 see how $x_1 = x$ relates to u through the transfer function matrix. In every case this yields an element of $(s\mathbf{I} - \mathbf{F})^{-1}\mathbf{G}$ which is $1/[(s+1)(s+2)]$, but this is actually (3.3.20) with $a = 1$ and $b = 2$. Thus any one of (3.3.22) will yield the same information for $x(t)$.

This equivalence between the various state variable representations and the original differential equation can be illustrated here because the equations are simple enough to manipulate. In a general case, given the transfer function matrix, it may be impossible to recover the correct original equation.

3.4 Discrete System Equations and the z-Transform

Up to now in this chapter we have discussed the solution of O.D.E.s and P.D.E.s by means of the L.T.; that is, we have dealt with functions continuous in time. We shall now turn to the problem of the solution of difference equations using the L.T. As soon as one considers discrete experimental data or the use of a digital computer, it is necessary to consider systems that are in some manner discrete. The theory of the analysis of linear discrete systems has resulted in a transform analogous to the L.T., called the *z-transform*. In this section we shall introduce the z-transform and apply it to a variety of situations involving discrete systems or sampled continuous systems.

3.4.1 Sampling and discrete systems

Because a digital computer is a machine operating with *numbers*, it cannot accept information which is a continuous function of time, such as the voltage from a thermocouple or the pressure from a flowmeter. For this reason, continuous physical measurements must be converted into numbers in some fashion. This operation is called *sampling*. An instant of time when a measurement is converted into a number is a *sampling point*, and the corresponding number is a *sample*.

No matter how often one samples, some information is always discarded in the process. However, this loss is often insignificant. In fact, if the signal $x(t)$ to be sampled can be presented in the form

$$x(t) = \sum_{-N}^{+N} a_n \exp(ni\omega t)$$

then we can prove the

SAMPLING THEOREM. If the average number of sampling points per unit time is more than $\pi/N\omega$—i.e., half the period of the highest frequency component signal of $x(t)$—then the original signal $x(t)$ can be reconstructed as accurately as desired from the samples

$$x(t_0), x(t_1), \cdots, x(t_k), \cdots$$

provided one is willing to wait for a sufficiently long period of time.

In less precise language, the sampling theorem says that the behavior of a "smooth" function between sampling points can be always reconstructed with arbitrary accuracy by means of some scheme of interpolation, provided one knows the values of the function at a large number of sampling points.

Figure 3.12 shows two curves each of which when sampled at the times shown will yield identical values of $x(t)$ vs. t. The solid line is a low-frequency curve, while the dashed line is a high-frequency curve. As the sampling frequency is increased, the regions of signal response are spread out. If the sampling frequency is decreased, however, distortion starts to appear as soon as the sampling frequency is less than twice the highest frequency in the signal. Now spurious components tend to ruin the reconstruction of the signal.

Note that the sampling theorem does not require that samples be taken in any kind of a uniform pattern. In practice, however, it is often most con-

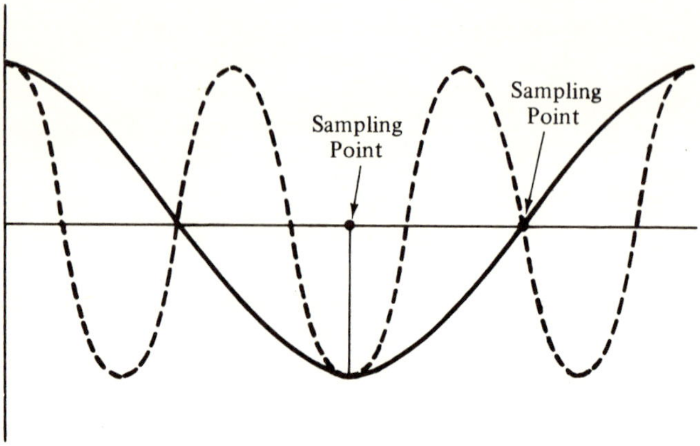

Figure 3.12. Two different sampled curves.

venient to sample periodically, so that the sampling points are given by

$$t_k = k\tau, \qquad k = 0, 1, 2, \ldots$$

where τ is called the sampling period; the integer k denotes the kth sample.

Whether sampling is introduced into a system deliberately (so as to be able to use a digital computer for control purposes) or unavoidably (when a measurement on the process is represented by a physical "sample" which is analyzed in the laboratory), we are dealing with a *sampled-data system*. In view of the sampling theorem, it is intuitively obvious that *sampled-data systems need not be inferior to systems in which there is no sampling*.

If the sampling operation takes place somewhat more rapidly than the changes in the process behavior which the control system is designed to counteract, then sampling does not impose a serious limitation on the overall system. For instance, in designing a control system for an airplane, rather rapid sampling is essential since a modern aircraft can move an appreciable distance in a fraction of a second. On the other hand, some chemical reactions may have a response time of the order of several hours, so that even very crude methods of sampling (a laboratory analysis every half hour or so) may provide an adequate amount of information. Thus the desirable sampling rate depends primarily on the dynamics of the process to be controlled.

Since the output of a digital computer operating in a control system is a sequence of numbers, these numbers must be converted into an actuating signal which affects the process, such as a change in valve opening which affects the flow rate. In other words, a continuous signal must be reconstructed from the numbers which are provided by the digital computer. Such an operation is known as *holding*. Two common methods are the *zero-order hold* (or staircase interpolation) and the *first-order hold* (or trapezoidal interpolation).

In the zero-order hold the signal, $F(t)$ [or if control is being used, $u(t)$] is held constant at the previous sampled value during the entire sampling period. Thus

$$F(t) = F(k\tau) = F_k(t), \qquad k\tau \leq t < (k+1)\tau,\ k = 0, 1, 2, \ldots$$

where the sequence of numbers $F(0), F(\tau), F(2\tau), \ldots$ is the output of the digital computer or the holding device. If, for example, a series of unit impulses (or any real continuous time functions) is the input, the output is a series of rectangular pulses of width τ. For a single impulse the response is merely equivalent to a square wave and thus has a transfer function of [see (3.2.6)]

$$g(s) = \frac{1 - e^{-s\tau}}{s} \tag{3.4.1}$$

This type of sampling device is also referred to as a *sample-and-hold* device.

For first-order holding, the output operates on the slope of the function sampled rather than its level. Thus

$$F(t) = F(k\tau) + (t - k\tau)\dot{F}(k\tau), \qquad k = 0, 1, 2, \ldots$$

However, since

$$\dot{F}(k\tau) \simeq \frac{F(k\tau) - F[(k-1)\tau]}{\tau}$$

we can replace $\dot{F}(k\tau)$ by the above. For a unit impulse at $t = 0$, $F(0) = 1$ and $F(-\tau) = 0$ and we can generate the transfer function as

$$g(s) = \frac{1 + \tau s}{\tau s^2}[1 - e^{-s\tau}]^2 \tag{3.4.2}$$

Suppose that a process is in a state of equilibrium; that is, to a given constant input F_{ss} there corresponds a constant output y_{ss}. If the input to the process is slightly perturbed, then the output usually undergoes similar small perturbations about its equilibrium value. If the output perturbations are sufficiently small, the system can be considered as approximately linear. If so, let us write

$$F(t) = F_{ss} + \delta F(t)$$
$$y(t) = y_{ss} + \delta y(t)$$

where $\delta F(t)$ and $\delta y(t)$ denote the small perturbations about equilibrium. Let us assume now that $\delta F(t)$ is a *unit pulse:*

$$\delta F(t) \begin{cases} = 1, & 0 \leq t \leq \tau \\ = 0, & \text{otherwise} \end{cases} \tag{3.4.3}$$

We observe the time function $\delta y(t)$ which results from such a perturbation and denote it by $G(t)$, which is the unit pulse response of the process. Since linearity has been assumed, we may invoke the principle of superposition and therefore it follows that for any input signal operating in a zero hold we have

$$\delta y(t) = \sum_{l=-\infty}^{l\tau<t} G(t - l\tau)\delta F(l\tau) \tag{3.4.4}$$

Thus, as long as the assumption of linearity remains valid, we can obtain the response of the process to any such function by considering it as the sum of separate pulses of height $\delta F(k\tau)$ and adding up the individual effects. Since $y(t)$ is of interest only at the sampling points, we write

$$\delta y_k = \sum_{l=-\infty}^{k} G_{k-l}\,\delta F_l \tag{3.4.5}$$

where for simplicity we have written $F(k\tau) = F_k$, etc.

Equation (3.4.5) shows that the dynamic behavior of the process can be thought of as transforming the sequence of numbers $\delta F_0, \delta F_1, \ldots$ into the sequence of numbers $\delta y_0, \delta y_1, \ldots$. To calculate the δy_k from the knowledge of the δF_k we must know, of course, the numbers $G_0, G_1, \ldots$. ($G_k = 0$ when $k < 0$ since the process cannot respond before the input is applied.) The latter numbers, in turn, can be obtained by a variety of tests, the simplest being the application of a unit pulse to the process.

To illustrate this point suppose that the response to a unit step

$$\delta F(t) = \begin{cases} 1, & t \geq 0 \\ 0, & t < 0 \end{cases}$$

is $H_0, H_1, \ldots$ at the sampling points. Then, since any signal may be decomposed into unit steps, we can write

$$\delta y_k = \sum_{l=-\infty}^{k} H_{k-l}(\delta F_l - \delta F_{l-1}) \tag{3.4.6}$$

Since $H_k = 0$ when $k < 0$, by rearranging the summation in (3.4.6) we find

$$\delta y_k = \sum_{l=-\infty}^{k} (H_{k-l} - H_{k-(l+1)})\,\delta F_l$$

Hence

$$G_k = H_k - H_{k-1}$$

which shows how the unit pulse response may be determined from the unit step response by a simple arithmetic calculation.

The use of the unit pulse response is frequently impractical in practice because of the very large (theoretically, infinite) number of G_k which may have to be taken into account. This difficulty may be avoided by assuming that $\delta y(t)$ is related to $\delta F(t)$ by means of an ordinary differential equation with constant coefficients. It can readily be shown that this assumption is equivalent to assuming that δy_k is related to δF_k by means of the *linear difference equation with constant coefficients:*

$$b_0\,\delta y_k + b_1\,\delta y_{k-1} + \cdots + b_n\,\delta y_{k-n} = a_0\delta F_k + a_1\delta F_{k-1} + \cdots + a_q\,\delta F_{k-q} \tag{3.4.7}$$

Instead of the constants G_k, the process dynamics is now represented by the constants $a_0, \ldots, b_n$.

For convenience, let $b_0 = 1$. Then (3.4.7) may be "solved"; i.e., the set of numbers δy_k may be evaluated iteratively by rearranging (3.4.7) to give the recurrence formula

$$\begin{aligned} \delta y_k = a_0\,\delta F_k + \cdots + a_q\,\delta F_{k-q} \\ - (b_1\,\delta y_{k-1} + \cdots + b_n\,\delta y_{k-n}), \qquad k = 0, 1, 2, \ldots \end{aligned} \tag{3.4.8}$$

The *initial conditions* for the calculation are the values of δy_k prior to $k = 0$:

$$\delta y_{-1}, \cdots, \delta y_{-n}$$

If we assume that $\delta y_k = 0$ when $k < 0$ and if $F(t)$ is a unit pulse, then using (3.4.8) repeatedly we find

$$\delta y_0 = G_0, \quad \delta y_1 - G_1, \ldots$$

However, if we expand the fraction (with z a constant)

$$\frac{\delta \bar{y}(z)}{\delta f(z)} = g(z) = \frac{a_0 + a_1 z^{-1} + \cdots + a_q z^{-q}}{b_0 + b_1 z^{-1} + \cdots + b_n z^{-n}} \tag{3.4.9}$$

by formal long division according to ascending powers of z^{-1}, we find that the same numerical computations must be performed as are required by (3.4.8) and we are led to

$$G_0 + G_1 z^{-1} + G_2 z^{-2} + \cdots \tag{3.4.10}$$

Note that if we define $z = \exp[s\tau]$ in (3.4.9), then $g(z)$ is simply the L.T. of the zero-order hold and process combination.

The expression (3.4.9) is called the *z-transform* of the combination zero-order hold and the process. It is simply a short hand notation for (3.4.8), obtained by writing symbolically $\delta y_{k-n} = z^{-n}\,\delta y_k$. The *inverse z-transform* is simply the number series $G_0, G_1, \ldots$.

3.4.2 Explicit z-transforms

We can develop the specific z-transform in a more formal fashion. Thus we shall consider a continuous function $F(t)$ which is converted into a sequence of numbers $\{F(k\tau)\}$, where the k refers to equally spaced points sampled from the continuous function and τ is the time interval between samples. For simplicity we shall call $F(k\tau) = F(k)$ or F_k. Then the z-transform of F_k is denoted by

$$Z[F_k] = f(z) = \sum_{k=0}^{\infty} F(k\tau) z^{-k} \tag{3.4.11}$$

In other words,

$$F_k \xrightarrow{Z} f(z)$$

and

$$F_k \xleftarrow{Z^{-1}} f(z) \tag{3.4.12}$$

Here the operator Z means "the z-transform of." Alternatively, we may let

$$L[F_k] = f(s) = \sum_{k=0}^{\infty} F(k\tau)e^{-ks\tau} \tag{3.4.13}$$

However, if we specify that

$$z = \exp[s\tau]$$

or

$$s = \frac{1}{\tau} \ln z$$

we can recover (3.4.11). Thus the z-transform is equivalent to the Laplace transform but with a specific value of s.

To illustrate use of these formulas (and the construction of a table of z-transforms) we can analyze some simple functions. To start, note that we have previously defined that the L.T. for repetitive unit impulses is (3.2.26)

$$L[|||\,..] = 1 + e^{-s\tau} + e^{-2s\tau} + e^{-3s\tau} + \cdots$$

By using $z = e^{s\tau}$, this becomes

$$Z[|||\,..] = 1 + z^{-1} + z^{-2} + z^{-3} + \cdots$$

or

$$g(z) = \frac{z}{z - 1} \tag{3.4.14}$$

for a train of impulses sampled at the time of the impulses. For the case where $F(t)$ is a unit step function, $U(t)$, the sampled sequence will be merely 1, 1, 1, ... Thus from (3.4.11),

$$f(z) = \sum_{k=0}^{\infty} z^{-k}$$

which is a geometric series. For $|1/z| < 1$ this can be summed and converges to

$$f(z) = \frac{z}{z - 1} \tag{3.4.15}$$

Note the correspondence between (3.4.14) and (3.4.15); this occurs because when the unit step is sampled, the output is a unit impulse train.

Next we shall consider the case $F(t) = e^{-at}$. The sampled values are $F(k\tau) = e^{-ak\tau}$ and substituting into (3.4.11) there results

$$Z[e^{-ak\tau}] = f(z) = \sum_{n=0}^{\infty} (e^{-a}z^{-1})^k$$

which is another geometric series. This then leads to

$$Z[e^{-ak\tau}] = \frac{z}{z - e^{-a\tau}} \tag{3.4.16}$$

Obviously, we may continue this procedure for a variety of $F(t)$. Table 3.1 includes a convenient small set of such functions as well as the corresponding L.T. Further results may be found in Kuo [17]. Also shown in Table 3.1 are the final and initial value theorems, etc., analogous to those for the L.T.

Of specific interest, however, in Table 3.1 are the following:

1. The translation property indicates that each shift or time delay of a sequence corresponds to a multiplication by z^{-1} in the z-domain. This is useful for solving difference equations.
2. The shifting theorem is analogous to the L.T. of a derivative in a continuous system. This is also useful for solving difference equations.

There are, however, certain limitations or restrictions on the z-transform, which include:

1. While the L.T. pair is unique, the z-transform does not necessarily have this property. Thus (3.4.15) and (3.4.16) are the same but $F_k(t)$ for each is different.
2. The transform is applicable only to problems where the sampling is periodic.
3. Only values at the sampling instants have any meaning. No explicit information about the behavior of a function or a system is available between sampling points.

It would also seem apparent from the analogy to the L.T. that there is a z-transform transfer function, also called a pulse transfer function. Further, it would be expected that the form

$$\bar{c}(z) = g(z)f(z) \tag{3.4.17}$$

relating the input $f(z)$ and the output $\bar{c}(z)$, still holds This is quite true with a few qualifications. Thus if a discrete input $F_k(t)$ is applied to $G(t)$, the response $c(t)$ can be continuous and not discrete. Only if one synchronizes a sampling of $c(t)$ with $F_k(t)$ will (3.4.17) hold; in this case, as usual, $\bar{c}(z)$ gives information only at the sampling instants.

In the same sense, if $G(t)$ is made up of a series of cascades $G_1(t)$, $G_2(t)$, . . . , a synchronized sampler (discretizer) must be placed between each successive $G_i(t)$ in order for

$$g(z) = g_1(z)g_2(z)\cdots$$

to hold in analogy to the Laplace transform. With this proviso of special sampling or discretizing, almost all the material in Section 3.2.5 follows over directly to the z-transform.

EXAMPLE 3.9. To illustrate some of the features of the z-transform, we shall first solve the linear, constant coefficient difference equation

$$x(k + 2) + \tfrac{5}{6}x(k + 1) + \tfrac{1}{6}x(k) = F(k)$$

with

$$x(0) = 1$$
$$x(1) = 1$$

Letting

$$x(k) \xrightarrow{Z} \bar{x}(z)$$
$$F(k) \xrightarrow{Z} f(z)$$

the use of the shifting theorem (Table 3.1) yields

$$z^2\bar{x}(z) - [z^2x(0) + zx(1)] + \tfrac{5}{6}z\bar{x}(z) - \tfrac{5}{6}zx(0) + \tfrac{1}{6}\bar{x}(z) = f(z)$$

Solving for $\bar{x}(z)$ with the initial conditions above yields

$$\bar{x}(z) = \frac{1}{(z + \frac{1}{2})(z + \frac{1}{3})}\left[\left(z^2 + \frac{5}{6}z\right) + z + f(z)\right]$$

Note that the z-transfer function is merely

$$g(z) = \frac{1}{(z + \frac{1}{2})(z + \frac{1}{3})}$$

associated with $x(0) = x(1) = 0$, the system initially at rest. In this case a partial-fraction expansion and the use of Table 3.1 show that the response to $F(k) = 1$, a step function, is

$$x(k) = 8[(-\tfrac{1}{2})^{k+1} - 9(-\tfrac{1}{3})^k + 1]$$

EXAMPLE 3.10. As a second example consider the case of a CSTR, followed by a constant multiplier stage and a zero-order hold device. In series these devices have s-domain transfer functions of

$$g_1(s) = \frac{1}{1 + \theta s} \qquad \text{(CSTR)}$$

$$g_2(s) = K \qquad \text{(multiplier)}$$

$$g_3(s) = \frac{1 - e^{-s\tau}}{s} \qquad \text{[zero-order hold; see (3.4.1)]}$$

We desire to find $g_1g_2g_3 = g(z)$. From the translation theorem (Table 3.1) we see that $e^{-s\tau}$ is equivalent to z^{-1}; thus we merely need to consider $1/s(1 + \theta s)$. This may be expanded in partial functions:

$$\frac{1}{s(1 + \theta s)} = \frac{1}{s} + \frac{1}{(1/\theta) + s}$$

From Table 3.1 we note that $1/s \longrightarrow z/(z - 1)$ and $1/[(1/\theta) + s] \longrightarrow z/(z - e^{-\tau/\theta})$. Thus we have

$$g(z) = (1 - z^{-1})K\left[\frac{z}{z - 1} + \frac{z}{z - e^{-\tau/\theta}}\right]$$
$$= K\left[\frac{1 - e^{-\tau/\theta}}{z - e^{-\tau/\theta}}\right]$$

EXAMPLE 3.11. As a final example we shall consider (3.3.19), which has a transfer function

$$g(s) = \frac{1}{(s + 2)(s + 1)}$$

for $a = 1$, $b = 2$. To operate in the z-domain we shall introduce a zero-order hold in series with the transfer function. Thus we now have

$$\frac{1 - e^{-s\tau}}{s} g(s)$$

However, since

$$1 - e^{-s\tau} \longrightarrow (1 - z^{-1})$$

and

$$\frac{ab}{s(s + a)(s + b)} \longrightarrow \frac{1}{z - 1} + \frac{b}{a - b}\frac{z}{z - e^{-a\tau}} - \frac{a}{a - b}\frac{z}{z - e^{-b\tau}}$$

substituting $a = 1$, $b = 2$, and $\tau = 1$ and expanding the result in partial fractions yields

$$g(z) = \frac{0.63}{z - 0.37} - \frac{0.43}{z - 0.14}$$

3.4.3 Relationship to the state representation

Just as we were able to develop a transfer function matrix in the s-domain, we can accomplish the equivalent in terms of the z-transform. Now, however, we start from a linear vector difference equation

$$\begin{aligned} \mathbf{x}(k + 1) &= \mathbf{\Phi}(\tau)\mathbf{x}(k) + \mathbf{\Delta}(\tau)\mathbf{u}(k) \\ \mathbf{y}(k) &= \mathbf{H}\mathbf{x}(k), \qquad k = 0, 1, 2, \ldots \end{aligned} \tag{3.4.18}$$

The state equation may arise normally from a discrete system or it may be obtained from a sampled version of the continuous equation with [20]

$$\begin{aligned} \mathbf{\Phi}(\tau) &= \exp[\mathbf{F}\tau] \\ \mathbf{\Delta}(\tau) &= \int_0^t \exp[\mathbf{F}(\tau - \lambda)]\mathbf{G}\, d\lambda \end{aligned} \tag{3.4.19}$$

Defining

$$\mathbf{x}(k) \xrightarrow{z} \bar{\mathbf{x}}(z)$$

$$\mathbf{u}(k) \xrightarrow{z} \bar{\mathbf{u}}(z)$$

$$\mathbf{y}(k) \xrightarrow{z} \bar{\mathbf{y}}(z)$$

we apply the z-transform to (3.4.18). This yields with $\mathbf{x}(0)$ as a starting value:

$$\begin{aligned} z\bar{\mathbf{x}}(z) - z\mathbf{x}(0) &= \mathbf{\Phi}(\tau)\bar{\mathbf{x}}(z) + \mathbf{\Delta}(\tau)\bar{\mathbf{u}}(z) \\ \bar{\mathbf{y}}(z) &= \mathbf{H}\bar{\mathbf{x}}(z) \end{aligned}$$

However, for $\mathbf{x}(0) = 0$, a zero initial condition, these lead to

$$\bar{\mathbf{x}}(z) = [z\mathbf{I} - \mathbf{\Phi}(\tau)]^{-1}\,\mathbf{\Delta}(\tau)\bar{\mathbf{u}}(z)$$
$$\bar{\mathbf{y}}(z) = \mathbf{H}[z\mathbf{I} - \mathbf{\Phi}(\tau)]^{-1}\,\mathbf{\Delta}(\tau)\bar{\mathbf{u}}(z)$$

and the state and output transfer matrices (relating the state and input and the output and input) are, respectively,

$$\begin{aligned}\mathbf{g}(z) &= [z\mathbf{I} - \mathbf{\Phi}(\tau)]^{-1}\,\mathbf{\Delta}(\tau)\\ \mathbf{g}(z) &= \mathbf{H}[z\mathbf{I} - \mathbf{\Phi}(\tau)]^{-1}\,\mathbf{\Delta}(\tau)\end{aligned} \tag{3.4.20}$$

3.4.4 z-transforms and numerical integration techniques

We now wish to show how the z-transform may be applied to the numerical solution of O.D.E.s. The rationale here is that the differential equation is approximated by a specific difference equation; this form is then ideal for manipulation via the z-transform. Further details, especially in terms of stability and accuracy, can be found in Gilbert [11].

Consider the following first-order ordinary differential equation:

$$\begin{aligned}\dot{y} &= f(x, y)\\ y &= y(0) \qquad \text{at } x = x(0)\end{aligned} \tag{3.4.21}$$

Euler's method for integrating this equation is given by [21]

$$\begin{aligned}y(k+1) &= y(k) + \tau f[x(k), y(k)]\\ &= y(k) + \tau f(k)\end{aligned} \tag{3.4.22}$$

Defining

$$y(k) \xrightarrow{z} \bar{y}(z)$$
$$f(k) \xrightarrow{z} \bar{f}(z)$$

the z-transform of (3.4.22) yields

$$z\bar{y}(z) - zy(0) = \bar{y}(z) + \tau\bar{f}(z)$$

or

$$\bar{y}(z) = \frac{\tau}{z-1} f(z) + \frac{zy(0)}{z-1} \tag{3.4.23}$$

Note that the transfer function for the Euler integration is given by

$$g(z) = \frac{\tau}{z-1} \tag{3.4.24}$$

The use of another integrating technique would lead to analogous results. For example, the Adams-Moulton two-step implicit method is [21]

$$y(k+1) = y(k-1) + \frac{\tau}{3}[f(k+1) + 4f(k) + f(k-1)]$$

Using the translation theorem, the z-transform is [for $y(0) = 0$]

$$z\bar{y}(z) = z^{-1}\bar{y}(z) + \frac{\tau}{3}[z + 4 + z^{-1}]\bar{f}(z)$$

or

$$\bar{y}(z) = \frac{\tau}{3}\frac{(z^2 + 4z + 1)}{(z^2 - 1)}\bar{f}(z)$$

The transfer function for this integration technique is

$$g(z) = \frac{\tau}{3}\frac{(z^2 + 4z + 1)}{(z^2 - 1)}$$

In this way transfer functions can be obtained for any numerical integration method.

EXAMPLE 3.12. Let us consider the numerical solution of a specific differential equation. We take either

$$\dot{y} = ay + bu \tag{A}$$

or, more simply,

$$\dot{y} = -y, \qquad a = -1, \qquad b = 0 \tag{B}$$

In the first case $f(x, y) = ay + bu$, while in the second case $f(x, y) = -y$. Using an Euler integration, (3.4.23) and (B) yield

$$\bar{y}(z) = \frac{-\tau}{z - 1}\bar{y}(z) + \frac{zy(0)}{z - 1}$$

or

$$\bar{y}(z) = \frac{zy(0)}{z - (1 - \tau)} \tag{C}$$

For (3.4.23) and (A) the analogous result is

$$\bar{y}(z) = \frac{zy(0)}{z - (1 + \tau a)} + \frac{\tau b}{z - (1 + \tau a)}\bar{u}(z)$$

In other words, these equations represent the z-transform of the numerical Euler solution of the appropriate differential equation.

For (C) the exact solution is known to be

$$y(t) = y(0)e^{-t}$$

and the z-transform of this exact solution is

$$\bar{y}(z) = \frac{zy(0)}{z - e^{-\tau}} \tag{D}$$

This follows from

$$y(k\tau) = y_0 e^{-k\tau}$$

With the approximate solution, (C), and the exact solution, (D), given, it is possible to investigate such items as the error (or the accuracy) of the Euler integration. Further, the question of stability of the Euler integration, e.g., the range of values of τ for which errors remain bounded, may be ascertained [21].

Thus the z-transform provides a convenient unifying approach for analyzing the accuracy and stability of finite difference approximations to O.D.E.s.

3.5 Numerical Inversion Methods

As we have already mentioned, the utility of the L.T. lies in the simplifying feature of transferring from the t-domain to the s-domain. Thus the solution of O.D.E.s is converted to the solution of algebraic equations. However, the inverse transformation, from the s-domain to the t-domain, may be an overpowering problem even in the case where extensive tables (such as Table 3.1) exist. In Section 3.6 we shall explicitly show some of the degrees of difficulty associated with a variety of different problems.

With this inverse problem often being so difficult and complicated and with the availability of digital computers, it seems apparent that a numerical approach may have significant merit. In particular we shall outline here the approach used by Bellman et al. [4] for numerical inversion of the L.T. This uses Gauss-Legendre-type quadrature formulas to evaluate $F(t)$ given a specific $f(s)$ (note the work of Piessens in Section 3.5.4, however). Further, we shall point out other possibilities which use Laguerre and Jacobi polynomials in place of Legendre polynomials.

As Bellman et al. point out, this numerical procedure is one which is subject to many difficulties. In fact, the inversion is one of a class of so-called improper problems of modern mathematics, because the inversion is not solvable for all values of the numerical or functional parameters of the problem and small perturbations of these parameters may induce large perturbations in the solution. More specifically, for

$$f(s) = \int_0^\infty e^{-st} F(t)\, dt$$

the problem is that given an $f(s)$ in analytic or even tabular form to find a corresponding $F(t)$. As such this is the inverse of the usual easy problem of evaluating an integral. Further, small changes in $f(s)$ may result in large changes in $F(t)$; small perturbations generate large perturbations. This is equivalent to the well-known difficulties of inverting certain matrices. Because of these features, the numerical solution must be approached with a considerable degree of caution.

3.5.1 Orthogonal polynomials

Since the basic feature of the method to be described lies in the use of orthogonal polynomials, we shall briefly outline certain features of these polynomials. Further details may be found in Davis and Rabinowitz [7].

Consider $w(t) \geq 0$ as a given function on $\{a, b\}$ and the integral

$$\int_a^b w(t)h(t)g(t)\, dt = (h, g) \tag{3.5.1}$$

Two functions $h(t)$ and $g(t)$ are called *orthogonal* over $\{a, b\}$ with respect to the weighting function $w(t)$ if

$$(h, g) = \int_a^b w(t)h(t)g(t)\,dt = 0 \tag{3.5.2}$$

Polynomials $P_m(t)$ and $P_n(t)$ of degree m and n, respectively, are orthogonal if

$$(P_m, P_n) = \int_a^b w(t)P_m(t)P_n(t)\,dt = 0, \qquad m \neq n \tag{3.5.3}$$

In the same sense, but normalized, a set of *orthonormal* polynomials P_m^* is defined by the relation

$$(P_m^*, P_n^*) = \int_a^b w(t)P_m^*(t)P_n^*(t)\,dt = \begin{cases} 0, & m \neq n \\ 1, & m = n \end{cases} \tag{3.5.4}$$

Some examples of well-known polynomials are:

1. $a = -1, b = 1, w(t) = 1$: Legendre polynomials.
2. $a = 0, b = 1, w(t) = 1$: shifted Legendre polynomials.
3. $a = -1, b = 1, w(t) = (1 - t)^a(1 + t)^b, a, b > -1$: Jacobi polynomials.
4. $a = 0, b = \infty, w(t) = e^{-t}$: Laguerre polynomials.

Other special polynomials are due to Chebyshev and Hermite, but we shall not present them here. It is interesting to note that the Laguerre polynomials have the limits of $\{0, \infty\}$ and a weight factor of e^{-t} corresponding most closely to that of the Laplace transform formula.

Each of these polynomials has certain common properties. We shall list them using the Legendre polynomials to present specific forms.

1. Each orthogonal or orthonormal polynomial satisfies a three-term recurrence relation

$$P_n^*(t) = (a_n t + b_n)P_{n-1}^*(t) - c_n P_{n-2}^*(t), \qquad n = 2, 3, \ldots$$

with

$$P_{-1}^*(t) = 0$$
$$P_0^*(t) = 1$$

as starting values. The Legendre polynomials are

$$\begin{aligned} P_0(t) &= 1 \\ P_1(t) &= t \\ P_2(t) &= \frac{3t^2 - 1}{2} \\ P_3(t) &= \frac{5t^3 - 3t}{2} \\ P_4(t) &= \frac{35t^4 - 30t^2 + 3}{8} \end{aligned} \tag{3.5.5}$$

2. The zeros (or roots) of the orthogonal polynomials are real, simple, and located in the interior of $\{a, b\}$. From (3.5.5) it is obvious that succeeding polynomials exhibit such behavior and that $P_3(t)$ has three roots, etc. These roots, and those of the other polynomials, are extensively tabulated in the literature [7].

3. When $n = m$ the normal polynomials are not zero and the resultant value of the integral is known. Thus for the Legendre polynomials

$$\int_{-1}^{+1} P_n^2(t)\, dt = \frac{2}{2n+1} \tag{3.5.6}$$

4. If N distinct points $t_1, t_2, \ldots, t_N$ of the interval $\{a, b\}$ are specified in advance, coefficients $\omega_1, \omega_2 \ldots, \omega_N$ exist such that

$$\int_a^b w(t)F(t)\, dt = \sum_{k=1}^{N} \omega_k F(t_k) \tag{3.5.7}$$

will be exact for all polynomials of degree $N - 1$ or lower. If the ts and the ωs are treated as unknowns (adjustable parameters), it will be exact for polynomials of degree $2N - 1$ or lower.

It was Gauss who showed that for the Legendre polynomials, as an example, the t_k are the roots of $P_N(t)$ and the

$$\omega_k = \int_{-1}^{+1} \frac{P_N(t)\, dt}{(t - t_k)\dot{P}_N(t_k)}, \qquad k = 1, 2, \ldots, N \tag{3.5.8}$$

In this form we have a Legendre-Gauss approximation or quadrature to an integral (3.5.7), the points and weights of which have been tabulated in the literature [7].

3.5.2 Change of variables

Now we are in a position to attack our original problem, namely, to find $F(t)$ given the integral

$$f(s) = \int_0^{\infty} e^{-st}F(t)\, dt$$

To carry this out with the Legendre-Gauss quadrature we first need to change the limits of integration so as not to encounter a divergent integral. We do this with the change

$$x = e^{-t} \tag{3.5.9}$$

yielding

$$f(s) = \int_0^1 x^{s-1}F(-\ln x)\, dx$$

or, letting $G(x) = F(-\text{not} \log x)$, as

$$f(s) = \int_0^1 x^{s-1}G(x)\, dx \tag{3.5.10}$$

However, since the Legendre orthogonal polynomials are specified in $\{-1, 1\}$ and (3.5.10) is in $\{0, 1\}$, we change over to the shifted Legendre polynomials which have $\{0, 1\}$ limits (see above). The change in variable of

$$\frac{1+t}{2} = x \quad \text{or} \quad t = 2x - 1 \tag{3.5.11}$$

converts (3.5.7) to

$$\int_0^1 2F(2x-1)dx \simeq \sum_{k=1}^{N} \omega_k F(2x_k - 1)$$

or letting $G(x) = F(2x - 1)$

$$\int_0^1 G(x)\,dx = \sum_{k=1}^{N} \omega_k G(x_k) \tag{3.5.12}$$

where the x_k are the roots of the shifted Legendre polynomials $P_n(x) = P_n(2x - 1)$ and the ω_k are the new set of weights analogous to (3.5.8). Bellman et al. have tabulated these x_k and ω_k to 16 decimal places for $N = 1, 2, \ldots, 15$ [4].

With limits of (3.5.10) and (3.5.12) properly set we now apply (3.5.12) to (3.5.10) to yield

$$f(s) = \sum_{k=1}^{N} \omega_k x_k{}^{s-1} G(x_k) \tag{3.5.13}$$

Letting s assume N values yields a linear system of N equations in the N unknown $G(x_k)$

$$f(i+1) = \sum_{k=1}^{N} \omega_k x_k^i G(x_k), \qquad i = 0, 1, \ldots, N-1 \tag{3.5.14}$$

or allowing $y_k = \omega_k G(x_k)$,

$$f(i+1) = \sum_{k=1}^{N} x_k^i y_k, \qquad i = 0, 1, \ldots, N-1 \tag{3.5.15}$$

There is also one further change of variable which may be used. This follows from the new variable $x = e^{-t}$ or $t_k = -\ln x_k$, where the x_k are the roots of $P_n(x) = 0$. Because of the behavior of the log, the roots yield only a corresponding small range of t_k. Thus it is necessary to devise a procedure for evaluating $F(t)$ over a wider range of ts. We shall mention such possible methods shortly.

3.5.3 The matrix inversion

Calling $f(i + 1) = a_i$, we can write (3.5.15) as

$$a_i = \sum_{k=1}^{N} x_k^i y_k, \qquad k = 0, 1, \ldots, N-1 \tag{3.5.16}$$

which is merely a set of linear algebraic equations in the y_k. The coefficient matrix is called a *Vandermonde matrix*, **V**,

$$\mathbf{V} = \begin{bmatrix} 1 & 1 & \cdots & 1 \\ x_1 & x_2 & \cdots & x_N \\ \cdot & & & \cdot \\ \cdot & & & \cdot \\ \cdot & & & \cdot \\ x_1^{N-1} & x_2^{N-1} & \cdots & x_N^{N-1} \end{bmatrix}$$

whose elements are the roots [translated by (3.5.9)] of the Legendre polynomial [13]. Bellman et al. show a convenient method for inverting **V** and even tabulate multiples of $\mathbf{V}^{-1}$ for converting $f(i+1)$ into $G(x_k)$ for $N = 3, 4, \ldots, 15$. Piessens [25] has, however, indicated that these tabulated values may include significant round off errors. As a result he has recomputed many of the necessary values [26].

It should be restated again that roundoff errors (or ill-conditioning of **V** such that **V** is almost singular) propagate as N increases. Thus it is probably wise to use double precision to protect the calculations; in fact, Han [13] suggests that one retain at least N digits in any calculation.

3.5.4 Further considerations

As we have already pointed out, the change of variable given by (3.5.9) is such as to yield only small values of t_k. In addition, a large value of s implies a small value of t and vice versa. Thus it may be necessary to expand the range of t. This can be done by noting that

$$L[F(at)] = \frac{f(s/a)}{a}$$

and instead of using $f(1), \ldots, f(N)$, the values of $f(1/a), \ldots, f(N/a)$ are used. Fortunately this does not change the coefficient matrix or its inverse. Alternatively one could use

$$L\left[F\left(\frac{t}{b}\right)\right] = bf(bs)$$

in the same sense.

With these features Bellman et al. present a set of FORTRAN programs which require only a specific $f(s)$ subroutine to complete. As previously mentioned, Piessens [25, 26] has suggested that the roundoff problems associated with this method and the fact that $F(t)$ is obtained at a restricted number of nonequidistant points are severe constraints on the utility of the algorithm. As a result he presents an alternative procedure in which a Lagrangian or Newtonian type of interpolation is used. In contrast to the approximation in t-space of Bellman et al., this approach begins with an approximation in s-space. An inversion of this approximation leads to t-space equations with unknown coefficients. This has the feature that equidistant points in t-space may be generated and thus error values can be estimated as N is increased. Further, the method no longer requires the solution of highly ill-conditioned matrices. As such this procedure of Piessens is highly recommended.

3.5.5 Laguerre quadrature

Since the Legendre (or shifted Legendre) polynomials are only one set of such polynomials which can be used for quadrature, other authors have extended the above ideas to include other orthogonal polynomials. In particular, Weeks [29] and Cook et al. [6] have examined the features of the Laguerre polynomials.

In this case we have

$$\int_0^\infty e^{-t}F(t)\,dt \simeq \sum_{k=1}^{N} \omega_k F(t_k) \tag{3.5.17}$$

where the ω_k is the kth root of the Nth Laguerre polynomial. Just as the roots and the weights of the Legendre polynomial are tabulated in the literature, so are those for the Laguerre polynomials; i.e., see Krylov and Skoblya [16]. The Laguerre polynomials are

$$\begin{aligned}
L_0(t) &= 1\\
L_1(t) &= -t + 1\\
L_2(t) &= t^2 - 4t + 2\\
L_3(t) &= -t^3 + 9t^2 - 19t + 6\\
&\;\;\vdots\\
L_{n+1} &= (1 + 2n - t)L_n - n^2 L_{n-1}
\end{aligned}$$

To apply this idea to the Laplace transform we write

$$f(s) = \int_0^\infty e^{-st}F(t)\,dt = \int_0^\infty [F(t)e^{-t(s-1)}]e^{-t}\,dt$$

and thus (3.5.17) leads to

$$f(s) \simeq \sum_{k=1}^{N} \{\omega_k e^{-t_k(s-1)}\}F(t_k) \tag{3.5.18}$$

By allowing s to assume N values, a set of N linear equations in the N unknowns $F(t_k)$, $k = 1, 2, \ldots, N$, results. Further, these equations result without the change of variable $x = e^{-t}$, as previously required with the Legendre polynomials.

3.5.6 Alternative approaches

A number of authors including Luke [22] and Miller and Guy [23] have used Jacobi orthogonal polynomials in much the same fashion. Here the limits on the orthogonality are $\{-1, +1\}$ and $w(t) = (1 - t)^a(1 + t)^b$, $a, b > -1$. Luke, as an example, used the substitution $x = e^{-vb}$ $(v > 0)$ and the shifted Jacobi polynomials (shifted to $\{0, 1\}$) to generate the linear equations. While numerical results show much the same problems of ill conditioning, this approach seems quite competitive with the Legendre approach.

Further, Stehfest [28] has outlined a finite and convergent series approach to $F(t)$. Of decided importance is that an ALGOL program as well as many numerical results is presented. In a direct comparison with the Legendre polynomial the method of Stehfest seems to restrict the ill conditioning and thus yield more accurate answers.

The use of a finite Fourier cosine transform has been advocated by Dubner and Abate [9]. The attractiveness of this approach lies in its simplicity. Further, they make use of the computationally important fast Fourier transform to simplify the computation.

EXAMPLE 3.13. To illustrate some of the features of this section, we have chosen four s-domain functions, $f(s)$, and numerically calculated the corresponding time functions, $F(t)$. In each case, as shown in Table 3.2, we know the true $F(t)$ and thus the numerical inversion can be checked for accuracy. Also shown in Table 3.2 are references for other equivalent calculations on the same functions.

TABLE 3.2
TRIAL FUNCTIONS FOR NUMERICAL INVERSION

$f(s)$	$F(t)$	Reference
1. $\exp\left[-\frac{\sqrt{s}}{2}\right]$	$\frac{\exp(-1/16t)}{4(\pi t^3)^{1/2}}$	Miller and Guy [23]
2. $\frac{1}{s^2 + s + 1}$	$\frac{2}{\sqrt{3}} \exp[-t/2] \sin \frac{t\sqrt{3}}{2}$	Piessens [26], Weeks [29]
3. $\frac{\exp[-25s]}{s}$	$\begin{cases} 0, & t < 25 \\ 1, & t > 25 \end{cases}$	Miller and Guy [23], Weeks [29]

In all cases we have used revised versions of the Bellman and the Piessens programs previously described. Since the Piessens program is only available in a report, we show our revised version in Table 3.3.

Before explicitly commenting on the results of the calculations we need to point out a few factors. In the Bellman program the parameter N refers to the number of Legendre points; in the Piessens program the points $s = a, 2a, \ldots, Na$ are used for interpolation and inversion, where a is an arbitrary constant. Thus the first method, Bellman's, has one adjustable parameter N, while the second, Piessens', has two parameters, N and a. In the first method the inversion was carried out with $N = 7$ and $N = 15$; in the second method $N = 3$, 7 and 15 were used with $a = 0.01$, 0.1 at each value of N.

To minimize the amount of tabular information, we generally show the end result of a computation rather than the intermediate answers. This means that the percentage error between the numerical inversion value of $F(t)$ and the exact value of $F(t)$ is usually presented at different values of t. In all the results to be quoted, the computation time was less than 0.1 second on an IBM 360/91.

TABLE 3.3
REVISED VERSION OF PIESSENS' PROGRAM FOR NUMERICAL INVERSION OF LAPLACE TRANSFORM

```
THIS SUBROUTINE DEFINES FF(X), THE LAPLACE TRANSFORM TO BE
INVERTED, SETS THE VALUES OF THE INPUT PARAMETERS, AND CALLS
LAPINV WHICH PRODUCES THE APPROXIMATE INVERSE OF FF(X).

***INPUT***

FF(X) MUST BE DEFINED AS AN EXTERNAL DOUBLE PRECISION FUNCTION.

THE OTHER INPUT PARAMETERS ARE:

     NUM: THE DESIRED ORDER OF THE APPROXIMATION

     A: DISTANCE BETWEEN THE INTERPOLATION POINTS, USUALLY 0.1 TO
        2.0 . THE INTERPOLATION POINTS ARE: A, 2*A, 3*A,..., NUM*A

THE FOLLOWING PARAMETERS DETERMINE THE VALUES OF T FOR WHICH
THE INVERSE WILL BE APPROXIMATED.

     TO: THE INITIAL VALUE OF T

     H: SPACING BETWEEN VALUES OF T

     NUMM: NUMBER OF T VALUES

THE INVERSE WILL THEN BE APPROXIMATED AT: TO, TO+H, TO+2*H,...,
TO+(NUMM-1)*H

***OUTPUT***

THE ONLY PARAMETERS RETURNED BY LAPINV TO THE CALLING PROGRAM
ARE THE COEFFICIENTS OF THE LEGENDRE SERIES . THESE ARE STORED IN
THE VECTOR C.

***PRINT***

IF NUMM.NE.0 THE COEFFICIENTS C AND THE APPROXIMATIONS OF ORDER
(NUM-1) AND NUM FOR THE INVERSE ARE PRINTED. IF NUMM = 0 ONLY
THE COEFFICIENTS ARE CALCULATED AND THERE IS NO PRINTOUT.

***NOTE***

BECAUSE OF POSSIBLE LOSS OF SIGNIFICANT FIGURES NUM MUST BE
LESS THAN 21.
```

TABLE 3.3 (Continued)

```
      DOUBLE PRECISION FUNCTION FF(X)
      DOUBLE PRECISION X
      THIS STATEMENT DEFINES THE LAPLACE TRANSFORM TO BE INVERTED.
      FF = 0.1D+01/(X*X + X + 0.1D+01)
      RETURN
      END

      DOUBLE PRECISION FF,A,TO,H,C(30)
      EXTERNAL FF
      NUM = 15
      A = 0.1D+00
      TO = 0.0D+00
      H = 0.5D+00
      NUMM = 21
      CALL LAPINV (FF,NUM,A,TO,H,NUMM,C)
      END

      SUBROUTINE LAPINV (FF,NUM,A,TO,H,NUMM,C)
      THIS SUBROUTINE PRODUCES A NUMERICAL APPROXIMATION TO THE
      INVERSE OF THE LAPLACE TRANSFORM DEFINED BY FF.
      DOUBLE PRECISION C(30),FF,A,TERM,DEN,AJ,T,E,FIO,FI1,AINV,FI2,TO,
     1 H,AINVV,DEXP
      CALCULATION OF THE COEFFICIENTS OF THE LEGENDRE SERIES
      C(1) = A*FF(A)
      DO 100 K=2,NUM
      T = K*A
      TERM = 0.1D+01/T
      DEN = 0.0D+00
      DO 102 J=2,K
      AJ = (J-1)*A
      DEN = TERM*C(J-1) + DEN
  102 TERM = TERM*(T-AJ)/(T+AJ)
  100 C(K) = (FF(T)-DEN)/TERM
      IF(NUMM)205,205,201
  201 WRITE(6,779) (C(K),K=1,NUM)
      EVALUATION OF THE LEGENDRE SERIES
      T = TO-H
      NUM1 = NUM-1
      WRITE (6,107)NUM1,NUM
      DO 200 J= 1,NUMM
      T= T+H
      E = DEXP(-A*T)
      FIO = 0.1D+01
      FI1 = 0.2D+01*E-0.1D+01
```

TABLE 3.3 (Continued)

```
    AINV = C(1) +C(2)*FI1
    DO 105 K=3,NUM
    AJ = K-2
    FI2 = ((E+E-0.1D+01)*(0.1D+01 +AJ+AJ)*FI1- AJ*FIO)/(AJ+0.1D+01)
    AINVV = AINV
    AINV = AINV +C(K)*FI2
    FIO = FI1
105 FI1 = FI2
    WRITE (6,106) T,AINVV,AINV
200 CONTINUE
779 FORMAT(35H1INVERSION OF THE LAPLACE TRANSFORM/1X,34(1H-)//1X,
   1 35HCOEFFICIENTS OF THE LEGENDRE SERIES//4(3X,D17.10))
107 FORMAT(//5X,1HT,2(6X,I2,17H-TH ORDER APPROX.)//)
106 FORMAT (2X,F6.2,4X,D17.10,8X,D17.10)
205 RETURN
    END
```

Table 3.4 shows a selected set of computational results. For function 1, as an example, we show Bellman's $N = 7$ and 15 results and Piessens' results for $N = 3$, 7, and 15 with $a = 0.1$ and $N = 15$ with $a = 0.01$ and 1.0. In none of the cases is the fit superb, and the two methods yield about the same accuracy. Note that in this case Piessens' method yields a more accurate answer as the value of the parameter a increases. In other functions this is not true, and the value of $a = 0.1$ may be the best.

Function 2 shows the influence of increasing the value of N in Piessens' method; the higher N, the more accurate the results. In this case Piessens' method seems more accurate than Bellman's.

Function 3 shows how the step function up at $t = 25.0$ is approximated by Piessens' method. The a priori supposed s-shaped approximation to the true discontinuity with an overshoot is achieved.

TABLE 3.4
SELECTED RESULTS ON NUMERICAL INVERSION OF FUNCTIONS IN TABLE 3.2
Function 1

t	% error	t	% error
0.03	−25.00	0.03	−9.37
0.14	23.48	0.24	5.22
0.35	−34.89	0.36	−6.23
0.69	53.21	0.51	7.55
1.21	−79.61	1.19	−13.40
2.05	117.64	1.54	16.13
3.67	−173.93	1.98	−19.35
		2.58	23.12
Bellman, $N = 7$		Bellman, $N = 15$	

TABLE 3.4 (Continued)

t	% error	% error	% error	% error	% error
0.03	−81.91	−43.38	−23.24	−63.99	−9.37
0.09	−77.08	−34.74	−20.00	−56.03	4.12
0.15	−63.48	−5.71	3.42	−32.53	−4.63
0.21	−47.51	22.43	19.01	−6.69	1.97
0.27	−30.00	46.89	25.42	19.61	4.82
0.33	−11.39	66.51	23.94	45.39	−2.98
0.39	8.02	80.84	16.77	70.01	−6.76
	$a = 0.1$ $N = 3$	$a = 0.1$ $N = 7$	$a = 0.1$ $N = 15$	$a = 0.01$ $N = 15$	$a = 1.0$ $N = 15$

Piessens

Function 2

t	% error	t	% error	t	% error	% error	% error
0.03	0.96	0.03	−0.93	0.5	11.18	10.38	−0.09
0.14	−0.32	0.24	0.26	2.0	−41.09	−8.36	−0.07
0.35	0.22	0.69	−0.21	3.5	489.53	97.21	0.59
0.69	−0.22	1.54	0.38	5.0	−161.19	−21.26	0.13
1.21	0.36	3.46	−26.46	6.5	−119.74	86.83	−0.91
2.05	−1.26	5.12	−19.60	8.0	−283.43	−168.16	−2.90
3.67	−329.54			9.5	−512.85	70.47	1.74
Bellman, $N = 7$		Bellman, $N = 15$		10.0	−865.07	254.08	6.47
					$a = 0.10$ $N = 3$	$a = 0.10$ $N = 7$	$a = 1.0$ $N = 15$

Piessens

Function 3

t	Approximation	t	Approximation
1.0	−0.1589 D − 01	21.0	−0.1513 D − 01
3.0	−0.1896 D − 01	23.0	0.2126
5.0	0.1613 D − 01	25.0	0.5030
7.0	−0.1434 D − 01		*Step occurs*
9.0	0.2484 D − 01	26.6	0.6414
11.0	−0.3383 D − 01	28.0	0.8666
13.0	0.7252 D − 02	30.0	0.1007 D + 01
15.0	0.4990 D − 01	32.0	0.1071 D + 01
17.0	−0.2299 D − 01	34.0	0.1084 D + 01
19.0	−0.9171 D − 01	36.0	0.1067 D + 01
		38.0	0.1040 D + 01
		40.0	0.1012 D + 01

$a = 0.1$, $N = 15$, Piessens

From these results and others not shown, Piessens' approach shows more versatility and in some cases better accuracy than Bellman's. However, this is achieved only at a cost, namely, the selection of two parameters rather than one. Thus the value of a selected can be crucial in obtaining a reasonable approximation to $F(t)$. However, in the usual case where $F(t)$ is unknown the question of how to determine the best value of a is a moot point.

In either case, however, the methods seem viable for digital computation (not desk computation). Our limited experience reveals that it is best to calculate quantities as needed, i.e., $\mathbf{V}^{-1}$, rather than try to read items in as input. However, this is a personal factor and is a function of the number of times the program is to be used.

3.6 Examples of Applications of the Laplace Transform

To illustrate several of the concepts developed in the previous sections we shall now analyze a number of different systems.

3.6.1 Exchange of ions in a porous adsorbent in a CSTR

Consider a CSTR containing a known weight of sodium-type spherical ion-exchange particles or resin of uniform size in contact with a solution of uniform composition (see Figure 3.13). Entering the reactor at a constant flow rate is a solution of known hydrogen ion concentration and leaving at the same rate is solution at the composition of the reactor. The entering hydrogen ions distribute themselves in the following way: Some diffuse into the internal voids of the resin, and a certain fraction of these, depending on the equilibrium relation, are picked up by the solid structure and displace the sodium ions on the resin; some go toward increasing the hydrogen ion concentration in the free volume of the reactor; the rest leave in the exit stream. It is desired to obtain an expression for the effluent concentration of hydrogen ions at any time of operation.

To obtain the system equations for this physical problem we shall first make a material balance for hydrogen ions for the CSTR:

$$qC_0 = qC + \frac{3W\alpha}{\rho R} D \frac{\partial c}{\partial r}\bigg|_{r=R} + V\frac{dC}{dt} \tag{3.6.1}$$

in = out + into particle + accumulation

where C = concentration in reactor volume; exit concentration
C_0 = concentration in feed to CSTR; assumed constant

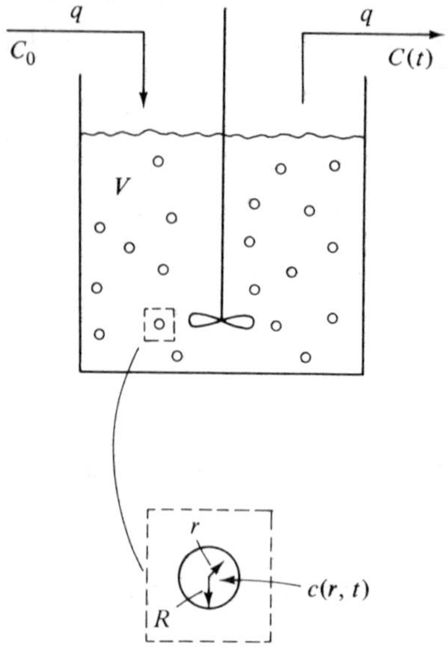

Figure 3.13. Exchange of ions in a porous adsorbent in a CSTR system.

c = concentration inside particles
t = time of operation
q = flow rate
W = weight of the dry resin particles
ρ = density of the dry resin particles
r = radius variable in spheres
R = radius of resin particles
α = void volume of resin
D = diffusivity of the ion inside the resin
V = free volume of the reactor

The second term on the right-hand side of (3.6.1) follows by noting that $\frac{4}{3}\pi R^3 \rho = W$ and that the flux of ions entering a single sphere is

$$-4\pi R^2 D\alpha \frac{\partial c}{\partial r}\bigg|_{r=R}$$

Consider now the interior of an individual particle of resin. The depletion of the hydrogen ions diffusing through the region enclosed between two concentric spherical shells occurs in two ways. Some is adsorbed on the solid, while the rest goes toward reducing the concentration gradient in this section, appearing as an increase in the hydrogen ion concentration in the void volume. A balance on a finite shell of radius r and thickness Δr and the limiting of this result, lim $\Delta r \longrightarrow 0$, yields

$$\alpha D\left[\frac{\partial^2 c}{\partial r^2} + \frac{2}{r}\frac{\partial c}{\partial r}\right] = \alpha\frac{\partial c}{\partial t} + \frac{\partial n}{\partial t} \tag{3.6.2}$$

flux = accumulation + adsorption

where n = concentration of adsorbed ions.

The assumption of linear equilibrium between c and n is also added (more complicated relations could be used):

$$n = kc \tag{3.6.3}$$

This converts (3.6.2) to

$$\frac{\partial^2 c}{\partial r^2} + \frac{2}{r}\frac{\partial c}{\partial r} - \gamma\frac{\partial c}{\partial t} \tag{3.6.4}$$

where $\gamma = (k + \alpha)/\alpha D$.

Boundary conditions are now necessary. If a film of liquid surrounds a particle, there is resistance to mass transfer from the main fluid mass to the particle surface which is embodied in a boundary condition at the surface

$$-D\left(\frac{\partial c}{\partial r}\right)_{r=R} = k_f(c - C) \qquad \text{at } r = R \tag{3.6.5}$$

By allowing the mass transfer coefficient $k_f \longrightarrow \infty$ this boundary condition reduces to

$$c = C, \qquad r = R \tag{3.6.6}$$

indicating zero resistance to mass transfer at the surface. Finally, we add an initial condition on the spheres and solution:

$$c = C - 0, \qquad t = 0 \tag{3.6.7}$$

The system of equations (3.6.1), (3.6.4), (3.6.5), and (3.6.7) completely describes the physical system. To simplify the analysis we change variables to

$$y = \frac{r}{R}, \qquad \tau = \frac{t}{\gamma R^2}, \qquad u = \frac{rc}{RC_0}, \qquad U = \frac{C}{C_0}$$

and obtain

$$m(1 - U) - n\frac{dU}{d\tau} - P\left(\frac{\partial u}{\partial y} - u\right)_{y=1} = 0 \tag{3.6.8}$$

$$\frac{\partial^2 u}{\partial y^2} = \frac{\partial u}{\partial \tau} \tag{3.6.9}$$

$$\frac{\partial u}{\partial y} + (l - 1)u = lU, \qquad y = 1 \tag{3.6.10}$$

$$u = U = 0, \qquad \tau = 0 \tag{3.6.11}$$

where

$$m = \frac{qR^2}{\alpha D}, \qquad n = \frac{V}{k + \alpha}, \qquad P = \frac{3W}{\rho}, \qquad l = \frac{k_f R}{D}$$

Defining

$$U(\tau) \xrightarrow{L} H(s)$$

$$u(y, \tau) \xrightarrow{L} h(y, s)$$

(3.6.8)–(3.6.11) become

$$\frac{m}{s} - mH - nsH - P\left(\frac{dh}{dy} - h\right)_{y=1} = 0 \tag{3.6.12}$$

$$\frac{d^2h}{dy^2} - sh = 0 \tag{3.6.13}$$

$$\frac{dh}{dy} + (l - 1)h = lH, \qquad y = 1 \tag{3.6.14}$$

where (3.6.11) has already been used. The solution to (3.6.13) is

$$h = A \sin\sqrt{-s}y + B \cos\sqrt{-s}y$$

with A and B as arbitrary constants. However,

$$h = L[u] = L\left[\frac{yc}{C_0}\right] = yL\left[\frac{c}{C_0}\right]$$

and so

$$L\left[\frac{c}{C_0}\right] = \frac{A}{y} \sin \sqrt{-s}y + \frac{B}{y} \cos \sqrt{-s}y$$

For the concentration c to remain finite at $y = 0$ (or $r = 0$), B must equal zero and

$$h = A \sin \sqrt{-s}y \tag{3.6.15}$$

Eliminating H from (3.6.12) and (3.6.14) and solving with (3.6.15) for the constant A, (3.6.15) becomes

$$h = \frac{ml \sin \sqrt{-s}y}{s\{[m + ns + Pl]\sqrt{-s} \cos \sqrt{-s} + [(m + ns)(l - 1) - Pl] \sin \sqrt{-s}\}} \tag{3.6.16}$$

Now we have completed the first half of our analysis, namely, to carry the solution through the s-domain. All that remains is to return to the t-domain by inverting (3.6.16). To do this we shall use (3.1.44) with h in the form

$$h = \frac{p(s)}{q(s)} \tag{3.6.17}$$

where there are an infinite number of roots of $q(s) = 0$. Thus

$$\sum_{n=1}^{\infty} \frac{p(s_n)}{\dot{q}(s_n)} e^{s_n\tau} \xleftarrow{L^{-1}} h(y, s) \tag{3.6.18}$$

The roots of $q(s) = 0$ are $s = 0$ and the zeros of

$$\{(m + ns + Pl)\sqrt{-s}\cos\sqrt{-s} + [(m + ns)(l - 1) - Pl]\sin\sqrt{-s}\}$$

Considering the root $s = 0$ first, the inverse can be obtained by the formula

$$u = \lim_{s \to 0} sh$$

which corresponds to using (3.1.44) for the particular value $s = 0$. Therefore

$$u = \lim_{s \to 0}\left\{\frac{ml \sin\sqrt{-s}y}{(m + ns + Pl)\sqrt{-s}\cos\sqrt{-s} + [(m + ns)(l - 1) - Pl]\sin\sqrt{-s}}\right\}$$
$$= y$$

Using (3.1.44), the result for the other roots may be found where the s_n are the roots of the equation

$$\sqrt{-s}\cos\sqrt{-s} = \frac{(m + ns)(1 - l) + Pl}{m + ns + Pl} \tag{3.6.19}$$

and

$$p(s_n) = ml \sin\sqrt{-s_n}y$$
$$q(s) = s\{(m + ns + Pl)\sqrt{-s}\cos\sqrt{-s} + [(m + ns)(l - 1) - Pl]\sin\sqrt{-s}\}$$

Differentiating $q(s)$ yields

$$\dot{q}(s) = \frac{\sin\sqrt{-s}}{2G}\phi(s)$$

where

$$\phi(s) = -(2ln^2 + l^2n^2)s^2 + (G^2 + 3nlG - 2lmn - 2l^2mn)s + lm(G - lm)$$
$$G = m + ns + Pl$$

Thus the complete inverse transform is

$$u = y + \sum_{n=1}^{\infty} \frac{2Gml \sin\sqrt{-s_n}\,y}{\phi(s_n)\sin\sqrt{-s_n}} e^{s_n\tau} \tag{3.6.20}$$

Implicit in this derivation is that the roots of (3.6.19) are all real and distinct. This may be proved but involves concepts beyond the scope of the present discussion.

This equation relates the concentration changes inside the resin particles to the time of contact of the particles with the solution. However, the concentration which can most easily be measured experimentally is the solution concentration. By using (3.6.20) in conjunction with (3.6.10) and carrying out the necessary integrations, there results

$$U = \frac{C}{C_0} = 1 + \sum_{n=1}^{\infty} \frac{2mPl^2}{\phi(s_n)} e^{s_n\tau} \tag{3.6.21}$$

This is the general solution of the problem under investigation.

While this solution includes a mass transfer resistance, we can obtain a simpler form by allowing k_f (or l) to approach infinity. Thus

$$\lim_{l\to\infty} U = 1 + \lim_{l\to\infty} \sum_{n=1}^{\infty} \frac{2mPl^2}{\phi(s_n)} e^{s_n\tau}$$
$$= 1 + \sum_{n=1}^{\infty} \frac{2mP}{\psi(s_n)} e^{s_n\tau}$$

where

$$\psi(s_n) = \lim_{l\to\infty} \frac{\phi(s_n)}{l^2}$$

Performing the indicated limit process on the explicit form of $\phi(s_n)$, one obtains

$$\psi(s_n) = -n^2 s_n{}^2 + (P^2 + 3nP - 2mn)s_n + m(P - m)$$

where the s_n are now solutions of the equation

$$\sqrt{-s}\cot\sqrt{-s_n} = \lim_{l\to\infty} \frac{(m + ns_n)(1 - l) + Pl}{m + ns_n + Pl}$$
$$= \frac{P - m - ns_n}{P}$$

The final set of equations is thus

$$U = \frac{C}{C_0} = 1 - 2mP\sum_{n=1}^{\infty} \frac{e^{s_n\tau}}{n^2 s_n{}^2 - (P^2 + 3nP - 2mn)s_n + m(m - P)} \tag{3.6.22}$$

with

$$\sqrt{s_n}\cot\sqrt{-s_n} = \frac{P - m - ns_n}{P} \tag{3.6.23}$$

Equation (3.6.22) is the final result for the proposed problem. Back substitution into the original set of equations shows that all are satisfied. Further, as W, the weight of resin, approaches zero, the solution reduces to

$$\frac{C}{C_0} = 1 - e^{-qt/V}$$

the equation for a simple CSTR with no adsorption.

3.6.2 Mass and/or heat transfer with local equilibrium

In this section we shall analyze the behavior of three essentially equivalent physical systems: the saturation phase of a chromatographic, adsorption, or ion-exchange column; the flow of a hot gas through a bed packed with small particles; and the heat exchanger with a constant-temperature second phase and no heat transfer to the environment (see Figure 3.14).

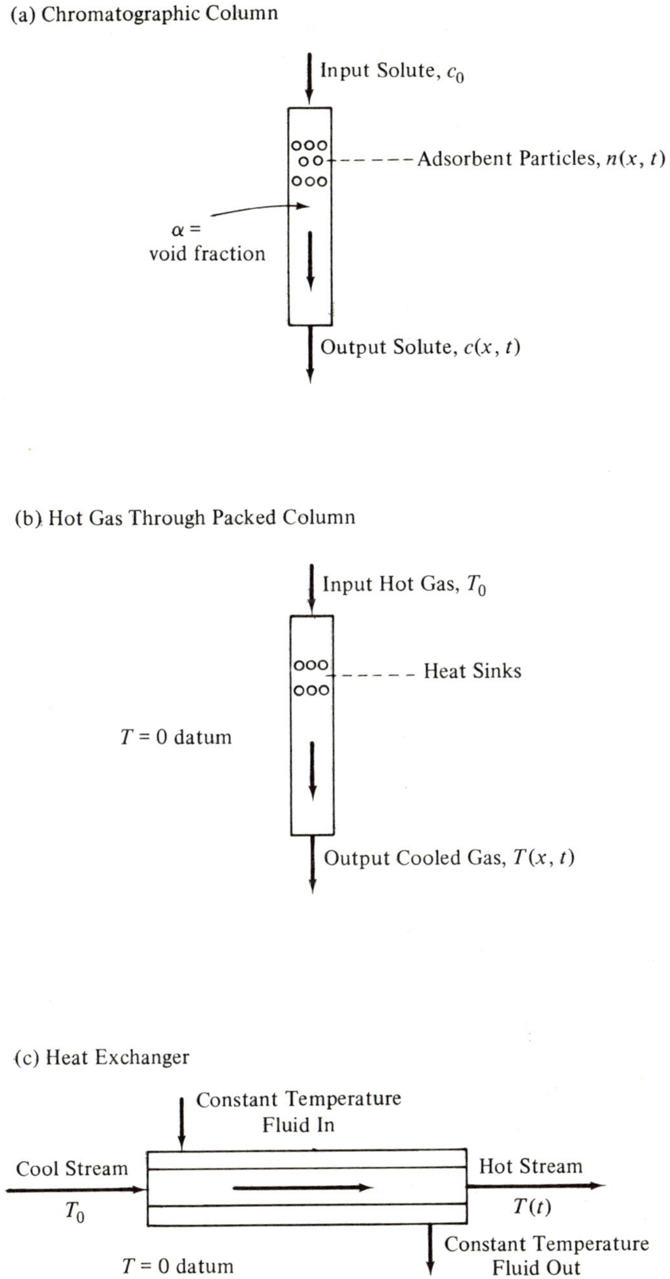

Figure 3.14. Three different systems with the same mathematical characteristics.

While not necessary, we shall for simplicity and convenience assume negligible axial gradients, constant system properties, and an infinite length system. Certain of these assumptions will be removed in later sections. For the interested reader, in this area we recommend a number of references [2, 12, 19].

Consider an infinitely long column filled with adsorbent particles through which is being passed, at a constant rate, a solution containing a single adsorbable component (see Figure 3.15). As the solution is introduced into the

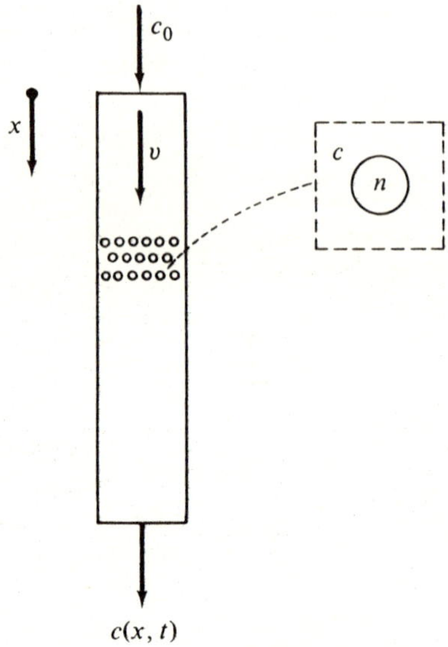

Figure 3.15. Adsorption column.

column the adsorbate composition in the solution changes with time of operation as well as with the specific level in the bed. During this transient period, calculate the composition of the adsorbate at any point in the column and at any time. Assume that equilibrium is instantaneously reached between the adsorbate concentration in the solution and on the solid; i.e., mass transfer steps are neglected.

With c and n as concentrations of adsorbate in the solution and on the solid, v the solution velocity through the bed, α the void volume of the bed, a material balance on an elemental section of bed length Δx and a subsequent limiting process yield

$$v\frac{\partial c}{\partial x} + \frac{\partial c}{\partial t} + \frac{1}{\alpha}\frac{\partial n}{\partial t} = 0 \qquad (3.6.24)$$

This is the mass balance for the process.

Since equilibrium is assumed at any level in the bed, a relationship can be specified between c and n. In the present case a linear equilibrium relationship is used:

$$n = bc \tag{3.6.25}$$

In addition the following initial and boundary conditions are specified:

$$c = c_0, \qquad x = 0 \tag{3.6.26}$$

$$n = 0, \qquad t \leq \frac{x}{v} \tag{3.6.27}$$

The first condition states that the incoming solution to the bed, $x = 0$, is a constant, c_0. The second condition indicates that the solid has zero adsorbate on it at $t = 0$; since x/v is the time required for the solution to reach a distance x units down the column, this equation also states that any level in the bed has zero adsorbate until the solution reaches it. Equations (3.6.24)–(3.6.27) must be solved to obtain $c(x, t)$ or $n(x, t)$ in terms of the system parameters.

Making the change of variables,

$$y = \alpha(t - x/v)$$

$$z = \frac{x}{v}$$

these equations may be combined to yield

$$\begin{aligned} &b\frac{\partial c}{\partial y} + \frac{\partial c}{\partial z} = 0 \\ &c = c_0, \qquad x = 0 \\ &n = 0, \qquad c = 0, \qquad y = 0 \end{aligned} \tag{3.6.28}$$

Before explicitly carrying out the solution, let us present the analogous equations for heat transfer in a packed bed. For a hot gas flowing through a packed bed, assuming no radial gradients in the bed or particle and no axial mixing, the equations analogous to (3.6.24)–(3.6.27) are

$$v\alpha c_f \rho_f \frac{\partial T}{\partial x} + \alpha c_f \rho_f \frac{\partial T}{\partial t} + (1 - \alpha)c_s \rho_s \frac{\partial T_s}{\partial t} = 0 \tag{3.6.24a}$$

$$T = T_s \tag{3.6.25a}$$

$$T = T_0, \qquad x = 0 \tag{3.6.26a}$$

$$T = 0, \qquad t = 0 \tag{3.6.27a}$$

where T and T_s are the fluid and solid temperatures; c_f, ρ_f and c_s, ρ_s are fluid and solid heat capacities and densities; and an initial datum temperature of $T = 0$ is assumed.

Since time is unrestricted and the packed bed is assumed infinite in length, the L.T. can be applied to either the distance or the time variable. To illustrate the procedure we shall solve the chromatographic column transforming z and the heat transfer problem transforming t. Thus let

$$c(z, y) \xrightarrow{L} \bar{c}(s, y) = \bar{c}$$

so that (3.6.28) becomes

$$b\frac{d\bar{c}}{dy} + s\bar{c} - c_0 = 0$$

$$\bar{c} = 0, \qquad y = 0$$

The solution of this first-order O.D.E. is

$$\bar{c} = Ke^{-(s/b)y} + \frac{c_0}{s}$$

and using $\bar{c} = 0$, $y = 0$, $K = -c_0/s$. Thus

$$\bar{c} = \frac{c_0}{s}[1 - e^{-(y/b)s}] \tag{3.6.29}$$

However, the right-hand side is merely the transform of a step function of height c_0 and duration y/b. Thus inverting (3.6.29) yields

$$c(z, y) = \begin{cases} c_0, & x < \dfrac{y}{b} \\ 0, & x > \dfrac{y}{b} \end{cases} \tag{3.6.30}$$

For an adsorbate which obeys a linear equilibrium, a sharp-edged band of adsorbate moves down the packed bed.

Turning now to (3.6.24a)–(3.6.27a), we define

$$T(x, t) \xrightarrow{L} \bar{T}(x, s) = \bar{T}$$

and obtain

$$v\alpha c_f\rho_f\frac{d\bar{T}}{dx} + [\alpha c_f\rho_f + (1 - \alpha)c_s\rho_s]s\bar{T} = 0$$

$$\bar{T} = \frac{T_0}{s}, \qquad x = 0$$

The solution of this O.D.E. is

$$\bar{T} = \frac{T_0}{s}e^{-(\bar{\alpha}x/v)s} \tag{3.6.31}$$

where

$$\bar{\alpha} = \frac{\alpha c_f\rho_f + (1 - \alpha)c_s\rho_s}{\alpha c_f\rho_f}$$

However, (3.6.31) is merely the L.T. of a delayed step function and thus, using (3.2.2),

$$T(x, t) = \begin{cases} T_0, & t > \dfrac{\bar{\alpha} x}{v} \quad \text{or} \quad x < \dfrac{vt}{\bar{\alpha}} \\ 0, & t < \dfrac{\bar{\alpha} x}{v} \quad \text{or} \quad x > \dfrac{vt}{\bar{\alpha}} \end{cases} \tag{3.6.32}$$

Except for nomenclature, (3.6.30) and (3.6.32) are equivalent to each other. The influent temperature moves through the bed as a sharp front with height T_0. The position of the front is given by $x_f = vt/\bar{\alpha}$ with a speed of propagation of $v/\bar{\alpha}$. Since $\bar{\alpha}$ is, in general, large, the speed is low and is essentially given by

$$\frac{v \alpha c_f \rho_f}{c_s \rho_s}$$

It is important to note that in this case where either independent variable could be used for the L.T. the proper way to make a choice is to use that variable which will yield a zero term (and thus a homogeneous equation) when the derivative is transformed. On this basis the transformation used for the heat transfer system is the simpler.

Finally, we note from (3.6.31) that

$$\bar{T}(1, s) = \frac{T_0}{s} e^{-(\bar{\alpha}/v)s}$$

$$\bar{T}(0, s) = \frac{T_0}{s}$$

and thus

$$\bar{T}(1, s) = e^{-(\bar{\alpha}/v)s} \bar{T}(0, s) \tag{3.6.33}$$

The transfer function of the temperature at $x = 1$ to changes in influent temperature is

$$g(s) = e^{-(\bar{\alpha}/v)s} \tag{3.6.34}$$

which is equivalent to a pure delay or dead time as indicated by (E) of Example 3.6.

3.6.3 Heat and/or mass transfer with local resistance

Next we shall consider an equivalent type of problem but with a local resistance rather than equilibrium between phases. This means that we shall need a mass transfer or heat transfer coefficient equation to replace (3.6.25) or (3.6.25a). The analysis will be carried out in terms of the heat transfer in a packed bed, but essentially identical results can be obtained for the chromatography system and for a heat exchanger system with heat loss (or gain) to the environment. The paper of Klinkenberg [15] details this analogy, while the papers of Cohen and Johnson [5] and Goldstein [12] further show the type of solution obtained.

We now wish to solve the problem of

$$v\alpha c_f\rho_f\frac{\partial T}{\partial x} + \alpha c_f\rho_f\frac{\partial T}{\partial t} + (1-\alpha)c_s\rho_s\frac{\partial T_s}{\partial t} = 0$$

$$c_s\rho_s\frac{\partial T_s}{\partial t} = \frac{3h}{R}[T - T_s] \tag{3.6.35}$$

$$T = T_0, \qquad x = 0$$

$$T = 0, \qquad t = 0$$

Let

$$T(x, t) \xrightarrow{L} \bar{T}(x, s) = \bar{T}$$

$$T_s(x, t) \xrightarrow{L} \bar{T}_s(x, s) = \bar{T}_s$$

and we obtain from the equations above

$$v\alpha c_f\rho_f\frac{d\bar{T}}{dx} + \alpha c_f\rho_f s\bar{T} + (1-\alpha)c_s\rho_s s\bar{T}_s = 0 \tag{3.6.36}$$

$$c_s\rho_s s\bar{T}_s = \frac{3h}{R}[\bar{T} - \bar{T}_s] \tag{3.6.37}$$

$$\bar{T} = \frac{T_0}{s}, \qquad x = 0 \tag{3.6.38}$$

Solving (3.6.37) for $\bar{T}_s$ and substituting into (3.6.36) yields

$$v\alpha c_f\rho_f\frac{d\bar{T}}{dx} + \left[\alpha c_f\rho_f s + \frac{(1-\alpha)c_s\rho_s sh}{(R/3)c_s\rho_s s + h}\right]\bar{T} = 0$$

whose solution is, using (3.6.38),

$$\bar{T} = \frac{T_0}{s}e^{-(x/v)s}e^{-(\beta x/v)[s/(s+\Delta)]} \tag{3.6.39}$$

where

$$\beta = \frac{3(1-\alpha)h}{R\alpha c_f\rho_f}, \qquad \Delta = \frac{3h}{Rc_s\rho_s}$$

This, then, is the solution of the L.T. equations (in the s-domain) and we need to invert the solution to obtain $T(x, t)$.

Before we actually carry out the inversion, note that

$$\bar{T}(1, s) = e^{-(s/v)}e^{-[\beta s/v(s+\Delta)]}\bar{T}(0, s)$$

$$= g(s)\bar{T}(0, s) \tag{3.6.40}$$

Thus $g(s)$ is the transfer function for the temperature at $x = 1$ to changes in influent temperature. Further, we can find from the transform of (3.6.37) that

$$\bar{T}_s(x, s) = \frac{\Delta}{s + \Delta}\bar{T}(x, s)$$

and thus that

$$\bar{T}_s(1, s) = \left(\frac{\Delta}{s + \Delta}\right)e^{-s/v}e^{-[\beta s/v(s+\Delta)]}\bar{T}(0, s)$$

$$= g_1(s)\bar{T}(0, s) \qquad (3.6.41)$$

$g_1(s)$ is now the transfer function between the solid temperature response at $x = 1$ to a change in fluid inlet temperature.

Returning to (3.6.39), we see that the first exponential is merely the translation operator and that the second is of the form

$$e^{-b[s/(s+\Delta)]} \qquad (3.6.42)$$

where $b = \beta x/v$. Unfortunately, this form is not readily available in tables of transforms and thus we must manipulate its form. Convert this to

$$e^{-[bs/(s+\Delta)]+b}e^{-b} = e^{b\Delta/(s+\Delta)}e^{-b} \qquad (3.6.43)$$

and note that

$$\frac{e^{b\Delta/(s+\Delta)}}{s(s+\Delta)}(s + \Delta) = \frac{e^{b\Delta/(s+\Delta)}}{s + \Delta} + \Delta\frac{e^{b\Delta/(s+\Delta)}}{s(s + \Delta)} \qquad (3.6.44)$$

However, these changes merely recast (3.6.39) into

$$\bar{T} = T_0 e^{-sx/v}e^{-\beta x/v}e^{-b}\left[\frac{e^{b\Delta/(s+\Delta)}}{s + \Delta} + \Delta\frac{e^{b\Delta/(s+\Delta)}}{s(s + \Delta)}\right]$$

The first term in the brackets has an inverse transform which we can obtain from Table 3.1, while the second, because of the s in the denominator, is the integral of the first. Using

$$I_0(2\sqrt{kt}) \xleftarrow{L^{-1}} \frac{e^{k/s}}{s}$$

where $I_0(\cdot)$ is the modified Bessel function of the first kind of order zero, it follows that

$$e^{-\Delta t}I_0(2\sqrt{b\,\Delta t}) \xleftarrow{L^{-1}} \frac{e^{b\Delta/(s+\Delta)}}{s + \Delta}$$

Thus we may now completely invert (3.6.39) to yield

$$T(x, t) = \begin{cases} T_0 e^{-\beta x/v}e^{-b}\left[e^{-\Delta[t-(x/v)]}I_0\left(2\sqrt{b\,\Delta\left(t - \frac{x}{v}\right)}\right) \right. \\ \left. \quad + \Delta\int_0^{t-x/v} e^{-\Delta}I_0(2\sqrt{b\,\Delta\lambda})\,d\lambda\right], & t > \frac{x}{v} \\ 0, & t < \frac{x}{v} \end{cases} \qquad (3.6.45)$$

There are three extensions of this problem which are interesting. First, we shall comment on the problem of a nonadiabatic packed bed with surface heat transfer. This means that there will now be a radial (in the packed bed) transfer of heat. If the radial mixing may be characterized by a diffusion coefficient K and the ambient temperature is zero, the heat balance equation becomes

$$K\left[\frac{\partial^2 T}{\partial r^2} + \frac{1}{r}\frac{\partial T}{\partial r}\right] - v\alpha c_f \rho_f \frac{\partial T}{\partial x} + \alpha c_f \rho_f \frac{\partial T}{\partial t} + (1-\gamma)c_s\rho_s\frac{\partial T_s}{\partial t} = 0 \quad (3.6.46)$$

The second equation corresponding now to a resistance at the wall of the packed bed has the same form as (3.6.35).

Obviously, the problem is complicated by terms containing the partial derivatives with respect to the radial variable r. Such terms can be removed directly from (3.6.46) by means of a finite Bessel transform, leaving essentially the problem we have just solved. The inverse finite Bessel transform can then be used (this is extremely simple to do) to solve the entire problem [2].

The second interesting extension of the current problem is the countercurrent (or parallel flow) heat exchanger or the heat or mass transfer problem but with a moving solid. In this case we would get heat (or mass) balances of the form

$$\begin{aligned} \frac{\partial T}{\partial t} + \frac{\partial T}{\partial x} + K_1(T - T_1) &= 0 \\ \gamma\frac{\partial T_1}{\partial t} + \frac{\partial T_1}{\partial x} + K_2(T_1 - T) &= 0 \end{aligned} \quad (3.6.47)$$

where γ, K_1, and K_2 are combinations of constants or parameters. Applying the L.T. yields

$$\begin{aligned} (s + K_1)\bar{T} + \frac{d\bar{T}}{dx} &= K_1\bar{T}_1 \\ (\gamma s + K_2)\bar{T}_1 + \frac{d\bar{T}_1}{dx} &= K_2\bar{T} \end{aligned} \quad (3.6.48)$$

In addition to initial conditions, say $T = T_1 = 0$, there is one inlet condition on T, $T(x, t) = T(0, t)$, and another on T_1, $T_1(x, t) = T_1(1, t)$ (assuming countercurrent flow). The solution of (3.6.48) subject to these boundary conditions yields

$$\begin{bmatrix} \bar{T}(x, s) \\ \bar{T}_1(x, s) \end{bmatrix} = \begin{bmatrix} g_{11}(x, s) & g_{12}(x, s) \\ g_{21}(x, s) & g_{22}(x, s) \end{bmatrix} \begin{bmatrix} T(0, s) \\ T_1(1, s) \end{bmatrix}$$

Here the elements $g_{11}(x, s), \ldots, g_{22}(x, s)$ constitute the transfer function matrix for this system. Thus $g_{11}(1, s)$ and $g_{22}(0, s)$ are the transfer functions for one end response to disturbances at the other end of the unit, while $g_{12}(1, s)$ and $g_{21}(0, s)$ give one end response to disturbances at the same end of the unit, i.e., the input/output transfer functions.

Finally, we shall mention that we have considered constant initial and boundary conditions, (3.6.26) and (3.6.27), only for ease in subsequent presentation. Actually we could have used the most general conditions of

$$T = T_0(0, t), \qquad x = 0 \tag{3.6.49}$$

(a general time-varying input) and

$$T = T(x, 0), \qquad t = 0 \tag{3.6.50}$$

(a general initial profile). Using the convolution integral, (3.6.49) would only affect the last inversion step. Equation (3.6.50) could also be handled without any immediate problems other than the length and explicit complexity of the equations. We shall illustrate these features in the next two example systems.

3.6.4 Heat and/or mass transfer with axial mixing and local equilibrium

As the illustrative example we shall consider the chromatographic column of Section 3.6.2 but with axial mixing (local equilibrium is assumed). Equation (3.6.24) now becomes

$$D\frac{\partial^2 c}{\partial x^2} = v\frac{\partial c}{\partial x} + \frac{\partial c}{\partial t} + \frac{1}{\alpha}\frac{\partial n}{\partial t} \tag{3.6.51}$$

where D is the axial diffusivity. Adding a linear equilibrium

$$n = bc \tag{3.6.52}$$

and completely general initial and boundary conditions

$$c = c_0(t), \qquad x = 0 \tag{3.6.53}$$

$$\left.\begin{aligned} c &= c_i(x), \qquad t = 0 \\ n &= n_i(x), \qquad t = 0 \end{aligned}\right. \tag{3.6.54}$$

completes the specification. Note that (3.6.53) specifies any input to the bed and (3.6.54) any initial distribution. Combining (3.6.51) and (3.6.52) there results

$$D\frac{\partial^2 c}{\partial x^2} = v\frac{\partial c}{\partial x} + \gamma\frac{\partial c}{\partial t} \tag{3.6.55}$$

where $\gamma = (\alpha + b)/\alpha$.

Whereas the previous systems with no axial mixing were hyperbolic P.D.E.s, the inclusion of the axial mixing leads to a parabolic P.D.E. To solve this case it is convenient to first make the change of variable

$$c(x, t) = g(x, t)\exp\left[\frac{vx}{2D} - \frac{v^2 t}{4D\gamma}\right] \tag{3.6.56}$$

This converts (3.6.55), (3.6.53), and (3.6.54) to

$$\frac{D}{\gamma}\frac{\partial^2 g}{\partial x^2} = \frac{\partial g}{\partial t}$$

$$g = c_0(t)\exp\left(\frac{vt^2}{4D\gamma}\right), \qquad x = 0$$

$$g = c_i(x)\exp\left(-\frac{vx}{2D}\right), \qquad t = 0$$

Equation (3.6.56) is a standard transformation used to remove a first derivative from a parabolic P.D.E. It complicates the initial and boundary conditions, but since these were general to start with, this causes no serious subsequent difficulties.

Next we split the single problem into two problems:

Problem 1		Problem 2	
$\frac{D}{\gamma}\frac{\partial^2 f_1}{\partial x^2} = \frac{\partial f_1}{\partial t}$		$\frac{D}{\gamma}\frac{\partial^2 f_2}{\partial x^2} = \frac{\partial f_2}{\partial t}$	
$f_1 = c_i(x)\exp\left(-\frac{vx}{2D}\right)$,	$t = 0$	$f_2 = 0$,	$t = 0$
$f_1 = 0$,	$x = 0$	$f_2 = c_0(t)\exp\left(\frac{vt}{4D\gamma}\right)$,	$x = 0$

where it may be seen that

$$g = f_1 + f_2$$

and

$$c(x, t) = [f_1(x, t) + f_2(x, t)]\exp\left[\frac{vx}{2D} - \frac{v^2 t}{4D\gamma}\right] \tag{3.6.57}$$

Now if we solve the first problem for f_1 and the second problem for f_2, we shall have the solution to our original problem. To conserve space we shall illustrate the solution of the second problem.

Let

$$f_2(x, t) \xrightarrow{L} \bar{f}_2(x, s) = \bar{f}_2$$

$$f_2(0, t) \xrightarrow{L} \bar{f}_{20}$$

and the second problem becomes

$$\frac{d^2\bar{f}_2}{dx^2} - \frac{\gamma}{D}s\bar{f}_2 = 0$$

$$\bar{f}_2 = \bar{f}_{20}, \qquad x = 0$$

The solution of this O.D.E. is

$$\bar{f}_2 = Ae^{-(\sqrt{s\gamma/D})x} + Be^{+(\sqrt{s\gamma/D})x}$$

but since $\bar{f}_2$ must remain finite as $x \longrightarrow \infty$, $B = 0$ and thus

$$\bar{f}_2 = Ae^{-(\sqrt{s\gamma/D})x}$$
$$= \bar{f}_{20}e^{-(\sqrt{s\gamma/D})x}$$

The specification that $\bar{f}_2$ be finite as $x \longrightarrow \infty$ is merely a consequence of the fact that an infinitely long column is being used. To invert this expression we can convolute $\bar{f}_{20}$ and the exponential after noting from Table 3.1 that

$$\frac{a}{2\sqrt{\pi t^3}} \exp\left(-\frac{a^2}{4t}\right) \xrightarrow{L} \exp[-a\sqrt{s}\,]$$

Carrying out the convolution yields

$$f_2(x, t) = \frac{x}{2}\sqrt{\frac{\gamma}{\pi D}} \int_0^t c_0(\lambda) \exp\left[\frac{v^2\lambda}{4D\gamma} - \frac{\gamma x^2}{4D(t-\lambda)}\right] \frac{d\lambda}{(t-\lambda)^{3/2}} \tag{3.6.58}$$

In a corresponding manner

$$f_1(x, t) = \frac{1}{2}\sqrt{\frac{\gamma}{\pi D t}} \int_0^\infty c_i(\lambda)\left[\exp\left(-\frac{v\lambda}{2D}\right)\right]\left[\exp\left(\frac{-\gamma(\lambda - x)^2}{4Dt}\right) - \exp\left(\frac{-\gamma(\lambda + x)^2}{4Dt}\right)\right] d\lambda \tag{3.6.59}$$

and the original problem has been solved.

Obviously, this solution is extremely complicated to use. We can simplify it by assuming that $c_0(t) = c_0 =$ constant and that $c_i(t) = c_i =$ constant; manipulation of the integrals using the

$$\text{error function:} \quad \text{erf}(t) = \frac{2}{\sqrt{\pi}} \int_0^t e^{-\lambda^2}\, d\lambda$$

and the

$$\text{complimentary error function:} \quad \text{erfc}(t) = 1 - \text{erf}(t) = \frac{2}{\sqrt{\pi}} \int_t^\infty e^{-\lambda^2}\, d\lambda$$

yields

$$\frac{c - c_i}{c_0 - c_i} = H(\Delta) \tag{3.6.60}$$

where

$$vt\alpha = \Delta \tag{3.6.61}$$

$$H(\Delta) = \frac{1}{2}\left[1 + \text{erf}\left(\sqrt{\frac{v\Delta}{4\gamma\alpha D}} - x\sqrt{\frac{\gamma\alpha v}{4D\Delta}}\right) + e^{vx/D}\, \text{erfc}\left(\sqrt{\frac{v\Delta}{4\gamma\alpha D}} + x\sqrt{\frac{\gamma\alpha v}{4D\Delta}}\right)\right] \tag{3.6.62}$$

Equations (3.6.60)–(3.6.62) may even be further simplified. Thus if $D = 0$, no axial mixing, and $c_i = 0$, they simplify to the solution previously given for (3.6.30); this is the sharp front solution. Further, if c_0 is a step up in concentration, i.e.,

$$c_0 = \begin{cases} c_0, & t < t_1 \\ c_{00}, & t > t_1 \end{cases}$$

where $c_i = 0$, (3.6.60) reduces to

$$c(x, t) = \begin{cases} c_0 H(\Delta) + (c_{00} - c_0) H(\Delta - \bar{\Delta}), & \bar{\Delta} < \Delta < \infty \\ c_0 H(\Delta), & \Delta < \bar{\Delta} \end{cases} \tag{3.6.63}$$

where $vt_1\alpha = \bar{\Delta}$. This solution also contains the square-wave input since $c_{00} = 0$ can be chosen.

Lapidus and Amundson [19] used both (3.6.60) and (3.6.63) to show the influence of the ratio v/D on $c(x, t)$. In the case of (3.6.60) with $c_i = 0$, a v/D of approximately 1–2 smears out the front, but as v/D approaches a large value the front approaches a square wave or sharp front. The same phenomenon occurs with (3.6.63) with solutions approaching a square wave as v/D becomes large. In other words, the effect of axial mixing is to cause the original sharp front to approximate a sigmoid or S-shaped front.

Finally, we shall point out that all the above is for an infinitely long packed bed. In the case of a finite bed of length L the boundary conditions would be

$$\begin{aligned} vc_0 &= vc - D\frac{\partial c}{\partial x}, & x &= 0 \\ \frac{\partial c}{\partial x} &= 0, & x &= L \end{aligned} \tag{3.6.64}$$

corresponding to the so-called Danckwerts or Wehner-Wilhelm conditions. In this case the arbitrary constant B in the solution above is no longer zero; rather, A and B must be evaluated from (3.6.64). Bastian and Lapidus [3] have obtained the solution to this problem, but the complexity prevents our developing it here.

3.6.5 Heat and/or mass transfer with axial mixing and local resistance

Now we shall investigate the extension of Section 3.6.3 to axial mixing with local resistance. Thus we have

$$\begin{aligned} D\frac{\partial^2 c}{\partial x^2} &= v\frac{\partial c}{\partial x} + \frac{\partial c}{\partial t} + \frac{1}{\alpha}\frac{\partial n}{\partial t} \\ \frac{\partial n}{\partial t} &= k_1 c - k_2 n \\ c &= c_0(t), \qquad x = 0 \\ c &= c_i(x), \qquad t = 0 \\ n &= n_i(x), \qquad t = 0 \end{aligned} \tag{3.6.65}$$

where (3.6.65) is a special form of (3.6.35). Because of the extensive details of the solution [19], we shall only present a limited portion of the analysis.

Let

$$c(x, t) \xrightarrow{L} \bar{c}(x, s) = \bar{c}$$

$$n(x, t) \xrightarrow{L} \bar{n}(x, s) = \bar{n}$$

Then the L.T.s of the first two equations are

$$D\frac{d^2\bar{c}}{dx^2} = v\frac{d\bar{c}}{dx} + s\bar{c} - c_i(x) + \frac{1}{\alpha}[s\bar{n} - n_i(x)]$$

$$s\bar{n} - n_i(x) = k_1\bar{c} - k_2\bar{n}$$

Elimination of $\bar{n}$ gives

$$\frac{d^2\bar{c}}{dx^2} - \frac{v}{D}\frac{d\bar{c}}{dx} - \frac{\bar{c}}{D}\left[\frac{\alpha s(s + k_2) + sk_1}{\alpha(s + k_2)}\right] = -\frac{1}{D}\left[c_i(x) + \frac{k_2 n_i(x)}{\alpha(s + k_2)}\right]$$

and the conditions

$$c_0(t) \xrightarrow{L} \bar{c}_0(s) = \bar{c}_0, \qquad x = 0$$

$$\bar{c} \text{ remains finite as } x \longrightarrow \infty$$

As before it is convenient to make a change of variable

$$y = \bar{c}\exp\left[-\frac{vx}{2D}\right]$$

and to split the resulting problem into two alternative problems such that

$$y(x, s) = y_1(x, s) + y_2(x, s)$$

where (after some manipulations)

$$\frac{d^2y_1}{dx^2} - gy_1 = p(x) \qquad\qquad \frac{d^2y_2}{dx^2} \quad gy_2 = 0$$

$$y_1 = 0, \quad x = 0 \qquad\qquad y_2 = \bar{c}_0, \quad x = 0$$

$$y_1 = 0, \quad \text{when } x \longrightarrow \infty \qquad\qquad y_2 = 0 \quad \text{when } x \longrightarrow \infty$$

where

$$g = \frac{1}{D}\frac{s^2 + As + B}{s + k_2}$$

$$A = \frac{v^2\alpha + 4D(k_1 + \alpha k_2)}{4D\alpha}$$

$$B = \frac{v^2k_2}{4D}$$

$$p(x) = \frac{\alpha sc_i(x) + k_2[n_i(x) + \alpha c_i(x)]}{\alpha D(s + k_2)}\exp\left[\frac{vx}{2D}\right]$$

We shall not pursue the solutions of y_1 and y_2 and the corresponding inverses here. However, in both cases the solution for y_1 and y_2 is relatively straightforward (note that $y_2 = \bar{c}_0 \exp[-x\sqrt{g}]$ and that the inverse L.T. can actually be obtained from tables [19]).

For the special case of $c_i = n_i = 0$ and $c_0(t) = c_0 = \text{constant}$, the solution reduces to

$$\frac{c}{c_0} = \exp\left[\frac{vx}{2D}\right]\left[F(t) + k_2 \int_0^t F(t)\, dt\right] \tag{3.6.66}$$

where

$$F(t) = e^{-k_2 t} \int_0^t I_0\left[2\sqrt{\frac{k_1 k_2 \lambda}{\alpha}(t - \lambda)}\right] \frac{x}{2\sqrt{\pi D \lambda^3}} \exp\left[-\frac{x^2}{4D\lambda} - \lambda d\right] d\lambda$$

and

$$d = \frac{v^2}{4D} + \frac{k_1}{\alpha} - k_2$$

This solution contains terms previously found for the axial mixing-equilibrium solution, (3.6.58), and from the zero axial mixing-local resistance solution (3.6.45). This, of course, is expected. For the special case where axial mixing is neglected, $D = 0$, (3.6.66) further reduces to (3.6.45) with the appropriate changes in terminology.

3.6.6 Frequency response of a heat exchanger

Consider the single pipe heat exchanger as shown in Figure 3.16. Assuming that the wall temperature $T_w(t)$ is uniform down the length of the exchanger, the dynamic behavior of the fluid in the exchanger is described by

$$vc_f\rho_f \frac{\partial T(x, t)}{\partial x} + c_f\rho_f \frac{\partial T(x, t)}{\partial t} = -hA[T(x, t) - T_w(t)] \tag{3.6.67}$$

This is merely a rephrased version of the first two equations in (3.6.35). At $t = 0$ the exchanger is at steady state, corresponding to a constant wall temperature $T_{w_{ss}}$ and inlet temperature $T_{0_{ss}}$:

$$T(x, 0) = T_{ss}(x) \tag{3.6.68}$$

The inlet fluid has a time-varying temperature $T_0(t)$:

$$T(0, t) = T_0(t) \tag{3.6.69}$$

We shall let $\beta = c_f\rho_f/hA$, and the steady-state temperature profile $T_{ss}(x)$

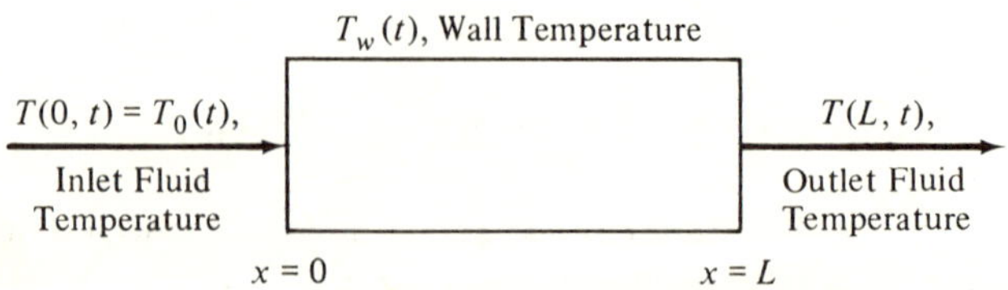

Figure 3.16. A single pipe heat exchanger.

is governed by

$$v\frac{dT_{ss}(x)}{dx} = -\frac{1}{\beta}(T_{ss}(x) - T_{w_{ss}}) \qquad T_{ss}(0) = T_{0_{ss}} \tag{3.6.70}$$

The solution of (3.6.70) is

$$T_{ss}(x) = T_{w_{ss}} + (T_{0_{ss}} - T_{w_{ss}})e^{-x/\beta v} \tag{3.6.71}$$

Let $y(x, t) = T(x, t) - T_{ss}(x)$, the deviation of the temperature from the initial steady state, and (3.6.67)–(3.6.69) become

$$\begin{gathered} v\frac{\partial y(x, t)}{\partial x} + \frac{\partial y(x, t)}{\partial t} = -\frac{1}{\beta}[y(x, t) - y_w(t)] \\ y(t, 0) = y_0(t) \\ y(0, x) = 0 \end{gathered} \tag{3.6.72}$$

Now define the L.T. with respect to t (since the x-variable is finite in length):

$$y(x, t) \xrightarrow{L} \bar{y}(x, s)$$

$$y_w(t) \xrightarrow{L} \bar{y}_w(s)$$

Applying this to (3.6.72) yields

$$s\bar{y}(x, s) + v\frac{d\bar{y}(x, s)}{dx} = -\frac{1}{\beta}[\bar{y}(x, s) - \bar{y}_w(s)]$$

$$\bar{y}(0, s) = \bar{y}_0(s)$$

Solving and evaluating at the outlet, $x = L$, we get

$$\frac{[(1/\beta) + s]\bar{y}(L, s) - (1/\beta)\bar{y}_w(s)}{[(1/\beta) + s]\bar{y}_0(s) - (1/\beta)\bar{y}_w(s)} = \exp\left[-\frac{L}{v}\left(\frac{1}{\beta} + s\right)\right] \tag{3.6.73}$$

We wish to study the frequency response of the outlet fluid temperature $T(L, t)$ to sinusoidal input variations in

1. The inlet temperature $T_0(t)$.
2. The wall temperature $T_w(t)$.
3. The velocity v.

1. *Inlet temperature frequency response.* Set $\bar{y}_w(s) = 0$ [i.e., $T_w(t) = T_{w_{ss}}$] and from (3.6.73) the transfer function between inlet and outlet temperatures is

$$g_1(s) = \frac{\bar{y}(L, s)}{\bar{y}_0(s)} = \exp\left(-\frac{L}{v\beta}\right)\exp\left(-\frac{L}{v}s\right) \tag{3.6.74}$$

Thus

$$\begin{gathered} |g_1(i\omega)| = e^{-L/v\beta} = e^{-\alpha/\beta}, \qquad \alpha = \frac{L}{v} \\ \angle g_1(i\omega) = -\frac{L\omega}{v} = -\alpha\omega \end{gathered} \tag{3.6.75}$$

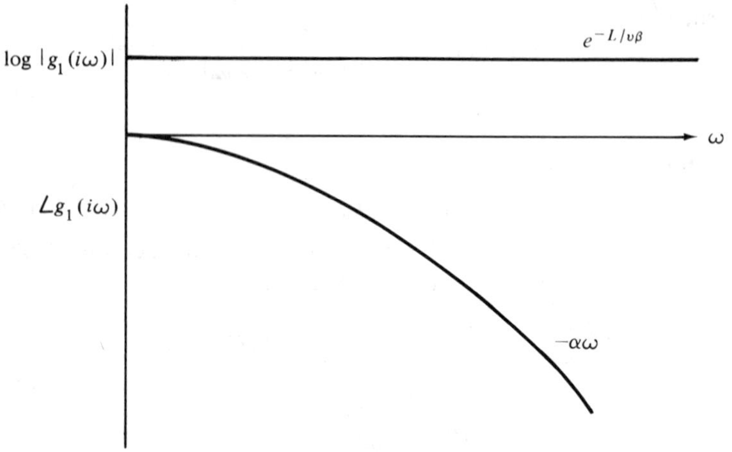

Figure 3.17. Bode plot for inlet temperature variation of heat exchanger.

The Bode plot is therefore as shown in Figure 3.17, exhibiting the characteristics of a pure time delay; that is, inlet temperature changes pass through the exchanger in plug flow, modified in amplitude by the factor $e^{-L/v\beta}$, which is a phase lag that increases linearly with ω. At the value of ω corresponding to a 45-degree phase lag, for example, the outlet temperature is exactly 45 degrees behind the inlet temperature at that instant.

2. *Wall temperature frequency response*. Set $\bar{y}_0(s) = 0$ [i.e., $T_0(t) = T_{0_{ss}}$]. The transfer function between the wall temperature and the outlet temperature is [from (3.6.73)]

$$g_2(s) = \frac{\bar{y}(L, s)}{\bar{y}_w(s)} = \left(\frac{1}{1 + \beta s}\right)\left\{1 - \exp\left(-\frac{\alpha}{\beta}\right)\exp(-\alpha s)\right\} \tag{3.6.76}$$

Setting $s = i\omega$ to determine the frequency response characteristics, we find

$$\begin{aligned} |g_2(i\omega)| &= \frac{1}{(1 + \beta^2\omega^2)^{1/2}}\{1 + e^{-2\alpha/\beta} - 2e^{-\alpha/\beta}\cos\alpha\omega\}^{1/2} \\ \angle\, g_2(i\omega) &= \tan^{-1}(-\alpha\omega) + \tan^{-1}\left(\frac{-e^{-\alpha/\beta}\sin\alpha\omega}{e^{-\alpha/\beta}\cos\alpha\omega - 1}\right) \end{aligned} \tag{3.6.77}$$

and the Bode plot is shown in Figure 3.18. The oscillations in both the magnitude and phase angle are called *resonance*. Resonance results because as the fluid moves through the exchanger, with a residence time of α, the wall temperature $T_w(t)$ is oscillating periodically with a frequency ω. As ω is increased, it is possible that the magnitude of the transfer function actually increases for certain values of ω. Keep in mind, of course, that each time ω is changed experimentally we must wait for all transients to die out, so that the periodic response is the steady-state response to the periodic forcing.

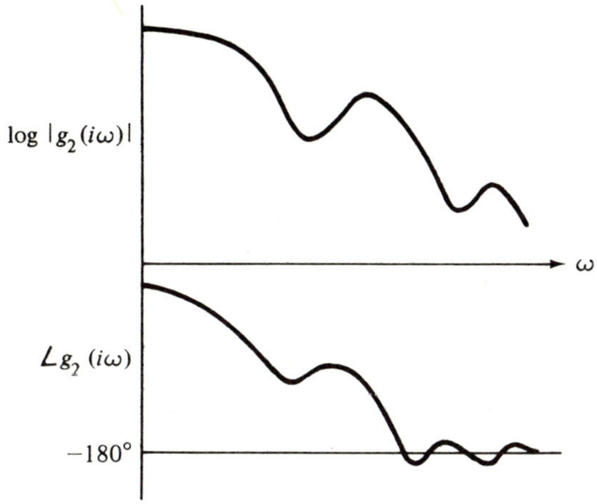

Figure 3.18. Bode plot for wall temperature variation of heat exchanger.

3. *Velocity frequency response*. Set $T_w(t) = T_{w_{ss}}$ and $T_0(t) = T_{0_{ss}}$ and consider the response of the outlet fluid to sinusoidal variation of v. The key problem is that v enters as a product with the derivative of $T(x, t)$ in the original P.D.E. for $T(x, t)$, (3.6.67). To handle this let

$$v(t) = v_{ss} + \dot{v}(t)$$

Then we have

$$\beta\frac{\partial y}{\partial t} + \beta v_{ss}\frac{dT_{ss}(x)}{dx} + \beta v_{ss}\frac{\partial y}{\partial x} + \beta\dot{v}(t)\frac{dT_{ss}(x)}{dx} + \beta\dot{v}(t)\frac{\partial y}{\partial x}$$
$$= T_{w_{ss}} - T_{ss}(x) - y(x, t)$$

Neglecting the last term on the left-hand side (L.H.S.) for small perturbations in v, we obtain

$$\beta\frac{\partial y}{\partial t} + \beta v_{ss}\frac{\partial y}{\partial x} = -y - \beta\dot{v}\frac{dT_{ss}(x)}{dx}$$

Using the L.T.,

$$\beta s\bar{y}(x, s) + \beta v_{ss}\frac{d\bar{y}(x, s)}{dx} = -\bar{y}(x, s) + \frac{\bar{\dot{v}}(s)}{v_{ss}}(T_{0_{ss}} - T_{w_{ss}})e^{-x/\beta v_{ss}}$$

$$\bar{y}(0, s) = \frac{T_{0_{ss}}}{s}$$

Solving this O.D.E., we find the transfer function between the outlet temperature and the velocity to be

$$g_3(s) = \frac{\bar{y}(L, s)}{\bar{\dot{v}}(s)} = \frac{T_{0_{ss}} - T_{w_{ss}}}{sv_{ss}\beta}e^{-\alpha/\beta}(1 - e^{-\alpha s})$$

The magnitude and phase angle of $g_3(i\omega)$ are

$$|g_3(i\omega)| = \frac{2(T_{0_{ss}} - T_{w_{ss}})}{\beta v_{ss}\omega} e^{-\alpha/\beta}(1 - \cos \alpha\omega)^{1/2}$$
$$\angle g_3(i\omega) = \tan^{-1}\left(\frac{1 - \cos \alpha\omega}{\sin \alpha\omega}\right) \tag{3.6.78}$$

The reader should decide whether or not we get resonance in this case.

3.6.7 Sequences of polymerization reactions

We have indicated that the z-transform can be applied advantageously to continuous time systems which are sampled and thus cast into a discrete time or difference equation form. In addition, many staged operations in chemical engineering are initially in a discrete form (the stage number) and the resulting difference equations are suitable for use of the z-transform. Here we wish to show how a sequence of polymer reactions lead to differential-difference equations whose solution can be facilitated by the transform technique. Murdoch [24], Abraham [1], and Kilkson [14], in particular, have used this approach.

There is a concept here which is worth mentioning before we carry out the actual operations. When the point in the analysis is reached where the differential-difference equations are available [see (3.6.84)] a number of paths are possible. These include the use of the Laplace transform to yield difference equations, which are then solved by the z-transform; the use of z-transforms to yield differential equations, which are then solved by the Laplace transform; and the use of either initial transform to make the first conversion and then solve the resulting equations in any manner desired. Here we shall use the z-transform first and then solve the resulting differential equations without the L.T. The work of Lapidus and Amundson [18] is indicative of how a two-step sequence may be used to solve the most complex type of problem.

We shall consider a batch polymerization conducted at constant temperature in which all chains in the reaction are linear. On this basis we may write the sequence of reactions

$$\begin{aligned} P + c &\longrightarrow P_1 \\ P_1 + c &\longrightarrow P_2 \\ P_2 + c &\longrightarrow P_3 \\ &\vdots \\ P_k + c &\longrightarrow P_{k+1} \\ &\vdots \end{aligned} \tag{3.6.79}$$

where the symbols $P, P_1, P_2, \ldots$ represent concentrations and c is a concentration treated as a constant. Associated with each equation is a rate equation, e.g.,

$$r_k = \text{rate of } k\text{th reaction step}$$

which might have the form

$$r_k = l_k P_{k-1} c \tag{3.6.80}$$

where l_k is a rate constant. Also, we specifically define r_f as the disappearance rate of P. The problem to be investigated is the following: Given pure P (feed concentration P_f) and c, what is the yield or concentration of P_k?

To simplify the discussion we shall let

$$\begin{aligned} v_n &= \frac{P_n}{P_f} \qquad \text{(a normalized concentration)} \\ b &= -\ln \frac{P}{P_f} \end{aligned} \tag{3.6.81}$$

or

$$\frac{P}{P_f} = e^{-b} \qquad \text{(a log transformation)} \tag{3.6.82}$$

It now follows that the following identities can be specified:

$$\frac{r_1}{r_f} = \frac{dP_1/dt}{-dP/dt} = -\frac{dP_1}{dP} = \frac{dv_1}{db} e^{+b}$$

where (3.6.80)–(3.6.82) have been used. Also,

$$\frac{-dP_1/dt}{dP/dt} = \frac{r_2 - r_1}{-r_1} = \frac{r_1 - r_2}{r_1} = 1 - \frac{l_2}{l_1} v_1 e^{+b}$$

Thus it follows that

$$\frac{dv_1}{db} + \frac{l_2}{l_1} v_1 = e^{-b}, \qquad k = 1 \tag{3.6.83}$$

In the same fashion

$$\frac{dv_{k+1}}{db} = \frac{l_2}{l_1}(v_k - v_{k+1}), \qquad k = 2, 3, 4, \ldots \tag{3.6.84}$$

Equations (3.6.83) and (3.6.84) are differential-difference equations relating the normalized concentration of P_k to the exponential or log variable b.

Since we are going to solve these equations with the z-transform and we note from (3.4.11) that $k = 0$ is defined in the z-transform, we first rescale (3.6.83) and (3.6.84) to

$$\frac{dv_0}{db} + \frac{l_2}{l_1} v_0 = e^{-b}, \qquad k = 0 \tag{3.6.83a}$$

$$\frac{dv_{k+1}}{db} = \frac{l_2}{l_1}(v_k - v_{k+1}), \qquad k = 1, 2, \ldots \tag{3.6.84a}$$

This does not really change anything except our numbering scheme.

Note that (3.6.83a) is a first-order O.D.E., while (3.6.84a) is a differential-difference equation. If we solve (3.6.83a) by the L.T. using (see Table 3.1)

$$v_0 \xrightarrow{L} \bar{v}_0$$

$$\frac{e^{at} - e^{dt}}{a - d} \xrightarrow{L} \frac{1}{(s-a)(s-d)}$$

$$e^{-t} \xrightarrow{L} \frac{1}{s+1}$$

we find that

$$v_0 = \frac{e^{-b} - e^{-l_2/l_1 b}}{(l_2 - l_1)/l_1} \tag{3.6.85}$$

where $v_0 = 0$ at $b = 0$ has been used. It then follows that

$$\frac{dv_0}{db} = \frac{-e^{-b} + l_2/l_1 e^{-l_2/l_1 b}}{(l_2 - l_1)/l_1} \tag{3.6.86}$$

With this information in hand we return to (3.6.84a) and from Table 3.1 note that

$$F(t + \tau) \xrightarrow{Z} zf(z) - zF(0)$$

$$v_k \xrightarrow{Z} \bar{v}(z)$$

Applied to (3.6.84a), there results

$$\frac{d}{db}(z\bar{v} - zv_0) = \frac{l_2}{l_1}\bar{v} - \frac{l_2}{l_1}[z\bar{v} - zv_0]$$

or, upon substitution of (3.6.85) and (3.6.86) and some algebraic manipulations,

$$\frac{d\bar{v}}{db} + \frac{l_2}{l_1}\frac{(z-1)}{z}\bar{v} = e^{-b} \tag{3.6.87}$$

Equation (3.6.87) is now only an O.D.E. in b-space in terms of the variable z. We could solve this equation by the L.T., but instead we shall solve by standard techniques. Thus

$$\bar{v}_H = K \exp\left[\frac{-l_2}{l_1}\frac{(z-1)}{z}b\right] \qquad \text{(homogeneous)}$$

$$\bar{v}_P = \exp[-b] \qquad \text{(particular)}$$

Substitution of these two equations into (3.6.87) and the use of $\bar{v} = 0$ at $b = 0$ to evaluate the arbitrary constant K leads to

$$\bar{v} = e^{-b}\left[\frac{\exp\{[(l_2/l_1)/z] - [(l_2/l_1) - 1]b\} - 1}{[(l_2/l_1)/z] - [(l_2/l_1) - 1]}\right] \tag{3.6.88}$$

The inverse z-transform for this function can be found in Table 3.1 and when applied yields (also shifting the variable k back up by 1)

$$v_k = \frac{(l_2/l_1)^{k-1} e^{-b}}{(k-1)!} \int_0^b e^{-[(l_2/l_1)-1]b} b^k \, db$$

This equation gives the yield of P_k as a function of the system parameters and the concentrations of P and P_f.

REFERENCES

1. Abraham, W. H., "Path-Dependent Distribution of Molecular Weight in Linear Polymers," *Ind. Eng. Chem. Fund. 2*, 221 (1963).
2. Amundson, N. R., "Solid-Fluid Interactions in Fixed and Moving Beds," *Ind. Eng. Chem. 48*, 26 (1956).
3. Bastian, W. C., and L. Lapidus, "Longitudinal Diffusion in Ion Exchange and Chromatographic Columns. Finite Column," *J. Phys. Chem. 60*, 816 (1956).
4. Bellman, R., R. E. Kalaba, and J. Lockett, *Numerical Inversion of the Laplace Transform*, American Elsevier, New York, 1966.
5. Cohen, W. C., and E. F. Johnson, "Dynamic Characteristics of Double-Pipe Heat Exchangers," *Ind. Eng. Chem. 48*, 1031 (1956).
6. Cook, G. E., E. D. Denman, and H. M. Carr, "Numerical Inversion of Laplace Transforms by the Laguerre-Gauss Quadrature Approximation," *IEEE Trans. Auto. Cont. 12*, 623 (1967).
7. Davis, P. J., and P. Rabinowitz, *Numerical Integration*, Ginn/Blaisdell, Waltham, Mass., 1967.
8. Ditkin, V. A., and A. P. Prodnikov, *Integral Transforms and Operational Calculus*, trans. I. N. Sneddon, Pergamon, Elmsford, N.Y., 1965.
9. Dubner, H., and J. Abate, "Numerical Inversion of Laplace Transforms and the Finite Fourier Cosine Transform," *J. ACM 15*, 115 (1968).
10. Gautschi, W., "On the Condition of a Matrix Arising in the Numerical Inversion of the Laplace Transform," *Math. Comp. 23*, 109 (1969).
11. Gilbert, E. G., "Dynamic Error Analysis of Digital and Combined Analog-Digital Computer Systems," *Simulation 6*, 241 (1966).
12. Goldstein, S., "On the Mathematics of Exchange Processes in Fixed Columns," *Proc. Roy. Soc. (London) A219*, 151 (1953).
13. Han, C. D., "A Note on Numerical Inversion of the Laplace Transform by Legendre Polynomials," *IEEE Trans. Auto. Cont. 12*, 230 (1967).
14. Kilkson, H., "Effect of Reaction Path and Initial Distribution on Molecular Weight Distribution of Irreversible Condensation Polymers," *Ind. Eng. Chem. Fund. 3*, 281 (1964).
15. Klinkenberg, A., "Heat Transfer in Cross-Flow Heat Exchangers and Packed Beds," *Ind. Eng. Chem. 46*, 2285 (1954).

16. Krylov, V. I., and N. S. Skoblya, *Handbook of Numerical Inversion of Laplace Transforms*, Israel Program for Scientific Translations, 1969.

17. Kuo, B. C., *Discrete-Data Control Systems*, Prentice-Hall, Englewood Cliffs, N.J., 1970.

18. Lapidus, L., and N. R. Amundson, "Stagewise Absorption and Extraction Equipment," *Ind. Eng. Chem. 42*, 1071 (1950).

19. Lapidus, L., and N. R. Amundson, "Mathematics of Adsorption in Beds. VI. The Effect of Longitudinal Diffusion," *J. Phys. Chem. 56*, 984 (1952).

20. Lapidus, L., and R. Luus, *Optimal Control of Engineering Processes*, Ginn/Blaisdell, Waltham, Mass., 1967.

21. Lapidus, L., and J. H. Seinfeld, *Numerical Solution of Ordinary Differential Equations*, Academic Press, New York, 1971.

22. Luke, Y. L., "Review 6," *Math. Comp. 22*, 215 (1968).

23. Miller, M. K., and W. T. Guy, "Numerical Inversion of the Laplace Transform By Use of Jacobi Polynomials," *SIAM J. Numer. Anal. 3*, 624 (1966).

24. Murdoch, P. G., "Finite-Difference Transforms for Application to Stage by Stage Processes," *AIChE J. 7*, 526 (1961).

25. Piessens, R., "Numerical Inversion of the Laplace Transform," *IEEE Trans. Auto. Cont. 14*, 299 (1969).

26. Piessens, R., *Tables for the Inversion of the Laplace Transform*, Report TW3, Applied Mathematics Section, University of Leuven, Belgium, Sept. 1969.

27. Roberts, G., and H. Kaufman, *Tables of Laplace Transforms*, Saunders, Philadelphia, 1966.

28. Stehfest, H., "Numerical Inversion of Laplace Transforms, Algorithm 368," *Commun. ACM 13*, 47 (1970).

29. Weeks, W. T., "Numerical Inversion of Laplace Transforms Using Laguerre Functions," *J. ACM 13*, 419 (1966).

30. Wylie, C. R., *Advanced Engineering Mathematics*, 2nd ed., McGraw-Hill, New York, 1960.

PROBLEMS

3.1 By use of the convolution integral, show the time functions for the transform functions

$$g(s) = \frac{1}{s+1}; \qquad f(s) = \frac{3}{s}$$

$$g(s) = \frac{1}{(s+2)^2}; \qquad f(s) = 4$$

$$g(s) = \frac{2}{(s+1)(s+5)}; \qquad f(s) = \frac{6}{s^2+4}$$

3.2 Show that the functions

$$F(t) = \begin{cases} t, & t \neq 3 \\ 5, & t = 3 \end{cases}$$

and

$$G(t) = t$$

have the same L.T.s. Discuss this result in terms of uniqueness of the inverse L.T.s.

3.3 Consider the time functions

$$F(t) = t^2 U(t = 0)$$

and

$$F(t) = t^2 U(t - \tau)$$

What are the corresponding L.T.s?

3.4 Consider the final value theorem for $F(t) = e^t$. What difficulties does one encounter here in applying the theorem?

3.5 If the input to a plant is zero, the dynamic response may be represented as

$$\mathbf{x}(t) = \mathbf{\Phi}(t)\mathbf{x}(0)$$

Show by means of the initial value theorem that

$$\mathbf{\Phi}(t = 0) = \mathbf{I}$$

and suggest how this may be used as a check on $\bar{\mathbf{\Phi}}(s)$. For the process

$$\dot{\mathbf{x}}(t) = \begin{bmatrix} 0 & 1 \\ -2 & -3 \end{bmatrix} \mathbf{x}$$

check your suggestion.

3.6 Consider a system whose transfer function is given by

$$g(s) = \frac{1}{1 + g_1(s)} = \alpha_0 + \alpha_1 s + \alpha_2 s^2 + \cdots$$

If the system is subjected to a ramp input, show by the final value theorem that

$$g(\infty) = \alpha_1$$

for $\alpha_0 = 0$.

3.7 Show that the two block diagrams in Fig. P3.7 have the same input/output transfer functions.

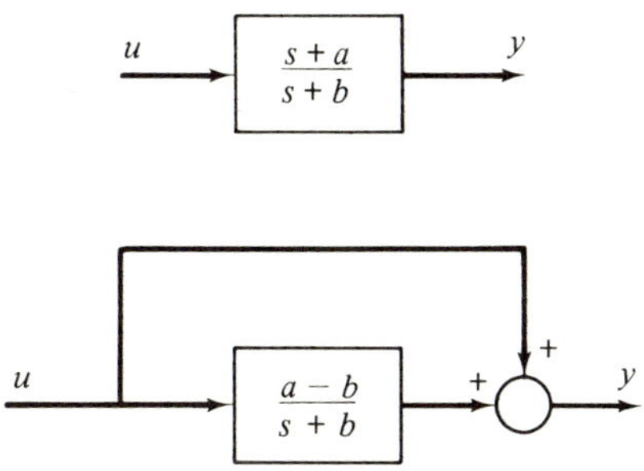

Figure P3.7

3.8 For a transfer function of the form

$$g(s) = \frac{(s+4)(s+5)}{(s+1)(s+2)(s+3)}$$

the zeros are $s = -4$ and $s = -5$; the poles are $s = -1$, $s = -2$, and $s = -3$. Make up an s-plane plot of imaginary vs. real roots and show where all the zeros and poles lie. Deduce that the system for which $g(s)$ is a transfer function has a damped (stable) solution.

3.9 Consider a system whose transfer function is

$$g(s) = \frac{1}{s-1}$$

To compensate for this function to make the system output damped, it is suggested that the two steps in Fig. P3.9 be used. Comment on the resulting system behavior, especially when a nonzero initial value $x(0)$ (or a roundoff error) occurs.

Figure P3.9.

3.10 Consider a system with a transfer function

$$g(s) = \exp[-\tau s]$$

What is the response of this system to a ramp input

$$u(t) = kt$$

By considering the series form

$$g(s) = \frac{1}{1 + \tau s + (\tau s)^2/2 + (\tau s)^3/3 + \cdots}$$

comment on truncated versions of the series form as a representation for the correct response.

3.11 Using the L.T., show that the solution of (3.6.87) is (3.6.88).

3.12 Consider the third-order differential equation

$$\dddot{x} + c_2\ddot{x} + c_1\dot{x} + c_0 x = u(t)$$

Show that

$$\bar{x}(s) = \frac{\bar{u}(s)}{s^3 + c_2 s^2 + c_1 s + c_0} + \frac{(s^2 + c_2 s + c_1)x(0) + (s + c_2)\dot{x}(0) + \ddot{x}(0)}{s^3 + c_2 s^2 + c_1 s + c_0}$$

In Figure P3.12, can you suggest what the three inputs should be to specify the behavior of this system?

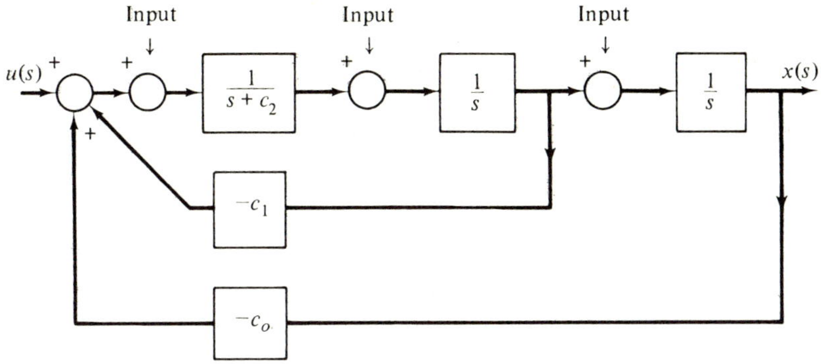

Figure P3.12

3.13 Show by partial fraction expansion that

$$\frac{1}{a-b}[a^k - b^k] \xleftarrow{Z^{-1}} \frac{z}{(z-a)(z-b)}$$

3.14 Given the z-transform

$$f(z) = \frac{0.792z^2}{(z-1)(z^2 - 0.416z + 0.208)}$$

determine that the value of $F_k(\tau) = 1$ by the final value theorem. Given the z-transform

$$f(z) = \frac{(1 - e^{-a\tau})z}{(z-1)(z - e^{-a\tau})}$$

where a is a constant show by a partial fraction expansion of $f(z)/z$, that

$$F_k(\tau) = 1 - e^{-an\tau}$$

3.15 The Adams-Bashforth two-step integration formula is given by

$$y(k+1) = y(k-1) + 2\tau f(k)$$

Show that the z-transform transfer function is

$$g(z) = \frac{2\tau z}{z^2 - 1}$$

For the backwards Euler integration formula

$$y(k+1) = y(k) + \tau f(k+1)$$

what is the transfer function?

3.16 Consider the four-compartment (or well-mixed stages) shown in Figure P3.16. The flow rates of material between compartments is

$$q_{ij} = k_{ij}c_i$$

where c_i is the compartment concentration of a tracer injected as $u(t)$. Each compartment is of volume V_i and thus contains Q_i moles of tracer. If $u(t)$ is an impulse of tracer, find the transfer function of c_1 as output for the external input.

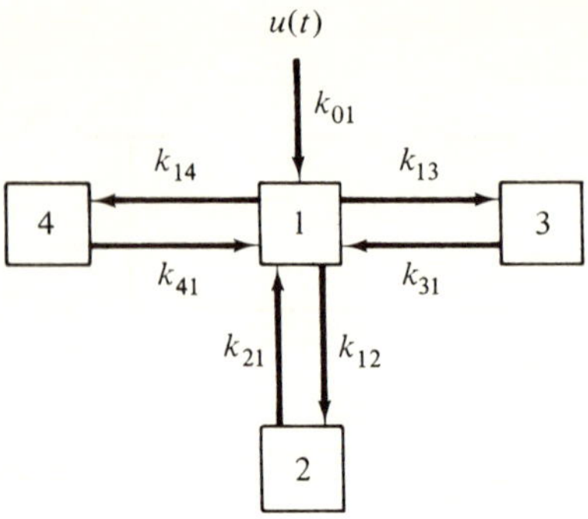

Figure P3.16

3.17 Consider a tank of the form shown in Figure P3.17. The orifice at the bottom of the tank has a linear head flow relation such that

$$q_1 = \frac{1}{R}h$$

where R is a constant parameter determined by the orifice. The output of this system is $q = q_0 - q_1$ with q_0, q_1, and h being functions of time; A is the cross-sectional area of the tank. Show that for the initial state zero, the overall transfer function relating $\bar{q}(s)$ and $\bar{q}_0(s)$ is

$$g(s) = \frac{RAs}{RAs + 1}$$

If the flow input is an ideal step function, show that

$$q = \exp\left[\frac{-t}{RA}\right]$$

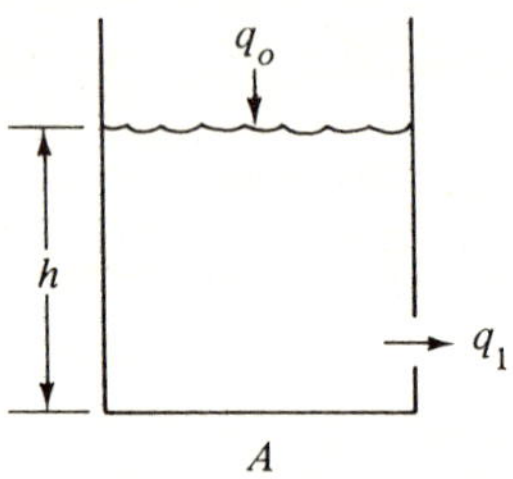

Figure P3.17

3.18 Consider two finite length countercurrent devices with interchange between them. The appropriate balances are

$$\frac{\partial x_1}{\partial t} + \frac{\partial x_1}{\partial z} + a_1(x_1 - x_2) = 0$$

$$\beta\frac{\partial x_2}{\partial t} + \frac{\partial x_2}{\partial z} + a_2(x_2 - x_1) = 0$$

$$x_1(0, t) = k_1(t)$$
$$x_2(1, t) = k_2(t)$$
$$x_1(z, 0) = x_2(z, 0) = 0$$

Show that the use of the L.T. leads to two linear O.D.E.s. Solve these equations and show that

$$\bar{x}_1(z, s) = g_{11}(z, s)\bar{k}_1(s) + g_{12}(z, s)\bar{k}_2(s)$$
$$\bar{x}_2(z, s) = g_{21}(z, s)\bar{k}_1(s) + g_{22}(z, s)\bar{k}_2(s)$$

where $g_{11}(z, s), \ldots, g_{22}(z, s)$ are component transfer functions. Find the explicit form of these transfer functions. Also find $x_1(t, z)$ and $x_2(t, z)$ for the case $k_1(t) = k_1$ and $k_2(t) = k_2$; k_1 and k_2 are constants.

3.19 A semi-infinite insulated bar which coincides with the x-axis, $x > 0$, is initially at zero temperature. At $t = 0$, a quantity of heat is instantaneously generated at the point $x = x_1$, $x_1 > 0$. We wish to find the temperature of the bar at any point and at any time $t > 0$. The governing equations are

$$\frac{\partial T}{\partial t} = k\frac{\partial^2 T}{\partial x^2}$$
$$T(x, 0) = 0$$
$$T(x_1, t) = Q\,\delta(t)$$

where k and Q are constants.

By taking the L.T. with

$$T(x, t) \Leftarrow \bar{T}(x, s) = \bar{T}(s)$$

show that

$$\bar{T}(s) = Ke^{-\sqrt{s/k}\,x}$$

where the fact that the temperature must be finite has been used. Then show that

$$T(x, t) = \frac{Q}{2\sqrt{\pi k t}}\, e^{-(x-x_1)^2/4kt}$$

3.20 For a first-order irreversible reaction in a wetted wall absorber,

$$-D\frac{\partial^2 c}{\partial x^2} + \frac{\partial c}{\partial t} + kc = 0 \qquad \text{(material balance)}$$

$$c = c^*, \quad x = 0, \quad t > 0$$
$$c = 0, \quad t = 0, \quad x > 0$$
$$c = \text{finite}, \quad x \longrightarrow \infty$$

where c = concentration of reactant A in liquid
c^* = equilibrium concentration at liquid surface
x = distance into liquid measured from surface
t = normalized time variable
D, k = constants

Show that

$$c(x, t) = \frac{c^*x}{2\sqrt{\pi D}} \int_0 e^{-k\lambda} e^{-x^2/4D\lambda} \lambda^{-3/2} \, d\lambda$$

3.21 Consider that the chromatographic column of Section 3.6.2 has established a band of solute a distance W down the column. At this time the solution flow is stopped. Pure solvent, not the same as in the band formation case, is then allowed to flow into the column at the same velocity as previously (this is termed *developing* the column). Assuming that the new linear equilibrium is

$$n = b'c$$

with $b \neq b'$ because of the different solvent, show that

$$n(z, y) = \begin{cases} 0, & 0 < z < \dfrac{y}{b'} \\ n_0, & \dfrac{y}{b'} < z < \dfrac{W}{v} + \dfrac{y}{b'} \\ 0, & \dfrac{W}{v} + \dfrac{y}{b'} < z \end{cases}$$

and that the absorbed band is developed as a square wave. n_0 is the adsorbed concentration in the original band. *Hint:* Convert the material balance equation in terms of $n(z, y)$ rather than $c(z, y)$. Also, use an initial condition corresponding to a step function on the solid.

Elements of Probability Theory 4

For our purposes the most fundamental classification of mathematical models is either deterministic or stochastic models. Up to this point we have developed the most prevalent analytical techniques for the solution of linear, deterministic process models. We shall now turn to the question of stochastic mathematical models.

The basic approach to developing a system model should certainly be to describe the changes taking place in the system by principles of conservation of mass, momentum, and energy. For a well-defined system, such as laminar flow in a circular tube, the application of these principles usually leads to a deterministic model of one of the forms in Chapter 1. For a *complex* system, for example, one in which turbulent flow exists, the application of conservation relations often does not lead directly to a soluble model, a common reason for which is the inherent random nature of the system. Therefore it may be advantageous to employ a probablistic or stochastic rather than a deterministic description for such a system. In this and the following chapter we shall develop the fundamental mathematical principles underlying stochastic modeling.

4.1 The Concept of a Random Variable

Think of a laboratory experiment the result of which is unknown until the experiment has been completed, as opposed to an experiment the result of which can be predicted precisely knowing all the conditions beforehand (of course, in the latter case we probably would not want to perform the experiment in the first place if we knew what the outcome would be). Fortunately (or maybe unfortunately), we usually do not know the result of an experiment until it is finished. Let us call the result of such an experiment an *event*. There may be a discrete number or a continuous spectrum of possible events, or outcomes, but the most we might be able to say about which event will result from a particular trial is the *probability* of observing each possible event.

We have now introduced a new concept, *probability*, which we must define in order to be more precise about the outcomes of our experiment. Let us write $\text{Prob}\{X\} = P(X)$ to signify the probability that the event X occurs. Even though X denotes an outcome rather than a number, $P(X)$ *is* a number, i.e., the *probability* that X occurs. Now, how do we define *probability*? We can only define probability on the basis of our physical reasoning about the process the outcome of which may be X. Nevertheless, we can set down some general rules to help us determine the outcome probabilities for a particular process.

Intuitively, we would say that the probability of a certain event is 1, while the probability of an event which cannot occur is zero. Also, if there are two *mutually exclusive* events X_1 and X_2, the probability of obtaining either X_1 or X_2 is the sum of the individual probabilities, $P(X_1)$ and $P(X_2)$. We can state the following axioms of probability theory formally:

1. $P(X) \geq 0$ (probability is nonnegative).
2. If X is certain, $P(X) = 1$.
3. If two events X_1 and X_2 are mutually exclusive,

$$\text{Prob}\{X_1 \text{ or } X_2\} = P(X_1) + P(X_2)$$

Consider for a moment the roll of a single die. Each of the outcomes corresponds to a number, the number being one of the integers 1, 2, . . . , 6. We can let X denote the number which will result from the roll of the die; that is, X is a random variable that can assume the integer values 1, 2, . . . , 6. If we throw the die n times and the value j occurs n_j times, it seems reasonable to define the probability of outcome j as

$$\text{Prob}\{X = j\} = P(X = j) = \lim_{n \to \infty} \frac{n_j}{n} \tag{4.1.1}$$

The random variable, which we have denoted by X, is a symbol for the outcome of the event in which we are interested. In the die example, the random variable X could only assume discrete values, the integers from 1 to 6. However, X could also denote the instantaneous value of a continuous variable, say the concentration in a reactor at a given location and time. Even if the outcome of an event has no obvious numerical value, we can assign numerical values to the random variable X corresponding to the possible outcomes. For example, in the tossing of a coin we can let $X = 1$ denote a head and $X = 2$ denote a tail.

It is customary to denote random variables by capital letters and to denote the values they may assume by the corresponding lowercase letters. Therefore x denotes one of the values that X may assume. In the case of the throw of the die we could have written $P(X = x)$, where x assumes the values 1 to 6.

The random variable X is called a *discrete random variable* if it may assume a finite or a denumerably infinite set of values, denoted as x_j, $j = 1, 2, \ldots, N$. The probability that X assumes the value x_j is denoted by $p(j)$. $p(j)$ is called the *probability mass function* of the random variable X, because the values of X may be thought of as being confined to the mass points $x_1, \ldots, x_N$. The probability mass function $p(j)$ has the following properties:

$$0 \leq p(j) \leq 1$$
$$\sum_{j=1}^{N} p(j) = 1 \tag{4.1.2}$$

If X is a *continuous random variable*, then we define the *probability density function* $p(x)$ such that

$$p(x)\,dx = \text{Prob}\{x < X \leq x + dx\} \tag{4.1.3}$$

where $p(x)\,dx$ may be thought of as the fraction of the total mass of X which lies in the range x to $x + dx$; that is, $p(x)$ is the *density* of X in this region. The probability density function has the following properties:

$$0 \leq p(x) \leq 1$$
$$\int_{-\infty}^{\infty} p(x)\,dx = 1 \tag{4.1.4}$$

We stress that $p(x)$ itself is not a probability; rather, $p(x)\,dx$ is a probability. In fact, $p(x)$ has the wrong units to be a probability. A probability, by assumption, is dimensionless. Let us suppose that X denotes a temperature measurement in degrees Fahrenheit (°F). Then dx also has units of °F. Thus for $p(x)\,dx$ to be dimensionless, $p(x)$ must have units of $(°\text{F})^{-1}$. In general, whatever the units of the random variable, the probability mass or density function is measured in $(\text{units})^{-1}$.

We shall now introduce the *distribution function*. Consider the probability that a random variable X will not exceed a given value x. Clearly, this probability is a function of the threshold value x. We call this function the distribution function and denote it by $F(x)$. For a discrete random variable $F(x_j)$ or $F(j)$ is the probability that $X \leq x_j$, where $x_1 < \cdots < x_N$. In terms of the probability mass function

$$F(j) = \sum_{i=1}^{j} p(i) \tag{4.1.5}$$

Let us return to the die-throwing experiment. For $x_1 = 1, p(1) = \frac{1}{6}$, and so $F(1)$, the probability that $X \leq 1$ (clearly X cannot be less than 1, but only exactly equal to 1), is $\frac{1}{6}$. At $x = 1$, $F(j)$ jumps from zero to $\frac{1}{6}$, just as at $x = 2$, $F(j)$ jumps from $\frac{1}{6}$ to $\frac{1}{3}$. Finally, $F(6) = 1$, since no numbers higher than 6 can be obtained. Thus the distribution function for a discrete random variable is a staircase function of discrete jumps.

For a continuous random variable, $F(x)$ is related to the probability density function by

$$F(x) = \int_{-\infty}^{x} p(\zeta)\, d\zeta \tag{4.1.6}$$

A distribution function $F(x)$ has the following general properties:

1. $0 \leq F(x) \leq 1$.
2. $\lim_{x \to -\infty} F(x) = 0$ and $\lim_{x \to \infty} F(x) = 1$.
3. As x increases, $F(x)$ must not decrease.
4. $\text{Prob}\{x_1 < X \leq x_2\} = F(x_2) - F(x_1)$.

Properties 1 and 2 follow since $F(x)$ is itself a probability and since the value of X must lie somewhere on the x-axis. Clearly, as x increases, the probability that $X \leq x$ must not decrease, since new intervals are continually being added to x. If these new intervals have probability zero, then $F(x)$ must at least remain constant. Property 4 follows from axiom 3, namely that

$$\text{Prob}\{X \leq x_2\} = \text{Prob}\{X \leq x_1\} + \text{Prob}\{x_1 < X \leq x_2\}$$

For a continuous random variable, this becomes

$$F(x_2) - F(x_1) = \int_{x_1}^{x_2} p(x)\, dx \tag{4.1.7}$$

At points x where the derivative of $F(x)$ exists, the derivative is equal to the density $p(x)$ at that point:

$$p(x) = \frac{dF(x)}{dx} \tag{4.1.8}$$

In Figure 4.1 we illustrate probability mass and density functions and their associated distribution functions. We also show the region corresponding to the probability that $x_1 < X \leq x_2$.

A discrete random variable may actually be considered as a special case of a continuous random variable in which the probability density is concentrated as impulses at N discrete points. For a discrete random variable assuming values $x_1, \ldots, x_N$ with probabilities $p(1), \ldots, p(N)$, the probability

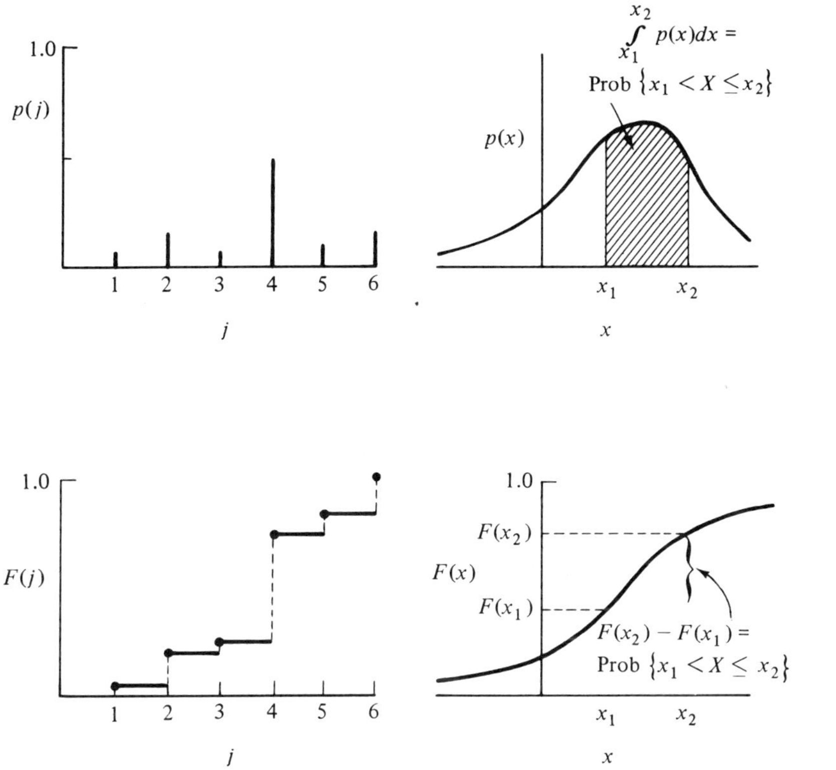

(a) Typical Probability Mass Function $p(j)$ and Distribution Function $F(j)$ for a Discrete Random Variable X.

(b) Typical Probability Density Function $p(x)$ and Distribution Function $F(x)$ for a Continuous Random Variable X.

Figure 4.1. Typical mass, density, and distribution functions.

density function of X can be expressed as a sum of impulses

$$p(x) = p(1)\delta(x - x_1) + p(2)\delta(x - x_2) + \cdots + p(N)\delta(x - x_N) \qquad (4.1.9)$$

In view of the second relation in (4.1.2), $p(x)$ satisfies the second relation in (4.1.4).

It is possible to have a mixed discrete and continuous probability density function in which the density function is a continuous curve but contains impulses. Consider the density function shown in Figure 4.2, where no value of X greater than x_1 can be encountered. Such a density could result, for example, if the outcome of the random event is passed through a device which takes all values of X greater than x_1 and converts them to $X = x_1$. The strength of the impulse at x_1 would then be

$$\int_{x_1}^{\infty} p(x)\, dx$$

which is the probability of observing $X = x_1$.

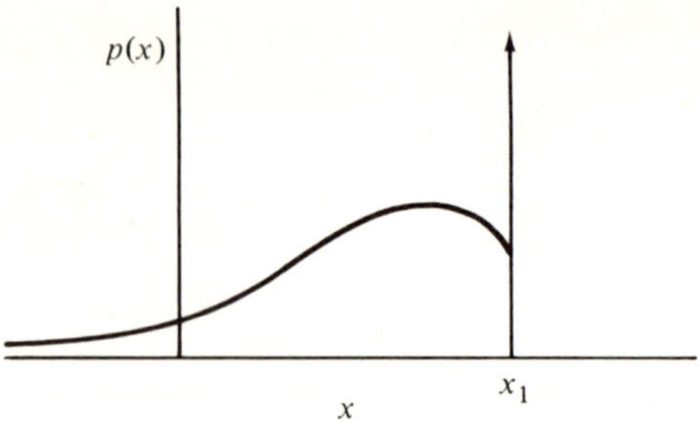

Figure 4.2. Mixed discrete and continuous probability density function.

4.2 Joint Distributions

Often in an experiment two or more quantities are measured; i.e., two or more random variables are observed. Let us consider first the case of two random variables, X and Y, both of which are discrete, X assuming N_1 possible values and Y assuming N_2 possible values. Thus when X and Y are measured, there are $N_1 N_2$ possible outcomes. For a complete characterization of X and Y, we need to know the *joint probability mass function* $p_{XY}(j_1, j_2)$, which is defined as

$$p_{XY}(j_1, j_2) = \text{Prob}\{X = x_{j_1}, Y = y_{j_2}\} \tag{4.2.1}$$

i.e., the probability that X assumes the value x_{j_1} and Y assumes the value y_{j_2}. If we are interested in the probability that X assumes the value x_{j_1} regardless of the value that Y assumes, we can define the *marginal probability mass function* by

$$\begin{aligned} p_X(j_1) &= \text{Prob}\{X = x_{j_1}\} \\ &= \sum_{j_2=1}^{N_2} p_{XY}(j_1, j_2) \end{aligned} \tag{4.2.2}$$

When dealing with two random variables, we shall use the subscripts XY and X and Y to denote the joint and marginal probability functions, respectively.

The joint probability mass function $p_{XY}(j_1, j_2)$ must satisfy the relation

$$\sum_{j_1=1}^{N_1} \sum_{j_2=1}^{N_2} p_{XY}(j_1, j_2) = 1 \tag{4.2.3}$$

Similarly, for continuous random variables X and Y, the *joint probability density function* $p_{XY}(x, y)$ is defined by

$$p_{XY}(x, y)\, dx\, dy = \text{Prob}\{x < X \leq x + dx; y < Y \leq y + dy\} \tag{4.2.4}$$

If we are interested in the probability that X lies in the region x to $x + dx$

regardless of the value that Y assumes, we can define the *marginal probability density function* by

$$p_X(x)\,dx = \int_{-\infty}^{\infty} P_{XY}(x, y)\,dy \tag{4.2.5}$$

The joint probability density function $p_{XY}(x, y)$ must satisfy the relation

$$\int_{-\infty}^{\infty}\int_{-\infty}^{\infty} p_{XY}(x, y)\,dx\,dy = 1 \tag{4.2.6}$$

We can also define distribution functions for two random variables. The *joint* distribution functions for discrete and continuous random variables X and Y are defined by

$$\begin{aligned} F_{XY}(j_1, j_2) &= \text{Prob}\{X \le x_{j_1};\, Y \le y_{j_2}\} \\ F_{XY}(x, y) &= \text{Prob}\{X \le x;\, Y \le y\} \end{aligned} \tag{4.2.7}$$

In terms of the joint probability mass and density functions

$$F_{XY}(j_1, j_2) = \sum_{k_1=1}^{j_1}\sum_{k_2=1}^{j_2} p_{XY}(k_1, k_2) \tag{4.2.8}$$

$$F_{XY}(x, y) = \int_{-\infty}^{x}\int_{-\infty}^{y} p_{XY}(\zeta, \eta)\,d\eta\,d\zeta \tag{4.2.9}$$

The joint distribution functions have the following properties:

$$\begin{aligned} F_{XY}(N_1, N_2) &= 1 \\ F_{XY}(1, j_2) = F_{XY}(j_1, 1) &= 0 \end{aligned} \tag{4.2.10}$$

and

$$\begin{aligned} F_{XY}(\infty, \infty) &= 1 \\ \lim_{x\to-\infty} F_{XY}(x, y) = \lim_{y\to-\infty} F_{XY}(x, y) &= 0 \end{aligned} \tag{4.2.11}$$

The first property of (4.2.11) simply states that X and Y must be located somewhere in the x, y-plane. The second states that if either value is restricted to be less than or equal to $-\infty$, the region of possible values in the x, y-plane is empty. We also observe that $F_{XY}(x, y)$ is a nondecreasing function of x and y.

We may also define the *marginal distribution functions* F_X and F_Y by

$$\begin{aligned} F_X(j_1) &= \text{Prob}\{X \le x_{j_1}\}, \qquad & F_Y(j_2) &= \text{Prob}\{Y \le y_{j_2}\} \\ F_X(x) &= \text{Prob}\{X \le x\}, \qquad & F_Y(y) &= \text{Prob}\{Y \le y\} \end{aligned} \tag{4.2.12}$$

where, for example, $F_X(x)$ is the probability that $X \le x$ regardless of the value of Y. Therefore $F_X(x) = F_{XY}(x, \infty)$ and $F_Y(y) = F_{XY}(\infty, y)$. The marginal distribution functions have the following properties:

$$\begin{aligned} F_X(N_1) = F_Y(N_2) &= 1 \\ F_X(1) = F_Y(1) &= 0 \end{aligned} \tag{4.2.13}$$

and

$$F_X(\infty) = F_Y(\infty) = 1$$
$$\lim_{x \to -\infty} F_X(x) = \lim_{y \to -\infty} F_Y(y) = 0 \tag{4.2.14}$$

We may obtain the joint density function by differentiating the joint distribution function if F_{XY} is continuous at the point x, y:

$$p_{XY}(x, y) = \frac{\partial^2 F_{XY}(x, y)}{\partial x \, \partial y} \tag{4.2.15}$$

We may obtain the marginal density functions $p_X(x)$ and $p_Y(y)$ by differentiating the marginal distribution functions:

$$p_X(x) = \frac{dF_X(x)}{dx}, \qquad p_Y(y) = \frac{dF_Y(y)}{dy} \tag{4.2.16}$$

Figure 4.3 shows typical joint mass and density functions for two random variables.

The two random variables X and Y are called *independent random variables* if the joint mass or density function is a product of the two marginal mass or density functions:

$$p_{XY}(j_1, j_2) = p_X(j_1)p_Y(j_2) \tag{4.2.17}$$

$$p_{XY}(x, y) = p_X(x)p_Y(y) \tag{4.2.18}$$

An alternative way of defining independence is that the events $X \leq x$ and $Y \leq y$ are independent for all values of x and y, namely

$$F_{XY}(j_1, j_2) = F_X(j_1)F_Y(j_2) \tag{4.2.19}$$

$$F_{XY}(x, y) = F_X(x)F_Y(y) \tag{4.2.20}$$

We have confined our attention in this section to two random variables and their associated distributions. The concepts of joint mass and density functions can easily be extended to n random variables. In such a case, it is common to drop the subscripts on the density and distribution functions. Thus for n random variables, it is understood that $p(j_1, j_2, \ldots, j_n)$ is the joint mass function and that $p(x_1, x_2, \ldots, x_n)$ is the joint density function. Note also that we let the random variables be denoted by $X_1, X_2, \ldots, X_n$, as opposed to above, where we have used X and Y when there are only two random variables. n random variables are independent if and only if their n-dimensional joint density function is equal to the product of the n marginal density functions. (This can also be stated in terms of the joint and marginal distribution functions.)

We can also have the case in which some of the random variables are independent of certain others but not all. In particular, for $n_1 + n_2$ random variables the first n_1 are independent of the last n_2 if the $n_1 + n_2$-dimensional joint density is equal to the product of the two n_1- and n_2-dimensional joint densities.

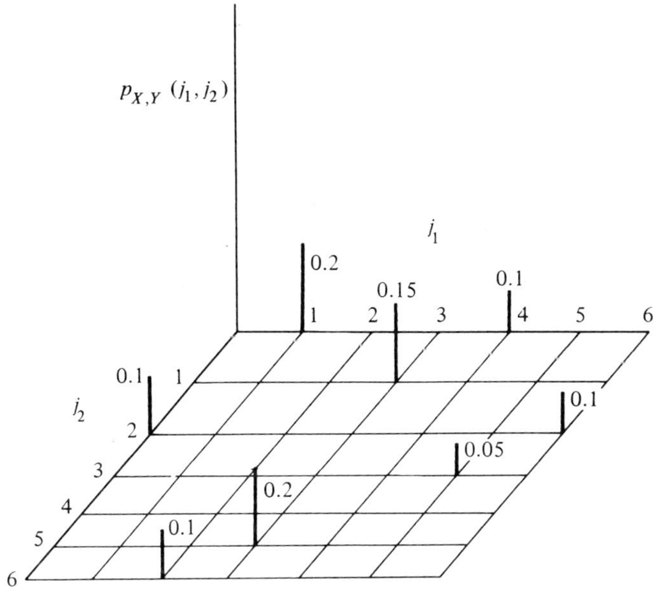

Example of a Joint Probability Mass Function $p_{X,Y}(j_1, j_2)$

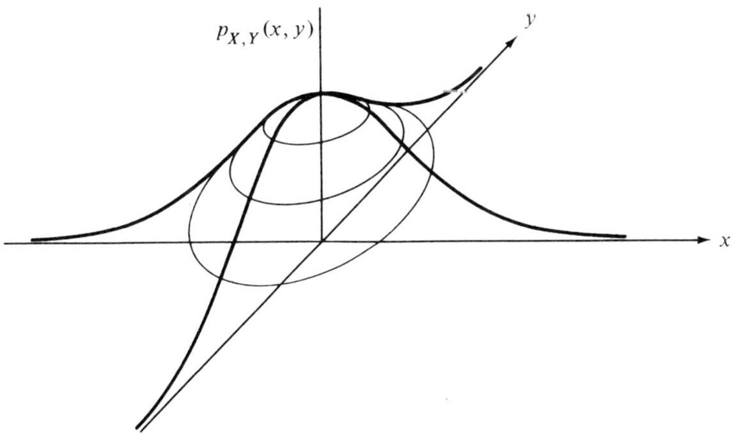

Example of a Joint Probability Density Function $p_{X,Y}(x, y)$

Figure 4.3. Examples of joint mass and density functions for two random variables X and Y.

4.3 Conditional Probabilities

A conditional probability is the probability that given two random variables, one random variable assumes some value if the other variable is fixed at a certain value. Consider first the case of two discrete random variables X and Y. We have already defined the joint probability mass function by (4.2.1) as $p_{XY}(j_1, j_2)$, the probability that $X = x_{j_1}$ and $Y = y_{j_2}$, and the marginal probability mass function by (4.2.2) as $p_X(j_1)$, the probability that $X = x_{j_1}$ regardless of the value of Y. Now, we want to know the probability that $X = x_{j_1}$ given that $Y = y_{j_2}$. This probability is called a *conditional probability* since it is the probability that $X = x_{j_1}$ conditioned on $Y = y_{j_2}$. We use the notation

$$p(j_1 \mid j_2) = \text{Prob}\{X = x_{j_1} \text{ if } Y = y_{j_2}\}$$

The relation between the joint, marginal, and conditional probability mass functions is

$$p(j_1 \mid j_2) = \frac{p_{XY}(j_1, j_2)}{p_Y(j_2)} \tag{4.3.1}$$

or, equivalently,

$$p(j_2 \mid j_1) = \frac{p_{XY}(j_1, j_2)}{p_X(j_1)} \tag{4.3.2}$$

where $p(j_2 \mid j_1) = P\{Y = y_{j_2} \mid X = x_{j_1}\}$. Equating $p_{XY}(j_1, j_2)$ in (4.3.1) and (4.3.2),

$$p(j_1 \mid j_2) = \frac{p(j_2 \mid j_1) p_X(j_1)}{p_Y(j_2)} \tag{4.3.3}$$

which is called *Bayes' rule.*

The concept of conditional probabilities can be easily extended to continuous random variables. We define the *conditional probability density* $p(x \mid y)$ as the probability density of the random variable X given that the random variable Y assumes the value of y. Thus $p(x \mid y)\, dx$ is the probability of finding X in the range $(x, x + dx)$ given that $y < Y \leq y + dy$. The density $p(x \mid y)$ can be envisioned as the intersection of a plane at $Y = y$ with the joint probability density function $p_{XY}(x, y)$.

It follows from the definition of conditional probability that

$$\text{Prob}\{X \leq x \mid y < Y \leq y + dy\} = \frac{\text{Prob}\{X \leq x, y < Y \leq y + dy\}}{\text{Prob}\{y < Y \leq y + dy\}}$$

or, equivalently,

$$F(x \mid y) = \frac{\int_{-\infty}^{x} p_{XY}(\zeta, y)\, d\zeta}{p_Y(y)} \tag{4.3.4}$$

where the conditional distribution function $F(x \mid y)$ is the probability that $X \leq x$ conditioned on $Y = y$. The derivative of $F(x \mid y)$ with respect to x is the conditional density $p(x \mid y)$. Differentiating both sides of (4.3.4) with respect to x, we obtain

$$p(x \mid y) = \frac{p_{XY}(x, y)}{p_Y(y)} \tag{4.3.5}$$

It follows likewise that

$$p(y \mid x) = \frac{p_{XY}(x, y)}{p_X(x)} \tag{4.3.6}$$

Equating (4.3.5) and (4.3.6), we obtain

$$p(y \mid x) = \frac{p(x \mid y) p_Y(y)}{p_X(x)} \tag{4.3.7}$$

which is Bayes' rule for continuous random variables.

EXAMPLE 4.1. Two machines are turning out a product. Machines 1 and 2 turn out 9 and 10 products per minute, respectively. The acceptability of the products from the two machines is not identical; on the average, out of 10 products from machine 1, 8 are acceptable, and from machine 2, only 7. What is the probability that an acceptable product has come from machine 1?

Let X_1 and X_2 denote the events that a product comes from machines 1 and 2, respectively, and let Y denote the event that it is acceptable. X_1 and X_2 are mutually exclusive events since a product must come from either machine 1 or 2 but not both. The probability of an acceptable product $p(Y)$ is then the probability that the product is acceptable if it has come from machine 1 times the probability it came from machine 1 plus the probability that it is acceptable if it has come from machine 2 times the probability it came from machine 2:

$$p(Y) = p(Y \mid X_1) p(X_1) + p(Y \mid X_2) p(X_2)$$

However, the probability of an acceptable product from machine 1 is also given by $p(X_1 \mid Y) p(Y)$, etc. Thus the probability that an acceptable product has come from machine 1, $p(X_1 \mid Y)$, is given by (another form of Bayes' rule)

$$p(X_1 \mid Y) = \frac{p(Y \mid X_1) p(X_1)}{p(Y \mid X_1) p(X_1) + p(Y \mid X_2) p(X_2)}$$

For our problem, $p(Y \mid X_1) = 0.8$, $p(Y \mid X_2) = 0.7$, and $p(X_1) = 0.9 p(X_2)$. Substituting into the above expression,

$$p(X_1 \mid Y) = 0.493$$

EXAMPLE 4.2. Assume that the random variables X and Y have the joint density (the jointly Gaussian density)

$$p_{XY}(x,y) = \frac{1}{2\pi\sqrt{M}} e^{-(P_{22}x^2 + P_{11}y^2 - 2P_{12}xy)/2M}$$

where $M = P_{11}P_{22} - P_{12}^2$. Find $p_X(x)$, $p_Y(y)$, $p(x \mid y)$, and $p(y \mid x)$.

First,

$$\begin{aligned} p_X(x) &= \int_{-\infty}^{\infty} p_{XY}(x, y)\, dy \\ &= \frac{1}{2\pi\sqrt{M}} \int_{-\infty}^{\infty} e^{-(P_{22}x^2 + P_{11}y^2 - 2P_{12}xy)/2M}\, dy \\ &= \frac{1}{2\pi\sqrt{M}} e^{-P_{22}x^2/2M} \int_{-\infty}^{\infty} e^{-(P_{11}y^2 - 2P_{12}xy)/2M}\, dy \\ &= \frac{1}{\sqrt{2\pi P_{11}}} e^{-x^2/2P_{11}} \end{aligned}$$

where

$$\int_{-\infty}^{\infty} e^{-py^2 + qy}\, dy = \sqrt{\frac{\pi}{p}}\, e^{q^2/4p}$$

Similarly,

$$p_Y(y) = \frac{1}{\sqrt{2\pi P_{22}}} e^{-y^2/2P_{22}}$$

Also,

$$p(x \mid y) = \frac{p_{XY}(x, y)}{p_Y(y)} = \sqrt{\frac{P_{22}}{2\pi M}}\, e^{-P_{22}[x - (P_{12}/P_{22})y]^2/2M}$$

and

$$p(y \mid x) = \frac{p_{XY}(x, y)}{p_X(x)} = \sqrt{\frac{P_{11}}{2\pi M}}\, e^{-P_{11}[y - (P_{12}/P_{11})x]^2/2M}$$

4.4 Properties of Random Variables

We shall now consider special properties of random variables, namely their *expectations*. The expected value of the random variable X is denoted by $E(X)$ and is also called the *mean* of X. The expected value is computed by

$$E(X) = \sum_{j=1}^{N} x_j p(j) \qquad \text{(discrete)} \tag{4.4.1}$$

$$= \int_{-\infty}^{\infty} xp(x)\, dx \qquad \text{(continuous)} \tag{4.4.2}$$

The expected value of a function of X, $f(X)$, is found from

$$E[f(X)] = \sum_{j=1}^{N} f(x_j)p(j) \qquad \text{(discrete)} \tag{4.4.3}$$

$$= \int_{-\infty}^{\infty} f(x)p(x)\, dx \qquad \text{(continuous)} \tag{4.4.4}$$

The expected values (4.4.3) and (4.4.4) exist provided that

$$\sum_{j=1}^{N} p(j) \mid f(x_j) \mid < \infty \tag{4.4.5}$$

$$\int_{-\infty}^{\infty} \mid f(x) \mid p(x)\, dx < \infty \tag{4.4.6}$$

EXAMPLE 4.3. Let the discrete random variable X assume the values $j = 0, 1, 2, \ldots$ with probability mass function (the *Poisson distribution*)

$$p(j) = \frac{\lambda^j}{j!}e^{-\lambda}, \qquad \lambda < 0$$

Compute the expected value of X. Using (4.4.1),

$$E(X) = \sum_{j=0}^{\infty} j\frac{\lambda^j}{j!}e^{-\lambda} = \lambda e^{-\lambda}\sum_{j=1}^{\infty}\frac{\lambda^{j-1}}{(j-1)!}$$

$$= \lambda e^{-\lambda}\sum_{k=0}^{\infty}\frac{\lambda^k}{k!} = \lambda e^{-\lambda}e^{\lambda} = \lambda$$

EXAMPLE 4.4 Let the discrete random variable X assume the values $j = 0, 1, 2, \ldots, N$ with probability mass function (the *binomial distribution*)

$$p(j) = \frac{N!}{j!(N-j)!}p^j(1-p)^{N-j}, \qquad 0 < p < 1$$

Compute the expected value of X.

$$E(X) = \sum_{j=0}^{N} j\frac{N!}{j!(N-j)!}p^j(1-p)^{N-j}$$

$$= \sum_{j=1}^{N} j\frac{N!}{j!(N-j)!}p^j(1-p)^{N-j}$$

$$= Np\sum_{j=1}^{N}\frac{(N-1)!}{(j-1)!(N-j)!}p^{j-1}(1-p)^{N-j}$$

$$= Np\sum_{j=0}^{N-1}\frac{(N-1)!}{j!(N-1-j)!}p^j(1-p)^{N-1-j}$$

$$= Np[p + (1-p)]^{N-1} = Np$$

Both the Poisson and binomial distributions will be discussed in more detail later.

The most important properties of probability distributions are the *moments* of the distribution, of which the mean is a special case. We shall define the expectation $E(X^r)$ as the *rth noncentral moment of* X. We shall use the notation μ'_r for this moment. Thus for discrete and continuous random variables,

$$\mu'_r = \sum_{j=0}^{N} x_j^r p(j) \tag{4.4.7}$$

$$= \int_{-\infty}^{\infty} x^r p(x)\,dx \tag{4.4.8}$$

By definition $\mu'_0 = 1$, and μ'_1 is the mean of X. We shall define the expectation $E[(X - \mu'_1)^r]$ as the *rth central moment of* X, that is, the rth moment about the mean μ'_1. Thus, denoting the central moments by μ_r,

$$\mu_r = \sum_{j=0}^{N} (x_j - \mu'_1)^r p(j) \tag{4.4.9}$$

$$= \int_{-\infty}^{\infty} (x - \mu'_1)^r p(x)\,dx \tag{4.4.10}$$

From the definition of the central moments, $\mu_0 = 1$ and $\mu_1 = 0$. The second central moment μ_2 is called the *variance* of the distribution and is often denoted by σ^2. μ_3 is a measure of the skewness of the distribution about the mean, and $\mu_3 = 0$ if the distribution is symmetric. μ_4 is a measure of the distortion from a normal (Gaussian) distribution resulting in a low center and high ends of the distribution or a high center and low ends, that is, a measure of the flatness of the distribution.

The noncentral and central moments can be related. Consider (4.4.10) for the rth central moment. Let us expand $(x - \mu'_1)^r$ by the binomial theorem:

$$(x - \mu'_1)^r = \sum_{i=0}^{r} (-1)^i \binom{r}{i} x^{r-i} \mu_1'^i \tag{4.4.11}$$

where

$$\binom{r}{i} = \frac{r!}{(r-i)!\,i!} \tag{4.4.12}$$

Using (4.4.11) in (4.4.10), we obtain

$$\mu_r = \int_{-\infty}^{\infty} \left[\sum_{i=0}^{r} (-1)^i \binom{r}{i} x^{r-i} \mu_1'^i \right] p(x)\, dx \tag{4.4.13}$$

Since the integrals $\int_{-\infty}^{\infty} x^r p(x)\, dx$ are simply the rth noncentral moments μ'_r, the various terms in the expansion (4.4.13) represent the various moments multiplied by numerical coefficients and powers of μ'_1. For $r = 1, 2, 3, 4$, (4.4.13) becomes

$$\mu_1 = 0 \tag{4.4.14}$$

$$\mu_2 = \mu'_2 - \mu_1'^2 \tag{4.4.15}$$

$$\mu_3 = \mu'_3 - 3\mu'_2\mu'_1 + 2\mu_1'^3 \tag{4.4.16}$$

$$\mu_4 = \mu'_4 - 4\mu'_3\mu'_1 + 6\mu'_2\mu_1'^2 - 3\mu_1'^4 \tag{4.4.17}$$

Similarly, the noncentral moments can be determined in terms of the central moments and the mean μ'_1:

$$\mu'_2 = \mu_2 + \mu_1'^2 \tag{4.4.18}$$

$$\mu'_3 = \mu_3 + 3\mu_2\mu'_1 + \mu_1'^3 \tag{4.4.19}$$

$$\mu'_4 = \mu_4 + 4\mu_3\mu'_1 + 6\mu_2\mu'_1 + \mu_1'^4 \tag{4.4.20}$$

For the moment μ'_r to exist it is necessary that the series (4.4.7) or the integral (4.4.8) be absolutely convergent. If the moment of order r exists, all the moments of order smaller than r will also exist.

The two most important moments of a distribution are the mean μ'_1 and the variance μ_2. The nonnegative square root of the variance, usually denoted σ, is called the *standard deviation* of the distribution. The standard deviation σ has the same units as the random variable X and is often used as a measure of the dispersion of a distribution about its mean value.

The following relationships can be easily verified:

$$E[(aX)^r] = a^r E(X^r) \tag{4.4.21}$$

$$E[aX + b] = aE(X) + b \tag{4.4.22}$$

$$E[(X - c)^2] = \mu_2 + (c - \mu'_1)^2 \tag{4.4.23}$$

Since $(c - \mu'_1)^2 > 0$, $E[(X - c)^2]$ is a minimum when $c = \mu'_1$. In mechanical terms, this result states that the moment of inertia about any point c is minimum when that point is the center of mass μ'_1. Next, consider the variance of $(aX + b)$:

$$E\{[(aX + b) - (a\mu'_1 + b)]^2\} = a^2\mu_2 \tag{4.4.24}$$

Thus the variance of $aX + b$ is unaffected by the displacement b and is simply a^2 times the variance of X.

We can define a normalized random variable Y by

$$Y = \frac{X - \mu'_1}{\sigma} \tag{4.4.25}$$

where μ'_1 and σ are the mean and standard deviation of X, respectively. We note, from (4.4.22) and (4.4.24), that $E(Y) = 0$ and $E(Y^2) = 1$.

Two other parameters often used to characterize distributions are the *coefficient of variation* ν, defined as the ratio of the standard deviation to the expected value,

$$\nu = \frac{\sigma}{\mu'_1} \tag{4.4.26}$$

and the *coefficient of skewness* γ, given by

$$\gamma = \frac{\mu_3}{\sigma^3} \tag{4.4.27}$$

The coefficient of skewness γ measures the extent to which a distribution departs from a symmetric distribution [one for which $F(a - x) = 1 - F(a + x) - \text{Prob}\{X = a + x\}$]. In a symmetric distribution all the central moments of odd order equal zero. Thus each central moment of odd order serves as a measure of asymmetry of the distribution. The coefficient of skewness expresses the third central moment relative to the standard deviation. γ may be positive or negative, giving rise to what is called positive or negative asymmetry.

Let us now consider two random variables X and Y. The expected value of any function $g(X, Y)$ is defined by

$$E[g(X, Y)] = \sum_{j_1=1}^{N_1} \sum_{j_2=1}^{N_2} g(x_{j_1}, y_{j_2}) p_{XY}(j_1, j_2) \tag{4.4.28}$$

in the discrete case, or

$$E[g(X, Y)] = \int_{-\infty}^{\infty} \int_{-\infty}^{\infty} g(x, y) p_{XY}(x, y)\, dx\, dy \tag{4.4.29}$$

These expected values exist, provided that

$$\sum_{j_1=1}^{N_1} \sum_{j_2=1}^{N_2} |g(x_{j_1}, y_{j_2})| p_{XY}(j_1, j_2) < \infty \tag{4.4.30}$$

$$\int_{-\infty}^{\infty} \int_{-\infty}^{\infty} |g(x, y)| p_{XY}(x, y)\, dx\, dy < \infty \tag{4.4.31}$$

If $g(X, Y) = X + Y$, we find that (from this point on we shall illustrate results for continuous random variables only; results for discrete random variables follow with the obvious modifications)

$$\begin{aligned} E(X + Y) &= \int_{-\infty}^{\infty} \int_{-\infty}^{\infty} (x + y) p_{XY}(x, y)\, dx\, dy \\ &= \int_{-\infty}^{\infty} \int_{-\infty}^{\infty} x p_{XY}(x, y)\, dx\, dy + \int_{-\infty}^{\infty} \int_{-\infty}^{\infty} y p_{XY}(x, y)\, dx\, dy \\ &= \int_{-\infty}^{\infty} x p_X(x) dx + \int_{-\infty}^{\infty} y p_Y(y)\, dy \\ &= E(X) + E(Y) \end{aligned} \tag{4.4.32}$$

Thus the expected value of a sum of n random variables is equal to the sum of their individual expectations. Note that we have not assumed independence of the variables to obtain (4.4.32).

If $g(X, Y) = X^l Y^k$, where $l \geq 0$ and $k \geq 0$, $\mu'_{lk} = E(X^l Y^k)$ is called the noncentral moment of order $l + k$ of the random variables X and Y. We note immediately that $\mu'_{10} = E(X)$ and that $\mu'_{01} = E(Y)$, that is, the expected values of the marginal distributions. Similarly, $\mu'_{20} = E(X^2)$ and $\mu'_{02} = E(Y^2)$. The central moments are defined by

$$\mu_{lk} = E[(X - \mu'_{10})^l (Y - \mu'_{01})^k] \tag{4.4.33}$$

The central moment μ_{11} is called the *covariance* of X and Y. The covariance is related to the other moments by

$$\begin{aligned} \mu_{11} &= E[(X - \mu'_{10})(Y - \mu'_{01})] \\ &= E(XY) - \mu'_{10} E(Y) - \mu'_{01} E(X) + \mu'_{10} \mu'_{01} \\ &= \mu'_{11} - \mu'_{10} \mu'_{01} \end{aligned} \tag{4.4.34}$$

We shall now consider the moments of the sum and product of two random variables X and Y. The expected value of the sum Z of two random variables has been given by (4.4.32). If X and Y are independent random variables, the expected value of the product XY is simply the product of the individual expected values:

$$E(XY) = \int_{-\infty}^{\infty} x p_X(x)\, dx \int_{-\infty}^{\infty} y p_Y(y)\, dy = E(X) E(Y) \tag{4.4.35}$$

Thus, if X and Y are independent, $\mu_{11} = 0$; however, the converse is not necessarily true.

Let us now consider the variance of the sum Z of two random variables:

$$\begin{aligned}\sigma_Z^2 &= E\{[(X - \mu'_{10}) + (Y - \mu'_{01})]^2\}\\ &= E[(X - \mu'_{10})^2] + 2E[(X - \mu'_{10})(Y - \mu'_{01})] + E[(Y - \mu'_{01})^2]\\ &= \sigma_X^2 + \sigma_Y^2 + 2\mu_{11} \qquad (4.4.36)\end{aligned}$$

If X and Y are independent,

$$\sigma_Z^2 = \sigma_X^2 + \sigma_Y^2 \qquad (4.4.37)$$

Both the covariance and the variance are second-order central moments; in fact, the variance is simply a special case of the covariance in which $X = Y$. One problem in using the covariance as a measure of the degree of dependence between X and Y is the question of units. If X is measured in moles per cubic foot (moles/ft³) and Y in °F, the covariance will have units of moles-°F/ft³, not a very workable dimension. Thus it is convenient to place the covariance in dimensionless form. A convenient way of doing this is to divide the covariance by the product of the individual standard deviations σ_X and σ_Y; calling this dimensionless form ρ, we have

$$\rho = \frac{E(XY) - \mu'_{01}\mu'_{10}}{\sigma_X\sigma_Y} = E\left[\left(\frac{X - \mu'_{10}}{\sigma_X}\right)\left(\frac{Y - \mu'_{01}}{\sigma_Y}\right)\right] \qquad (4.4.38)$$

where ρ is called the *correlation coefficient* between X and Y.

The correlation coefficient satisfies the inequality

$$-1 \leq \rho \leq 1 \qquad (4.4.39)$$

If X and Y are independent, $\rho = 0$. However, if $\rho = 0$, X and Y are not necessarily independent, since $\mu_{11} = 0$ does not necessarily imply that X and Y are independent. In such a case, we say that X and Y are *uncorrelated.*

EXAMPLE 4.5. Let us show that if $\rho = \pm 1$, the probability that $Y = aX + b$ is unity. Suppose that Prob $\{Y \neq aX + b\} = 0$. We can write

$$\begin{aligned}\mu'_{01} &= E(Y)\\ &= \text{Prob}\,\{Y = aX + b\}E(Y \mid Y = aX + b)\\ &\quad + \text{Prob}\,\{Y \neq aX + b\}E(Y \mid Y \neq aX + b)\\ &= E(aX + b) = a\mu'_{10} + b\\ \sigma_Y^2 &= E[(Y - \mu'_{01})^2]\\ &= \text{Prob}\,\{Y = aX + b\}E[(Y - \mu'_{01})^2 \mid Y = aX + b]\\ &\quad + \text{Prob}\,\{Y = aX + b\}E[(Y - \mu'_{01})^2 \mid Y = aX + b]\\ &= E[(aX + b - \mu'_{10})^2]\\ &= a^2\sigma_X^2\end{aligned}$$

In a similar way we obtain $\mu_{11} = a^2\sigma_X^2$. Using (4.4.38), we find that $\rho^2 = 1$.

We can extend these concepts to n random variables. Consider, for example, the n discrete random variables $X_1, X_2, \ldots, X_n$. Variable X_1 can assume values from zero to N_1, variable X_2 values from zero to N_2, etc. The mean of the vector $\mathbf{X} = [X_1, X_2, \ldots, X_n]^T$ is given by

$$\boldsymbol{\mu}_{\mathbf{x}} = \sum_{j_1=0}^{N_1} \cdots \sum_{j_n=0}^{N_n} \begin{bmatrix} X_1(j_1) \\ \cdot \\ \cdot \\ \cdot \\ X_n(j_n) \end{bmatrix} p(j_1, j_2, \ldots, j_n)$$

$$= \begin{bmatrix} \mu_{X_1} \\ \mu_{X_2} \\ \cdot \\ \cdot \\ \cdot \\ \mu_{X_n} \end{bmatrix} \tag{4.4.40}$$

where $\mu_{X_1} = E(X_1)$, etc. Thus the mean of a vector of n random variables is simply the vector of the individual mean values. A very important quantity associated with n random variables is the covariance. We have seen that a single random variable has a single variance $\mu_2(= \sigma^2)$ and that two random variables X and Y have the two individual variances σ_X^2 and σ_Y^2 as well as the single covariance $E[(X - \mu'_{10})(Y - \mu'_{01})]$. We shall now extend the concept of the covariance to n random variables. We shall define the $n \times n$ *covariance matrix* $\mathbf{P}$ as

$$\mathbf{P} = E[(\mathbf{X} - \boldsymbol{\mu}_{\mathbf{X}})(\mathbf{X} - \boldsymbol{\mu}_{\mathbf{X}})^T]$$
$$= E\begin{bmatrix} (X_1 - \mu_{X_1})^2 & \cdots & (X_1 - \mu_{X_1})(X_n - \mu_{X_n}) \\ \cdot & & \cdot \\ \cdot & & \cdot \\ \cdot & & \cdot \\ (X_n - \mu_{X_n})(X_1 - \mu_{X_1}) & \cdots & (X_n - \mu_{X_n})^2 \end{bmatrix} \tag{4.4.41}$$

where the expectation operator applies to each element of the matrix. The terms on the diagonal of $\mathbf{P}$ are the individual variances and the off-diagonal elements are the covariances. It is clear that the matrix is symmetric, so there are only $n(n + 1)/2$ distinct quantities.

EXAMPLE 4.6. Let Z denote the sum of n continuous random variables, $Z = X_1 + X_2 + \cdots + X_n$. Compute the variance of Z, that is, $E[(Z - E(Z))^2]$.

We begin by noting that $E(Z) = \sum_{i=1}^{n} E(X_i)$. Next we compute Z^2:

$$Z^2 = \sum_{i=1}^{n} \sum_{j=1}^{n} X_i X_j = \sum_{i=1}^{n} X_i^2 + \sum_{i=1}^{n} \sum_{j \neq i}^{n} X_i X_j$$

Then the second noncentral moment of Z is

$$E(Z^2) = \sum_{i=1}^{n} E(X_i^2) + \sum_{i=1}^{n} \sum_{j \neq i}^{n} E(X_i X_j)$$

We note that the desired variance can be written as $E(Z^2) - E(Z)^2$. Thus

$$E(Z)^2 = \sum_{i=1}^{n} E(X_i)^2 + \sum_{i=1}^{n} \sum_{j \neq i}^{n} E(X_i)E(X_j)$$

Finally,

$$E(Z^2) - E(Z)^2 = \sum_{i=1}^{n} [E(X_i^2) - E(X_i)^2] + \sum_{i=1}^{n} \sum_{j \neq i}^{n} [E(X_i X_j) - E(X_i)E(X_j)]$$
$$= \sum_{i=1}^{n} \sigma_{X_i}^2 + \sum_{i=1}^{n} \sum_{j \neq i}^{n} \sigma_{X_i X_j}^2$$

where $\sigma_{X_i}^2$ is the variance of X_i and $\sigma_{X_i X_j}^2$ is the covariance of X_i and X_j.

4.5 Generating and Characteristic Functions

In the theory of probability distributions two special functions play a very powerful role. These two functions are known as *generating functions* and *characteristic functions*. To introduce these two functions, let us begin by considering the sum of two continuous random variables X and Y, $Z = X + Y$, where $p_{XY}(x, y)$ is the joint probability density function for X and Y. We want to calculate the probability density function for Z in terms of $p_{XY}(x, y)$. We assume for the moment that both X and Y are nonnegative random variables.

Let $F_Z(z)$ be the distribution function for Z, i.e., the probability that $Z \leq z$, and $p_Z(z)$ be the probability density function for Z. Thus

$$F_Z(z) = \int_0^{\infty} \int_0^{z-x} p_{XY}(x, y)\, dy\, dx \tag{4.5.1}$$

which can be verified by drawing the line $Z = X + Y$ in the X, Y-plane and considering the area beneath it. Make the change of variable $y = \zeta - x$. For a given x, $dy = d\zeta$, and the range of ζ is from 0 to z. Then

$$\begin{aligned} F_Z(z) &= \int_0^{\infty} \int_0^{z} p_{XY}(x, \zeta - x)\, d\zeta\, dx \\ &= \int_0^{z} \int_0^{\infty} p_{XY}(x, \zeta - x)\, dx\, d\zeta \end{aligned} \tag{4.5.2}$$

where we have interchanged the order of integration since the range of ζ is independent of x. Since

$$p_Z(z) = \frac{dF_Z(z)}{dz} \tag{4.5.3}$$

then

$$p_Z(z) = \int_0^{\infty} p_{XY}(x, z - x)\, dx \tag{4.5.4}$$

If X and Y are independent,

$$p_Z(z) = \int_0^{\infty} p_X(x) p_Y(z - x)\, dx \tag{4.5.5}$$

which is recognizable as a convolution integral (Section 3.1.4).

We shall now take a very important step. We shall define the Laplace transforms of the three marginal density functions $p_X(x)$, $p_Y(y)$, and $p_Z(z)$ as

$$\psi_X(s) = \int_0^\infty e^{-sx} p_X(x)\, dx \tag{4.5.6}$$

$$\psi_Y(s) = \int_0^\infty e^{-sy} p_Y(y)\, dy \tag{4.5.7}$$

$$\psi_Z(s) = \int_0^\infty e^{-sz} p_Z(z)\, dz \tag{4.5.8}$$

The L.T. of the probability density functions for X, Y and Z are called the *generating functions* for X, Y and Z.†

First, we note that in the L.T. domain, (4.5.5) can be written as

$$\psi_Z(s) = \psi_X(s)\psi_Y(s) \tag{4.5.9}$$

Thus the generating function for the sum of two independent random variables is equal to the product of the individual generating functions. For example, if X and Y are independent random variables, distributed according to $p_X(x) = ae^{-ax}$ and $p_Y(y) = be^{-by}$ for $x \geq 0$, $y \geq 0$, then $\psi_X(s) = a/(s + a)$ and $\psi_Y(s) = b/(s + b)$ and

$$\psi_Z(s) = \frac{ab}{(s + a)(s + b)} \tag{4.5.10}$$

Inverting $\psi_Z(s)$, we obtain the density function for $Z = X + Y$,

$$p_Z(z) = \frac{ab}{b - a}(e^{-az} - e^{-bz}) \tag{4.5.11}$$

Thus, the probability density of the sum of two independent random variables can be obtained from inversion of the product of the L.T. of the two individual probability densities.

However, the generating functions play a much more important role than their use in computing the density of a sum of two independent random variables. Note that the L.T. of a density function, such as in (4.5.6)–(4.5.8), are also the expected values of exponentials of the random variables, for example,

$$\psi_X(s) = E(e^{-sX}) \tag{4.5.12}$$

Thus we can rederive (4.5.9) directly, using

$$\begin{aligned} E(e^{-sZ}) &= E(e^{-s(X+Y)}) = E(e^{-sX}e^{-sY}) \\ &= E(e^{-sX})E(e^{-sY}) \end{aligned} \tag{4.5.13}$$

†In probability theory it is more customary to refer to the z-transform of a probability mass function rather than the L. T. of a probability density function as the generating function (e.g., see Feller, Vol. I [1]). However, because of the direct correspondence between the Laplace and z-transforms, we have chosen to introduce the generating function as the Laplace transform of a probability density function. In Example 4.7 we illustrate the generating function for a discrete random variable.

The *most important* use of the generating function is in the calculation of the moments of a distribution. Consider

$$\psi_X(s) = E(e^{-sX}) = \int_0^\infty e^{-sx} p_X(x)\,dx \tag{4.5.14}$$

Let us expand e^{-sx} in a Maclaurin series:

$$e^{-sx} = 1 - sx + \frac{s^2}{2!}x^2 - \frac{s^3}{3!}x^3 + \cdots \tag{4.5.15}$$

Substituting (4.5.15) into (4.5.14), we find

$$\begin{aligned}\psi_X(s) &= 1 - sE(X) + \frac{s^2}{2!}E(X^2) - \frac{s^3}{3!}E(X^3) + \cdots \\ &= 1 - s\mu_1' + \frac{s^2}{2!}\mu_2' - \frac{s^3}{3!}\mu_3' + \cdots \end{aligned} \tag{4.5.16}$$

Thus, if the generating function of X is known explicitly, the coefficients of its Maclaurin series are the noncentral moments of X. Alternatively, the noncentral moments can be obtained by

$$\mu_r' = (-1)^r \frac{d^r \psi_X(s)}{ds^r}\bigg|_{s=0} \tag{4.5.17}$$

Thus the Maclaurin coefficients can be found by differentiation or the use of a known expansion for $\psi_X(s)$.

Consider the generating function for $Y = aX + b$. We note that

$$\begin{aligned} E(e^{-sY}) &= E(e^{-s(aX+b)}) \\ &= e^{-bs}E(e^{-asX}) \end{aligned} \tag{4.5.18}$$

Replacing s by as in (4.5.16) will enable calculation of the moments of Y. A particular form of linear transformation which is important is $X - \mu_X$, where the generating function of $X - \mu_X$ is

$$\begin{aligned} E(e^{-s(X-\mu_X)}) &= 1 - sE(X - \mu_X) + \frac{s^2}{2!}E\,(X - \mu_X)^2 \\ &\quad - \frac{s^3}{3!}E((X - \mu_X)^3) + \cdots \\ &= 1 - s\mu_1 + \frac{s^2}{2!}\mu_2 - \frac{s^3}{3!}\mu_3 + \cdots \end{aligned} \tag{4.5.19}$$

Therefore the central moments can be obtained from the generating function of $X - \mu_X$ or, more easily, from $e^{s\mu_X}$ times the generating function of X:

$$\begin{aligned} E(e^{-s(X-\mu_X)}) &= e^{s\mu_X}E(e^{-sX}) \\ &= e^{s\mu_X}\psi_X(s) \end{aligned} \tag{4.5.20}$$

Combining (4.5.19) and (4.5.20),

$$e^{s\mu_X}\psi_X(s) = 1 - s\mu_1 + \frac{s^2}{2!}\mu_2 - \frac{s^3}{3!}\mu_3 + \cdots \tag{4.5.21}$$

so that the central moments can be found from

$$\mu_r = (-1)^r \frac{d^r}{ds^r}[e^{s\mu_X}\psi_X(s)]\bigg|_{s=0} \tag{4.5.22}$$

obviating the need for the conversion relations (4.4.14)–(4.4.17).

EXAMPLE 4.7. When X is a discrete random variable, the generating function of X is defined as the z-transform of the probability mass function:

$$\psi(z) = \sum_{j=0}^{n} p(j)z^j$$

where $-1 \leq z \leq 1$. For example, the random variable X obeying the binomial distribution

$$p(j) = \binom{N}{j} p^j(1-p)^{N-j}, \qquad j = 0, 1, \ldots, N$$

has the generating function

$$\psi(z) = \sum_{j=0}^{N} \binom{N}{j}(pz)^j(1-p)^{N-j}$$
$$= (pz + q)^N$$

The random variable X obeying the Poisson distribution

$$p(j) = \frac{\lambda^j}{j!}e^{-\lambda}, \qquad j = 0, 1, 2, \ldots$$

has the generating function

$$\psi(z) = \sum_{j=0}^{\infty} \frac{(\lambda z)^j}{j!}e^{-\lambda} = e^{-\lambda}e^{\lambda z} = e^{-\lambda(1-z)}$$

The moments of X can be determined from the derivatives of $\psi(z)$ at $z = 1$. We note that

$$\dot{\psi}(z) = \sum_j jp(j)z^{j-1}$$

$$\ddot{\psi}(z) = \sum_j j(j-1)p(j)z^{j-2}$$

so that

$$\dot{\psi}(1) = \sum_j jp(j) = E(X)$$

$$\ddot{\psi}(1) = \sum_j j(j-1)p(j) = E(X^2) - E(X)$$

We have seen that the generating functions of *nonnegative* continuous and discrete random variables are the expected values of e^{-sX} and z^X, respectively, where $0 \leq z \leq 1$. The usefulness of the generating functions comes largely from the multiplicative properties, $e^{-s(X+Y)} = e^{-sX}e^{-sY}$ and $z^{X+Y} = z^X z^Y$. This same multiplicative property holds for the exponential function with a purely imaginary argument, i.e., e^{itX}, where X is real. However, since e^{itX} is bounded, its expectation exists under any circumstances [1]. Therefore, the use of $E[e^{itX}]$ as a substitute for generating functions provides a more powerful and general tool for analyzing sums and moments of random variables. In particular, for random variables defined on the entire real line ($-\infty$ to $+\infty$), it is necessary to employ this expectation rather than those associated with the generating function.

If X is a random variable with density $p(x)$, then the *characteristic function* of X is the function $\phi(t)$ defined for real t by

$$\phi(t) = E(e^{itX}) = \int_{-\infty}^{\infty} e^{itx}p(x)\,dx \tag{4.5.23}$$

We see that $\phi(t)$ is simply the Fourier transform of $p(x)$ (see Section 3.1.2). However, we note that $\phi(t)$ is defined in (4.5.23) with e^{itx} in the transform rather than e^{-itx} as we saw in Chapter 3. Since the limits of integration are from $-\infty$ to $+\infty$, this change of sign in the exponent does not have any effect on the intrinsic properties of the Fourier transform. It seems that the characteristic function in probability theory has evolved with the definition as in (4.5.23) while in virtually all other applications of the Fourier transform, e^{-itx} is employed in the integrand. In order to be consistent with the classic treatises on probability theory, e.g. Feller, we will conform to this distinction.

For a discrete random variable, the characteristic function is defined by

$$\phi(t) = E(e^{itX}) = \sum_{j=0}^{N} e^{itx_j}p(x_j) \tag{4.5.24}$$

The characteristic function has the following properties:

$$\phi(0) = 1$$
$$|\phi(t)| \leq 1$$
$$\phi(-t) = \phi^*(t)$$

where $\phi^*(t)$ is the complex conjugate of $\phi(t)$.

The characteristic function can also be used to determine the moments of a distribution. Consider the discrete random variable X for which $\phi(t)$ is given by (4.5.24). Differentiating (4.5.24) r times,

$$\begin{aligned}\frac{d^r\phi}{dt^r} &= \sum_{j=0}^{n} p(x_j)i^r x_j^r e^{itx_j} \\ &= E(i^r X^r e^{itX})\end{aligned} \tag{4.5.25}$$

Similarly, for a continuous random variable X,

$$\begin{aligned}\frac{d^r\phi}{dt^r} &= \int_{-\infty}^{\infty} i^r x^r p(x)e^{itx}\,dx \\ &= E(i^r X^r e^{itX})\end{aligned} \tag{4.5.26}$$

Thus the relation analogous to (4.5.17) is

$$\mu'_r = i^{-r}\frac{d^r\phi(t)}{dt^r}\bigg|_{t=0} \tag{4.5.27}$$

EXAMPLE 4.8. Determine the characteristic function of a variable which assumes the values ± 1 with probabilities $\frac{1}{2}$. Using (4.5.24),

$$\begin{aligned}\phi(t) &= 0.5e^{-it} + 0.5e^{it}\\ &= \cos t\end{aligned}$$

EXAMPLE 4.9. If X is a continuous random variable with density

$$p(x) = \begin{cases} 0, & x < 0 \\ 1, & 0 \leq x \leq 1 \\ 0, & x > 1 \end{cases}$$

determine its characteristic function. Using (4.5.23),

$$\phi(t) = \int_{-\infty}^{\infty} p(x)e^{itx}\,dx = \int_0^1 e^{itx}\,dx = \frac{e^{it} - 1}{it}$$

EXAMPLE 4.10. Consider a discrete random variable X which obeys the Poisson distribution; that is, X can assume the values $x_j = j$, where j is any integer ≥ 0, according to the mass function:

$$p(j) = \frac{\lambda^j}{j!}e^{-\lambda}, \qquad \lambda > 0$$

1. Determine the characteristic function of X:

$$\begin{aligned}\phi(t) &= \sum_{j=0}^{\infty} e^{itj}\frac{\lambda^j}{j!}e^{-\lambda}\\ &= e^{-\lambda}\sum_{j=0}^{\infty}\frac{(\lambda e^{it})^j}{j!}\\ &= e^{-\lambda}e^{\lambda e^{it}} = \exp[\lambda(e^{it} - 1)]\end{aligned}$$

2. Determine μ_1' and μ_2':

$$\mu_1' = \frac{1}{i}\frac{d\phi}{dt}\bigg|_{t=0}$$

$$\frac{d\phi}{dt} = \lambda i \exp(it)\exp[\lambda(e^{it} - 1)]$$

$$\frac{d\phi}{dt}\bigg|_{t=0} = \lambda i$$

$$\therefore \quad \mu_1' = \lambda$$

$$\mu_2' = -\frac{d^2\phi}{dt^2}\bigg|_{t=0}$$

$$\frac{d^2\phi}{dt^2} = \lambda i^2 \exp(it)\exp[\lambda(e^{it} - 1)][\lambda\exp(it) + 1]$$

$$\therefore \quad \mu_2' = \lambda(\lambda + 1)$$

EXAMPLE 4.11. Find the characteristic function and μ_1' and μ_2' of a continuous random variable X obeying a normal distribution:

$$p(x) = \frac{1}{\sqrt{2\pi}}e^{-x^2/2}$$

We have

$$\phi(t) = \frac{1}{\sqrt{2\pi}} \int_{-\infty}^{\infty} \exp(itx) \exp\left(\frac{-x^2}{2}\right) dx$$
$$= \exp\left(\frac{-t^2}{2}\right)$$

It can easily be shown that

$$\mu_1' = 0$$
$$\mu_2' = 1$$

Define the *joint characteristic function* for the two variables X and Y as the double Fourier transform of the joint density:

$$\begin{aligned}\phi_{XY}(t_1, t_2) &= E[e^{i(t_1X + t_2Y)}] \\ &= \int_{-\infty}^{\infty}\int_{-\infty}^{\infty} e^{i(t_1x + t_2y)} p_{XY}(x, y)\, dx\, dy\end{aligned} \tag{4.5.28}$$

If X and Y are independent, the joint characteristic function is equal to the product of the marginal characteristic functions:

$$\phi_{XY}(t_1, t_2) = \phi_X(t_1)\phi_Y(t_2) \tag{4.5.29}$$

If (4.5.29) holds for all values of t_1 and t_2, then X and Y must be independent. If we let $t_1 = t_2 = t$, then we can obtain the characteristic function for $Z = X + Y$:

$$\phi_{XY}(t, t) = E[e^{it(X+Y)}] = E[e^{itZ}] \tag{4.5.30}$$

Results (4.5.29) and (4.5.30) can easily be extended to n random variables. We can compute the marginal characteristic functions from the joint characteristic function by simply letting values of $t = 0$. Specifically,

$$\begin{aligned}\phi_{XY}(t_1, 0) &= \int_{-\infty}^{\infty} e^{it_1x} p_X(x)\, dx \\ &= E(e^{it_1X})\end{aligned} \tag{4.5.31}$$

since if $t_2 = 0$, the integration with respect to y in (4.5.28) can be performed.

As in the case of one random variable, the joint characteristic function can be used for the calculation of moments. Let the exponential in a double Maclaurin series in powers of t_1 and t_2 be

$$\begin{aligned}e^{i(t_1x + t_2y)} &= 1 + i(t_1x + t_2y) - \frac{1}{2!}(t_1x + t_2y)^2 \\ &\quad + \frac{i}{3!}(t_1x + t_2y)^3 - \cdots\end{aligned} \tag{4.5.32}$$

Then

$$\begin{aligned}\phi_{XY}(t_1, t_2) &= 1 + it_1E(X) + it_2E(Y) - \frac{1}{2!}t_1^2E(X^2) \\ &\quad - t_1t_2E(XY) - \frac{1}{2!}t_2^2E(Y^2) + \cdots\end{aligned} \tag{4.5.33}$$

Therefore, if $\phi_{XY}(t_1, t_2)$ is known explicitly in terms of t_1 and t_2, it can either be expanded in a double Maclaurin series or differentiated successively to determine both marginal moments such as $E(X^2)$ and $E(Y^2)$ and product moments such as $E(XY)$. A knowledge of the joint characteristic function will provide us with not only those moments obtainable from the marginal characteristic functions but also with the product moments $E(X^l Y^m)$, which are not generally obtainable from marginal properties.

Sometimes it is convenient to use a set of parameters other than the moments. Let us consider the function

$$\Psi(t) = \ln \phi(t) \tag{4.5.34}$$

Expanding $\phi(t)$ in a power series in the neighborhood of $t = 0$,

$$\phi(t) = 1 + \sum_{r=1}^{\infty} \frac{\mu'_r}{r!}(it)^r \tag{4.5.35}$$

Let us denote the right-hand side by $1 + v$. Then

$$\begin{aligned} \Psi(t) &= \ln \phi(t) = \ln (1 + v) \\ &= v - \frac{v^2}{2} + \frac{v^3}{3} - \cdots \end{aligned} \tag{4.5.36}$$

We write $\Psi(t)$ as

$$\Psi(t) = \sum_{r=1}^{\infty} \frac{\kappa_r}{r!}(it)^r \tag{4.5.37}$$

From (4.5.35) and (4.5.37) we obtain

$$\begin{aligned} \phi(t) &= 1 + \sum_{r=1}^{\infty} \frac{\mu'_r}{r!}(it)^r = \exp\left[\sum_{r=1}^{\infty} \frac{\kappa_r}{r!}(it)^r\right] \\ &= 1 + \sum_{r=1}^{\infty} \frac{\kappa_r}{r!}(it)^r + \frac{1}{2!}\left[\sum_{r=1}^{\infty} \frac{\kappa_r}{r!}(it)^r\right]^2 + \cdots \end{aligned} \tag{4.5.38}$$

The coefficients κ_r are called the *cumulants* or *semiinvariants* of the random variable X. To express the cumulants in terms of the moments or the moments in terms of the cumulants, we compare successively the coefficients of $(it)^r$ for particular values of r in (4.5.38). Doing this we obtain

$$\begin{aligned} \kappa_1 &= \mu'_1 & & \\ \kappa_2 &= \mu'_2 - \mu'^2_1 & \kappa_2 &= \mu_2 \\ \kappa_3 &= \mu'_3 - 3\mu'_1\mu'_2 + 2\mu'^3_1, & \kappa_3 &= \mu_3 \\ \kappa_4 &= \mu'_4 - 3\mu'^2_2 - 4\mu'_1\mu'_3 + 14\mu'^2_1\mu'_2 - 8\mu'^4_1, & \kappa_4 &= \mu_4 - 3\mu_2^2 \end{aligned} \tag{4.5.39}$$

and

$$\begin{aligned} \mu'_1 &= \kappa_1 \\ \mu'_2 &= \kappa_2 + \kappa_1^2 \\ \mu'_3 &= \kappa_3 + 3\kappa_1\kappa_2 + \kappa_1^3 \\ \mu'_4 &= \kappa_4 + 3\kappa_2^2 + 4\kappa_1\kappa_3 + 4\kappa_1^2\kappa_2 + \kappa_1^4 \end{aligned} \tag{4.5.40}$$

4.6 Common Probability Distributions

We shall now present several of the more common discrete and continuous probability distributions that arise in the analysis of physical systems. The distributions are summarized in Tables 4.1 and 4.2, and the distributions are shown in Figure 4.4. More complete discussions of these distributions may be found in the references at the end of the chapter. This section will be devoted to exercises verifying some of the properties in Tables 4.1 and 4.2 and exploring some of the general properties of the distributions.

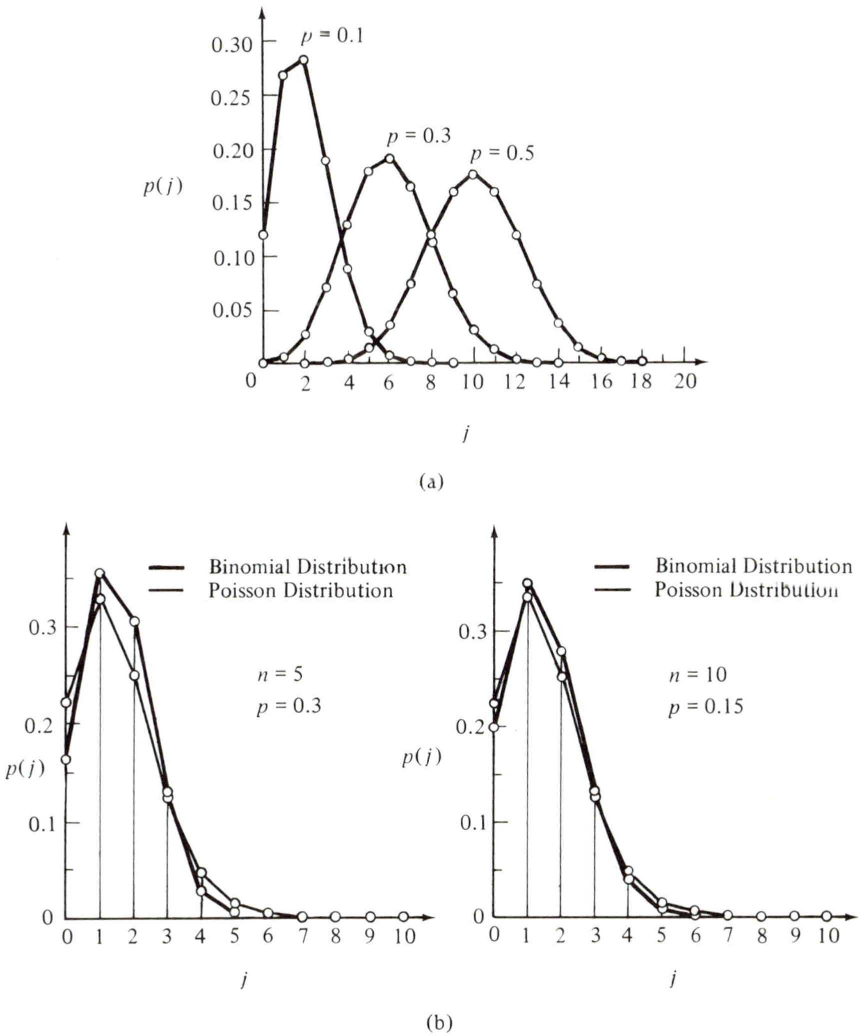

*See page 178 for complete caption for Figure 4.4.

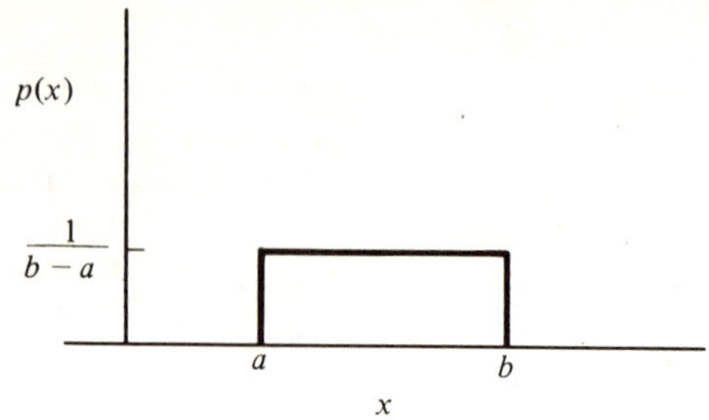

(c)

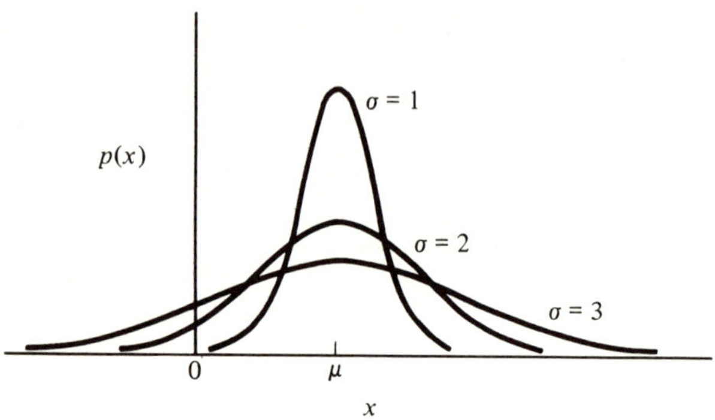

(d)

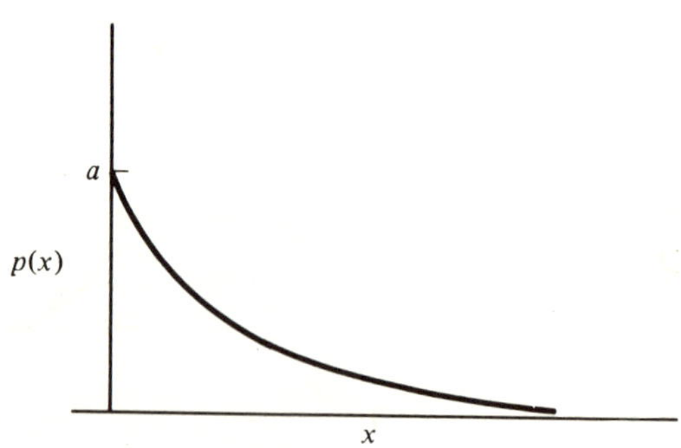

(e)

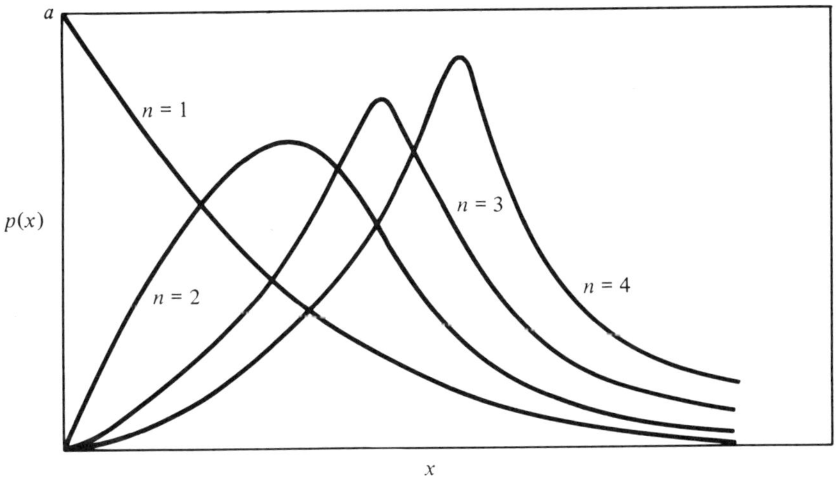

(f)

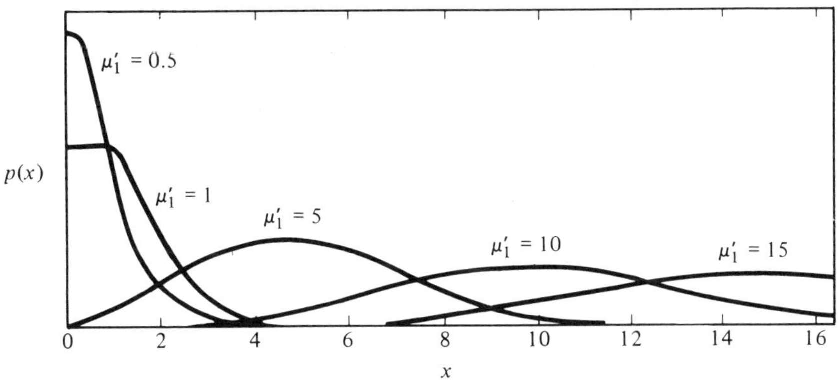

(g)

*See page 178 for complete caption for Figure 4.4.

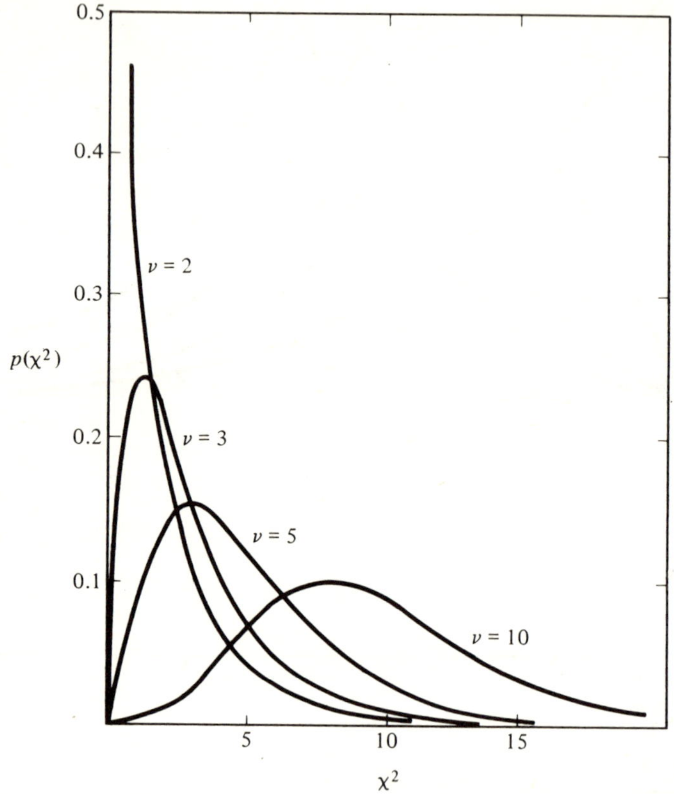

(h)

Figure 4.4. (a) Binomial distribution as a function of p. (Reprinted from M. Fisz, *Probability Theory and Mathematical Statistics*, 3rd ed., John Wiley & Sons, Inc., New York, 1963, p. 133. Reprinted by permission of the copyright owner.) (b) Comparison of Binomial and Poisson distributions for $n = 5$ and 10 with $p = 0.3$ and $p = 0.15$. (Reprinted from M. Fisz, *Probability Theory and Mathematical Statistics*, 3rd ed., John Wiley & Sons, Inc., New York, 1963, p. 142. Reprinted by permission of the copyright owner.) (c) Uniform distribution. (d) Normal distribution as a function of the standard deviation σ. (e) Exponential distribution. (f) Gamma distribution. (g) Poisson distribution for the continuous random variable X as a function of μ'_1. (h) Chi-square distribution.

TABLE 4.1
SOME COMMON PROBABILITY MASS FUNCTIONS

Name	Probability mass function	Moments	Remarks
Binomial	$p(m) = \binom{n}{m} p^m q^{n-m}$, $m = 0, 1 \dots, n$ $\binom{n}{m} = \frac{n!}{m!(n-m)!}$	$\mu_1' = np$ $\mu_2' = npq + n^2p^2$ $\mu_2 = npq$	The binomial distribution arises when an event has two possible outcomes, the probability of the first outcome (success) being p and the second (failure) being $q = 1 - p$, and the event is repeated n times. $p(m)$ represents the probability that in n trials there will be m successes. Each event is assumed independent of the others, and the probability of success and failure are the same for every trial. An experiment of this type is called a sequence of Bernoulli trials.
Poisson	$p(j) = \frac{\lambda^j}{j!} e^{-\lambda}, \lambda > 0$, $j = 0, 1, 2, \dots$	$\mu_1' = \lambda$ $\mu_2' = \lambda(\lambda + 1)$ $\mu_2 = \lambda$	Consider random events occurring in time, such as radioactive disintegrations or incoming telephone calls to an exchange. Each event is represented by a point on the time axis, and we are interested in the probability that we shall discover exactly j points in an interval of fixed length if the average rate of occurrence of points in that interval is λ. Events occur independently, obtained as the limit of the binomial distribution as n becomes large with $np = \lambda$.
Multinomial	$p(j_1, j_2, \dots, j_r)$ $= \frac{n!}{j_1! \cdots j_r!} p_1^{j_1} \cdots p_r^{j_r}$	$E(X_j) = np_j$ $\sigma^2(X_j) = np_j(1 - p_j)$ $E[(X_j - np_j)(X_l - np_l)]$ $= -np_jp_l$	Perform n random experiments. As a result of each experiment one of the events $A_j, j = 1, 2, \dots, r$, occurs. Let $p_j = \text{Prob}\{A_j\}$. The results of the n experiments are independent. We are interested in the probability that A_1 occurs j_1 times, A_2 occurs j_2 times, $\dots$, A_r occurs j_r times. This is represented by $p(j_1, j_2, \dots, j_r)$, the multidimensional form of the binomial distribution.

TABLE 4.2
SOME COMMON PROBABILITY DENSITY FUNCTIONS

Name	Probability density function	Moments	Remarks
Uniform	$p(x) = \begin{cases} \frac{1}{b-a}, & a \le x \le b \\ 0, & x < a,\ x > b \end{cases}$	$\mu'_r = \frac{b^{r+1} - a^{r+1}}{(r+1)(b-a)}$ $\mu_2 = (b-a)^2/12$ $\mu_3 = 0$ $\mu_4 = (b-a)^4/80$	X has a uniform likelihood of being found anywhere on the segment of the real line between $x = a$ and $x = b$. The simplest case of a continuous random variable.
Normal (Gaussian)	$p(x) = \frac{1}{\sqrt{2\pi}\sigma} \exp\left[-\frac{1}{2}\left(\frac{x-\mu}{\sigma}\right)^2\right]$, $-\infty < x < \infty$	$\mu'_1 = \mu$ $\mu'_2 = \sigma^2 + \mu^2$ $\mu'_3 = 3\sigma^2\mu + \mu^3$ $\mu'_4 = 3\sigma^4 + 6\mu^2\sigma^2 + \mu^4$ $\mu_{2k+1} = 0$, all k $\mu_{2k} = 1 \cdot 3 \cdots (2k-1)\sigma^{2k}$	The distribution approached by the sum of a large number of independent random variables under most conditions, and usually describing experimental errors. Approached as the limit of the binomial and Poisson distributions.
Exponential	$p(x) = ae^{-ax}$, $x \ge 0$	$\mu_r = \frac{r!}{a^r}$	Consider events occurring randomly in time, such as radioactive disintegration or incoming telephone calls to an exchange. Each event is represented by a point on the time axis, and we are interested in the density function for X, the time elapsed between occurrences, where the events occur at an average rate of a per time. Let $p(x)\,dx$ = probability that the time elapsed between two events is between x and $x + dx$. This distribution also describes the residence time density of a particle in an ideally mixed vessel.
Gamma	$p_n(z) = \frac{a^n z^{n-1} e^{-az}}{(n-1)!}$, $z \ge 0$	$\mu'_1 = \frac{n}{a}$	Consider the sum Z of n independent random variables, $Z = X_1 + X_2 + \cdots + X_n$, each exponentially distributed with average rate a.

		$\mu_2' = \dfrac{n(n+1)}{a^2}$ $\mu_2 = \dfrac{n}{a^2}$	Then $p_n(z)$, the distribution of Z, is given by the gamma distribution. For example, Z can represent the residence time of a particle in a series of n ideally mixed vessels.
Poisson	$p_n(x) = \dfrac{(ax)^n}{n!} e^{-ax}$ $x \geq 0$	$\mu_1' = ax$ $\mu_2' = (ax)^2 + ax$ $\mu_3' = (ax)^3 + 3(ax)^2 + ax$ $\mu_4' = (ax)^4 + 6(ax)^3 + 7(ax)^2 + ax$ $\mu_2 = ax$ $\mu_3 = ax$	Consider the probability that exactly n events will occur in a time interval x if the time between individual events is exponentially distributed. In terms of the gamma distribution, if Z_n is the total time taken for n events to occur, then $Z_n > x$ and $Z_{n+1} \leq x$. The probability that exactly n events have occurred in time x is $F_{Z_n}(x) - F_{Z_{n+1}}(x)$, where $F_Z(x)$ is the distribution function for the gamma density. Then $p_n(x)$ for the Poisson distribution is $p_n(x) = F_{Z_n}(x) - F_{Z_{n+1}}(x)$.
Multidimensional normal (multidimensional Gaussian)	$p(\mathbf{x}) = \dfrac{1}{(2\pi)^{n/2} \lvert \mathbf{P} \rvert^{1/2}} \cdot$ $\exp\left[-\dfrac{1}{2}(\mathbf{x} - \boldsymbol{\mu})^T \mathbf{P}^{-1} (\mathbf{x} - \boldsymbol{\mu})\right]$ $-\infty < \mathbf{x} < \infty$	$\boldsymbol{\mu}_1' = \boldsymbol{\mu}$ $\boldsymbol{\mu}_2 = \mathbf{P}$	A generalization of the normal distribution to n random variables $\mathbf{X} = [X_1, X_2, \ldots, X_n]$. If $\mathbf{P}$ is a diagonal matrix, the components of $\mathbf{x} - \boldsymbol{\mu}$ are statistically independent since $p(\mathbf{x})$ may be factored into a product of n scalar normal distributions.
χ^2 (Chi-square)	$p(\chi^2) = \dfrac{1}{(2)^{\nu/2}\Gamma(\nu/2)} (\chi^2)^{(\nu/2)-1} e^{-\chi^2/2}$, $0 < \chi^2 < \infty$	$\mu_1' = \nu$ $\mu_2' = 2\nu$	Consider the set of ν independent, normally distributed random variables $X_1, X_2, \ldots, X_\nu$, each of which is characterized by the set of parameters $(\mu_{X_1}, \sigma_{X_1}^2), \ldots, (\mu_{X_\nu}, \sigma_{X_\nu}^2)$. We form the new variables $Y_i^2 = (X_i - \mu_{X_i})^2/\sigma_{X_i}^3$ and let $\chi^2 = Y_1^2 + Y_2^2 + \cdots + Y_\nu^2$. Then the probability density for χ^2 is at the left. The χ^2 distribution is widely used in statistical analysis for testing independence of variables and testing the fit of probability distributions to experimental data.

EXAMPLE 4.12. Let us derive the mean and variance of the binomial distributions shown in Table 4.1. We can describe a sequence of n Bernoulli trials in terms of n independent random variables $X_1, X_2, \ldots, X_n$, where $X_i = 1$ if the ith trial is successful and $X_i = 0$ if unsuccessful. Each variable X_i has the probability mass function

$$\text{Prob}\{X_i = 1\} = p(1) = p$$
$$\text{Prob}\{X_i = 0\} = p(0) = q = 1 - p$$

The total number of successes in n trials is simply the sum of the X_is,

$$Z = X_1 + X_2 + \cdots + X_n$$

since $X_i = 0$ if the ith trial is a failure.

We can determine the mean and variance of the number of successes Z. The mean of Z is simply the sum of the means, which are all identical to that of X_1:

$$E(Z) = nE(X_1)$$

Since the X_is are independent, the variance of Z is the sum of the individual variances which are all equal to that of X_1. Since $E(X_1) = 0 \cdot q + 1 \cdot p = p$,

$$E(Z) = np$$

Also, $E[(X_1 - p)^2] = p^2q + q^2p = pq$. Thus

$$E[(Z - np)^2] = npq$$

Let us also obtain the moments of Z from a generating function, $E(e^{-sZ})$, where

$$E(e^{-sZ}) = E(e^{-s(X_1+X_2+\cdots+X_n)})$$
$$= E(e^{-sX_1})E(e^{-sX_2}) \cdots E(e^{-sX_n})$$

since the Xs are independent. However, since the generating functions for each of the variables are identical,

$$E(e^{-sZ}) = [E(e^{-sX_1})]^n$$

Now

$$E(e^{-sX_1}) = e^{-s\cdot 0}p(0) + e^{-s\cdot 1}p(1)$$
$$= q + pe^{-s}$$

so that

$$E(e^{-sZ}) = (q + pe^{-s})^n$$

Using the binomial theorem,

$$E(e^{-sZ}) = 1 - nps\left(-\frac{ps^2}{2}\right) + \frac{n(n-1)}{2}\left(ps - \frac{ps^2}{2}\right)^2 - \cdots$$
$$= 1 - nps + (npq + n^2p^2)\frac{s^2}{2} - \cdots$$

Identifying the coefficients of the terms as the noncentral moments, we confirm that

$$E(Z) = np$$
$$E(Z^2) = npq + n^2p^2$$

where $E[(Z - np)^2] = npq$.

EXAMPLE 4.13. Consider the random variable X_n satisfying the binomial distribution,

$$p(j) = \frac{n!}{j!(n-j)!}p^j(1-p)^{n-j}, \qquad j = 0, 1, 2, \ldots, n \tag{A}$$

if

$$\lambda = np, \qquad \lambda > 0 \tag{B}$$

Then as n becomes large, let us demonstrate that $p(j)$ approaches the *discrete Poisson distribution* (recall Example 4.3):

$$p(j) = \frac{\lambda^j}{j!}e^{-\lambda} \tag{C}$$

To demonstrate this, we write (A), using (B), as

$$\begin{aligned} p(j) &= \frac{n!}{j!(n-j)!}\left(\frac{\lambda}{n}\right)^j\left(1 - \frac{\lambda}{n}\right)^{n-j} \\ &= \frac{\lambda^j}{j!}\left(1 - \frac{\lambda}{n}\right)^n \frac{n(n-1)\cdots(n-j+1)}{n^j}\frac{1}{[1-(\lambda/n)]^j} \\ &= \frac{\lambda^j}{j!}\left(1 - \frac{\lambda}{n}\right)^n \frac{1\cdot[1-(1/n)]\cdots\{1-[(j-1)/n]\}}{[1-(\lambda/n)]^j} \end{aligned} \tag{D}$$

Noting that

$$\lim_{n\to\infty}\left(1 - \frac{\lambda}{n}\right)^n = e^{-\lambda}$$

and

$$\lim_{n\to\infty}\frac{1\cdot[1-(1/n)]\cdots\{1-[(j-1)/n]\}}{[1-(\lambda/n)]^j} = 1$$

we obtain (C). We refer the reader to Figure 4.4(b) in which the binomial and Poisson distributions are compared for $n = 5$ and 10 and $p = 0.3$ and 0.15, so that in each case $\lambda = np = 1.5$. As n increases, the two distributions will become almost indistinguishable.

EXAMPLE 4.14. *The Gamma Distribution.* Let us examine some of the properties of the gamma distribution. First, let us derive the density function for the gamma distribution, $p(z)$, assuming that Z is the sum of n independent, exponentially distributed random variables. Since

$$E(e^{-sX_1}) = \int_0^\infty ae^{-sx}e^{-ax}\,dx = \frac{a}{s+a} \tag{A}$$

then if $Z = X_1 + X_2 + \cdots + X_n$,

$$\psi_Z(s) = E(e^{-sZ}) = \frac{a^n}{(s+a)^n} \tag{B}$$

Inverting, we obtain

$$p(z) = \frac{a^n z^{n-1}e^{-az}}{(n-1)!} \tag{C}$$

To see why we call this the gamma distribution, we note that the gamma function is defined by

$$\Gamma(n) = \int_0^\infty x^{n-1}e^{-x}\,dx \tag{D}$$

The integral is uniformly convergent with respect to n (n need not be an integer) and therefore $\Gamma(n)$ is a continuous function. Integrating by parts, we find that

$$\Gamma(n + 1) = n\Gamma(n) \tag{E}$$

In particular, if n is an integer, $\Gamma(n + 1) = n!$ Thus an alternative form for $p(z)$ is

$$p(z) = \frac{a^n z^{n-1} e^{-az}}{\Gamma(n)} \tag{F}$$

4.7 Transformations of Random Variables

We are often interested in obtaining the probability density of a random variable Y which is some given function of a random variable X which has a known probability density $p_X(x)$. Let us say that we want to compute

$$Y = f(X) \tag{4.7.1}$$

The unknown density is written as $p_Y(y)$. If X changes its value by dx and the corresponding change in Y is dy, then the probability of finding X between x and $x + dx$ is the same as observing Y between the corresponding y and $y + dy$. Therefore

$$p_X(x)\,dx = p_Y(y)\,dy \tag{4.7.2}$$

which really should be written as

$$p_X(x)\,|dx| = p_Y(y)\,|dy| \tag{4.7.3}$$

since the probabilities are equal to the magnitudes of the areas under dx and dy. Thus

$$p_Y(y) = \frac{p_X(x)}{|dy/dx|} = \frac{p_X(x)}{|df/dx|} \tag{4.7.4}$$

Since $p_Y(y)$ is a function of y, the right-hand side of (4.7.4) must be expressed in terms of y. Assuming that (4.7.1) can be inverted, we can express x in terms of y by $x = f^{-1}(y)$ in (4.7.4). However, if a given value of Y corresponds to more than one value of X, (4.7.4) must be modified.

Consider the case in which $y = f(x)$ has two roots, x_1 and x_2. Let us then write

$$y = f_1(x_1), \qquad y = f_2(x_2)$$

where there is a one-to-one correspondence between x and y in each of these functions. Thus x_1 and x_2 are single-valued functions of y. Let the inverse relationship be denoted by $x_1 = f_1^{-1}(y)$ and $x_2 = f_2^{-1}(y)$. It follows that Y assumes values between y and $y + dy$ when x_1 lies in the interval $(x_1, x_1 + dx_1)$ and x_2 lies in the interval $(x_2, x_2 + dx_2)$. Thus

$$p_Y(y)\,|dy| = p_X(x_1)\,|dx_1| + p_X(x_2)\,|dx_2| \tag{4.7.5}$$

and

$$p_Y(y) = \frac{p_X(x_1)}{|df_1/dx_1|} + \frac{p_X(x_2)}{|df_2/dx_2|} \tag{4.7.6}$$

We can extend these relations to joint probability densities. Assume that we have the two random variables X and Y with a joint density function $p_{XY}(x, y)$. Let U and V be two other random variables related to X and Y by

$$U = g(X, Y), \qquad V = h(X, Y) \tag{4.7.7}$$

If U and V are both single-valued functions of X and Y,

$$p_{UV}(u, v)|du\,dv| = p_{XY}(x, y)|dx\,dy| \tag{4.7.8}$$

or

$$p_{UV}(u, v) = \frac{p_{XY}(x, y)}{|du\,dv/dx\,dy|} \tag{4.7.9}$$

The relation between the area elements in the two systems can be expressed as

$$du\,dv = J\left(\frac{u, v}{x, y}\right) dx\,dy \tag{4.7.10}$$

where J is called the Jacobian of the transformation and is given by

$$J\left(\frac{u, v}{x, y}\right) = \det \begin{bmatrix} \dfrac{\partial g}{\partial x} & \dfrac{\partial g}{\partial y} \\ \dfrac{\partial h}{\partial x} & \dfrac{\partial h}{\partial y} \end{bmatrix} \tag{4.7.11}$$

Therefore

$$p_{UV}(u, v) = \frac{p_{XY}(x, y)}{|J(u, v/x, y)|} \tag{4.7.12}$$

If x and y are multiple valued functions, namely, if $(x_1, y_1), \ldots, (x_k, y_k)$ are solutions of $u = g(x, y)$ and $v = h(x, y)$ for given u and v, then

$$p_{UV}(u, v) = \frac{p_{XY}(x_1, y_1)}{|J(u, v/x_1, y_1)|} + \cdots + \frac{p_{XY}(x_k, y_k)}{|J(u, v/x_k, y_k)|} \tag{4.7.13}$$

These results can, of course, be extended to n random variables.

Assume that we have n random variables $X_1, \ldots, X_n$ with joint density function $p_X(x_1, \ldots, x_n)$. We desire to find the joint density function for the n random variables $Y_1, \ldots, Y_n$, $p_Y(y_1, \ldots, y_n)$, which are related to the Xs by

$$Y_i = f_i(X_1, \ldots, X_n), \qquad i = 1, 2, \ldots, n \tag{4.7.14}$$

We assume that there is a single-valued relation between the Xs and the Ys so that

$$X_i = g_i(Y_1, \ldots, Y_n), \qquad i = 1, 2, \ldots, n \tag{4.7.15}$$

Generalizing (4.7.12),

$$\begin{aligned} p_Y(y_1, \ldots, y_n) &= \frac{p_X(x_1, \ldots, x_n)}{[dy_1 \cdots dy_n/dx_1 \cdots dx_n]} \\ &= \frac{p_X(x_1, \ldots, x_n)}{[J(y_1, \ldots, y_n/x_1, \ldots, x_n)]} \end{aligned} \tag{4.7.16}$$

where the Jacobian

$$J\left(\frac{y_1, \ldots, y_n}{x_1, \ldots, x_n}\right) = \det \begin{bmatrix} \frac{\partial f_1}{\partial x_1} & \frac{\partial f_1}{\partial x_2} & \cdots & \frac{\partial f_1}{\partial x_n} \\ \cdot & & & \\ \cdot & & & \\ \cdot & & & \\ \frac{\partial f_n}{\partial x_1} & \frac{\partial f_n}{\partial x_2} & \cdots & \frac{\partial f_n}{\partial x_n} \end{bmatrix} \tag{4.7.17}$$

EXAMPLE 4.15. Consider the problem of throwing a dart. Assume that the coordinates of a point where the dart hits (X, Y) are independent random variables, normally distributed according to

$$p_X(x) = \frac{1}{\sqrt{2\pi}\sigma} e^{-x^2/2\sigma^2}$$

$$p_Y(y) = \frac{1}{\sqrt{2\pi}\sigma} e^{-y^2/2\sigma^2}$$

Find the probability density $p_{R\Theta}(r, \theta)$ for the two random variables R and Θ, the distance of the hit from the bulls-eye, and the angle of the hit with respect to the x-axis. Thus

$$R = (X^2 + Y^2)^{1/2}, \qquad \Theta = \tan^{-1}\left(\frac{Y}{X}\right)$$

From (4.7.12),

$$p_{R\Theta}(r, \theta) = \frac{p_{XY}(x, y)}{|J(r, \theta/x, y)|}$$

where

$$J\left(\frac{r, \theta}{x, y}\right) = \det \begin{bmatrix} \frac{\partial R}{\partial X} & \frac{\partial R}{\partial Y} \\ \frac{\partial \Theta}{\partial X} & \frac{\partial \Theta}{\partial Y} \end{bmatrix} = \frac{1}{\sqrt{x^2 + y^2}} = \frac{1}{r}$$

Thus

$$p_{R\Theta}(r, \theta) = \frac{r}{2\pi\sigma^2} e^{-(x^2+y^2)/2\sigma^2} = \frac{r}{2\pi\sigma^2} e^{-r^2/2\sigma^2}$$

Since Θ does not appear in $p_{R\Theta}(r, \theta)$, R and Θ must be independent and $p_{R\Theta}(r, \theta) = p_\Theta(\theta) p_R(r)$. Also, $p_\Theta(\theta)$ must be a constant. Since θ varies over 0 to 2π, $p_\Theta(\theta)$ is a constant equal to $1/2\pi$ since

$$\int_0^{2\pi} p_\Theta(\theta)\, d\theta = 1$$

must be satisfied. Thus $p_R(r) = (r/\sigma^2) e^{-r^2/2\sigma^2}$, which is known as the Rayleigh density function, shown in Figure 4.5.

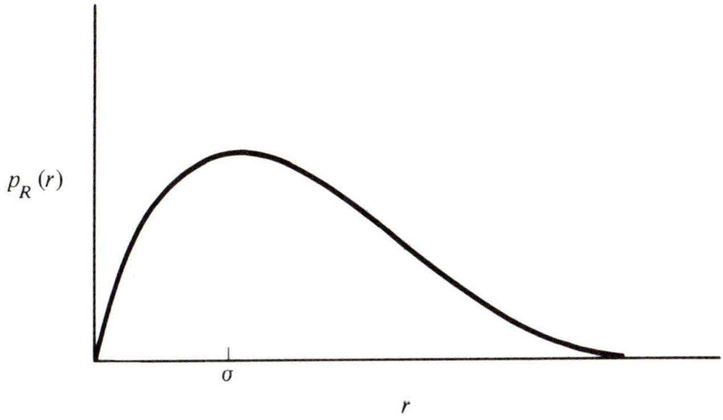

Figure 4.5. Rayleigh density function, $p_R(r) = (r/\sigma^2)\exp[-r^2/2\sigma^2]$.

4.8 The Central Limit Theorem

In this section we shall present the elements of some important results in probability theory pertaining to the distribution of the sum of a large number of random variables. We start by quoting two theorems, the proofs of which we shall not include. These may be found, for example, in Feller [1] and Fisz [2].

THEOREM 4.1 (Lévy-Cramér). Consider a sequence of continuous random variables $X_1, X_2, \ldots$. If the sequence $\{F_n(x), n = 1, 2, \ldots\}$ of distribution functions is convergent† to the distribution function $F(x)$, then the corresponding sequence of characteristic functions $\{\phi_n(t)\}$ converges at every point t $(-\infty < t < +\infty)$ to the function $\phi(t)$, which is the characteristic function of the limit distribution function $F(x)$, and the convergence to $\phi(t)$ is uniform with respect to t in every finite interval on the t-axis.

THEOREM 4.2 (Lévy-Cramér). If the sequence of characteristic functions $\{\phi_n(t)\}$ converges at every point t $(-\infty < t < +\infty)$ to a function $\phi(t)$ continuous in some interval $|t| < \tau$, then the sequence $\{F_n(x)\}$ of corresponding distribution functions converges to $F(x)$, which corresponds to $\phi(t)$.

†The sequence $\{F_n(x)\}$ of distribution functions of the random variables $\{X_n\}$ is called *convergent* if there exists a distribution function $F(x)$ such that at every continuity point of $F(x)$

$$\lim_{n\to\infty} F_n(x) = F(x)$$

$F(x)$ is called the *limit distribution function.*

From Theorems 4.1 and 4.2 we can further state that $\{F_n(x)\}$ is convergent to $F(x)$ if the sequence $\{\phi_n(t)\}$ is convergent at every point t $(-\infty < t < +\infty)$ to a function $\phi(t)$ continuous in some neighborhood $|t| < \tau$ of the origin. The limit function $\phi(t)$ is then the characteristic function of the limit distribution function $F(x)$.

We shall now present the statement and proof of the principal theorem of this section, called the *central limit theorem*. Consider a sequence $\{X_k\}$ of equally distributed, independent random variables whose second-order moment exists. Let $E(X_k) = m$ and $E[(X_k - m)^2] = \text{variance}(X_k) = \sigma^2$, for every k. Consider the random variable Z_n defined by

$$Z_n = X_1 + X_2 + \cdots + X_n \tag{4.8.1}$$

where $E(Z_n) = nm$ and, by the independence of the X_k, $E[(Z_n - nm)^2] = n\sigma^2$. Let

$$\hat{Z}_n = \frac{Z_n - nm}{\sigma\sqrt{n}} \tag{4.8.2}$$

THEOREM 4.3 (Lindeberg-Lévy). If $X_1, X_2, \ldots$ are independent random variables with the same distribution whose standard deviation $\sigma \neq 0$ exists, then the sequence of distribution functions $\{F_n(x)\}$ of $\hat{Z}_n$, given by (4.8.2), satisfies, for every Z,

$$\lim_{n\to\infty} F_n(z) = \frac{1}{2\pi}\int_{-\infty}^{z} e^{-z^2/2}\,dz \tag{4.8.3}$$

that is, the distribution of $\hat{Z}_n$ approaches the normal distribution.

PROOF Let us write (4.8.2) in the form

$$\hat{Z}_n = \frac{1}{\sigma\sqrt{n}}\sum_{k=1}^{n}(X_k - m) \tag{4.8.4}$$

All the random variables $(X_k - m)$ have the same distribution by assumption and thus the same characteristic function $\phi_X(t)$. Thus

$$\phi_{\hat{Z}}(t) = \phi_X\left(\frac{t}{\sigma\sqrt{n}}\right)^n \tag{4.8.5}$$

Since $E(X_k - m) = 0$ and $E[(X_k - m)^2] = \sigma^2$, we can expand $\phi_X(t)$ in a Maclaurin series about $t = 0$ as

$$\phi_X(t) = 1 - \tfrac{1}{2}\sigma^2 t^2 + o(t^2) \tag{4.8.6}$$

where $o(t^2)$ denotes terms that obey the condition $\lim_{t\to 0}[o(t^2)/t^2] \longrightarrow 0$. Therefore, substituting (4.8.6) into (4.8.5),

$$\phi_{\hat{Z}}(t) = \left[1 - \frac{t^2}{2n} + o\left(\frac{t^2}{n}\right)\right]^n \tag{4.8.7}$$

where

$$\lim_{n\to\infty} n\cdot o\left(\frac{t^2}{n}\right) = 0 \tag{4.8.8}$$

Taking the logarithm of (4.8.7), we obtain

$$\begin{aligned}\ln \phi_{\hat{Z}}(t) &= n \ln\left[1 - \frac{t^2}{2n} + o\left(\frac{t^2}{n}\right)\right]\\ &\cong n\left[-\frac{t^2}{2n} + o\left(\frac{t^2}{n}\right)\right]\\ &= -\frac{t^2}{2} + n\cdot o\left(\frac{t^2}{n}\right)\end{aligned} \tag{4.8.9}$$

Thus $\lim_{n\to\infty} \ln \phi_{\hat{Z}}(t) = -t^2/2$ or

$$\lim_{n\to\infty} \phi_{\hat{Z}}(t) = e^{-t^2/2} \tag{4.8.10}$$

However, $e^{-t^2/2}$ is simply the characteristic function of a normally distributed random variable. From Theorem 4.2, if $\{\phi_n(t)\} \longrightarrow \phi(t)$, then $\{F_n(x)\} \longrightarrow F(x)$, corresponding to $\phi(t)$, confirming (4.8.3). This completes the proof of Theorem 4.3.

EXAMPLE 4.16. Verify Theorem 4.3 for the binomial distribution.

Let $\{X_n\}$ be a sequence of discrete random variables which obeys the binomial distribution; that is, $Z_n = X_1 + X_2 + \cdots + X_n$, and

$$\text{Prob}\,\{Z_n = r\} = \binom{n}{r} p^r q^{n-r}$$

We know that $E(Z_n) = np$ and $E[(Z_n - np)^2] = npq$. Consider the sequence

$$\hat{Z}_n = \frac{Z_n - np}{\sqrt{npq}}$$

Now, the characteristic function of a single variable X is

$$\phi_X = pe^{it} + q$$

and that of the sum is therefore

$$\phi_Z = (pe^{it} + q)^n$$

To determine $\phi_{\hat{Z}}$, we shall consider for the moment the transformation $Y = aX + b$. We want to relate ϕ_X and ϕ_Y. Then $\phi_Y = e^{ibt}\phi_X(at)$. In particular, if $Y = (1/\sigma)X - (m/\sigma)$, $\phi_Y = \exp\,[-(mit/\sigma)]\phi(t/\sigma)$. Thus

$$\phi_{\hat{Z}} = \exp\left(\frac{-npit}{\sqrt{npq}}\right)\left[p\exp\left(\frac{it}{\sqrt{npq}}\right) + q\right]^n$$

Rewriting,

$$\phi_{\hat{Z}} = \left[q\exp\left(-\frac{pit}{\sqrt{npq}}\right) + p\exp\left(\frac{qit}{\sqrt{npq}}\right)\right]^n$$

Let us expand the two exponentials of the form e^{iz} in the neighborhood of $z = 0$

according to

$$e^{iz} = \sum_{j=0}^{k} \frac{(iz)^j}{j^k} + o(z^k)$$

Thus

$$p \exp\left(\frac{qit}{\sqrt{npq}}\right) = p + it\sqrt{\frac{pq}{n}} - \frac{qt^2}{2n} + o\left(\frac{t^2}{n}\right)$$

$$q \exp\left(\frac{-pit}{\sqrt{npq}}\right) = q - it\sqrt{\frac{pq}{n}} - \frac{pt^2}{2n} + o\left(\frac{t^2}{n}\right)$$

and we have

$$\phi_{\hat{Z}}(t) = \left[1 - \frac{t^2}{2n} + o\left(\frac{t^2}{n}\right)\right]^n$$

or

$$\ln \phi_{\hat{Z}}(t) = n \ln\left[1 - \frac{t^2}{2n} + o\left(\frac{t^2}{n}\right)\right]$$

$$\cong -\frac{t^2}{2} + n \cdot o\left(\frac{t^2}{n}\right)$$

Finally,

$$\lim_{n\to\infty} \ln \phi_{\hat{Z}}(t) = \frac{-t^2}{2}$$

or

$$\lim_{n\to\infty} \phi_{\hat{Z}}(t) = e^{-t^2/2}$$

EXAMPLE 4.17. Show that the gamma distribution, given in Example 4.14, tends to the normal distribution as $n \longrightarrow \infty$. The density function for the gamma distribution is that of the sum of n exponentially distributed random variables:

$$p_Z(z) = \frac{a^n z^{n-1} e^{-az}}{\Gamma(n)}$$

Let us redefine

$$\hat{Z}_n = \frac{Z_n - \mu'_1}{\sigma}$$

$$= \frac{Z_n - n/a}{\sqrt{n/a}}$$

where $\mu'_1 = n/a$ and $\mu_2 = n/a^2$ for the gamma distribution. We shall prove the convergence to the normal distribution in two ways. First, we shall do it directly. We note that $Z = (1/a)\,(n + \hat{Z}\sqrt{n})$, so that

$$p_Z(\hat{z}) = \frac{a}{\Gamma(n)}(n + \hat{z}\sqrt{n})^{n-1} e^{-(n+\hat{z}\sqrt{n})}$$

or

$$\ln p_Z(\hat{z}) = \ln a - \ln \Gamma(n) + (n-1)\ln(n + \hat{z}\sqrt{n}) - (n + \hat{z}\sqrt{n})$$

Recall Stirling's approximation,

$$\ln[(n-1)!] = \ln(2\pi)^{1/2} + (n - \tfrac{1}{2})\ln(n-1) - (n-1)$$

and $\Gamma(n) = (n-1)!$. Thus

$$\ln p_Z(\hat{z}) = \ln a - \ln (2\pi)^{1/2} - (n - \tfrac{1}{2}) \ln (n-1) + (n-1) \ln (n + \hat{z}\sqrt{n}) - (1 + \hat{z}\sqrt{n})$$

We note that

$$-(n - \tfrac{1}{2}) \ln (n-1) = -n \ln (n-1) + \tfrac{1}{2} \ln (n-1)$$

$$(n+1) \ln (n + \hat{z}\sqrt{n}) = n \ln n + n \ln (1 + \hat{z}\sqrt{n}) - \ln n - \ln (1 + \hat{z}\sqrt{n})$$

Using the relation

$$\ln (1+x) = x - \frac{x^2}{2} + \frac{x^3}{3} - \cdots, \qquad |x| < 1$$

where possible, we obtain, after collecting terms,

$$\ln p_Z(\hat{z}) = \ln a - \ln (2\pi)^{1/2} + \frac{1}{2} \ln (n-1) - \ln n - \frac{1}{2}\frac{1}{n} - \frac{1}{2}\hat{z}^2 + \frac{1}{3}\frac{\hat{z}^3}{\sqrt{n}} - \frac{\hat{z}}{\sqrt{n}} + \frac{1}{2}\frac{\hat{z}^2}{n}$$

Now, let $n \longrightarrow \infty$, noting that $\ln [(n-1)/n] \longrightarrow 0$,

$$\lim_{n\to\infty} \ln p_Z(\hat{z}) = \ln a - \ln (2\pi)^{1/2} - \tfrac{1}{2}\hat{z}^2$$

However, since $\sigma = \sqrt{n/a}$, $\ln \sigma = \frac{1}{2} \ln n - \ln a$,

$$\lim_{n\to\infty} \ln (\sigma p_Z(\hat{z})) = -\ln (2\pi)^{1/2} - \tfrac{1}{2}\hat{z}^2$$

Thus

$$\lim_{n\to\infty} \sigma p_Z(\hat{z}) = \frac{1}{\sqrt{2\pi}} e^{-\hat{z}^2/2}$$

completing the demonstration by direct arguments.

Now let us obtain the same result via the route of Theorem 4.3. The characteristic function for the exponential distribution is

$$\phi_X(t) = \frac{a}{a - it}$$

and for the gamma distribution is

$$\phi_Z(t) = \frac{a^n}{(a - it)^n}$$

Therefore

$$\phi_{\hat{Z}}(t) = e^{-it\mu_1'/\sigma} \phi_X\left(\frac{t}{\sigma}\right)$$

$$= e^{-i\sqrt{n}t} \phi_X\left(\frac{at}{\sqrt{n}}\right)$$

$$= e^{-i\sqrt{n}t} \frac{1}{(1 - it/\sqrt{n})^n}$$

Take the limit as $n \longrightarrow \infty$:

$$\lim_{n\to\infty} \ln \phi_{\hat{Z}}(t) = -i\sqrt{n}\,t - n \ln \left(1 - \frac{it}{\sqrt{n}}\right)$$

Note that for large n

$$\ln\left(1 - \frac{it}{\sqrt{n}}\right) = -\frac{it}{\sqrt{n}} + \frac{1}{2}\left(\frac{it}{\sqrt{n}}\right)^2 - \cdots$$

Therefore

$$\lim_{n\to\infty} \ln \phi_{\hat{z}}(t) = -i\sqrt{n}\,t + i\sqrt{n}\,t - \tfrac{1}{2}t^2 + \cdots$$
$$= -\tfrac{1}{2}t^2$$

or

$$\lim_{n\to\infty} \phi_{\hat{z}}(t) = e^{-t^2/2}$$

completing the demonstration. We note the relative ease of using the characteristic function as opposed to the direct method.

The result of this example will be important to us subsequently when we consider chemical reactor models consisting of sequences of well-mixed vessels in series.

Figure 4.6 shows the interrelationships among the distributions we have considered.

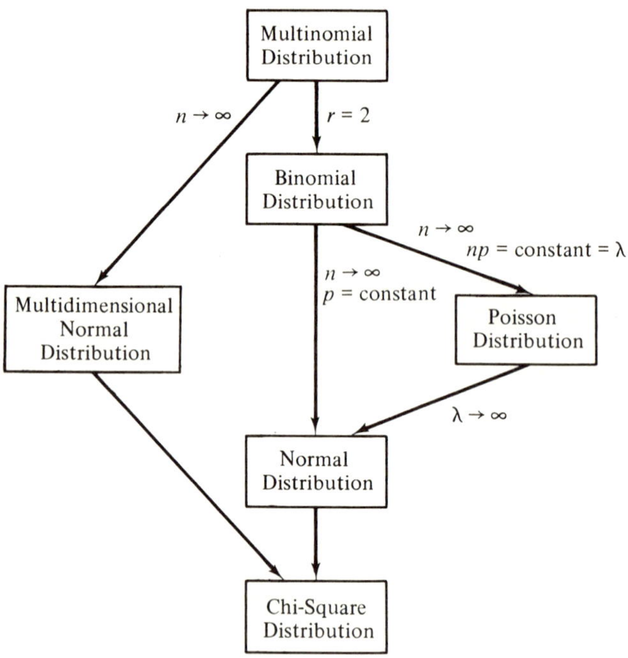

Figure 4.6. Relationships among the common distributions. (Reprinted from H. W. Alexander, *Elements of Mathematical Statistics*, John Wiley & Sons, Inc., New York, 1961, p. 212. Reprinted by permission of the copyright owner.)

REFERENCES

1. Feller, W., *An Introduction to Probability Theory and Its Applications*, vol. I, 3rd ed., Wiley, New York, 1968.

2. Fisz, M., *Probability Theory and Mathematical Statistics*, 3rd ed., Wiley, New York, 1963.

Two additional references which have excellent treatments of probability theory are

McFadden, J. A., *Physical Concepts of Probability*, Van Nostrand Reinhold, New York, 1971.

Papoulis, A., *Probability, Random Variables, and Stochastic Processes*, McGraw-Hill, New York, 1965.

PROBLEMS

4.1 The joint probability density function of two random variables is given by

$$p_{XY}(x, y) = xye^{-(x^2+y^2)/2}, \qquad x \geq 0, y \geq 0$$

a. Find $p_X(x)$, $p_Y(y)$, $p(x \mid y)$, and $p(y \mid x)$.
b. Are X and Y independent?

4.2 The joint probability density function of two random variables is given by

$$p_{XY}(x, y) = \beta \exp\left[-(x^2 + 2xy + 2y^2)\right]$$

a. Determine β.
b. Determine $p_X(x)$, $p_Y(y)$, $p(x \mid y)$, and $p(y \mid x)$.
c. Are X and Y independent?

4.3 Two light bulbs are burning in a room. For bulb 1 we assume that the remaining lifetime obeys an exponential distribution ae^{-bt}, $t \geq 0$, where $b = 2$ days^{-1}. For bulb 2 we know that it will burn out sometime in the next day, with a uniform distribution over the day. The two bulbs operate independently. What is the probability that bulb 1 will burn out before bulb 2?

4.4 For the joint density

$$p_{XY}(x, y) = \frac{1}{2} \exp\left[-\frac{x^2 - 2xy + y^2}{2}\right]$$

determine μ'_{10}, μ'_{01}, μ_{11}, μ_{20}, μ_{02}, and ρ.

4.5 Consider the beta distribution for a continuous random variable X in the range $0 \leq X \leq 1$, which has the density function

$$p(x) = \frac{\Gamma(p+q)}{\Gamma(p)\Gamma(q)} x^{p-1}(1-x)^{q-1}, \qquad 0 \leq x \leq 1$$
$$p > 0; q > 0$$

where p and q are parameters, and

$$\frac{\Gamma(p)\Gamma(q)}{\Gamma(p+q)} = B(p,q) = \int_0^1 x^{p-1}(1-x)^{q-1}\,dx$$

$B(p,q)$ is called the beta function. Show that the mean and variance of the beta distribution are

$$\mu'_1 = \frac{p}{p+q}$$

$$\mu_2 = \frac{pq}{(p+q)^2(p+q+1)}$$

This distribution takes on many shapes depending on the values of p and q. For this reason, it is often used for fitting experimental data.

4.6 The probability density functions for X and Y are

$$p_X(x) = ae^{-x}U(x), \qquad p_Y(y) = be^{y}U(-y)$$

$$p_X(x) = ae^{-|x|}, \qquad p_Y(y) = be^{y}U(-y)$$

Determine the density function for $Z = X + Y$.

4.7 Consider the Cauchy density:

$$p(x) = \frac{1}{\pi}\frac{\lambda}{\lambda^2 + (x-u)^2}, \qquad -\infty < x < \infty, \quad \lambda > 0$$

a. Show that $\int_{-\infty}^{\infty} p(x)\,dx = 1$. *Hint:* Let $y = (x-u)/\lambda$.

b. Find the characteristic function for X when governed by the Cauchy density. First, find the characteristic function for the normalized Cauchy density:

$$p(y) = \frac{1}{\pi}\frac{1}{1+y^2}$$

To do this, consider first the bilateral exponential density

$$p(y) = \tfrac{1}{2}e^{-|y|}$$

and show that the characteristic function corresponding to this density is

$$\phi_1(t) = \frac{1}{1+t^2}$$

We note that since

$$e^{-|y|} = \frac{1}{\pi}\int_{-\infty}^{\infty} \frac{e^{ity}}{1+y^2}\,dy$$

we can write

$$e^{-|t|} = \frac{1}{\pi}\int_{-\infty}^{\infty} \frac{e^{ity}}{1+y^2}\,dy$$

and the characteristic function of the normalized Cauchy density is

$$\phi(t) = e^{-|t|}$$

c. Determine $\phi_Y(t)$ in terms of $\phi_X(t)$ if

$$Y = \frac{X-\mu}{\lambda}$$

d. Use this result to show that for the original Cauchy density

$$\phi_2(t) = \exp(i\mu t - \lambda|t|)$$

e. Show that moments do not exist for the Cauchy distribution.

4.8 The joint distribution of the random variables X and Y is given by the density

$$p_{XY}(x, y) = \begin{cases} \frac{1}{4}[1 + xy(x^2 - y^2)], & \text{for } |x| \leq 1 \text{ and } |y| \leq 1 \\ 0, & \text{for all other points} \end{cases}$$

a. Show that X and Y are dependent. *Hint:* Show that they are not independent.

b. Show that the density of $Z = X + Y$ is

$$p_Z(z) = \begin{cases} \frac{1}{4}(2 + z), & -2 \leq z \leq 0 \\ \frac{1}{4}(2 - z), & 0 < z \leq 2 \\ 0, & |z| > 2 \end{cases}$$

This is called a triangular distribution.

c. Show that the characteristic functions of X, Y, and Z are

$$\phi_X(t) = \phi_Y(t) = \frac{\sin t}{t}$$

$$\phi_Z(t) = \left(\frac{\sin t}{t}\right)^2$$

Note that even though X and Y are not independent,

$$\phi_Z(t) = \phi_X(t)\phi_Y(t)$$

4.9 The expression

$$m_l = \sum_k k(k-1)\cdots(k-1+l)\,p(k)$$

where $p(k) = \text{Prob}\{X = k\}$ is called the *factorial moment* of order l of X.

a. Express m_l by means of the generating function of X.

b. Find the relations between the factorial moments and the ordinary moments.

c. Define the central factorial moment of order l.

4.10 If the random variable X obeys the normal distribution

$$p_X(x) = \frac{1}{\sqrt{2\pi}\sigma} e^{-x^2/2\sigma^2}$$

and $Y = X^2$, determine $p_Y(y)$.

4.11 The joint probability density function of two random variables is given by

$$p_{XY}(x, y) = \beta \exp[-(x^2 + 2xy + 2y^2)]$$

Determine the mean and variance of $Z = XY$.

4.12 If A, B, and C are independent random variables, uniformly distributed over [0, 1], show that the probability that the equation $Ax^2 + 2Bx + C = 0$ will have real roots is $\frac{5}{9}$.

4.13 A chalk mark is made on the outer edge of a tire. When the car stops at random, if we let X denote the height of the mark above the pavement, show that the distribution function for X is

$$\text{Prob}\{X \leq x\} = F(x) = \frac{1}{2} + \frac{1}{\pi}\sin^{-1}(x-1), \qquad 0 \leq x \leq 2$$

What is the density function for X?

4.14 A random variable X is uniformly distributed between zero and 1.

a. Using a generating function, show that the variance and fourth central moment of X are $\mu_2 = \frac{1}{12}$ and $\mu_4 = \frac{1}{80}$ and thus $\mu_4/\mu_2^2 = 1.8$.

b. Let X_1 and X_2 be two independent measures of X and let $Y = \frac{1}{2}(X_1 + X_2)$. Show by a generating function that the corresponding central moments of Y are $\frac{1}{24}$ and $\frac{1}{240}$, respectively, and that the corresponding ratio is 2.4. (Note that using the central limit theorem this ratio should approach 3.)

4.15 "Buffon's Famous Needle Problem." Compute the probability that a needle of length l will fall across a line when dropped at random on a plane ruled with equally spaced parallel lines (spacing $h \geq l$).

4.16 "Game Theory" Payoff Matrix. Calculate the probability that an $m \times n$ matrix $A = (A_{ij})$, $i = 1, \ldots, m$, $j = 1, \ldots, n$, with random entries has a saddle point. A saddle point is an element A_{kl} such that

$$A_{kl} < A_{kj}, \qquad l \neq j \text{ all } j$$

$$A_{kl} > A_{il}, \qquad k \neq i \text{ all } i$$

i.e., the saddle element is the smallest in a row and the largest in a column. Assume at first that the A_{ij}s are uniformly distributed in $[-1, 1]$. Then show that the result is essentially independent of the distribution. Find an asymptotic formula for large m, n.

4.17 Waiting Times. Let the Poisson distribution

$$p_k(t) = \frac{(\lambda t)^k}{k!}e^{-\lambda t}$$

be the probability of k successes in time t. Let T_r, a random variable, be the time of the rth occurrence. Find the probability density for T_r, i.e., p_{T_r}, where

$$\text{Prob}\{t < T_r \leq t + dt\} = p_{T_r}(t)\,dt$$

4.18 Consider two jointly normal random variables X_1 and X_2. Assume that each has zero mean. The joint density function

$$p(x_1, x_2) = \frac{1}{2\pi|\mathbf{P}|^{1/2}}\exp\left(-\frac{1}{2}\mathbf{x}^T\mathbf{P}^{-1}\mathbf{x}\right)$$

where $\mathbf{x} = [X_1, X_2]^T$ and $\mathbf{P}$ is the 2×2 covariance matrix. Show that the new random vector $\mathbf{y} = [Y_1, Y_2]$ given by

$$\mathbf{y} = \mathbf{Ax}$$

where $\mathbf{A}$ is an arbitrary 2×2 matrix, is also normal. Compute the elements of the covariance matrix of $\mathbf{y}$ in terms of the elements of $\mathbf{P}$ and $\mathbf{A}$.

Stochastic Mathematical Models 5

A process may exhibit random characteristics for a number of reasons:

1. The actual physical process is stochastic, as when molecular phenomena are observed on a macroscopic scale, such as radioactive decay.
2. The process is based on macroscopic fluctuations, our lack of information of which makes the process appear random, such as turbulent flow.
3. Our concept of a model for the process does not adequately represent the process.
4. The process depends on human behavior, such as the diffusion of news in a social group.

By a *stochastic mathematical model* we mean a mathematical representation of a stochastic phenomenon. Stochastic models will vary depending on the nature of the random variables which characterize the phenomenon. The variables may be discrete or continuous and they may vary continuously with time or only assume values at certain instants of time. Our objective in formulating a stochastic model is to be able to represent any regular features of the process as concisely and exactly as possible. Although most of the applications of stochastic models have been in the physical and biological sciences and engineering, problems in the social sciences are now being modeled as stochastic processes. In the question of labor mobility, for instance, we cannot predict when a man will decide to change his job. The process must be described probabilistically. A similar example involves the diffusion of news and rumors through a community, a process depending on chance encounters between members of the group.

The primary desired quantities from a stochastic model are (1) the mean values of the state variables as a function of time, (2) the variance of the state variables as a function of time, and (3) the degree of interdependence existing among the various state variables. Let us now proceed to be more specific about the types of systems of interest.

Consider a set of random variables $X_1, X_2, \ldots$. Instead of treating these as separate random variables, we may consider them as the values of a single random variable X at a sequence of times. For example, if X denotes the concentration of tracer emerging from a chemical reactor, X_1, X_2, X_3 could represent the concentrations measured at $t = t_1$, $t = t_2$, and $t = t_3$. This sequence of random variables can be called a *random sequence* or a *stochastic process*. Note that we now associate the concept of *time* with the evolution of the random variable X. In the sequence, $X_1, X_2, \ldots$, the index can be envisioned to play the role of time, since the successive values of X are ordered by the index. Thus we now think of the random variable X as a function of the parameter t, which, for convenience, we call time.

There are two possibilities for the dependence of X on t. First, X may assume values at only discrete values of t, such as illustrated above, X_1, X_2, X_3, etc. In this case, X_t is called a *discrete time* stochastic process. If the random variable X can be observed continuously as a function of time t, $X(t)$ is called a *continuous time* stochastic process, the second possibility. In addition, the random variable X may itself be a discrete or continuous random variable. Since X represents the *state* of the stochastic process, these two cases are referred to as *discrete state* and *continuous state* stochastic processes, respectively. Therefore there are four possible combinations of X (state) and t (index parameter) classifications:

1. Discrete state, discrete time stochastic processes.
2. Discrete state, continuous time stochastic processes.
3. Continuous state, discrete time stochastic processes.
4. Continuous state, continuous time stochastic processes.

These are illustrated in Table 5.1.

An example of a discrete state, discrete time stochastic process is the following. An unbiased coin is tossed repeatedly, one point awarded for each head and one point subtracted for each tail. Let X_k denote the number of points accumulated after k tosses. The random variable X_k can assume only integer values, and so X_k is a discrete random variable. Also, the index k

TABLE 5.1
CLASSIFICATION AND COMMON EXAMPLES OF STOCHASTIC PROCESSES

	Discrete State	Continuous State
Discrete Time	Sequence of coin tosses Random walk on a line, in the plane, and in three dimensions Number of persons in a waiting line at 1-minute intervals Social mobility in a family line	Measurement of wind velocity at specified time intervals Position of a particle in a turbulent field at specified time intervals
Continuous Time	Emission of radioactive particles Molecules impinging on a surface Births and deaths in a population Incoming telephone calls to an exchange Number of persons in a waiting line Movement of a tracer particle through a sequence of ideally mixed vessels Labor mobility of a class of people Survival after treatment for cancer Diffusion of news and rumors	Continuous measurement of wind velocity Position of a particle in a turbulent field as a continuous function of time

represents the accumulated number of tosses (and not an actual measure of time) and assumes only integer values. Clearly, the distribution of X_k is dependent on k. For example, the probabilities of $X_1 = 1$ and $X_1 = -1$ are each $\frac{1}{2}$, whereas the probability of $X_2 = 1$ is zero, since it is not possible to have one point after two tosses. The complete probabilistic description of the sequence $X_1, X_2, \ldots, X_n$ is given only by the n-dimensional joint density $p(j_1, j_2, \ldots, j_n)$. If n is very large, the corresponding joint density must be just as large. If the tossing of the coin governs whether we take a step to the right (if a head is tossed) or to the left (if a tail is tossed), the process is called a *random walk on a line* and is described identically to the accumulation of points.

A discrete state, continuous time stochastic process is exemplified by the process of molecules hitting the surface of a catalyst. The molecules arrive at random times and may stick or not, say with equal probability. We can let $X(t)$ denote the total number of molecules on the surface at time t. Therefore X can assume only integer values, whereas time t runs continuously.

If we observe the value of a continuous variable at discrete times, the resulting process is a continuous state, discrete time stochastic process. For example, say we observe the displacement of a particle in a turbulent field at 1-minute intervals. We can let X_k represent the X-coordinate of the displacement of the particle at $t = t_k$, $k = 1, 2, 3, \ldots$.

Finally, if we observe this particle's displacement $X(t)$ continuously as a function of time t, the displacement can be represented as a continuous state, continuous time stochastic process.

To describe continuous time stochastic processes, we must extend the concept of the random variable X to a *random function* $X(t)$. If we observe a random function $X(t)$ over a time interval $[t_a, t_b]$, there are an infinite number of possible values or realizations of $X(t)$. The various possible realizations of $X(t)$ are called *sample functions*, three of which, for $X(t)$ a continuous random variable, are shown in Figure 5.1. The sample functions shown are only three of those which $X(t)$ may assume, in the same way that the values the random variable X may assume are sample values. At any $t = t_1$, the

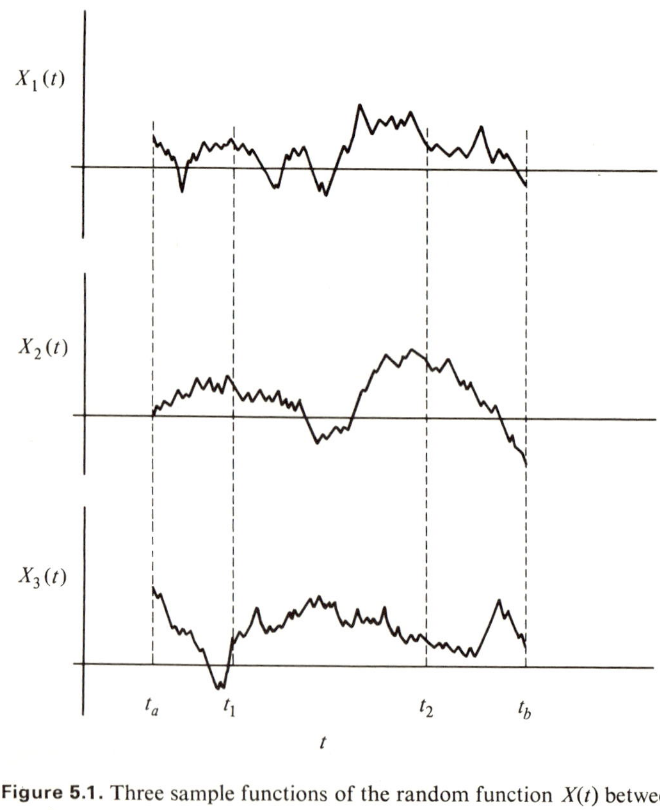

Figure 5.1. Three sample functions of the random function $X(t)$ between t_a and t_b.

value of $X(t_1)$ is a random variable, as shown in Figure 5.1. If we observe $X(t)$ continuously, then we are measuring an infinity of random variables $X(t_k)$, one for each instant of time. Our sample space, then, has infinite dimension. For a complete probabilistic description of a random function $X(t)$, $t_a \leq t \leq t_b$, we must specify all the appropriate joint distributions of instantaneous values. For any sequence of m times in the given interval, $t_a \leq t_1 \leq t_2 \leq \cdots \leq t_m \leq t_b$, the joint distribution of the instantaneous values $X(t_1)$, $X(t_2), \ldots, X(t_m)$ must be given. We shall see that it is important to know the distributions of the instantaneous values in order to reduce the description of such a process to a manageable level.

In this chapter, we shall consider the above four categories of stochastic processes. Of particular importance will be special classes of processes in each category which have been successful in representing physical phenomena.

5.1 Discrete State, Discrete Time Stochastic Processes

Let us consider the stochastic process $X_0, X_1, \ldots, X_t, \ldots,$ where the index parameter t assumes the values 0, 1, 2, . . . and the random variable X assumes the values 1, 2, ... (Note that even if X assumes discrete values other than the integers, we can always relabel these values with integer values.) Thus we say that X_n is in state j if $X_n = j$.

To describe the sequence $X_0, X_1, \ldots, X_{n-1}, X_n$ completely, the joint probability mass function $p(j_0, j_1, \ldots, j_n)$ of all elements in the sequence must in general be specified. Obviously, this involves an enormous amount of information. Fortunately, if the sequence satisfies what is called the *Markov property*, the situation is greatly simplified. In this section we shall consider only those processes which satisfy the Markov property; these processes are called *Markov chains*.

Denote by $p_{ij}^{(n)}$ the conditional probability that at the nth step the system is in state j $(X_n = j)$ given that the system was in state i at the $(n-1)$st step $(X_{n-1} = i)$:

$$p_{ij}^{(n)} = \text{Prob}\{X_n = j \mid X_{n-1} = i\} \tag{5.1.1}$$

A sequence of discrete random variables forms a Markov chain if for any $i, j, n = 1, 2, 3, \ldots,$ the equalities

$$\begin{aligned} p_{ij}^{(n)} &= \text{Prob}\{X_n = j \mid X_{n-1} = i\} \\ &= \text{Prob}\{X_n = j \mid X_{n-1} = i, X_{n-2} = i_{n-2}, \ldots, X_0 = i_0\} \end{aligned} \tag{5.1.2}$$

are satisfied for arbitrary $i_{n-2}, \ldots, i_0$. Thus the Markovian assumption is that only the last state occupied by the process is relevant in determining its future behavior; that is, the probability of making a transition to any state depends only on the state presently occupied. The Markovian assumption is equivalent to specifying that the system has a very limited memory and that events only in the immediate past influence the future. Few real systems possess this property. Nevertheless, the Markov assumption has proved extremely useful in treating systems as diverse as genetics and inventory control [9]. There is no direct way to establish whether a given system is truly Markovian. The ultimate test of the Markov model is its validity in representing real systems.

To define the M-state Markov chain we must specify for each state i and time n the probability of making a transition to any other state j, $p_{ij}^{(n)}$, $1 \leq i, j \leq M$, $n = 0, 1, 2, \ldots$, the *transition probability*. If $p_{ij}^{(n)}$ is independent of n, i.e., the transition probabilities do not depend on the time of the transition, the Markov chain resulting is called *homogeneous*. We shall restrict our attention to this case. Clearly, the transition probabilities must satisfy

$$\begin{aligned} &0 \leq p_{ij} \leq 1, && 1 \leq i, j \leq M \\ &\sum_{j=1}^{M} p_{ij} = 1, && i = 1, 2, \ldots, M \end{aligned} \tag{5.1.3}$$

The $M \times M$ matrix of transition probabilities is called the transition matrix and is denoted by $\mathbf{P}$:

$$\mathbf{P} = \begin{bmatrix} p_{11} & p_{12} & \cdots & p_{1M} \\ p_{21} & p_{22} & \cdots & p_{2M} \\ \vdots & & & \vdots \\ p_{M1} & p_{M2} & \cdots & p_{MM} \end{bmatrix} \tag{5.1.4}$$

Since, from (5.1.3), the sum of the elements of each row of $\mathbf{P}$ must be 1, only $M - 1$ transition probabilities per row need be specified. Thus only $M(M - 1)$ transition probabilities need be specified for the entire M-state Markov chain.

Starting from a known state, say at $n = 0$, at $n = 1$ the state switches randomly to another state; at $n = 2$, it changes again; etc. An important question is, What is the probability $p_{ij}(n)$ that starting from state i the process has reached state j in exactly n transitions? Note carefully the distinction between $p_{ij}^{(n)}$ and $p_{ij}(n)$. Since we shall consider only homogeneous Markov chains, we shall have no further need for $p_{ij}^{(n)}$, and so we shall use only p_{ij} and $p_{ij}(n)$. We want to obtain a general relation for $p_{ij}(n)$ in terms of the transition probabilities p_{ij}. Start by considering a one-step transition between any two states. By definition, $p_{ij}(1) = p_{ij}$. Next, consider two steps that transfer the system from state i to state j. The transition can be envisioned as from state i to state k and then from state k to state j:

$$p_{ij}(2) = \sum_{k=1}^{M} p_{ik} p_{kj} \tag{5.1.5}$$

Continuing, it is easy to generalize (5.1.5) to n steps:

$$p_{ij}(n) = \sum_{k=1}^{M} p_{ik}(m)p_{kj}(n-m), \qquad n = 1, 2, 3, \ldots \tag{5.1.6}$$

where m is an integer, $1 \leq m \leq n - 1$. For recursive calculation of $p_{ij}(n)$ it is convenient to use (5.1.6) with $m = n - 1$. If the state at $n = 0$ is known as i, then $p_{ij}(0) = \delta_{ij}$, the Kronicker delta.

The n-step transition probabilities clearly must satisfy

$$\begin{aligned} &0 \leq p_{ij}(n) \leq 1, \qquad 1 \leq i, j \leq M, \quad n = 0, 1, 2, \ldots \\ &\sum_{j=1}^{M} p_{ij}(n) = 1, \qquad i = 1, 2, \ldots, M, \quad n = 0, 1, 2, \ldots \end{aligned} \tag{5.1.7}$$

The transition probabilities $p_{ij}(n)$ can be considered to be the elements of an $M \times M$ matrix $\mathbf{P}(n)$, called the n-step transition probability matrix. The matrix form of (5.1.6) for $m = n - 1$ is

$$\mathbf{P}(n) = \mathbf{P}(n-1)\mathbf{P}, \qquad n = 1, 2, 3, \ldots \tag{5.1.8}$$

with $\mathbf{P}(0) = \mathbf{I}$, the identity matrix. However, it is easy to see from (5.1.8) that

$$\mathbf{P}(n) = \mathbf{P}^n \tag{5.1.9}$$

The rows of $\mathbf{P}(n)$ specify the probabilities that starting in the particular state corresponding to the row, after n transitions, the process will exist in any of the M states.

If the probability of the initial state being $1, 2, \ldots, M$ is given by the elements of the row vector $\boldsymbol{\pi}(0)$ and that of the final state being $1, 2, \ldots, M$ by the row vector $\boldsymbol{\pi}(n)$,

$$\begin{aligned} \boldsymbol{\pi}(0) &= [\pi_1(0), \ldots, \pi_M(0)] \\ \boldsymbol{\pi}(n) &= [\pi_1(n), \ldots, \pi_M(n)] \end{aligned} \tag{5.1.10}$$

then the fundamental Markov chain evolution relation is

$$\boldsymbol{\pi}(n) = \boldsymbol{\pi}(0)\mathbf{P}^n \tag{5.1.11}$$

The individual probabilities $\pi_i(n)$ are called *state probabilities.*

EXAMPLE 5.1. *Two-State Markov Chains.* Consider a sequence of Bernoulli trials, which we can always represent as a two-state Markov chain. The two states are 1 and 2, say the head or tail of a coin. If the last toss was a head, the probability of achieving a head or tail on the next toss is p or q. If the last toss were a tail, the probability of achieving a head or tail on the next toss is also p or q. Thus $p_{11} = p_{21} = p$ and $p_{12} = p_{22} = q$, and the current state has *no* effect on the outcome of the transition. The transition matrix for a sequence of Bernoulli trials is

$$\mathbf{P} = \begin{bmatrix} p & q \\ p & q \end{bmatrix}$$

Next consider the behavior of a family in its television watching. In a certain time period there are two programs, 1 and 2. If it watched program 1 last week, the family will watch program 1 this week with probability p_{11}, while if it watched program 2 last week, it will watch program 2 this week with probability p_{22}. The history of the viewing behavior of the family forms a two-state Markov chain. Let us see what the probabilities are of the family watching either program after 5 weeks if $p_{11} = 0.8$ and $p_{22} = 0.7$. Then

$$\mathbf{P} = \begin{bmatrix} 0.8 & 0.2 \\ 0.3 & 0.7 \end{bmatrix}$$

$$\mathbf{P}^2 = \begin{bmatrix} 0.7 & 0.3 \\ 0.45 & 0.55 \end{bmatrix}$$

$$\mathbf{P}^3 = \begin{bmatrix} 0.65 & 0.35 \\ 0.525 & 0.475 \end{bmatrix}$$

$$\mathbf{P}^4 = \begin{bmatrix} 0.625 & 0.375 \\ 0.5625 & 0.4375 \end{bmatrix}$$

$$\mathbf{P}^5 = \begin{bmatrix} 0.6125 & 0.3875 \\ 0.58125 & 0.41875 \end{bmatrix}$$

After 5 weeks, the probability that the family will watch program 1 if it originally watched program 2 is 0.58125. We notice that as n increases, the two rows of $\mathbf{P}^n$ appear to be approaching each other. Thus as n gets large, the influence of the initial state is diminishing. As $n \longrightarrow \infty$, it appears that each row will approach 0.6, 0.4. As we shall soon see, the behavior of a Markov chain in the limit of a very large number of transitions is quite an important aspect of Markov chains.

Suppose at week 0 that the probability that the family watched show 1 was 0.1 (reception on this channel was poor that night). Then we would like to ask, What are the probabilities that at the fifth week they will watch programs 1 and 2? At week 0, $\boldsymbol{\pi}(0) = [0.1 \quad 0.9]$, and we want $\boldsymbol{\pi}(5)$. Using (5.1.11),

$$\boldsymbol{\pi}(5) = \boldsymbol{\pi}(0)\mathbf{P}^5$$

we find

$$\begin{aligned} \boldsymbol{\pi}(5) &= [0.1 \quad 0.9] \begin{bmatrix} 0.6125 & 0.3875 \\ 0.58125 & 0.41875 \end{bmatrix} \\ &= [0.584375 \quad 0.415625] \end{aligned}$$

As $n \longrightarrow \infty$, if $\mathbf{P}^n$ approaches a constant value, as it does in this example, then $\boldsymbol{\pi}(n)$ also approaches a constant value. We shall examine this asymptotic behavior shortly.

EXAMPLE 5.2. Determine the transition matrix for a random walk with absorbing barriers. A particle may be at one of the points $1, 2, 3, \ldots, M$ on the x-axis. If it arrives at $x = 1$ or $x = M$, it will remain there forever—thus the term *absorbing barriers*. If a particle is at point i at moment t, it can move to $i + 1$ with probability p and to $i - 1$ with probability $q = 1 - p$. It cannot remain at its current location. Thus the locations of the particle form a homogeneous Markov chain with M

states. The transition probabilities are

$$p_{11} = p_{MM} = 1$$

$$p_{ij} = \begin{cases} p, & j = i + 1 \\ q, & j = i - 1 \\ 0, & \text{otherwise} \end{cases}$$

Thus **P** has the form

$$\mathbf{P} = \begin{bmatrix} 1 & 0 & 0 & 0 & \cdots & 0 & 0 & 0 \\ q & 0 & p & 0 & \cdots & 0 & 0 & 0 \\ 0 & q & 0 & p & \cdots & 0 & 0 & 0 \\ & & \cdot & & & & \cdot & \\ & & \cdot & & & & \cdot & \\ & & \cdot & & & & \cdot & \\ 0 & 0 & 0 & 0 & \cdots & 0 & 0 & 1 \end{bmatrix}$$

EXAMPLE 5.3. Determine the transition matrix for a random walk with two reflecting barriers. The transition probabilities are

$$p_{11} = q = 1 - p$$

$$p_{ij} = \begin{cases} p, & i = 1, 2, 3, \ldots, \quad j = i + 1 \\ q, & i = 2, 3, \ldots, \quad j = i - i \\ 0, & \text{otherwise} \end{cases}$$

and so the transition matrix is

$$\mathbf{P} = \begin{bmatrix} q & p & 0 & 0 & 0 & & & & 0 \\ q & 0 & p & 0 & 0 & \cdots & & & 0 \\ 0 & q & 0 & p & 0 & & & & 0 \\ & & \cdot & & & & & & \cdot \\ & & \cdot & & & & & & \cdot \\ & & \cdot & & & & 0 & q & p \end{bmatrix}$$

The states of a Markov chain can be classified in various ways. State j is *accessible* from state i if for some integer $n \geq 0$, $p_{ij}(n) > 0$, namely, that in a finite number of transitions state j can be reached from state i. If two states are not accessible, it must be true that either $p_{ij}(n) = 0$ or $p_{ji}(n) = 0$, or both relations hold. Two states are said to *communicate* if they are each accessible to each other. We can partition the elements of **P** into classes, called equivalence classes [9]. The states in an equivalence class are those which communicate with each other. It may be possible, starting in one class, to enter another class with positive probability, but it would not be possible to return to the original class. If we could, the two classes would have to form a single class. We call the Markov chain *irreducible* if **P** contains only one class, that is, if all states communicate with each other. The random walk with absorbing barriers has

$$\mathbf{P} = \begin{bmatrix} 1 & 0 & 0 & \cdots & & & 0 \\ q & 0 & p & \cdots & & & 0 \\ \cdot & & & & & & \cdot \\ & & & & & & \cdot \\ \cdot & & & & & & \cdot \\ \cdot & & & & q & 0 & p \\ 0 & \cdots & & & 0 & 0 & 1 \end{bmatrix}$$

and has three classes $\{i = 0\}$, $\{i = 1, 2, \ldots, M - 1\}$, and $\{i = M\}$. We see that it is possible to reach the first and third classes from the second class but not vice versa.

If $p_{ii}(n) > 0$, the state can return to state i after n steps. We then define the period of state i to be the greatest common divisor of all integers $n \geq 1$ for which $p_{ii}(n) > 0$. A Markov chain in which each state has period 1 is called *aperiodic*. We might note that most Markov chains we deal with in practice are aperiodic.

Let us now study the structure of a Markov chain as $n \longrightarrow \infty$. Return for a moment to Example 5.1. We saw that the rows of $\mathbf{P}^n$ were becoming identical as $n \longrightarrow \infty$. This is a common asymptotic behavior of Markov chains, although it does not always happen that the rows become identical. A sufficient condition for such asymptotic behavior is that the process be able to make a transition (have a nonzero transition probability) from any state to any other state. In short, if all the elements of $\mathbf{P}^n$ are nonzero for some n, then the entire Markov chain will converge to a $\mathbf{P}$ with identical rows. A necessary condition is that there exist only one subset of states that must be occupied as $n \longrightarrow \infty$. Formally stated, there must be a single aperiodic recurrent class.

After n steps

$$\pi_j(n) = \sum_{i=1}^{M} \pi_i(0) p_{ij}(n), \qquad j = 1, 2, \ldots, M \tag{5.1.12}$$

However, if, as $n \longrightarrow \infty$, the rows of $\mathbf{P}^n$ become identical, the elements of $\mathbf{P}^n$, $p_{ij}(n)$, become equal to a value p_j that depends only on the column j. Then

$$\begin{aligned} \pi_j &= \sum_{i=1}^{M} \pi_i(0) p_j \\ &= p_j \end{aligned} \tag{5.1.13}$$

The so-called *ergodic probabilities* π_j can be computed easily. Since

$$\boldsymbol{\pi}(n) = \boldsymbol{\pi}(n-1)\mathbf{P} \tag{5.1.14}$$

must always hold, the ergodic probabilities can be computed from

$$\boldsymbol{\pi} = \boldsymbol{\pi}\mathbf{P} \tag{5.1.15}$$

which gives rise to the M simultaneous equations

$$\pi_j = \sum_{i=1}^{M} \pi_i p_{ij}, \qquad j = 1, 2, \ldots, M \tag{5.1.16}$$

which must be solved to determine the π_j subject to

$$\sum_{i=1}^{M} \pi_i = 1 \tag{5.1.17}$$

Given the transition matrix $\mathbf{P}$ for a Markov chain we can readily write down the equations that the ergodic or stationary probabilities must satisfy if they exist. If the Markov chain is such that all the rows of $\mathbf{P}^n$ do not become identical as $n \longrightarrow \infty$, (5.1.16) and (5.1.17) will lead to an indeterminate set of equations.

We can now actually compute the ergodic probabilities for the television-watching family. Equation (5.1.16) leads to the two simultaneous equations

$$\pi_1 = 0.8\pi_1 + 0.3\pi_2$$
$$\pi_2 = 0.2\pi_1 + 0.7\pi_2$$

each of which reduces to

$$0.2\pi_1 = 0.3\pi_2$$

To obtain a unique solution we must invoke (5.1.17), $\pi_1 + \pi_2 = 1$, thereby giving the values $\pi_1 = 0.6$, $\pi_2 = 0.4$.

EXAMPLE 5.4. Let us examine the limiting behavior of a random walk with absorbing barriers as in Example 5.2. Let $M = 3$, so that

$$\mathbf{P} = \begin{bmatrix} 1 & 0 & 0 \\ q & 0 & p \\ 0 & 0 & 1 \end{bmatrix}$$

It is easy to show that $\mathbf{P}^n = \mathbf{P}$. The process must either end up in state 1 or 3. Beginning in state 2, it will go to states 1 and 3 with probabilities p and q, respectively, and remain there forever. We see that

$$\lim_{n \to \infty} \begin{bmatrix} p_{11}(n) = 1 & p_{12}(n) = 0 & p_{13}(n) = 0 \\ p_{21}(n) = q & p_{22}(n) = 0 & p_{23}(n) = p \\ p_{31}(n) = 0 & p_{32}(n) = 0 & p_{33}(n) = 1 \end{bmatrix}$$

Now let us see what happens when we try to compute the asymptotic probabilities for this process. The equations are

$$\pi_1 = \pi_1 + \pi_2 q$$
$$\pi_2 = 0$$
$$\pi_3 = \pi_2 p + \pi_3$$
$$\pi_1 + \pi_2 + \pi_3 = 1$$

These tell us only that $\pi_2 = 0$. The other two equations tell us only that $\pi_1 + \pi_3$

= 1. We cannot obtain a unique solution for $\boldsymbol{\pi}$, which we knew since the rows of the limiting form of **P** are not identical.

A state which has zero probability of occupancy after a large number of transitions is called a *transient state*. The elements of the corresponding column of the limiting n-step transition matrix for a transient state are all zero. States which once entered can never be left are called *absorbing* or *trapping states*.

Let us also examine the limiting behavior of a three-state random walk with reflecting barriers, as in Example 5.3. We have

$$\mathbf{P} = \begin{bmatrix} q & p & 0 \\ q & 0 & p \\ 0 & q & p \end{bmatrix}$$

The ergodic probabilities are computed from

$$\begin{aligned} \pi_1 &= \pi_1 q + \pi_2 q \\ \pi_2 &= \pi_1 p + \pi_3 q \\ \pi_3 &= \pi_2 p + \pi_3 p \\ \pi_1 &+ \pi_2 + \pi_3 = 1 \end{aligned}$$

The result is

$$\begin{aligned} \pi_1 &= \left[1 + \frac{p}{q} + \left(\frac{p}{q}\right)^2\right]^{-1} \\ \pi_2 &= \frac{p}{q}\pi_1 \\ \pi_3 &= \left(\frac{p}{q}\right)^2 \pi_1 \end{aligned}$$

We conclude that the asymptotic state probability vector $\boldsymbol{\pi}$ is unique only when the Markov process contains a single class, such as in the case of the random walk with reflecting barriers. Transient states are excluded from inclusion in a class, and so all states within a class must have nonzero probability of being occupied after many transitions, provided the system has entered that class. These are called *recurrent* states. All states belonging to the same class will have identical rows in $\mathbf{P}^n$ as $n \longrightarrow \infty$. Clearly, an absorbing state constitutes a class by itself. If we cannot obtain a unique solution for $\boldsymbol{\pi}$, then the Markov chain has more than one class.

EXAMPLE 5.5. *A Periodic Markov Chain.* Consider the transition matrix

$$\mathbf{P} = \begin{bmatrix} 0 & 1 & 0 \\ 0 & 0 & 1 \\ 1 & 0 & 0 \end{bmatrix}$$

Thus the process goes from 1 to 2, 2 to 3, and 3 to 1 indefinitely. Intuitively, this is a *periodic* process. It is easy to show that

$$\mathbf{P}^{n+3} = \mathbf{P}^n$$

The period of the process is thus 3.

EXAMPLE 5.6. *A Model for Social Mobility.* Bartholomew [2], in his interesting monograph, considers the problem of social mobility. Let us say we are interested in changes in the occupational makeup of a closed society with time. We measure time in terms of generations, which take positive integer values. Consider the development of a single family line which might be taken as representative of any in the society. Let M be the number of social classes in the society, and let p_{ij} denote the probability that the son of a father in class i is in class j. Thus p_{ij} is the transition probability for change of social class between two generations in the family. We assume that, on the average, each father has only one son, which is consistent with our assumption of a society with a fixed number of men, i.e., a closed society.

Let the probability that the initial progenitor of a family line is in class j at time (generation) zero be $\pi_j(0)$ and the probability that the nth generation is in class j be $\pi_j(n)$. Thus we have a Markov chain describing the dynamics of the society with time. In many applications, the population has been in existence for many generations as the "present" state corresponds to a large number n. It is therefore of considerable practical interest to investigate the behavior of $\pi_j(n)$ as $n \longrightarrow \infty$. As long as all the elements of $\mathbf{P}^n$ are nonzero for some n, the $\pi_j(n)$ approach the single-class ergodic probabilities. All transition matrices we would encounter in this problem would possess this property, since there should always be a nonzero probability that a son could go to any social level regardless of the initial level of his family line. The ergodic probabilities can then be computed from (5.1.16) and (5.1.17).

We have assumed that the development of a society depends only on its initial state and the transition matrix. Of these two features, the initial state has a diminishing influence as time passes. In the long run, the structure of the society is determined only by the transition matrix.

What is now required is a test of the adequacy of the model using actual data on mobility. Glass and Hall, in a study described by Bartholomew, sampled randomly 3500 pairs of fathers and sons in Britain. They classified the members of their sample according to the seven occupational groups listed in Table 5.2, which also gives the estimated transition probabilities. If the Markov chain model is adequate

TABLE 5.2
ESTIMATED TRANSITION PROBABILITIES FOR ENGLAND AND WALES IN 1949
(GLASS AND HALL'S DATA [2])

	Son's class						
Father's class	1	2	3	4	5	6	7
1. Professional and higher administrative	0.388	0.146	0.202	0.062	0.140	0.047	0.015
2. Managerial and executive	0.107	0.267	0.227	0.120	0.206	0.053	0.020
3. Higher-grade supervisory and nonmanual	0.035	0.101	0.188	0.191	0.357	0.067	0.061
4. Lower-grade supervisory and nonmanual	0.021	0.039	0.112	0.212	0.430	0.124	0.062
5. Skilled manual and routine nonmanual	0.009	0.024	0.075	0.123	0.473	0.171	0.125
6. Semiskilled manual	0.000	0.013	0.041	0.088	0.391	0.312	0.155
7. Unskilled manual	0.000	0.008	0.036	0.083	0.364	0.235	0.274

TABLE 5.3
ACTUAL AND EQUILIBRIUM DISTRIBUTIONS OF THE SOCIAL CLASSES IN ENGLAND AND WALES (1949), ESTIMATED FROM GLASS AND HALL'S DATA [2]

Class	Fathers	Sons	Predicted equilibrium
1	0.037	0.029	0.023
2	0.043	0.046	0.042
3	0.098	0.094	0.088
4	0.148	0.131	0.127
5	0.432	0.409	0.409
6	0.131	0.170	0.182
7	0.111	0.121	0.129

and the society has reached equilibrium, we would expect the actual distribution of fathers and sons to be close to the ergodic distribution obtained from (5.1.16). The comparisons appear in Table 5.3.

The differences between the three distributions are not large, although there does appear to be a shift toward the lower classes as we move across the table. If this trend is genuine and not merely the result of sampling fluctuations, it might indicate that the society has not yet reached equilibrium.

A more general concept than periodicity is recurrence. For $n \geq 1$ we let $f_{ii}(n)$ be the probability that, starting from state i, the first return to state i occurs after n transitions. We note that $f_{ii}(1) = p_{ii}$ and that

$$f_{ii}(n) = \sum_{k=0}^{n} f_{ii}(k) p_{ii}(n-k), \qquad n \geq 1$$

where $f_{ii}(0) = 0$. A state i is *recurrent* if and only if the probability of returning to state i at some time is unity:

$$\sum_{n=0}^{\infty} f_{ii}(n) = 1$$

EXAMPLE 5.7. *Absorption Probabilities* [11]. We have established that if states i and j are in the same aperiodic, recurrent class, then $p_{ij}(n) \longrightarrow \pi_j \geq 0$ as $n \longrightarrow \infty$. We now want to consider the special case in which i is nonrecurrent (transient) and j is recurrent. The problem with which we are already familiar that includes both transient and recurrent states is the random walk with absorbing barriers. Clearly, the absorbing barriers are recurrent states, whereas the interior points are transient states, since the probability that a particle will return to states 0 and M (the barriers) once it has reached them is unity. In general, we are interested in the probability that a process beginning in some arbitrary initial state will ultimately end up at 0 or M.

Let $\bar{\pi}_i(n)$ be the probability that the process will enter and thus be absorbed in recurrent class C for the first time at the nth transition, given that the initial state is transient state i in transient class T. Then the probability $\bar{\pi}_i$ that the process will ultimately be absorbed in class C is related to $\bar{\pi}_i(n)$ by

$$\bar{\pi}_i = \sum_{n=1}^{\infty} \bar{\pi}_i(n) \leq 1 \tag{A}$$

where the probability of being absorbed on the first transition is

$$\bar{\pi}_i(1) = \sum_j p_{ij}, \qquad j \text{ in class } C\ (j \in C) \tag{B}$$

and after two or more transitions,

$$\bar{\pi}_i(n) = \sum_j p_{ij}\bar{\pi}_j(n-1), \qquad n \geq 2,\ j \text{ in transient class } T\ (j \in T) \tag{C}$$

Using (C) in (A), we find for absorption in class C that

$$\begin{aligned}\bar{\pi}_i &= \bar{\pi}_i(1) + \sum_{n=2}^{\infty} \bar{\pi}_i(n) \\ &= \bar{\pi}_i(1) + \sum_{n=2}^{\infty} \sum_{j \in T} p_{ij}\bar{\pi}_j(n-1) \\ &= \bar{\pi}_i(1) + \sum_{j \in T} p_{ij} \sum_{n=2}^{\infty} \bar{\pi}_j(n-1) \\ &= \bar{\pi}_i(1) + \sum_{j \in T} p_{ij}\bar{\pi}_j \end{aligned} \tag{D}$$

As an example, consider the gambler's ruin problem, which is essentially a one-dimensional random walk with two absorbing barriers, at 0 and M. Two gamblers compete with state 0 representing bankruptcy for gambler 1 and state M representing bankruptcy for gambler 2. At each time the probability of gambler 1 winning (advancing to the right) is p, while the probability of gambler 2 winning is $q(q = 1 - p)$. We are interested in the probability that starting from state i gamblers 1 and 2 are ruined. We therefore want

$$\alpha_i = \bar{\pi}_i^0 = \text{probability of the process ultimately reaching state } 0$$

$$\beta_i = \bar{\pi}_i^M = \text{probability of the process ultimately reaching state } M$$

Employing (D), we have

$$\begin{aligned}\alpha_1 &= q + p\alpha_2 \\ \alpha_i &= q\alpha_{i-1} + p\alpha_{i+1}, \qquad 2 \leq i \leq M-2 \\ \alpha_{M-1} &= q\alpha_{M-2}\end{aligned} \tag{E}$$

This is a system of $M - 1$ linear difference equations. Assume a solution of the form $\alpha_r = x^r$. Substituting in the middle equations leads to

$$px^2 + q - x = 0 \tag{F}$$

which has the two solutions $x = 1$ and $x = q/p$. Therefore we form the sum

$$\alpha_r = A + B\left(\frac{q}{p}\right)^r \tag{G}$$

the arbitrary constants of which (A and B) are evaluated from the α_1 and α_{M-1} equation of (E). For $q \neq p$ we obtain

$$A = \frac{q^n}{q^n - p^n}, \qquad B = \frac{-p^n}{q^n - p^n}$$

and for $q = p$, $A = 1$ and $B = -1/n$, and so

$$\alpha_r = \begin{cases} \dfrac{(q/p)^n - (q/p)^r}{(q/p)^n - 1}, & q \neq p \\ \dfrac{n-r}{n}, & q = p \end{cases} \tag{H}$$

To determine $\boldsymbol{\pi}(n)$ for a general Markov chain it is necessary to compute $\mathbf{P}^n$. Let us now show how this can be carried out. Assume that the M eigenvalues of $\mathbf{P}$ are distinct and denoted by $\lambda_1, \lambda_2, \ldots, \lambda_M$. Define the right eigenvector $\boldsymbol{\alpha}_j$ of $\mathbf{P}$ by (each $\boldsymbol{\alpha}_j$ is an M-dimensional column vector)

$$\mathbf{P}\boldsymbol{\alpha}_j = \lambda_j \boldsymbol{\alpha}_j, \qquad j = 1, 2, \ldots, M$$

Then define the left eigenvector $\boldsymbol{\beta}_j$ of $\mathbf{P}$ by (each $\boldsymbol{\beta}_j$ is an M-dimensional column vector)

$$\boldsymbol{\beta}_j^T \mathbf{P} = \lambda_j \boldsymbol{\beta}_j^T, \qquad j = 1, 2, \ldots, M$$

where $\boldsymbol{\alpha}_j$ and $\boldsymbol{\beta}_j$ are orthogonal, that is,

$$\boldsymbol{\beta}_j^T \boldsymbol{\alpha}_k = \begin{cases} 1, & j = k \\ 0, & j \neq k \end{cases}$$

Let $\mathbf{H}$ be the matrix whose columns are the right eigenvectors of $\mathbf{P}$:

$$\mathbf{H} = [\boldsymbol{\alpha}_1, \boldsymbol{\alpha}_2, \ldots, \boldsymbol{\alpha}_M]$$

Then $\mathbf{H}^{-1}$ is the matrix whose rows are the left eigenvectors of $\mathbf{P}$:

$$\mathbf{H}^{-1} = \begin{bmatrix} \boldsymbol{\beta}_1^T \\ \boldsymbol{\beta}_2^T \\ \cdot \\ \cdot \\ \cdot \\ \boldsymbol{\beta}_M^T \end{bmatrix}$$

The diagonal matrix whose elements are the eigenvalues of $\mathbf{P}$ is found from

$$\boldsymbol{\Lambda} = \mathbf{H}^{-1}\mathbf{P}\mathbf{H} = \begin{bmatrix} \lambda_1 & & & & 0 \\ & \lambda_2 & & & \\ & & \cdot & & \\ & & & \cdot & \\ & 0 & & & \cdot \\ & & & & \lambda_M \end{bmatrix} \tag{5.1.18}$$

or

$$\mathbf{P} = \mathbf{H}\boldsymbol{\Lambda}\mathbf{H}^{-1}$$

For n steps

$$\mathbf{P}^n = \mathbf{H}\boldsymbol{\Lambda}^n\mathbf{H}^{-1} \tag{5.1.19}$$

Using (5.1.18), we can write $\mathbf{P}^n$ in the form

$$\mathbf{P}^n = \sum_{j=1}^{n} C_j \lambda_j^n (\boldsymbol{\alpha}_j \boldsymbol{\beta}_j^T) \tag{5.1.20}$$

where

$$C_j = (\boldsymbol{\beta}_j^T \boldsymbol{\alpha}_j)^{-1}$$

We can also write (5.1.20) in terms of the individual elements of **P**. Let

$$\boldsymbol{\alpha}_j = \begin{bmatrix} \alpha_{1j} \\ \alpha_{2j} \\ \cdot \\ \cdot \\ \cdot \\ \alpha_{Mj} \end{bmatrix}, \qquad \boldsymbol{\beta}_j = \begin{bmatrix} \beta_{j1} \\ \beta_{j2} \\ \cdot \\ \cdot \\ \cdot \\ \beta_{jM} \end{bmatrix} \tag{5.1.21}$$

and (5.1.20) becomes

$$p_{ik}(n) = \sum_{j=1}^{M} C_j \lambda_j^n \alpha_{ij} \beta_{jk} \tag{5.1.22}$$

where

$$C_j = \frac{1}{\sum_{l=1}^{M} \beta_{jl} \alpha_{lj}} \tag{5.1.23}$$

5.2 Discrete State, Continuous Time Stochastic Processes

We shall now consider stochastic processes, denoted $X(t)$, which depend on a continuous index parameter t but the values of which can assume only discrete values. Therefore we shall extend the concepts of the previous section to the case in which the discrete stochastic process X depends continuously on t. As before, we shall restrict our attention to processes possessing the Markov property.

A discrete process $X(t)$ is a Markov process if for $n = 1, 2, 3, \ldots,$ and for arbitrary $t_m (m = 0, 1, 2, \ldots, n)$, where $t_0 < t_1 \cdots < t_n$, and for arbitrary integers i, j the equality

$$\begin{aligned} \text{Prob}\{X(t_n) = j \mid X(t_{n-1}) = i, X(t_{n-2}) = i_{n-2}, \ldots, X(t_0) = i_0\} \\ = \text{Prob}\{X(t_n) = j \mid X(t_{n-1}) = i\} \end{aligned}$$

is satisfied for any integers $i_{n-2}, \ldots, i_0$. Note that we now refer to $X(t)$ as a Markov *process*, where in the last section X_t was a Markov *chain*.

For $t_1 < t_2$ we define the transition probability function

$$p_{ij}(t_1, t_2) = \text{Prob}\{X(t_2) = j \mid X(t_1) = i\} \tag{5.2.1}$$

where $p_{ij}(t_1, t_2)$ does not depend on the values of $X(t)$ for $t < t_1$.

The function $p_{ij}(t_1, t_2)$ satisfies

$$\sum_j p_{ij}(t_1, t_2) = 1 \tag{5.2.2}$$

$$p_{ij}(t_1, t_2) = \sum_k p_{ik}(t_1, \tau) p_{kj}(\tau, t_2) \tag{5.2.3}$$

where $t_1 < \tau < t_2$. This relation is analogous to (5.1.6) for the Markov chain and is commonly known as the *Chapman -Kolmogorov* equation.

Using the defining property of the Markov process, we obtain for $n = 1, 2, 3, \ldots$ and arbitrary t_m, $t_0 < t_1 < \cdots < t_m$, the following relation for the joint probability mass function of the random variables $X(t_0), X(t_1), \ldots, X(t_n)$:

$$\begin{aligned} p(i_0, i_1, \ldots, i_n) &= \text{Prob}\{X(t_0) = i_0, X(t_1) = i_1, \ldots, X(t_n) = i_n\} \\ &= p_{i_0} p_{i_0 i_1}(t_0, t_1) \cdots p_{i_{n-1} i_n}(t_{n-1}, t_n) \end{aligned} \tag{5.2.4}$$

where p_{i_0} is the probability mass function for $X(t_0)$. This relation is to be compared to (5.1.2) for the Markov chain.

A discrete Markov process is *homogeneous* if for arbitrary i, j and for arbitrary t_1, t_2, the transition probability function $p_{ij}(t_1, t_2)$ depends only on the difference $t_2 - t_1$. Thus $p_{ij}(t_1, t_2) = p_{ij}(t_2 - t_1)$. The properties (5.2.2) and (5.2.3) become

$$\sum_j p_{ij}(t) = 1 \tag{5.2.5}$$

$$p_{ij}(t + \tau) = \sum_k p_{ik}(t) p_{kj}(\tau) \tag{5.2.6}$$

A process is called a *process with independent increments* if for $n = 2, 3, \ldots$ and arbitrary $t_m (m = 0, 1, 2, \ldots, n)$, $t_0 < t_1 < \cdots < t_m$, the random variables $X(t_1) - X(t_0), X(t_2) - X(t_1), \ldots, X(t_n) - X(t_{n-1})$ are independent. If a stochastic process with independent increments has $\text{Prob}\{X(t_0) = i_0\} = 1$, then this process is a Markov process. With probability unity,

$$X(t_n) = i_0 + \sum_{m=1}^{n} (X(t_m) - X(t_{m-1})) \tag{5.2.7}$$

that is, $X(t_n)$ is the sum of $n + 1$ independent random variables and can be represented as a Markov process. However, a Markov process is not necessarily a process with independent increments. For a general Markov process we have that

$$\begin{aligned} &\text{Prob}\{X(t_1) = x_1, X(t_2) = x_1 + x_2\} \\ &\quad = \text{Prob}\{X(t_1) = x_1, X(t_2) - X(t_1) = x_2\} \\ &\quad = \text{Prob}\{X(t_1) = x_1\}\text{Prob}\{X(t_2) - X(t_1) = x_2 \mid X(t_1) = x_1\} \end{aligned} \tag{5.2.8}$$

whereas if the process has independent increments,

$$\begin{aligned} &\text{Prob}\{X(t_1) = x_1, X(t_2) - X(t_1) = x_2\} \\ &\quad = \text{Prob}\{X(t_1) = x_1\} \text{Prob}\{X(t_2) - X(t_1) = x_2\} \end{aligned} \tag{5.2.9}$$

A stochastic process $X(t)$ is called a *process with homogeneous increments* if for any t_1 and t_2, $t_1 < t_2$, the distribution function of $X(t_2) - X(t_1)$ depends only on $t_2 - t_1$. A process with independent increments which is a homogeneous Markov process is a process with homogeneous increments.

Consider the homogeneous Markov process $X(t)$ with $M + 1$ states. If there exists a $t^* < \infty$ such that $p_{ij}(t^*) > 0$, $i, j = 0, 1, \ldots, M$, then

$$\lim_{t \to \infty} p_{ij}(t) = \pi_j > 0 \tag{5.2.10}$$

Let $c_j(t) = \text{Prob}\{X(t) = j\}$, namely the probability that at time t the process is in state j. For a homogeneous Markov process

$$c_j(t) = \sum_{i=0}^{M} c_i(0) p_{ij}(t) \tag{5.2.11}$$

The limit

$$\lim_{t \to \infty} c_j(t) = c_j \tag{5.2.12}$$

exists independently of the initial distribution if the ergodic probabilities π_j defined by (5.2.10) exist. In that case,

$$\pi_j = c_j \tag{5.2.13}$$

5.2.1 The Poisson process

We shall now consider a special class of discrete state, continuous time Markov processes that represents a wide variety of physical phenomena. Let $X(t)$ denote the number of observed random events in the time interval $(0, t)$, $0 \leq t < \infty$, where $X(t)$ can assume integer values $i = 0, 1, 2, \ldots$. Also, for any t_1 and $t_2 (t_1 < t_2)$, the increment $X(t_2) - X(t_1)$ may equal $0, 1, 2, \ldots$. Let us assume that $X(t)$ satisfies the following conditions:

1. $X(t)$ is a process with independent increments. Thus the number of events in the intervals (t_0, t_1), (t_1, t_2), ... are independent random variables.
2. $X(t)$ is a process with homogeneous increments. Thus the probability of appearance of a given number of events is the same for all time intervals of the same length, or

$$\text{Prob}\{X(t) - X(0) = i\} = W_i(t)$$

where $W_i(t)$ is the probability that i events will occur in time interval t.

3.

$$\lim_{\Delta t \to 0} \frac{W_1(\Delta t)}{\Delta t} = \lambda, \qquad \lambda > 0$$

$$\lim_{\Delta t \to 0} \frac{1 - W_0(\Delta t) - W_1(\Delta t)}{\Delta t} = 0$$

The first condition in condition 3 states that the probability of occurrence of one event during a small interval of time Δt is $\lambda \Delta t + o(\Delta t)$. The second condition in 3 says that the probability of occurrence of more than one signal during Δt, as $\Delta t \to 0$, is of the order $o(\Delta t)$. Therefore the events occur suddenly, instantaneously, and singly, not in pairs or other multiple combinations. A stochastic process $\{X(t), 0 \leq t < \infty\}$, where $X(t)$ satisfies conditions 1–3 and $\text{Prob}\{X(0) = 0\} = 1$, is called a homogeneous *Poisson process*.

We shall now show that for the Poisson process

$$W_i(t) = \text{Prob}\{X(t) = i\} = \frac{(\lambda t)^i}{i!} e^{-\lambda t}, \qquad \lambda > 0 \tag{5.2.14}$$

From condition 2 we can write, for $\Delta t > 0$,

$$W_0(t + \Delta t) = W_0(t) W_0(\Delta t) \tag{5.2.15}$$

Thus, if no event occurs in the interval $(0, t + \Delta t)$, no event can occur in either of the intervals $(0, t)$ or $(t, t + \Delta t)$. If, however, during the interval $(0, t + \Delta t)\, i$ events occur, the event could occur $i - k$ times in the interval $(0, t)$ and k times in the interval $(t, t + \Delta t)$, $k = 0, 1, 2, \ldots, i$. Then

$$W_i(t + \Delta t) = \sum_{k=0}^{i} W_{i-k}(t) W_k(\Delta t) \tag{5.2.16}$$

Let us express condition 3 as

$$W_1(\Delta t) = \lambda\, \Delta t + o(\Delta t) \tag{5.2.17}$$

$$\begin{aligned} 1 - W_0(\Delta t) - W_1(\Delta t) &= \sum_{k=2}^{\infty} W_k(\Delta t) \\ &= o(\Delta t) \end{aligned} \tag{5.2.18}$$

Therefore

$$W_0(\Delta t) = 1 - \lambda\, \Delta t + o(\Delta t) \tag{5.2.19}$$

Substituting (5.2.19) into (5.2.15),

$$\frac{W_0(t + \Delta t) - W_0(t)}{\Delta t} = -\lambda W_0(t) - W_0(t) \frac{o(\Delta t)}{\Delta t} \tag{5.2.20}$$

Similarly, using (5.2.16)–(5.2.19),

$$\frac{W_i(t + \Delta t) - W_i(t)}{\Delta t} = -\lambda W_i(t) + \lambda W_{i-1}(t) + \frac{o(\Delta t)}{\Delta t} \sum_{k=0}^{i} W_k(t) \tag{5.2.21}$$

Letting $\Delta t \to 0$ in (5.2.20) and (5.2.21), we obtain the system of linear differential equations

$$\dot{W}_0(t) = -\lambda W_0(t) \tag{5.2.22}$$

$$\dot{W}_i(t) = -\lambda W_i(t) + \lambda W_{i-1}(t), \qquad i = 1, 2, \ldots \tag{5.2.23}$$

By assumption, $W_0(0) = 1$, and so

$$W_0(t) = e^{-\lambda t} \tag{5.2.24}$$

To solve (5.2.23) we assume that $W_i(t) = U_i(t) e^{-\lambda t}$, where $U_0(0) = 1$ from (5.2.24). It can be shown that $U_i(t)$ satisfies

$$\dot{U}_i(t) = \lambda U_{i-1}(t), \qquad U_i(0) = 0, \qquad i = 1, 2, \ldots \tag{5.2.25}$$

or

$$U_i(t) = \lambda \int_0^t U_{i-1}(t)\, dt \tag{5.2.26}$$

Evaluating (5.2.26) for $i = 1, 2, \ldots,$ we obtain by induction

$$U_i(t) = \frac{(\lambda t)^i}{i!} \tag{5.2.27}$$

Thus we obtain (5.2.14),

$$W_i(t) = \frac{(\lambda t)^i}{i!} e^{-\lambda t} \tag{5.2.28}$$

Let us compute the transition probabilities $p_{ij}(t)$:

$$\begin{aligned} p_{ij}(t) &= \text{Prob}\{X(t_2) = j \mid X(t_1) = i\} \\ &= \frac{\text{Prob}\{X(t_2) = i, X(t_2) - X(t_1) = j - i\}}{\text{Prob}\{X(t_1) = i\}} \\ &= \text{Prob}\{X(t_2) - X(t_1) = j - i\} \\ &= \frac{(\lambda t)^{j-i}}{(j-i)!} e^{-\lambda t} \end{aligned} \tag{5.2.29}$$

where $t = t_2 - t_1$ and $j \geq i$. Clearly, $p_{ij}(t) = 0$ for $j < i$. From (5.2.29) we can see that

$$\lim_{t \to \infty} p_{ij}(t) = 0 \tag{5.2.30}$$

A typical Poisson process is shown in Figure 5.2. We note that $X(t)$ is simply a series of step functions.

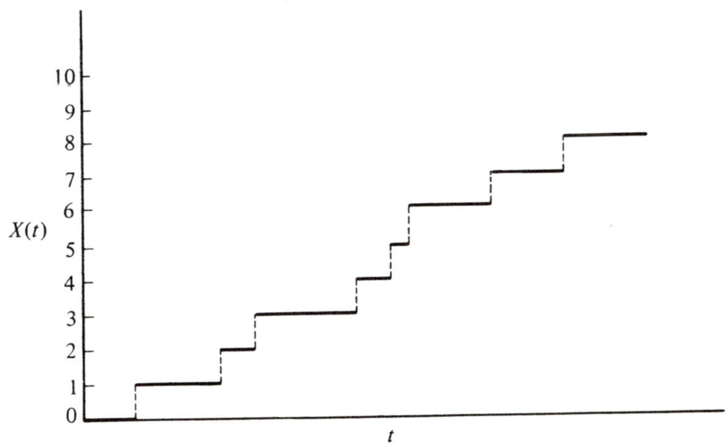

Figure 5.2. Poisson process.

5.2.2 The birth process

The Poisson process can be generalized by allowing the frequency λ to depend on the state of the process. Suppose that $X(t)$ is a homogeneous Markov process for which λ is a function of the current state j of the process; that is, instead of a constant λ we now write λ_j. Then

$$\lim_{\Delta t \to 0} \frac{p_{j,j+1}(\Delta t)}{\Delta t} = \lambda_j \tag{5.2.31}$$

$$\lim_{\Delta t \to 0} \frac{1 - p_{jj}(\Delta t) - p_{j,j+1}(\Delta t)}{\Delta t} = 0 \tag{5.2.32}$$

Such a process is called a *birth process*. The probability of the occurrence of an event during Δt equals $\lambda_j \, \Delta t + o(\Delta t)$ if at t, j events have already occurred. The terminology "birth process" arises because of the ability now to account for the appearance of new events as a function of the number of events which have already occurred, or, in other words, of the current population of the system. The birth process has been used to model the growth of biological populations in which new elements appear by the splitting of elements already in the population.

Let $V_m(t) = \text{Prob}\{X(t) = i + m\}$ and $\text{Prob}\{X(0) = i\} = 1$. Therefore

$$V_0(t + \Delta t) = V_0(t)V_0(\Delta t) \tag{5.2.33}$$

$$V_m(t + \Delta t) = \sum_{k=0}^{m} V_{m-k}(t)V_k(\Delta t) \tag{5.2.34}$$

and by arguments similar to the Poisson process,

$$V_0(\Delta t) = 1 - \lambda_i \, \Delta t + o(\Delta t) \tag{5.2.35}$$

$$V_1(\Delta t) = \lambda_i \, \Delta t + o(\Delta t) \tag{5.2.36}$$

Using (5.2.33)–(5.2.36), we generate the following system of differential equations:

$$\dot{V}_0(t) = -\lambda_i V_0(t) \tag{5.2.37}$$

$$\dot{V}_m(t) = -\lambda_{i+m} V_m(t) + \lambda_{i+m-1} V_{m-1}(t), \qquad m = 1, 2, \ldots \tag{5.2.38}$$

subject to

$$V_m(0) = \begin{cases} 1, & m = 0 \\ 0, & m \neq 0 \end{cases} \tag{5.2.39}$$

The solution of (5.2.37) is

$$V_0(t) = e^{-\lambda_i t} \tag{5.2.40}$$

The solution of (5.2.38) can be obtained successively for larger values of m.

The particular birth process with

$$\lambda_j = \lambda j \tag{5.2.41}$$

is often used to represent phenomena associated with cosmic radiation and the growth of biological populations and is called a *Furry-Yule process*.

Physically, the Furry-Yule process represents the growth of elements such that the probability that a given element will give birth to a new element in Δt is $\lambda \, \Delta t + o(\Delta t)$. Thus, if at some time there are j elements and the random events that give birth to new elements are independent, the probability of the birth of a new element during Δt is equal to $\lambda j\,\Delta t + o(\Delta t)$.

Equations (5.2.37) and (5.2.38) in the case of the Furry-Yule process take the form

$$\dot{V}_0(t) = -\lambda i V_0(t) \tag{5.2.42}$$

$$\dot{V}_m(t) = -\lambda(i+m)V_m(t) + \lambda(i+m-1)V_{m-1}(t), \qquad m = 1, 2, \ldots \tag{5.2.43}$$

It can be shown (see Problem 5.9) that the solution of (5.2.42) and (5.2.43) is

$$V_m(t) = \binom{m+i-1}{m} e^{-i\lambda t}(1-e^{-\lambda t})^m, \qquad m = 0, 1, 2, \ldots \tag{5.2.44}$$

Let us compute the mean and variance of $X(t)$, where $X(t)$ is a Furry-Yule process. First,

$$E(t) = E[X(t)] = \sum_{m=0}^{\infty} (i+m)V_m(t) \tag{5.2.45}$$

Differentiating both sides of (5.2.45) with respect to time and using (5.2.42) and (5.2.43),

$$\begin{aligned}\dot{E}(t) &= \sum_{m=0}^{\infty} (i+m)\dot{V}_m(t) \\ &= \lambda \sum_{m=0}^{\infty} (i+m)V_m(t) \\ &= \lambda E(t)\end{aligned} \tag{5.2.46}$$

Since the initial condition for (5.2.46) is $E(0) = i$, the solution is

$$E(t) = ie^{\lambda t} \tag{5.2.47}$$

Next, we compute the second noncentral moment of $X(t)$, $E[X(t)^2]$, which we denote by $E_2(t)$:

$$E_2(t) = \sum_{m=0}^{\infty} (i+m)^2 V_m(t) \tag{5.2.48}$$

Differentiating both sides with respect to t and using (5.2.42) and (5.2.43),

$$\begin{aligned}\dot{E}_2(t) &= \sum_{m=0}^{\infty} (i+m)^2 \dot{V}_m(t) \\ &= \lambda \sum_{m=0}^{\infty} (i+m)(2i+2m+1)V_m(t) \\ &= 2\lambda E_2(t) + \lambda E(t)\end{aligned} \tag{5.2.49}$$

The initial condition for (5.2.49) is $E_2(0) = i^2$, and so we must solve

$$\dot{E}_2(t) = 2\lambda E_2(t) + \lambda i e^{\lambda t}, \qquad E_2(0) = i^2 \tag{5.2.50}$$

The solution is

$$E_2(t) = e^{2\lambda t}[i^2 + i(1-e^{-\lambda t})] \tag{5.2.51}$$

Finally, the variance of $X(t)$, $\sigma^2[X(t)] = E_2(t)^2$, is

$$\sigma^2[X(t)] = i(e^{2\lambda t} - e^{\lambda t}) \tag{5.2.52}$$

5.2.3 Birth and death processes

Let us now consider processes in which events may not only appear but also disappear. In particular, consider a process $\{X(t), 0 \leq t < \infty\}$ satisfying conditions 1–3 of the Poisson process. We now want to include not only the appearance of an event but also its life, that is, the time interval between its "birth" and "death." The transition probabilities satisfy

1. $p_{j,j+1}(\Delta t) = \lambda_j \Delta t + o(\Delta t), \qquad j \geq 0$
2. $p_{j,j-1}(\Delta t) = \mu_j \Delta t + o(\Delta t), \qquad j \geq 1$
3. $p_{jj}(\Delta t) = 1 - (\lambda_j + \mu_j)\Delta t + o(\Delta t), \qquad j \geq 0$

where λ_j and μ_j are the birth and death rates when the process is in state j. Properties 1 and 2 state that if the process is in state j, then in a small interval of time the probabilities of the state increasing or decreasing by 1 are proportional to the length of the interval.

Let us denote by $X(t)$ the number of events present at time t and $V_j(t)$ as Prob $\{X(t) = j\}$. We can write

$$\begin{aligned} V_0(t + \Delta t) = {} & V_0(t)[1 - \lambda_0 \Delta t + o(\Delta t)] \\ & + V_1(t)[\mu_1 \Delta t + o(\Delta t)] + o(\Delta t) \end{aligned} \tag{5.2.53}$$

which states that the state $j = 0$ at $t + \Delta t$ can be attained either by having the process be in the $j = 0$ state at t and having no events occur in $(t, t + \Delta t)$ or by having the process be in the $j = 1$ state at t and having the existing signal disappear in $(t, t + \Delta t)$. The process is in the state j at $t + \Delta t$ either if it was at the state j at t and no signals arrived during $(t, t + \Delta t)$, if it was at the state $j - 1$ at t and one signal arrived in $(t, t + \Delta t)$, or if it was in the state $j + 1$ and one signal disappeared in $(t, t + \Delta t)$. Thus

$$\begin{aligned} V_j(t + \Delta t) = {} & V_j(t)[1 - \lambda_j \Delta t + o(\Delta t)][1 - \mu_j \Delta t + o(\Delta t)] \\ & + V_{j-1}(t)[\lambda_{j-1} \Delta t + o(\Delta t)] \\ & + V_{j+1}(t)[\mu_{j+1} \Delta t + o(\Delta t)] + o(\Delta t) \end{aligned} \tag{5.2.54}$$

Dividing by Δt and letting $\Delta t \to 0$, (5.2.53) and (5.2.54) become

$$\dot{V}_0(t) = -\lambda_0 V_0(t) + \mu_1 V_1(t) \tag{5.2.55}$$

$$\dot{V}_j(t) = \lambda_{j-1} V_{j-1}(t) - (\lambda_j + \mu_j) V_j(t) + \mu_{j+1} V_{j+1}(t) \tag{5.2.56}$$

Again the result is an infinite system of linear differential-difference equations. However, these equations are not recursive and require rather sophisticated solution procedures. The initial conditions depend on the state of the system at time zero. If the system is in state i at $t = 0$, we have

$$V_j(0) = \begin{cases} 1, & j = i \\ 0, & j \neq i \end{cases} \tag{5.2.57}$$

Now let us determine the distribution of the random variable T_j, the waiting time of $X(t)$ in state j. Letting $G_j(t) = \text{Prob}\{T_j \geq t\}$, we can write

$$G_j(t + \Delta t) = G_j(t)G_j(\Delta t) \tag{5.2.58}$$

since the births or deaths of signals are independent events.

We assert that the probability that $T_j \geq \Delta t$, $G_j(\Delta t)$, is equivalent to $p_{jj}(\Delta t) + o(\Delta t)$. Then, from property 3, (5.2.58) becomes

$$G_j(t + \Delta t) = G_j(t)[1 - (\lambda_j + \mu_j)\Delta t] + o(\Delta t) \tag{5.2.59}$$

or

$$\frac{G_j(t + \Delta t) - G_j(t)}{\Delta t} = -(\lambda_j + \mu_j)\Delta t + \frac{o(\Delta t)}{\Delta t} \tag{5.2.60}$$

As $\Delta t \longrightarrow 0$, then

$$\dot{G}_j(t) = -(\lambda_j + \mu_j)G_j(t) \tag{5.2.61}$$

with initial condition $G_j(0) = 1$. The solution is

$$G_j(t) = \exp[-(\lambda_j + \mu_j)t] \tag{5.2.62}$$

Therefore T_j is exponentially distributed with mean $(\lambda_j + \mu_j)^{-1}$.

The birth and death process can be described as follows. The state of the process remains in a given state j for a random length of time T_j which obeys an exponential distribution with parameter $(\lambda_j + \mu_j)$. When leaving state j the process can proceed to either state $j + 1$ or $j - 1$. Although we shall not prove it, the probability that the state advances to $j + 1$ is $\lambda_j/(\lambda_j + \mu_j)$, whereas the probability that it retreats to $j - 1$ is $\mu_j/(\lambda_j + \mu_j)$. The motion of the process could be described as a random walk with steps being taken either to the right or left at random times rather than at fixed time periods.

We might note that we could have dealt directly with the transition probabilities $p_{ij}(t)$ rather than introducing the probability $V_j(t)$. For example, in terms of transition probabilities, (5.2.55) and (5.2.56) are

$$\dot{p}_{i0}(t) = -\lambda_0 p_{i0}(t) + \mu_1 p_{i1}(t) \tag{5.2.63}$$

$$\dot{p}_{ij}(t) = \lambda_{j-1}p_{i,j-1}(t) - (\lambda_j + \mu_j)p_{ij}(t) + \mu_{j+1}p_{i,j+1}(t) \tag{5.2.64}$$

with $p_{ij}(0) = \delta_{ij}$. The notation is simpler if we use $V_j(t)$ since transitions occur only between adjacent states.

It is of interest to examine the behavior of the birth and death process as $t \longrightarrow \infty$ [11]. Let us determine the conditions under which the limits

$$\lim_{t\to\infty} p_{ij}(t) = \bar{p}_j \tag{5.2.65}$$

exist and are independent of the initial state i. If (5.2.65) holds, then the ergodic probabilities $\bar{p}_j$ satisfy the steady-state forms of (5.2.63) and (5.2.64):

$$-\lambda_0\bar{p}_0 + \mu_1\bar{p}_1 = 0 \tag{5.2.66}$$

$$\lambda_{j-1}\bar{p}_{j-1} - (\lambda_j + \mu_j)\bar{p}_j + \mu_{j+1}\bar{p}_{j+1} = 0 \tag{5.2.67}$$

The solution of (5.2.66) and (5.2.67) is obtained by induction. Letting

$$\pi_0 = 1, \qquad \pi_j = \frac{\lambda_0 \lambda_1 \cdots \lambda_{j-1}}{\mu_1 \mu_2 \cdots \mu_j}$$

we find that $\bar{p}_1 = \lambda_0 \mu_1^{-1} \bar{p}_0 = \pi_1 \bar{p}_0$. Then, if $\bar{p}_k = \pi_k \bar{p}_0$ for $k = 1, 2, \ldots, j$, we obtain

$$\begin{aligned} \mu_{j+1} \bar{p}_{j+1} &= (\lambda_j + \mu_j)\pi_j \bar{p}_0 - \lambda_{j-1}\pi_{j-1}\bar{p}_0 \\ &= \lambda_j \pi_j \bar{p}_0 + (\mu_j \pi_j - \lambda_{j-1}\pi_{j-1})\bar{p}_0 \\ &= \lambda_j \pi_j \bar{p}_0 \end{aligned}$$

Thus

$$\bar{p}_{j+1} = \pi_{j+1} \bar{p}_0 \tag{5.2.68}$$

So that the sequence $\{\bar{p}_j\}$ defines a distribution, we must have $\sum_j \bar{p}_j = 1$. If $\sum \pi_k < \infty$, we see that we obtain

$$\bar{p}_j = \frac{\pi_j}{\sum \pi_k}, \qquad j = 0, 1, 2, \ldots \tag{5.2.69}$$

However, if $\sum \pi_k = \infty$, $\bar{p}_0 = 0$ and we do not obtain a limiting stationary distribution.

EXAMPLE 5.8. Consider the process of molecules impinging on the surface of a catalyst particle. Assume that there are a finite number M of active sites that can be occupied at any time. The molecules arrive at the surface with frequency λ independently of the number of molecules already on the surface. However, the rate at which molecules leave the surface depends on the number on the surface. Specifically, the probability that a molecule will leave the surface in a time Δt is $\mu \, \Delta t + o(\Delta t)$. We assume that the residence times of molecules on the surface are independent of each other and are independent of the number of incoming molecules. Thus, if at time t we have j molecules on the surface, the probability of departure of one of them in the next t is $\mu j \, \Delta t + o(\Delta t)$, and of two of them, $o(\Delta t)$. This is, therefore, a birth and death process in which $X(t) =$ the number of sites occupied at time t, where $X(t)$ can assume the values $0, 1, 2, \ldots, M$. The parameters are $\lambda_j = \lambda$ and $\mu_j = \mu j$. Clearly, the individual residence times of molecules are exponentially distributed with parameter μ.

We let $V_j(t)$ be the probability that j sites are occupied at time t. We desire $V_j(t)$ as $t \longrightarrow \infty$, the stationary distribution of site occupation. The dynamics of the process are governed by

$$\begin{aligned} \dot{V}_0(t) &= -\lambda V_0(t) + \mu V_1(t) \\ \dot{V}_j(t) &= \lambda V_{j-1}(t) - (\lambda + \mu_j) V_j(t) + \mu(j+1) V_{j+1}(t), \qquad j = 1, 2, \ldots, M-1 \\ \dot{V}_M(t) &= \lambda V_{M-1}(t) - \mu M V_M(t) \end{aligned}$$

At steady state,

$$\begin{gathered} -\lambda \bar{V}_0 + \bar{V}_1 = 0 \\ \lambda \bar{V}_{j-1} - (\lambda + \mu_j)\bar{V}_j + \mu(j+1)\bar{V}_{j+1} = 0, \qquad j = 1, 2, \ldots, M-1 \\ \lambda \bar{V}_{M-1} - \mu M \bar{V}_M = 0 \end{gathered}$$

Solving,

$$\bar{V}_j = \frac{1}{j!}\left(\frac{\lambda}{\mu}\right)^j \bar{V}_0, \qquad j = 1, 2, \ldots, M$$

where

$$\bar{V}_0 = \frac{1}{\sum_{k=0}^{M} (1/k!)(\lambda/\mu)^k}$$

We could then compute $\bar{V}_j$ for various values of M and λ/μ.

This problem is identical to the description of a telephone exchange with M channels and no waiting lines in which individual calls have exponentially distributed durations.

EXAMPLE 5.9. Karlin [11] has considered a particular model of biological reproduction and population growth in which the population has not only births and deaths but also immigration. In particular, we let the birth parameter λ_j be $\lambda j + \alpha$, where the average instantaneous rate of growth of the population is λj from births and α from outside immigration and the death parameter $\mu_j = \mu j$. Let us compute the rate of change of the expected value $E[X(t)]$ of the size of the population. Substituting for λ_j and μ_j in (5.2.55) and (5.2.56),

$$\dot{V}_0(t) = -\alpha V_0(t) + \mu V_1(t)$$

$$\dot{V}_j(t) = [\lambda(j-1) + \alpha]V_{j-1}(t) - [(\lambda + \mu)j + \alpha]V_j(t) + \mu(j+1)V_{j+1}(t)$$

The expected value of $X(t)$ is

$$E[X(t)] = E(t) = \sum_{j=1}^{\infty} jV_j(t)$$

Multiplying the differential equation for $V_j(t)$ by j and summing, we obtain

$$\dot{E}(t) = \alpha + (\lambda - \mu)E(t)$$

the solution of which, subject to $E(0) = i$, if $X(0) = i$, is

$$E(t) = \begin{cases} \alpha t + i, & \lambda = \mu \\ \dfrac{\alpha}{\lambda - \mu}[e^{(\lambda-\mu)t} - 1] + ie^{(\lambda-\mu)t}, & \lambda \neq \mu \end{cases}$$

We note that $E(t) \longrightarrow \infty$ as $t \longrightarrow \infty$ if $\lambda \geq \mu$, whereas if $\lambda < \mu$, $E(t)$ approaches $\alpha/(\mu - \lambda)$ as $t \longrightarrow \infty$.

5.2.4 Kolmogorov equations

Up to this point we have considered discrete state, continuous time Markov processes in which a change of state may occur only by direct passage to a neighboring state. We can generalize to processes in which the change of state may occur by direct passage from any state to any other state. Let $p_{ij}(\tau, t)$, $0 < \tau < t$, be the transition probability function for the process. In this case the Chapman-Kolmogorov equations (5.2.3) become

$$p_{ij}(\tau, t) = \sum_k p_{ik}(\tau, s)p_{kj}(s, t), \qquad \tau < s < t \tag{5.2.70}$$

The objective is to derive a system of differential equations for $p_{ij}(\tau, t)$. Let

$$q_i(t) = \lim_{\Delta t \to 0} \frac{1 - p_{ii}(t, t + \Delta t)}{\Delta t} \tag{5.2.71}$$

$$q_{ij}(t) = \lim_{\Delta t \to 0} \frac{p_{ij}(t, t + \Delta t)}{\Delta t} \tag{5.2.72}$$

or, equivalently,

$$p_{ii}(t, t + \Delta t) = 1 - q_i(t) + o(\Delta t) \tag{5.2.73}$$

$$p_{ij}(t, t + \Delta t) = q_{ij}(t)\,\Delta t + o(\Delta t) \tag{5.2.74}$$

where $q_i(t)$ and $q_{ij}(t)$ are called *intensity functions*. Let us write (5.2.70) as

$$p_{ij}(\tau, t + \Delta t) = \sum_k p_{ik}(\tau, t) p_{kj}(t, t + \Delta t) \tag{5.2.75}$$

or

$$\begin{aligned} \frac{p_{ij}(\tau, t + \Delta t) - p_{ij}(\tau, t)}{\Delta t} &= \frac{1}{\Delta t}\Big[\sum_k p_{ik}(\tau, t) p_{kj}(t, t + \Delta t) - p_{ij}(\tau, t)\Big] \\ &= \sum_{k \neq j} p_{ik}(\tau, t) \frac{p_{kj}(t, t + \Delta t)}{\Delta t} \\ &\quad - p_{ij}(\tau, t) \frac{1 - p_{jj}(t, t + \Delta t)}{\Delta t} \end{aligned} \tag{5.2.76}$$

Using (5.2.71) and (5.2.72) and the fact that $\sum_k p_{ik}(\tau, t)$ converges absolutely, as well as the uniform convergence of (5.2.72) with respect to the first index when the second index is fixed,

$$\begin{aligned} \frac{\partial p_{ij}(\tau, t)}{\partial t} &= \lim_{\Delta t \to 0} \frac{p_{ij}(\tau, t + \Delta t) - p_{ij}(\tau, t)}{\Delta t} \\ &= \sum_{k \neq j} p_{ik}(\tau, t) \lim_{\Delta t \to 0} \frac{p_{kj}(t, t + \Delta t)}{\Delta t} \\ &\quad - p_{ij}(\tau, t) \lim_{\Delta t \to 0} \frac{1 - p_{jj}(t, t + \Delta t)}{\Delta t} \\ &= -p_{ij}(\tau, t) q_j(t) + \sum_{k \neq j} p_{ik}(\tau, t) q_{kj}(t) \end{aligned} \tag{5.2.77}$$

The initial conditions for (5.2.77) are

$$p_{ij}(\tau, \tau) = \begin{cases} 1, & i = j \\ 0, & i \neq j \end{cases} \tag{5.2.78}$$

Summarizing, we have treated i and τ as constants and j and t as variables in deriving (5.2.77) and (5.2.78). We could have followed an analogous procedure and kept j and t as constants and treated i and τ as variables. The resulting differential equation is

$$\frac{\partial p_{ij}(\tau, t)}{\partial \tau} = q_i(\tau) p_{ij}(\tau, t) - \sum_{k \neq i} q_{ik}(\tau) p_{kj}(\tau, t) \tag{5.2.79}$$

$$p_{ij}(t, t) = \begin{cases} 1, & j = i \\ 0, & j \neq i \end{cases} \tag{5.2.80}$$

It can be shown that (5.2.77) and (5.2.78) have solutions identical to (5.2.79) and (5.2.80). Equation (5.2.77) is called the *Kolmogorov forward equation* and (5.2.79) is called the *Kolmogorov backward equation.*

If the process is homogeneous, then $q_i(t)$ and $q_{ij}(t)$ are independent of t. Also, p_{ij} depends only on $t - \tau$. Thus (5.2.77)–(5.2.80) become

$$\dot{p}_{ij}(t) = -p_{ij}(t)q_j + \sum_{k \neq j} p_{ik}(t)q_{kj} \tag{5.2.81}$$

$$\dot{p}_{ij}(t) = q_i p_{ij}(t) - \sum_{k \neq i} q_{ik}p_{kj}(t) \tag{5.2.82}$$

$$p_{ij}(0) = \begin{cases} 1, & j = i \\ 0, & j \neq i \end{cases} \tag{5.2.83}$$

The differential equations for the Poisson, birth, and birth and death processes are all special cases of (5.2.81).

5.3 Continuous State Stochastic Processes

Up to this point, the random variables X_t and $X(t)$ could assume only discrete values; we shall now consider the case in which X_t or $X(t)$ can assume arbitrary real values, that is, the case in which they are continuous random variables. We shall discuss only continuous state, continuous time processes of which the continuous state, discrete time process is just a special case.

Consider a scalar stochastic process $X(t)$ as pictured, for example, in Figure 5.1. At any time t', $X(t')$ is a random variable. If we were to connect all possible values of $X(t)$ for all t, we would generate an infinite family of $X(t)$ curves, called the *ensemble* of $X(t)$. Three of these curves are shown in Figure 5.1. A complete probabilistic description of the stochastic process is obtained if one knows the nth-order joint probability density function $p(x_1, t_1; x_2, t_2; \ldots ; x_n, t_n)$ for any n. In dealing with discrete state stochastic processes, we noted that knowledge of the joint density of states at all times is almost never available. This statement can be made even more strongly in the case of continuous state stochastic processes.

We can write the nth-order joint density function for a process $X(t)$ in terms of its conditional densities as follows:

$$p(x_1, t_1; \ldots ; x_n, t_n) = p(x_2, t_2; \ldots ; x_n, t_n \mid x_1, t_1)p(x_1, t_1) \tag{5.3.1}$$

where $p(x_1, t_1)$ is called the first-order probability density function. Thus $p(x_1, t_1)\,dx_1$ is the probability that $x_1 \leq X(t_1) < x_1 + dx_1$. Continuing the process begun in (5.3.1), we find that

$$\begin{aligned} p(x_1, t_1; \ldots ; x_n, t_n) &= p(x_3, t_3; \ldots ; x_n, t_n \mid x_1, t_1; x_2, t_2) \\ &\quad \cdot p(x_1, t_1; x_2, t_2) \\ &= p(x_n, t_n \mid x_{n-1}, t_{n-1}; \ldots ; x_1, t_1) \\ &\quad \cdot p(x_{n-1}, t_{n-1}; \ldots ; x_1, t_1) \end{aligned} \tag{5.3.2}$$

It turns out that conditional densities, such as in (5.3.1) and (5.3.2), are of great interest in the study of stochastic processes. In particular, the concept of a Markov process is of tremendous utility in studying continuous stochastic processes. If

$$p(x_n, t_n | x_{n-1}, t_{n-1}; \ldots ; x_1, t_1) = p(x_n, t_n | x_{n-1}, t_{n-1}) \tag{5.3.3}$$

for every integer n, for every choice of $t_1, t_2, \ldots, t_n$, provided only that $t_1 < t_2 < \cdots < t_n$, the process is a Markov process. Thus the single most recently observed value of a Markov process contains as much information about the future evolution of the process as does knowledge of the entire past history of the process up to and including the last observed value.

The conditional densities (called also the transition probabilities) for a continuous Markov process obey the Chapman-Kolmogorov equation:

$$p(x, t | x_0, t_0) = \int_{-\infty}^{\infty} p(x, t | \eta, s) p(\eta, s | x_0, t_0)\, d\eta, \qquad t > s > t_0 \tag{5.3.4}$$

Thus the passage of $X(t)$ from state x_0 at t_0 to a state x at t is a product of two passages: (1) passage from state x_0 at t_0 to a state η at time s, and (2) passage from state η at time s to state x at time t. Note that the intermediate time s is fixed, η being the state, $\eta \leq X(s) \leq \eta + d\eta$, at time s.

Often we have some initial first-order probability density $p(x_0, t_0)$ and we desire to see how the density function $p(x, t)$ develops with time. For a Markov process we can relate its first-order density at any time to its initial density through the transition probability

$$p(x, t) = \int_{-\infty}^{\infty} p(x, t | x_0, t_0) p(x_0, t_0)\, dx_0 \tag{5.3.5}$$

The key question now is, How do we determine the transition probability $p(x, t | x_0, t_0)$ for a Markov process? As has been our custom, we would like to obtain a differential equation for $p(x, t | x_0, t_0)$. To derive this differential equation we begin with the Chapman-Kolmogorov equation (5.3.4) written in the form

$$p(x, t + \Delta t | x_0, t_0) = \int_{-\infty}^{\infty} p(x, t + \Delta t | x - \Delta x, t) p(x - \Delta x, t | x_0, t_0)\, d(\Delta x) \tag{5.3.6}$$

We want to expand the integrand in a Taylor series about $p(x + \Delta x, t + \Delta t | x, t) p(x, t | x_0, t_0)$. To simplify the notation somewhat, we note that all we want to do is expand the product $f(x)g(x - \Delta x)$ in a Taylor series about $f(x + \Delta x)g(x)$. In fact, if we let $h(x) = f(x)g(x - \Delta x)$ and $h(x + \Delta x) = f(x + \Delta x)g(x)$, the Taylor series is simply

$$h(x) = h(x + \Delta x) - \left.\frac{dh}{dx}\right|_{x+\Delta x} \Delta x + \frac{1}{2} \left.\frac{d^2 h}{dx^2}\right|_{x+\Delta x} \Delta x^2 - \cdots \tag{5.3.7}$$

Letting $h(x) = p(x, t + \Delta t \mid x - \Delta x, t)p(x - \Delta x, t \mid x_0, t_0)$ and substituting (5.3.7) into (5.3.6), we obtain

$$p(x, t + \Delta t \mid x_0, t_0) = \int_{-\infty}^{\infty} p(x + \Delta x, t + \Delta t \mid x, t)p(x, t \mid x_0, t_0)\, d(\Delta x) + \int_{-\infty}^{\infty} \sum_{n=1}^{\infty} \frac{(-1)^n}{n!} (\Delta x)^n \frac{\partial^n}{\partial x^n} [p(x + \Delta x, t + \Delta t \mid x, t)p(x, t \mid x_0, t_0)]\, d\,\Delta x \tag{5.3.8}$$

This equation can be rewritten as

$$p(x, t + \Delta t \mid x_0, t_0) = p(x, t \mid x_0, t_0) \int_{-\infty}^{\infty} p(x + \Delta x, t + \Delta t \mid x, t)\, d(\Delta x) + \sum_{n=1}^{\infty} \frac{(-1)^n}{n!} \frac{\partial^n}{\partial x^n} p(x, t \mid x_0, t_0) \cdot \int_{-\infty}^{\infty} (\Delta x)^n p(x + \Delta x, t + \Delta t \mid x, t)\, d(\Delta x) \tag{5.3.9}$$

By definition

$$\int_{-\infty}^{\infty} p(\xi, t \mid \eta, s)\, d\xi = 1 \tag{5.3.10}$$

so that (5.3.9) becomes

$$p(x, t + \Delta t \mid x_0, t_0) = p(x, t \mid x_0, t_0) + \sum_{n=1}^{\infty} \frac{(-1)^n}{n!} \frac{\partial^n}{\partial x^n} p(x, t \mid x_0, t_0) \cdot \int_{-\infty}^{\infty} (\Delta x)^n p(x + \Delta x, t + \Delta t \mid x, t)\, d(\Delta x) \tag{5.3.11}$$

We now define

$$\frac{\partial p(x, t \mid x_0, t_0)}{\partial t} = \lim_{\Delta t \to 0} \frac{p(x, t + \Delta t \mid x_0, t_0) - p(x, t \mid x_0, t_0)}{\Delta t} \tag{5.3.12}$$

and

$$\alpha_n(x, t) = \lim_{\Delta t \to 0} \frac{1}{\Delta t} \int_{-\infty}^{\infty} (\Delta x)^n p(x + \Delta x, t + \Delta t \mid x, t)\, d(\Delta x) \tag{5.3.13}$$

In effect, the $\alpha_n(x, t)$ are the moments of the increment, $X(t + \Delta t) - X(t)$, conditioned on $X(t) = x$, namely

$$\alpha_n(x, t) = \lim_{\Delta t \to 0} \frac{1}{\Delta t} E\{[X(t + \Delta t) - X(t)]^n \mid X(t) = x\} \tag{5.3.14}$$

Dividing (5.3.11) by Δt and taking the limit as $\Delta t \longrightarrow 0$, we obtain the differential equation governing $p(x, t \mid x_0, t_0)$:

$$\frac{\partial p(x, t \mid x_0, t_0)}{\partial t} = \sum_{n=1}^{\infty} \frac{(-1)^n}{n!} \frac{\partial^n}{\partial x^n} (\alpha_n(x, t)p(x, t \mid x_0, t_0)), \quad -\infty < x < \infty,\ t > t_0 \tag{5.3.15}$$

This is the general form of the differential equation governing the transition density $p(x, t \mid x_0, t_0)$ for a scalar Markov process $X(t)$. This equation is called the *forward Fokker-Planck equation* (also called the *forward Kolmogorov equation*), and, as we shall see, it is one of the most important equations of classic mathematical physics. The boundary conditions for the forward Fokker-Planck equation are

$$\begin{aligned} \lim_{t \to t_0} p(x, t \mid x_0, t_0) &= \delta(x - x_0) \\ \lim_{|x| \to \infty} p(x, t \mid x_0, t_0) &= 0 \end{aligned} \tag{5.3.16}$$

We call (5.3.15) the forward equation since it traces the development of the transition forward in time t.

A particularly important class of continuous Markov processes are those for which $\alpha_n(x, t) = 0$ for $n \geq 3$. These are called *diffusion processes* and obey the forward equation

$$\frac{\partial p(x, t \mid x_0, t_0)}{\partial t} = -\frac{\partial}{\partial x}(\alpha_1(x, t) p(x, t \mid x_0, t_0)) + \frac{1}{2} \frac{\partial^2}{\partial x^2}(\alpha_2(x, t) p(x, t \mid x_0, t_0)) \tag{5.3.17}$$

We, of course, recognize (5.3.17) as a general form of the convective diffusion equation. This suggests that a process consisting of a very large number of small displacements can be described as a diffusion process. This is indeed the case, and we shall develop this correspondence in some detail in the pages that follow.

By analogy to (5.2.79) and (5.2.80) we might expect that there exists a backward equation corresponding to (5.3.15). There does exist such an equation and it can be derived very similarly to (5.3.15). Limiting our attention henceforth to processes for which $\alpha_n(x, t) = 0$, $n \geq 3$, we write the backward equation corresponding to (5.3.17) as

$$-\frac{\partial p(x, t \mid x_0, t_0)}{\partial t_0} = \alpha_1(x_0, t_0) \frac{\partial p(x, t \mid x_0, t_0)}{\partial x_0} + \frac{1}{2} \alpha_2(x_0, t_0) \frac{\partial^2 p(x, t \mid x_0, t_0)}{\partial x_0^2}$$
$$-\infty < x_0 < \infty, \; t_0 < t \tag{5.3.18}$$

with the boundary conditions

$$\begin{aligned} \lim_{t_0 \to t} p(x, t \mid x_0, t_0) &= \delta(x - x_0) \\ \lim_{|x_0| \to \infty} p(x, t \mid x_0, t_0) &= 0 \end{aligned} \tag{5.3.19}$$

Equation (5.3.17) is a partial differential equation for p considered a function of the independent variables x and t; the variables x_0 and t_0 are merely parameters which enter through the boundary conditions (5.3.16). On the other hand, (5.3.18) is a partial differential equation for p as a function of the independent variables x_0 and t_0; here x and t are merely parameters which enter through the boundary conditions (5.3.19). In either equation, the functions which are related to the stochastic process $X(t)$ are the moments

α_1 and α_2. A complete probabilistic analysis of the properties of a Markov process may be obtained by finding the transition density $p(x, t \mid x_0, t_0)$ as a solution to one of the Fokker-Planck equations (if it satisfies one, it necessarily satisfies the other).

It is often of interest to determine only the first-order density $p(x, t)$ for a Markov process. We have seen in (5.3.5) that $p(x, t)$ can be computed once $p(x, t \mid x_0, t_0)$ is known. However, what would be useful is a partial differential equation for $p(x, t)$ itself. A little reflection on (5.3.5) and on the procedure used to derive (5.3.15) will quickly show that $p(x, t)$ satisfies an equation identical to (5.3.15). Thus for diffusion processes we have

$$\frac{\partial p(x,t)}{\partial t} = -\frac{\partial}{\partial x}(\alpha_1(x, t)p(x, t)) + \frac{1}{2}\frac{\partial^2}{\partial x^2}(\alpha_2(x, t)p(x, t)) \qquad (5.3.20)$$

5.3.1 Discontinuous Markov processes

In the above development we have assumed that the process $X(t)$ undergoes only small displacements in an infinitesimal time interval. It is possible, however, that $X(t)$ can undergo jumps as well as continuous changes. We can say that if at time t the process is in the state x, the probability that at $t + \Delta t$ it will suffer a jump is $q(x, t)\,\Delta t + o(\Delta t)$. Similarly, there is a probability $1 - q(x, t)\,\Delta t + o(\Delta t)$ that no discontinuous change will occur over the interval $(t, t + \Delta t)$. This assumption is, of course, the same as we made in studying discrete state, continuous time Markov processes. Now, however, we must take into account the fact that $X(t)$ may assume a continuous spectrum of values if it does make a jump. Therefore we let $\pi(t, x, \eta)\,d\eta$ be the probability that if a jump occurs at time t with the process at state x, the state after the jump will be in the interval $\eta \leq X(t) < \eta + d\eta$. A process in which changes of state occur *only* by jumps is called *purely discontinuous*.

The transition density of a purely discontinuous process is

$$p(x + \Delta x, t + \Delta t \mid x, t) = [1 - q(x, t)\,\Delta t]\,\delta(\Delta x) + q(x, t)\,\Delta t \pi(t, x, x + \Delta x) + o(\Delta t) \qquad (5.3.21)$$

Thus the probability is small that in a small interval of time a change of state will occur, but if a change does occur, it may be big. On the other hand, for a purely continuous Markov process, the probability of a big change in state in a small time interval is very small (of order less than Δt), although for any Δt, no matter how small, the probability that $X(t + \Delta t) - X(t) \neq 0$ is positive.

We now want to determine a differential equation governing $p(x, t \mid x_0, t_0)$ for a purely discontinuous process. We begin by substituting (5.3.21) into the Chapman-Kolmogorov equation (5.3.6):

$$p(x, t + \Delta t \mid x_0, t_0) = \int_{-\infty}^{\infty} \{[1 - q(x - \Delta x, t)\,\Delta t]\,\delta(\Delta x) + q(x - \Delta x, t) \cdot \Delta t \pi(t, x - \Delta x, x)\} p(x - \Delta x, t \mid x_0, t_0)\,d(\Delta x) \qquad (5.3.22)$$

This equation reduces to

$$p(x, t + \Delta t \mid x_0, t_0) = p(x, t \mid x_0, t_0) - q(x, t)\, \Delta t\, p(x, t \mid x_0, t_0) + \Delta t \int_{-\infty}^{\infty} q(\eta, t)\pi(t, \eta, x) p(\eta, t \mid x_0, t_0)\, d\eta \tag{5.3.23}$$

Dividing by Δt and using (5.3.12), we obtain the forward equation governing $p(x, t \mid x_0, t_0)$ for a purely discontinuous Markov process:

$$\frac{\partial p(x, t \mid x_0, t_0)}{\partial t} = -q(x, t) p(x, t \mid x_0, t_0) + \int_{-\infty}^{\infty} q(\eta, t)\pi(t, \eta, x) p(\eta, t \mid x_0, t_0)\, d\eta \tag{5.3.24}$$

As in the case of purely continuous Markov processes, the first-order density $p(x, t)$ satisfies the same equation as $p(x, t \mid x_0, t_0)$. The initial condition for (5.3.24) is

$$\lim_{t \to t_0} p(x, t \mid x_0, t_0) = \delta(x - x_0) \tag{5.3.25}$$

For a process which can undergo both continuous and discontinuous changes, the forward equation is

$$\frac{\partial p}{\partial t} = -\frac{\partial}{\partial x}(\alpha_1(x, t)p) + \frac{1}{2}\frac{\partial^2}{\partial x^2}(\alpha_2(x, t)p) - q(x, t)p + \int_{-\infty}^{\infty} q(\eta, t)\pi(t, \eta, x) p(\eta, t \mid x_0, t_0)\, d\eta \tag{5.3.26}$$

and the boundary conditions are (5.3.16).

EXAMPLE 5.10. Let us consider a purely continuous Markov process for which the transition density $p(x, t \mid x_0, t_0) = p(u, t \mid t_0)$, where $u = x - x_0$. Thus the process $X(t)$ evolves in such a way that the probability that the increment of the state in the time interval lies in the interval $u \leq X(t) - X(t_0) < u + \Delta u$ is independent of the value of the state at time t_0. The Chapman-Kolmogorov equation becomes

$$p(u, t \mid t_0) = \int_{-\infty}^{\infty} p(u - v, t \mid s) p(v, s \mid t_0)\, dv, \qquad t > s > t_0 \tag{A}$$

Defining

$$m(t) = \lim_{\Delta t \to 0} \frac{1}{\Delta t} \int_{-\infty}^{\infty} u p(u, t + \Delta t \mid t)\, du$$

$$\sigma^2(t) = \lim_{\Delta t \to 0} \frac{1}{\Delta t} \int_{-\infty}^{\infty} u^2 p(u, t + \Delta t \mid t)\, du$$

it can be shown easily that $p(u, t \mid t_0)$ satisfies the forward and backward Fokker-Planck equations:

$$\frac{\partial p}{\partial t} = -m(t)\frac{\partial p}{\partial u} + \frac{1}{2}\sigma^2(t)\frac{\partial^2 p}{\partial u^2} \tag{B}$$

$$-\frac{\partial p}{\partial t_0} = m(t_0)\frac{\partial p}{\partial u} + \frac{1}{2}\sigma^2(t_0)\frac{\partial^2 p}{\partial u^2} \tag{C}$$

We now want to solve either (B) or (C) to obtain the transition density $p(u, t \mid t_0)$. We let

$$\rho = u - \int_{t_0}^{t} m(v)\, dv \tag{D}$$

$$\theta = \int_{t_0}^{t} \sigma^2(v)\, dv \tag{E}$$

so that $p(u, t \mid t_0)$ becomes $p(\rho, \theta \mid t_0)$. Using (D) and (E), the forward and backward equations become

$$\frac{\partial p}{\partial \theta} = \frac{1}{2} \frac{\partial^2 p}{\partial \rho^2} \tag{F}$$

$$-\frac{\partial p}{\partial t_0} = \frac{1}{2} \frac{\partial^2 p}{\partial \rho^2} \tag{G}$$

The solution of these equations corresponding to the conditions

$$\lim_{\theta \to t_0} p(\rho, \theta \mid t_0) = \delta(\rho)$$

$$\lim_{|\rho| \to \infty} p(\rho, \theta \mid t_0) = 0$$

is just the Gaussian density

$$p(\rho, \theta \mid t_0) = \frac{1}{\sqrt{2\pi\theta}} \exp\left(-\frac{\rho^2}{2\theta}\right) \tag{H}$$

To convert back to the original variables, we let

$$M(t, t_0) = \int_{t_0}^{t} m(v)\, dv$$

$$B^2(t, t_0) = \int_{t_0}^{t} \sigma^2(v)\, dv = \theta$$

and (H) becomes

$$p(u, t \mid t_0) = \frac{1}{\sqrt{2\pi} B(t, t_0)} \exp\left\{-\frac{[u - M(t, t_0)]^2}{2B^2(t, t_0)}\right\} \tag{I}$$

An important special case of the particular type of process we are considering here is when $m(t) = 0$ and $\sigma^2(t) = \sigma^2$. In that case, (I) becomes simply

$$p(u, t \mid t_0) = p(u, \tau) = \frac{1}{\sqrt{2\pi\sigma^2\tau}} \exp\left(\frac{-u^2}{2\sigma^2\tau}\right), \qquad \tau = t - t_0 \tag{J}$$

5.3.2 The Wiener process

One particular type of purely continuous Markov process has a wide number of applications to physical problems. This is the Wiener process, customarily denoted by $W(t)$, which has the property that $W(t) - W(s)$ is a real number, normally distributed with

$$E[W(t) - W(s)] = 0$$

$$E\{[W(t) - W(s)]^2\} = \sigma^2 |t - s|, \qquad \sigma > 0$$

In addition, if $t_1 < t_2 < \cdots < t_n$, then $W(t_2) - W(t_1)$, $W(t_3) - W(t_2), \ldots,$ $W(t_n) - W(t_{n-1})$ are independent random variables. We can demonstrate this property as follows [7].

Let Δt be a small increment of time. Define $\Delta W(t) = W(t + \Delta t) - W(t)$. For fixed t and Δt, $W(t + \Delta t)$ and $W(t)$ are both Gaussian random variables, and so $\Delta W(t)$ is also a Gaussian random variable. The transition density for a Wiener process is thus given by

$$p(w, t \mid w_0, t_0) = \frac{1}{[2\pi\sigma^2(t - t_0)]^{1/2}} \exp\left[-\frac{(w - w_0)^2}{2\sigma^2(t - t_0)}\right] \tag{5.3.27}$$

Let us define the conditional probability density

$$p_{\Delta W}\left(\frac{\Delta\xi}{\xi}\right) d(\Delta\xi) = \text{Prob}\,\{\Delta\xi \leq \Delta W(t) < \Delta\xi + d(\Delta\xi) \mid W(t) = \xi\}$$

Since we are conditioning on the fixed event that $W(t) = \xi$, we observe that

$$\begin{aligned}
&\text{Prob}\{\Delta\xi \leq \Delta W(t) < \Delta\xi + d(\Delta\xi) \mid W(t) = \xi\} \\
&\quad = \text{Prob}\{\xi + \Delta\xi \leq W(t) + \Delta W(t) < \xi + \Delta\xi + d(\Delta\xi) \mid W(t) = \xi\} \\
&\quad = \text{Prob}\{\xi + \Delta\xi \leq W(t + \Delta t) < \xi + \Delta\xi + d(\Delta\xi) \mid W(t) = \xi\}
\end{aligned}$$

It follows from (5.3.27) that

$$\begin{aligned}
p_{\Delta W}(\Delta\xi \mid \xi) &= p(\xi + \Delta\xi, t + \Delta t \mid \xi, t) \\
&= \frac{1}{\{2\pi\sigma^2[(t + \Delta t) - t]\}^{1/2}} \exp\left\{-\frac{[(\xi + \Delta\xi) - \xi]^2}{2\sigma^2[(t + \Delta t) - t]}\right\} \\
&= \frac{1}{(2\pi\sigma^2\,\Delta t)^{1/2}} \exp\left[-\frac{(\Delta\xi)^2}{2\sigma^2\,\Delta t}\right] \\
&= p_{\Delta W}(\Delta\xi)
\end{aligned} \tag{5.3.28}$$

where

$$p_{\Delta W}(\Delta\xi)\, d(\Delta\xi) = \text{Prob}\{\Delta\xi \leq \Delta W(t) < \Delta\xi + d(\Delta\xi)\}$$

The key point is that

$$p_{\Delta W}(\Delta\xi \mid \xi) = p_{\Delta W}(\Delta\xi) \tag{5.3.29}$$

namely that the distribution of the increment $\Delta W(t)$ is independent of $W(t)$. Thus we have shown that the Wiener process is a process with *independent increments*. This property is not generally true of stochastic processes or even of Markov processes. From (5.3.28) we see that the random variable $\Delta W(t)$ is Gaussian with mean zero and variance Δt. The fact that the variance of $\Delta W(t)$ is of order Δt and not $(\Delta t)^2$ is an important property of the Wiener process.

5.3.3 Random movements of a particle

One of the classic problems of mathematical physics is that of describing the random motion of a particle in space [3]. In this problem, a particle undergoes a series of displacements $\mathbf{R}_1, \mathbf{R}_2, \ldots, \mathbf{R}_n$, the magnitude and direction of each displacement being independent of all the preceding ones. The probability that displacement i lies between $\mathbf{r}_i$ and $\mathbf{r}_i + d\mathbf{r}_i$ is given by a density function $p_i(\mathbf{r}_i)$. We then ask, What is the probability $p(\mathbf{r})\, d\mathbf{r}$ that after N displacements the coordinates of the particle lie in the interval $(\mathbf{r}, \mathbf{r} + d\mathbf{r})$? The ultimate position $\mathbf{R}$ of the particle is just the result of N independent displacements, $\mathbf{R}_1, \mathbf{R}_2, \ldots, \mathbf{R}_N$. We use the notation $\mathbf{R} = [X, Y, Z]$ to indicate that the displacement is a three-dimensional vector. In essence, the description of the scalar stochastic process in Sections 5.3.1 and 5.3.2 can be envisioned as the wandering of a particle, so that what we suggest here is not new. The only new aspect is that we now generalize to where the state of the process is a three-dimensional vector. In addition, we shall be interested in the state of the process after a large number of displacements. The material in this section has been developed following Chandrasekhar [3], to which the reader may refer for more elaboration.

To get started, let us return to the problem of random walk along a line. In particular, our particle suffers displacements along a straight line in the form of a series of steps of equal length, each step being taken either in the forward or backward direction with equal probability $\frac{1}{2}$. After taking N steps the particle could be at any of the points $-N, -N+1, \ldots, -1, 0, 1, \ldots, N-1$ and N. We ask, What is the probability $p(m, N)$ that the particle is at point m after N steps? Of course, this is just the problem of Bernoulli trials and $p(m, N)$ is simply

$$p(m, N) = \frac{N!}{[\frac{1}{2}(N+m)]!\,[\frac{1}{2}(N-m)]!}\left(\frac{1}{2}\right)^N \tag{5.3.30}$$

for which the mean and mean square displacement are zero and N, respectively. Of interest to us is the case when N is very large and $N \gg m$. However, again, we have already considered this case in Chapter 4 when discussing the central limit theorem. Using Stirling's approximation, as in Example 4.17, we find that

$$\ln p(m, N) \cong -\frac{1}{2}\ln N + \ln 2 - \frac{1}{2}\ln 2\pi - \frac{m^2}{2N}$$

In other words, for large N

$$p(m, N) = \left(\frac{2}{\pi N}\right)^{1/2} \exp\left(\frac{-m^2}{2N}\right) \tag{5.3.31}$$

Now, when N is large it is convenient to introduce instead of the integer position m the net displacement X from the starting point by $X = ml$, where l is the length of a step. Further, if we consider intervals ΔX along the straight line which are large compared with the length of a single step, we can ask for the probability $p(x, N)\,\Delta x$ that the particle will be in the interval $(x, x + \Delta x)$ after N displacements. Clearly,

$$p(x, N)\,\Delta x = p\left(\frac{x}{l}, N\right)\left(\frac{\Delta x}{2l}\right) \tag{5.3.32}$$

since $x/l = m$ can assume only even or odd values depending on whether N is even or odd. Combining (5.3.31) and (5.3.32), we obtain

$$p(x, N) = \frac{1}{(2\pi N l^2)^{1/2}} \exp\left(\frac{-x^2}{2Nl^2}\right) \tag{5.3.33}$$

Finally, suppose that the particle suffers n displacements per time, so that $N/n = t$, a time variable. Then (5.3.33) becomes

$$p(x, t) = \frac{1}{(4\pi D t)^{1/2}} \exp\left(\frac{-x^2}{4Dt}\right) \tag{5.3.34}$$

where $D = \frac{1}{2}nl^2$. This density clearly looks familiar, as it is precisely the density for the scalar (one-dimensional) Wiener process derived in the last section. Reviewing our assumptions for this random walk example, we see that they are essentially the same as for the Wiener process.

We could return to the three-dimensional problem and determine $p(\mathbf{r})$ for the displacements of the particle in space by studying the limit of the three-dimensional symmetric random walk as $N \longrightarrow \infty$. As expected, the computations are considerably more involved than for the one-dimensional case. Nevertheless, the final result is that

$$p(\mathbf{r}, t) = \frac{1}{(4\pi D t)^{1/2}} \exp\left(\frac{-|\mathbf{r}|^2}{4Dt}\right) \tag{5.3.35}$$

where $D = \frac{1}{6}n\sigma^2$, and σ^2 is the mean square displacement at any time.

Let us consider the motion of any body which is subject to random collisions with other bodies, for example, a microscopic particle in a liquid which moves as a result of collisions with the molecules of the liquid or a gas molecule at low pressure. We can let $X(t)$ denote the x-coordinate of the particle at time t which was at the origin at time zero. If the collisions which cause movement of the particle occur instantaneously and irregularly but often, that is, the increment $t - t_0$ is large in comparison with the time interval between two successive collisions, then $X(t) - X(t_0)$ is the sum of a large number of small changes. If the increments depend only on the length of the time interval $t - t_0$ and are independent, then $X(t)$ is a Wiener or so-called *Brownian motion* process. The mean of the process $M(t, t_0) = 0$, and the second moment $E[(X(t) - X(t_0))^2] = 2D^2(t - t_0)$, where D is called the *diffusion coefficient* of the process. The variance of the Brownian motion process is $2D^2$. In fact, since $X(t) - X(t_0)$ may be considered as the sum of a large number of independent random variables, the central limit theorem provides that $X(t) - X(t_0)$ has a normal distribution.

Now that we have seen the relation of these classic random walk problems to the Wiener process, we realize that we need not go through the difficult and tedious procedure of letting $N \longrightarrow \infty$ in the discrete problem to obtain probability densities in each case. We know that the transition density of the Wiener process satisfies the forward (or backward) equations, so that we expect that the problem of random particle movements can also be reduced to a boundary-value problem involving a partial differential equation. This is particularly important when restrictions on the motion of the particle in the form of reflecting and absorbing barriers are present.

Let Δt be an interval of time long enough for a particle to suffer a large number of displacements but still short enough for the net mean square increment $\langle |\Delta \mathbf{R}|^2 \rangle$ to be small (we use the brackets to denote the expected value). In this case, the probability that a particle suffers a net displacement $\Delta \mathbf{r}$ in time Δt is given by

$$\psi(\Delta \mathbf{r}, \Delta t) = \frac{1}{(4\pi D\,\Delta t)^{1/2}} \exp\left(\frac{-|\Delta \mathbf{r}|^2}{4D\,\Delta t}\right) \tag{5.3.36}$$

and is independent of $\mathbf{R}$. We still have $p(\mathbf{r}, t)$ as the probability density that after a total time t the particle lies in the interval $(\mathbf{r}, \mathbf{r} + \Delta \mathbf{r})$. The relation between $p(\mathbf{r}, t)$ and $\psi(\Delta \mathbf{r}, \Delta t)$ is given by [see (5.3.5)]

$$p(\mathbf{r}, t + \Delta t) = \int_{-\infty}^{\infty}\int_{-\infty}^{\infty}\int_{-\infty}^{\infty} p(\mathbf{r} - \Delta \mathbf{r}, t)\psi(\Delta \mathbf{r}, \Delta t)\, d(\Delta \mathbf{r}) \tag{5.3.37}$$

Our objective is to derive a differential equation for $p(\mathbf{r}, t)$. This problem is, of course, identical to that in the derivation of (5.3.20), the only difference being that here the stochastic process is a three-dimensional vector rather than a scalar. We can anticipate the result as a three-dimensional form of the forward Fokker-Planck equation, particular to this process.

We expand $p(\mathbf{r} - \Delta \mathbf{r}, t)$ in a Taylor series about $p(\mathbf{r}, t)$:

$$\begin{aligned} p(\mathbf{r} - \Delta \mathbf{r}, t) = p(\mathbf{r}, t) &- \frac{\partial p}{\partial x}\Delta x - \frac{\partial p}{\partial y}\Delta y - \frac{\partial p}{\partial z}\Delta z \\ &+ \frac{1}{2}\left[\frac{\partial^2 p}{\partial x^2}(\Delta x)^2 + \frac{\partial^2 p}{\partial y^2}(\Delta y)^2 + \frac{\partial^2 p}{\partial z^2}(\Delta z)^2\right. \\ &\left. + 2\Delta x\,\Delta y\frac{\partial^2 p}{\partial x\,\partial y} + 2\Delta x\,\Delta z\frac{\partial^2 p}{\partial x\,\partial z} + 2\Delta y\,\Delta z\frac{\partial^2 p}{\partial y\,\partial z}\right] - \cdots \end{aligned} \tag{5.3.38}$$

Substituting (5.3.36) and (5.3.38) into (5.3.37), we obtain, after dividing by Δt and taking the limit as $\Delta t \longrightarrow 0$, the forward equation for this process:

$$\frac{\partial p(\mathbf{r}, t)}{\partial t} = D\left(\frac{\partial^2 p}{\partial x^2} + \frac{\partial^2 p}{\partial y^2} + \frac{\partial^2 p}{\partial z^2}\right) \tag{5.3.39}$$

This is simply the diffusion equation. This result actually provides us with a molecular interpretation of the macroscopic diffusion equation. Up to this point we have been speaking of a single particle suffering displacements and have determined the probability of finding this particle in a certain volume of space at some time. However, we can imagine a very large number of particles starting under the same initial conditions and undergoing the displacements with no interference and ask for the *fraction* of the total number which will be found in a certain volume at some later time. The two problems are equivalent, so that $p(\mathbf{r}, t)$ is simply the *concentration* of the particles at position $\mathbf{r}$ at time t. Thus the motion of a large number of particles undergoing random displacements can be described macroscopically as a diffusion process with the diffusion coefficient $D = \frac{1}{6}n\langle R^2 \rangle$.

In the case of infinite boundaries the solution of (5.3.39) is just (5.3.35). If there are absorbing or reflecting barriers, then the boundary conditions on (5.3.39) are $p = 0$ and $\nabla p = 0$ (normal to the surface), respectively. Thus the problem of the random movement of a particle in space has been reduced, in the limit of a large number of steps, to the solution of a boundary-value problem involving the diffusion equation. This is a much simpler way to solve the problem than evaluating discrete densities in the limit as $N \longrightarrow \infty$.

We have considered only symmetric processes up to now. Clearly, there may be preferred directions for particle movement. A more general form of $\psi(\Delta\mathbf{r}, \Delta t)$ would be

$$\psi(\Delta\mathbf{r}, \Delta t) = \frac{1}{8(\pi\,\Delta t)^{1/2} D_X D_Y D_Z} \exp\left[-\frac{(\Delta x - u_X\,\Delta t)^2}{4D_X\,\Delta t} - \frac{(\Delta y - u_Y\,\Delta t)^2}{4D_Y\,\Delta t} - \frac{(\Delta z - u_Z\,\Delta t)^2}{4D_Z\,\Delta t}\right] \tag{5.3.40}$$

where the mean displacement in the x-direction is no longer zero but is $u_X\,\Delta t$, u_X being a constant. Substituting (5.3.40) and (5.3.38) into (5.3.37), dividing by Δt, and letting $\Delta t \longrightarrow 0$ yields the forward equation

$$\frac{\partial p(\mathbf{r}, t)}{\partial t} = -u_X\frac{\partial p}{\partial x} - u_Y\frac{\partial p}{\partial y} - u_Z\frac{\partial p}{\partial z} + D_X\frac{\partial^2 p}{\partial x^2} + D_Y\frac{\partial^2 p}{\partial y^2} + D_Z\frac{\partial^2 p}{\partial z^2} \tag{5.3.41}$$

where we see that the us can be interpreted as macroscopic velocities.

An example of the application of the problem of random particle motions to the analysis of chemical processing systems is the theoretical description of axial dispersion in packed beds. The problem of axial dispersion in randomly packed beds of solids has been a subject of considerable interest to chemical engineers for a number of years. Although the prediction of the degree of dispersion resulting for a specific type of packing cannot be determined exactly, the problem is amenable to a probabilistic analysis.

Dispersion in a packed bed results from the combined influence of convection and molecular diffusion and may be considered to be a stochastic process. A recent comprehensive theoretical analysis of packed bed dispersion from a stochastic viewpoint has been presented by Gunn [8].

Essentially, we are interested in the change of position with time of a tracer molecule introduced into the system at $t = 0$ at a point designated as the origin of our coordinate system. The coordinates of the molecule at time t, $X(t)$, $Y(t)$, and $Z(t)$, are continuous random variables. These random variables are functions of the distribution of solid surfaces and molecular interactions. Since both the packing distribution and the molecular motions are completely random, we assume that X, Y, and Z are independent. We do not know $p(x, y, z, t)$ in general, but we can investigate its behavior for long times $t \gg T$, where T is the time required for the average flow to move a distance equal to a characteristic length of the solid packing.

5.4 Characterization of Stochastic Processes

We have now introduced some of the common mathematical representations of stochastic processes. In process analysis we are often confronted with a random output signal from a process which we want to model. In most cases the signal varies continuously (such as a temperature or concentration recording) with time. Our first task is to characterize the random signal in terms of its statistical properties. Then we can proceed to suggest one of the stochastic models, if appropriate, for the system. Let us now consider the problem of describing or characterizing a continuous state, continuous time stochastic process, one for which we may not have a mathematical model. We begin by referring back to Figure 5.1, in which three possible sample functions for a random process $X(t)$ between t_a and t_b are shown. Instead of considering the instantaneous values of $X(t)$, let us look at the entire function $X(t)$ on (t_a, t_b). The observation of a random function such as $X(t)$ over an interval involves an infinity of measurements, and so the sample space of $X(t)$ can be said to have infinite dimension. Alternatively, we can define the sample space as the collection of all possible sample functions, that is, the collection of all possible $X(t)$ vs. t plots in (t_a, t_b), of which three are shown in Figure 5.1. A sample space so defined is called an *ensemble*. Thus an ensemble of sample functions can be visualized as the outputs of an infinite number of identical, but independent, systems. The randomness of the problem is associated with the occurrence of a particular sample function. The probability distribution over the ensemble is the probability distribution associated with observing a particular sample function.

The outcome of a physical experiment will be one sample function, but, of course, we do not know which one of the infinite number possible it will be. We can thus define a continuous time stochastic process in the following way. It is an infinite ensemble of sample functions, i.e., $X(t)$ functions, from which the outcome of a particular experiment can be envisioned as one of these chosen at random. The concept of an ensemble of processes is useful in characterizing a stochastic process, as we shall see.

5.4.1 Stationary processes

If the statistical properties of a stochastic process do not change with time, we refer to the process as *stationary*. Consider first a random sequence

$X_1, X_2, \ldots,$ as might result, for example, from regular observations of a random process. This sequence is stationary if

1. The marginal distribution for X_k is independent of k.
2. The joint distribution for X_k and X_{k+i} is independent of k for arbitrary values of i.
3. The joint distribution for X_k, X_{k+i}, and X_{k+j} is independent of k for arbitrary values of i and j, etc.

Thus stationarity of a random sequence implies that every member has the same marginal distribution and that the joint distribution of two members depends only on the difference in the indices of the two members. If the members are independent, as well as identically distributed, the sequence is stationary. In that case, the joint distribution for X_k and X_{k+i} is independent of both k and i. The mean value $E(X_k)$ as well as the higher-order moments are independent of k. The product moments, $E(X_k X_{k+i})$ are independent of k and depend only on i.

Consider now the definition of stationarity for continuous time stochastic processes. A continuous time stochastic process $X(t)$ is stationary if

1. The marginal distributions of $X(t_1)$, $X(t_2)$, $X(t_3)$, . . . are identical.
2. The joint distribution for $X(t_1)$ and $X(t_2)$ depends only on the difference $t_2 - t_1$.
3. The joint distribution for $X(t_1)$, $X(t_2)$, and $X(t_3)$ depends only on the differences $t_3 - t_2$ and $t_2 - t_1$, etc.

In short, a stationary process is one for which the marginal and joint probability densities are independent of a shift of the time axis; that is, the pair of random variables $X(t_1 + \tau)$ and $X(t_2 + \tau)$ have the same joint distribution as the variables $X(t_1)$ and $X(t_2)$.

All the instantaneous values $X(t)$ of a stationary process have the same mean, and the higher moments are independent of the time t of observation. The mean value of $X(t)$, at a particular time t, is an ensemble average, since it is the average value of $X(t)$ at time t for the infinite number of sample functions of $X(t)$ on (t_a, t_b). The ensemble average of the signals, or sample functions, at time t can be represented as the arithmetic average of the values of all the signals at time t:

$$\lim_{N\to\infty} \frac{1}{N} \sum_{i=1}^{N} X_i(t)$$

Thus the ensemble average is taken vertically along a line $t = t$ in Figure 5.1, not horizontally along a single sample function. For a stationary process, all vertical lines should give the same average value as the number of processes gets very large.

Next consider the product moment $E[X(t_1)X(t_2)]$. Given a large number of noise generators, we can observe all the instantaneous values at t_1 and t_2. Then for each generator we can compute $X(t_1)X(t_2)$. The arithmetic average of all the products

$$\lim_{N\to\infty} \frac{1}{N} \sum_{i=1}^{N} X_i(t_1)X_i(t_2)$$

is then the ensemble average $E[X(t_1)X(t_2)]$. For a stationary process, this product moment is a function only of $t_2 - t_1$. A process $X(t)$ is called *stationary in the wide sense* if $E[X(t)]$ is independent of t and for any t_1 and t_2 $E[X(t_1)X(t_2)]$ depends only on $t_2 - t_1$. Similarly, a random sequence X_1, X_2, X_3, . . . is stationary in the wide sense if $E(X_k)$ is independent of k and for any k and l, $E(X_k X_l)$ depends only on $l - k$. Note that the definition of wide stationarity *only considers the first- and second-order ensemble averages.* Thus a process stationary in the *wide sense* may not necessarily be stationary in the *strict sense*, as there may be higher moments that do depend on t.

We have been discussing ensemble averages, envisioned as being performed along vertical lines at constant t in Figure 5.1. In practice, we can never assemble an infinite number of random noise generators, so that ensemble averages can never really be measured. Usually we shall only have *one* sample function available to us for measurement (although, of course, we do not know a priori which of the infinite ones possible it will be). If we are to obtain averages for this single sample function, they will have to be made over time, that is, horizontally on Figure 5.1.

The time average of a random sequence is

$$\bar{X} = \frac{1}{N}(X_1 + X_2 + \cdots + X_N) \tag{5.4.1}$$

where $\bar{X}$ is a random variable itself since it is a linear combination of random variables. [In contrast, the ensemble average, denoted by $E(X_k)$, is not a random variable but a number. Also note that $E(X_k)$ depends in general on k except in the stationary case.] Normally, we let $N \longrightarrow \infty$ so that (5.4.1) becomes

$$\bar{X} = \lim_{N\to\infty} \frac{1}{N} \sum_{i=1}^{N} X_i \tag{5.4.2}$$

$\bar{X}$ does not necessarily equal $E(X_k)$; the former is a random variable, whereas the latter is not, and the former does not depend on k, whereas the latter may. A random sequence is called *ergodic* if

1. The ensemble average $E(X_k)$ is independent of k.
2. The time average $\bar{X}$ equals a single constant with unit probability.
3. $E(X_k) = \bar{X}$.
4. Conditions similar to 1–3 hold for all averages; e.g., $E(X_k^2)$ must be independent of k and must equal $\overline{X^2} = \lim_{N\to\infty} (1/N) \sum_{i=1}^{N} X_i^2$, etc.

Condition 1 is satisfied if the sequence is stationary. Condition 2 states that the time average must be the same for all the sample functions. If conditions 1 and 2 are fulfilled, then condition 3 states that the ensemble average can be obtained from a single sample function of infinite length.

Suppose that the elements of the sequence are independent and identically distributed and thus stationary. Then it does not matter whether the mean value of X_k is computed from an ensemble average of all the kth elements (which, of course, are identical to all the other elements) or a time average of all the elements in a single sample function. Even in the case in which the random sequence is stationary with dependence among related members of the sequence, we can justify the ergodic property by taking the time average over a long enough period to make negligible the effect of short-range dependence.

The time average of a random process is defined by $\bar{X}(t)$ as in

$$\bar{X}(t) = \lim_{T\to\infty} \frac{1}{2T} \int_{t-T}^{t+T} X(\tau)\, d\tau \tag{5.4.3}$$

A random process $X(t)$ is ergodic if

1. The ensemble average $E[X(t)]$ is independent of t.
2. The time average $\bar{X}(t)$ equals a single constant with unit probability.
3. $E[X(t)] = \bar{X}(t)$.
4. Conditions similar to 1–3 hold for all averages, e.g., the mean square and the product moments.

As before, stationarity is a necessary condition for ergodicity.

The ensemble mean and mean square values of the process $X(t)$ at time t are computed from the first-order density function $p(x, t)$ by

$$E[X(t)] = \int_{-\infty}^{\infty} xp(x, t)\, dx \tag{5.4.4}$$

$$E[X^2(t)] = \int_{-\infty}^{\infty} x^2 p(x, t)\, dx \tag{5.4.5}$$

where we reiterate that $p(x, t)\, dx$ is the probability that $X(t)$ will lie in the interval $(x, x + dx)$ at time t. We stress again that $p(x, t)$ is *not* a joint density of two variables x and t; rather, t only indicates that $p(x, t)$ is the density function for the random variable X at time t.

We have also already introduced the second-order density function $p(x_1, t_1; x_2, t_2)$, such that $p(x_1, t_1; x_2, t_2)\, dx_1\, dx_2$ is the joint probability that $X(t)$ will lie in the interval $(x_1, x_1 + dx_1)$ at t_1 and in the interval $(x_2, x_2 + dx_2)$ at t_2. It is clear that

$$p(x_1, t_1) = \int_{-\infty}^{\infty} p(x_1, t_1; x_2, t_2)\, dx_2 \tag{5.4.6}$$

and that

$$p(x_2, t_2) = \int_{-\infty}^{\infty} p(x_1, t_1; x_2, t_2)\, dx_1 \tag{5.4.7}$$

A process is strictly stationary if *all* the probability density functions are independent of a shift in the time axis.

EXAMPLE 5.11. Let us determine the first-order probability density function $p(x, t)$ for the random process [16]

$$X(t) = a \cos(\omega_0 t + \Theta)$$

where Θ is a uniformly distributed random variable on the interval $(0, 2\pi)$. Therefore the random process $X(t)$ is a function of a random variable Θ. Thus, to determine $p(x, t)$ in terms of $p(\theta)$, we need

$$p(x, t) = \frac{p(\theta)}{dx/d\theta}$$

We note, however, that X is a double-valued function of Θ, since there are two values of Θ corresponding to each value of X. Thus we need

$$p(x, t) = \frac{p(\theta_1)}{|dx/d\theta_1|} + \frac{p(\theta_2)}{|dx/d\theta_2|}$$

For both roots θ_1 and θ_2, $p(\theta) = \frac{1}{2}\pi$ and $x = a\cos(\omega_0 t + \theta)$, and so

$$\begin{aligned}\frac{1}{dx/d\theta} &= a \sin(\omega_0 t + \theta) \\ &= a[1 - \cos^2(\omega_0 t + \theta)]^{1/2} \\ &= a\left[1 - \frac{x^2}{a^2}\right]^{1/2} \\ &= \sqrt{a^2 - x^2}\end{aligned}$$

Thus

$$p(x, t) = \frac{1}{\pi\sqrt{a^2 - x^2}}, \qquad -a \le x \le a$$

We note that $p(x, t)$ is independent of t; that is, the amplitude distribution of the ensemble is the same for all values of t. Therefore at any value of t the probability of finding the random process $X(t)$ lying between x and $x + dx$ is given by $p(x, t)$. Finally, we note that this is a case in which the random process can be described analytically. In many cases the process is too complicated to allow analytical determination of $p(x, t)$.

5.4.2 The autocorrelation function

We have mentioned in the last section the product moments $E(X_k X_l)$ and $E[X(t_1)X(t_2)]$. We shall now consider these in more depth.

The product moment $E(X_k X_l)$ is called the *autocorrelation function* of the random sequence $X_1, X_2, \ldots$. The prefix "auto" refers to correlation of the sequence with itself. The autocorrelation function depends on k and l. The autocorrelation function $E(X_k X_l)$ refers to the ensemble average of the products $X_k X_l$ for all the sample functions. In a stationary sequence, $E(X_k X_{k+i})$ is denoted by R_i, where

$$R_i = E(X_k X_{k+i}) = \lim_{N \to \infty} \frac{1}{N} \sum_{j=1}^{N} X_{j_k} X_{j_{k+1}} \tag{5.4.8}$$

where X_{j_k} is the kth member of the jth sequence of the ensemble. Clearly, R_0 is just the mean square value of X_k.

The autocorrelation function R_i of a stationary random sequence has the following properties:

1. R_i is a symmetric function of i; that is,
$$R_i = E(X_k X_{k+i}) = E(X_{k-i} X_k) = R_{-i}$$
2. $R_i \leq R_0$ for all i (see Problem 5.10).
3. If the mean $E(X_k)$ of the process is μ,
$$R_i \geq 2\mu^2 - R_0$$
4. If the variance $\sigma^2(X_k)$ of the process is finite, and as $i \longrightarrow \infty$, X_k and X_{k+i} become asymptotically uncorrelated:
$$\lim_{i \to \infty} R_i = \mu^2$$

The autocorrelation $R(t_1, t_2)$ of a continuous time process $X(t)$ is the joint moment $E[X(t_1) X(t_2)]$,

$$R(t_1, t_2) = \int_{-\infty}^{\infty} x_1 x_2 p(x_1, t_1; x_2, t_2)\, dx_1\, dx_2 \tag{5.4.9}$$

For a stationary process $R(t_1, t_2)$ depends only on the difference $t_2 - t_1 = \tau$.

The autocorrelation function $R(\tau)$ of a stationary random process has the following properties:

1. $R(\tau)$ is an even function of τ; i.e., $R(\tau) = R(-\tau)$.
2. $R(\tau) \leq R(0)$.
3. $R(\tau) \geq 2\mu^2 - R(0)$.
4. If the variance $\sigma^2_{X(t)}$ is finite, and $\tau \longrightarrow \infty$, the instantaneous measurements $X(t)$ and $X(t - \tau)$ become uncorrelated, $R(\tau) \longrightarrow \mu^2$.

If the random processes are ergodic, then the ensemble average autocorrelation function must be equal to the time average autocorrelation. For a random sequence and process, the time average autocorrelations are

$$R_i = \lim_{N \to \infty} \frac{1}{N} (X_1 X_{1+i} + \cdots + X_N X_{N+i})$$

and

$$R(\tau) = \lim_{T\to\infty} \frac{1}{2T} \int_{-T}^{T} X(t)X(t+\tau)\,dt$$

The question arises, Is there a test to determine whether a process is ergodic or not? First, it must be established that the process is stationary. This can be done only through extensive calculations of means and mean squares at several instants of time. If these are time-invariant, we can then calculate various time averages for several sample functions. If these agree, then the process can be classed as ergodic, at least to the order of the average calculated.

EXAMPLE 5.12. Determine the autocorrelation function for a random process

$$X(t) = a\cos(\omega_0 t + \Theta)$$

where Θ is a uniformly distributed random variable on the interval $(0, 2\pi)$. By definition, $R(t_1, t_2) = E[X(t_1)X(t_2)]$, and so

$$\begin{aligned} R(t_1, t_2) &= a^2 E[\cos(\omega_0 t_1 + \Theta)\cos(\omega_0 t_2 + \Theta)] \\ &= \frac{a^2}{2} E[\cos(\omega_0 t_1 - \omega_0 t_2) + \cos(\omega_0 t_1 + \omega_0 t_2 + 2\Theta)] \\ &= \frac{a^2}{2} E[\cos(\omega_0 t_1 - \omega_0 t_2)] + E[\cos(\omega_0 t_1 + \omega_0 t_2 + 2\Theta)] \end{aligned}$$

The first quantity on the right-hand side is not random, and so

$$E[\cos(\omega_0 t_1 - \omega_0 t_2)] = \cos[\omega_0(t_1 - t_2)]$$

Then

$$\begin{aligned} E\{\cos[\omega_0(t_1 + t_2) + 2\Theta]\} &= \int_{-\infty}^{\infty} \cos[\omega_0(t_1 + t_2) + 2\theta]p(\theta)\,d\theta \\ &= \frac{1}{2\pi}\int_0^{2\pi} \cos[\omega_0(t_1 + t_2) + 2\theta]\,d\theta = 0 \end{aligned}$$

Thus

$$R(t_1, t_2) = \frac{a^2}{2}\cos[\omega_0(t_1 - t_2)]$$

All that can be concluded about this process is that it is stationary in the wide sense.

Let us briefly summarize the classifications of stochastic processes. The most fundamental division is between stationary and nonstationary processes. A process is stationary if all orders of its probability density functions are invariant under a shift of the time axis. Clearly, for a stationary process $p(x, t)$ is independent of t, and the second-order density $p(x_1, t_1; x_2, t_2) = p(x_1, x_2; t_2 - t_1)$. For the nth-order density function of a stationary process we have that $p(x_1, x_2, \ldots, x_n; t_1, t_2, \ldots, t_n) = p(x_1, x_2, \ldots, x_n; \tau_1, \tau_2, \ldots, \tau_{n-1})$, where $\tau_i = t_{i+1} - t_1$. It is not an easy task to determine whether or

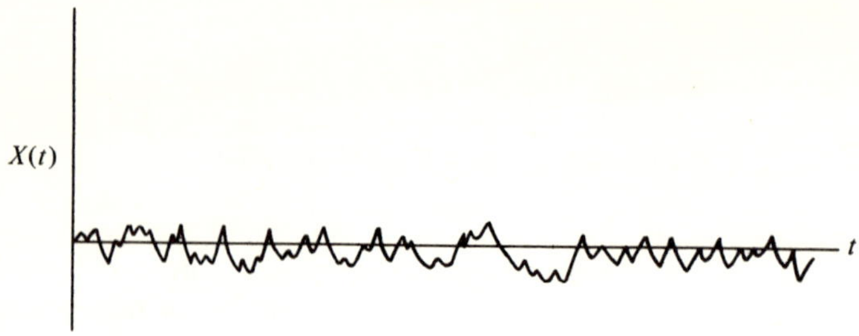

(a) A Stationary Random Process

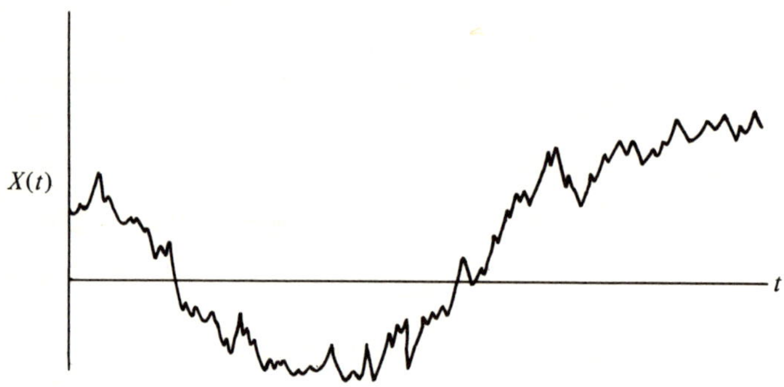

(b) A Nonstationary Random Process with Time-Varying Mean Value

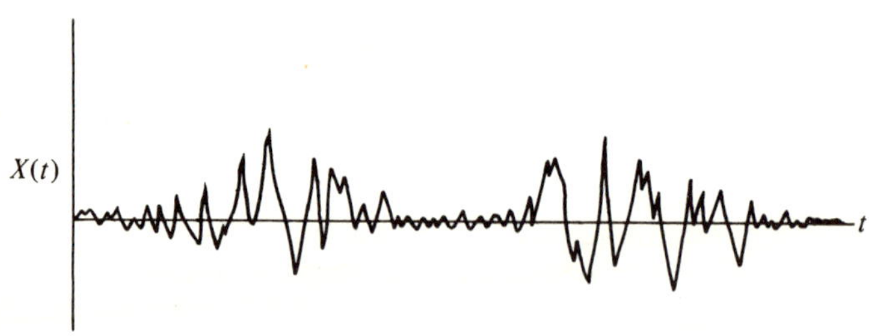

(c) A Nonstationary Random Process with Time-Varying Mean Square

Figure 5.3. Stationary and nonstationary random processes. (Reprinted from J. S. Bendat and A. G. Piersol, *Measurement and Analysis of Random Data*, John Wiley & Sons, Inc., New York, 1966, p. 334. Reprinted by permission of the copyright owner.)

not a process is stationary. It may be possible to assess whether a process is stationary or not from the underlying physical and chemical mechanisms governing the process. If there is no change in the basic nature of the process, its inputs, or its parameters with time, then it may be likely that the process is stationary. Stationary and nonstationary random processes are shown in Figure 5.3.

As we have noted, some processes are stationary in a limited sense, such as wide-sense stationary. For most practical applications, particularly with linear systems, wide-sense stationarity is sufficient for detailed analysis. The analysis of nonstationary processes is, on the other hand, quite difficult. We shall mainly be concerned with stationary or at least wide-sense stationary processes in this book.

The second fundamental division applies only to stationary processes and it is the division into ergodic and nonergodic processes. In an ergodic process the complete statistics can be determined from one sample function, since every sample function carries with it identical statistical information. Since it is only possible to determine time averages from a single sample function, it is clear that in an ergodic process the time averages must be identical to the ensemble averages. Thus the amplitude distribution of the ensemble at any time t is identical to the amplitude distribution of a single sample function. It should be clear that an ergodic process must be a stationary process, since there is only one set of probability density functions that can be obtained from a single sample function, and if this sample function contains all the statistical information on the process, the process must be stationary. A nonstationary process has an infinite set of density functions depending on the time origin.

An example of ergodicity can be illustrated by the process of injecting some tracer in an ideally mixed vessel with a constant, time-invariant through-

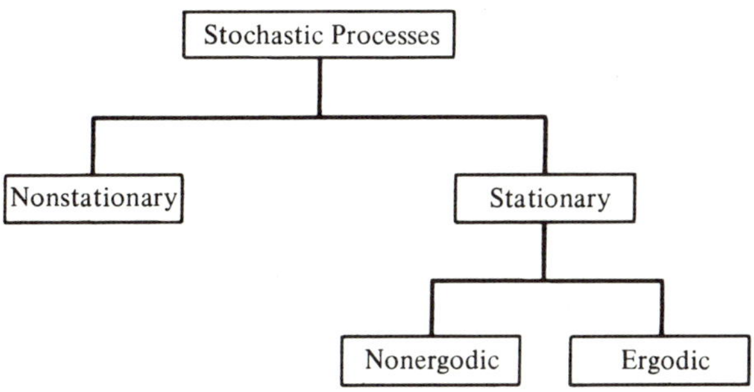

Figure 5.4. Classification of stochastic processes. (Reprinted from D. M. Himmelblau, *Process Analysis by Statistical Methods*, John Wiley & Sons, Inc., New York, 1970, p. 8. Reprinted by permission of the copyright owner.)

put. Let us take one tracer molecule and place it in the inlet stream and record its time of departure from the tank. Then let us repeat this experiment a large number of times to get a distribution of residence times. On the other hand, let us release a large number of molecules at the same time in the inlet (not enough to disturb the flow, however) and observe the residence time of each to obtain a distribution. Thus the statistical description of a single molecule going through the tank a large number of times is identical to that of a large number of molecules introduced simultaneously in the tank in one experiment.

We summarize the classification of stochastic processes in Figure 5.4.

5.4.3 The cross-correlation function

For two random sequences $X_1, X_2, \ldots$ and $Y_1, Y_2, \ldots$ we define the *cross-correlation function* $R_{XY}(k, l)$ as the ensemble average:

$$R_{XY}(k, l) = E(X_k Y_l) \tag{5.4.10}$$

Similarly, for two random processes, $X(t)$ and $Y(t)$, the cross-correlation function $R_{XY}(t_1, t_2)$ is defined by the ensemble average:

$$R_{XY}(t_1, t_2) = E[X(t_1)Y(t_2)] \tag{5.4.11}$$

The processes $X(t)$ and $Y(t)$ are jointly stationary if the joint density function of $X(t_1), \ldots, X(t_N), Y(t'_1), \ldots, Y(t'_M)$ is invariant under shift of the time axis for all values of $t_1, t_2, \ldots, t_N, t'_1, \ldots, t'_M$. The processes are jointly stationary in the wide sense if only their means and cross-correlation function are invariant under shift of the time axis, namely

$$R_{XY}(t_1, t_2) = R_{XY}(t_2 - t_1) \tag{5.4.12}$$

If $X(t)$ and $Y(t)$ are jointly ergodic, at least of order 2, then $R_{XY}(\tau)$ is equal to $\bar{R}_{XY}(\tau)$, the time-averaged cross-correlation function

$$\bar{R}_{XY}(\tau) = \lim_{T\to\infty} \frac{1}{2T} \int_{-T}^{T} X(t)Y(t + \tau)\, dt \tag{5.4.13}$$

Two processes $X(t)$ and $Y(t)$ are uncorrelated if their cross-correlation function is equal to the product of their means. The processes are called orthogonal if $R_{XY}(\tau) = 0$.

The cross-correlation function of stationary processes has the following properties:

1. $R_{XY}(\tau) = R_{YX}(-\tau)$
2. $|R_{XY}(\tau)| \leq [R_{XX}(0) R_{YY}(0)]^{1/2}$

From this point on we will not draw a distinction as to whether an auto- or cross-correlation function has been determined by an ensemble or a time average. Theoretically, an ensemble average would be desired, but practically, a time average is necessary. In general, we will confine our attention to stationary processes, for which, by the ergodic hypothesis, both methods of averaging should give the same result. Henceforth, we will omit overbars on the correlation functions.

5.4.4 Response of a linear system to a random input

We have already seen in Chapter 3 that the transfer function $W(s)$ is a basic characteristic of a linear dynamic system if in the analysis of its properties we employ the plane of the real or the complex variable s. If we conduct the analysis in the time domain, then the time function $w(t)$ which corresponds to $W(s)$ is given by L.T.

$$W(s) = \int_0^{\infty} e^{-st} w(t)\, dt$$

For ease in subsequent discussion and to conform to the literature we use here

$$w(t) \xrightarrow{L} W(s)$$

$$w(t) \xleftarrow{L^{-1}} W(s)$$

rather than the nomenclature of Chapter 3.

It is customary to refer to $w(t)$ as the *impulse response* or *system weighting* function. The response $y(t)$ of a linear system to an input $x(t)$ can be determined by inversion of $W(s)X(s)$, if $W(s)$ is known. Alternately, we can determine $y(t)$ produced by $x(t)$ if $w(t)$ is known. We represent $x(t)$ as the sum of a large number of pulses, each of which has a width Δt_i and height $x(t_i)$ Each pulse can be represented in the form $x(t_i)\, \Delta t_i\, \delta(t - t_i)$ and, consequently, $x(t)$ can be approximately expressed as

$$x(t) \cong \sum_{i=-n}^{n} x(t_i)\delta(t - t_i)\, \Delta t_i \tag{5.4.14}$$

The transient produced by the unit impulse $\delta(t - t_i)$ is the impulse response $w(t - t_i)$. Thus, the output produced by each impulse in the train is $x(t_i)\, \Delta t_i$ $w(t - t_i)$, and the output $y(t)$ produced by the series of impulses (5.4.14) is

$$y(t) = \sum_{i=-n}^{n} x(t_i) w(t - t_i)\, \Delta t_i$$

Taking the limit as $\Delta t_i \longrightarrow 0$, this becomes

$$y(t) = \int_{-\infty}^{\infty} w(t - \tau) x(\tau)\, d\tau = \int_{-\infty}^{\infty} w(\tau) x(t - \tau)\, d\tau \tag{5.4.15}$$

In physical systems $w(t - \tau) = 0$ for $\tau > t$, so that

$$y(t) = \int_{\infty}^{t} w(t - \tau) x(\tau)\, d\tau$$

Introducing a new integration variable $\lambda = t - \tau$, we have

$$y(t) = -\int_{\infty}^{0} w(\lambda)x(t-\lambda)\,d\lambda = \int_{0}^{\infty} w(\tau)x(t-\tau)\,d\tau \tag{5.4.16}$$

Equation (5.4.16) is the fundamental time domain relation for a scalar linear system. We see that it is merely a convolution integral (recall 3.2.44) which could have been obtained directly from the L.T. relations.

Now we suppose that the input $x(t)$ is a stationary random function with autocorrelation $R_{XX}(\tau)$. We wish to find the autocorrelation function of the output $R_{YY}(\tau)$ as a function of $R_{XX}(\tau)$ and $w(\tau)$. The autocorrelation function for $y(t)$ is

$$R_{YY}(\tau) = \lim_{T\to\infty} \frac{1}{2T} \int_{-T}^{T} y(t)y(t+\tau)\,dt \tag{5.4.17}$$

Using (5.4.16) in (5.4.15) gives (we need not assume that $w(t) = 0$ for $t < 0$)

$$R_{YY}(\tau) = \lim_{T\to\infty} \frac{1}{2T} \int_{-\infty}^{\infty} w(t_1)x(t-t_1)\,dt_1 \int_{-\infty}^{\infty} w(t_2)x(t+\tau-t_2)\,dt_2\,dt$$

Interchanging the order of the limit process and integration,

$$R_{YY}(\tau) = \int_{-\infty}^{\infty} w(t_1)\,dt_1 \int_{-\infty}^{\infty} w(t_2)\,dt_2 \lim_{T\to\infty} \frac{1}{2T} \int_{-T}^{T} x(t-t_1)x(t+\tau-t_2)\,dt \tag{5.4.18}$$

The term following the limit is just $R_{XX}(\tau + t_1 - t_2)$, so we obtain our desired result,

$$R_{YY}(\tau) = \int_{-\infty}^{\infty} w(t_1)\,dt_1 \int_{-\infty}^{\infty} w(t_2)\,R_{XX}(\tau + t_1 - t_2)\,dt_2 \tag{5.4.19}$$

The second fundamental relation is that between the autocorrelation function of the input and the cross-correlation function between the input and output. Using (5.4.13) and (5.4.15), we get

$$R_{XY}(\tau) = \lim_{T\to\infty} \frac{1}{2T} \int_{-T}^{T} x(t) \int_{-\infty}^{\infty} w(t_2)x(t+\tau-t_2)\,dt_2\,dt \tag{5.4.20}$$

Interchanging the order of integration and the limiting procedure, there results

$$R_{XY}(\tau) = \int_{-\infty}^{\infty} w(t_2)R_{XX}(\tau - t_2)\,dt_2 \tag{5.4.21}$$

We note that $R_{XY}(\tau)$ is simply a convolution of $w(t)$ and the autocorrelation of the input. *Thus $R_{XY}(\tau)$ may be interpreted as the response of a linear system when the input is the autocorrelation of the actual input.*

We can combine (5.4.19) and (5.4.21) to obtain the alternative relation

$$R_{YY}(\tau) = \int_{-\infty}^{\infty} w(t_1)R_{XY}(\tau + t_1)\,dt_1 \tag{5.4.22}$$

Thus $R_{YY}(\tau)$ may be interpreted as the convolution of $w(t)$ with the cross-correlation function, with τ running backwards to $-\infty$.

5.4.5 The power density spectrum

Consider for the moment a deterministic function $F(t)$. A useful characteristic of a signal $F(t)$ is often its *energy*, given by

$$E = \int_{-\infty}^{\infty} F^2(t)\,dt \tag{5.4.23}$$

If $f(\omega)$ is the Fourier transform of $F(t)$ (see Section 3.1.2), then (5.4.23) can be written as

$$E = \int_{-\infty}^{\infty} F(t)\left[\frac{1}{2\pi}\int_{-\infty}^{\infty} f(\omega)e^{i\omega t}\,d\omega\right]dt \tag{5.4.24}$$

Interchanging the order of integration,

$$E = \frac{1}{2\pi}\int_{-\infty}^{\infty} f(\omega)\left[\int_{-\infty}^{\infty} F(t)e^{i\omega t}\,dt\right]d\omega \tag{5.4.25}$$

The inner integral is just $f(-\omega)$, and so

$$E = \frac{1}{2\pi}\int_{-\infty}^{\infty} f(\omega)f(-\omega)\,d\omega \tag{5.4.26}$$

which may be written alternatively as

$$E = \frac{1}{2\pi}\int_{-\infty}^{\infty} |f(\omega)|^2\,d\omega \tag{5.4.27}$$

This result is known as *Parseval's theorem*. The quantity $|f(\omega)|^2(d\omega/2\pi)$ represents the fraction of the total energy of $F(t)$ that is concentrated in the frequency range $(\omega/2\pi, (\omega + d\omega)/2\pi)$. We can thus consider $|f(\omega)|^2$ as an *energy density*, and we commonly use the phrase *energy density spectrum*, denoting $|f(\omega)|^2$ by $\Phi_F(\omega)$.

Since $\Phi_F(\omega)$ is an even function of ω, we can simply write (5.4.27) as

$$E = \frac{1}{\pi}\int_{0}^{\infty} \Phi_F(\omega)\,d\omega \tag{5.4.28}$$

We can obtain the energy density spectrum of a function $F(t)$ in a somewhat different manner. Consider the integral

$$\int_{-\infty}^{\infty} F(t)F(t+\tau)\,dt \tag{5.4.29}$$

the Fourier transform of which is

$$\begin{aligned}\int_{-\infty}^{\infty} e^{-i\omega t}\,dt\int_{-\infty}^{\infty} F(t)F(t+\tau)\,dt &= \int_{-\infty}^{\infty} F(t)\,dt\int_{-\infty}^{\infty} F(t+\tau)e^{-i\omega t}\,dt\\ &= \int_{-\infty}^{\infty} F(t)e^{i\omega t}\,dt\int_{-\infty}^{\infty} F(\zeta)e^{-i\omega\zeta}\,d\zeta\\ &= f(\omega)f(-\omega)\\ &= |f(\omega)|^2\end{aligned} \tag{5.4.30}$$

Thus the energy density spectrum of $F(t)$ can be obtained by Fourier transform of the integral (5.4.29).

The concept of energy of a function $F(t)$ is meaningful only if (5.4.23) is finite. Some functions, such as periodic signals or noise signals, have infinite energy, and E as in (5.4.23) cannot be used as a measure of $F(t)$. In such a case, we define the *average power* P of the function $F(t)$ by

$$P = \lim_{T\to\infty} \frac{1}{2T} \int_{-T}^{T} F^2(t)\, dt \tag{5.4.31}$$

which is just the mean square value of $F(t)$.

Let us now proceed as in the case of the energy and obtain a power density spectrum. We form a new function $F_T(t)$ by truncating $F(t)$ outside the interval $|t| < T$, i.e.,

$$F_T(t) = \begin{cases} F(t), & |t| < T \\ 0, & |t| > T \end{cases} \tag{5.4.32}$$

As long as T is finite, $F_T(t)$ has finite energy, given by

$$\int_{-\infty}^{\infty} F_T^2(t)\, dt = \int_{-T}^{T} F^2(t)\, dt \tag{5.4.33}$$

Thus

$$\frac{1}{2T} \int_{-T}^{T} F^2(t)\, dt = \frac{1}{2\pi} \int_{-\infty}^{\infty} \frac{|f_T(\omega)|^2}{T}\, d\omega \tag{5.4.34}$$

The average power is given by

$$P = \lim_{T\to\infty} \frac{1}{2T} \int_{-T}^{T} F^2(t)\, dt = \frac{1}{2\pi} \int_{-\infty}^{\infty} \lim_{T\to\infty} \frac{|f_T(\omega)|^2}{T}\, d\omega \tag{5.4.35}$$

If the limit in the integral exists, we define $S_F(\omega)$ as

$$S_F(\omega) = \lim_{T\to\infty} \frac{|f_T(\omega)|^2}{T} \tag{5.4.36}$$

the *power density spectrum* of $F(t)$, so that

$$P = \frac{1}{2\pi} \int_{-\infty}^{\infty} S_F(\omega)\, d\omega \tag{5.4.37}$$

From (5.4.36) we see that $S_F(\omega)$ is a real and even function of ω. $S_F(\omega)\, d\omega$ can be interpreted as the fraction of the total average power of $F(t)$ in the frequency range $(\omega/2\pi, (\omega + d\omega)/2\pi)$.

As in the case of the energy density spectrum, the power density spectrum can be obtained in an alternative way. In particular, the power density spectrum can be obtained from the Fourier transform of

$$R_{FF}(\tau) = \lim_{T\to\infty} \frac{1}{2T} \int_{-T}^{T} F(t)F(t + \tau)\, dt \tag{5.4.38}$$

We can show that the Fourier transform of $R_{FF}(\tau)$ is $S_{FF}(\omega)$. $R_{FF}(\tau)$ is, of course, the time-averaged autocorrelation function of $F(t)$, $\overline{F(t)F(t+\tau)}$. [If $F(t)$ is a complex signal, then $R_{FF}(\tau)$ is defined as $\overline{F^*(t)F(t+\tau)}$, where $F^*(t)$ is the complex conjugate of $F(t)$.] Thus the power density spectrum of a function $X(t)$ is given by the Fourier transform of the time-averaged autocorrelation function $R_{XX}(\tau)$.

What if $X(t)$ is a random process? In this case we have already introduced the correlation functions. The Fourier transforms of the correlation functions, called the power density spectra, of random processes are thus important characteristics of the random processes. For ergodic processes all sample functions have the same averages and hence the power density spectra of all sample functions are identical.

We define $S_{XX}(\omega)$ and $S_{XY}(\omega)$ as the Fourier transforms of $R_{XX}(\tau)$ and $R_{XY}(\tau)$, respectively:

$$S_{XX}(\omega) = \int_{-\infty}^{\infty} R_{XX}(\tau)e^{-i\omega\tau}\,d\tau \tag{5.4.39}$$

$$S_{XY}(\omega) = \int_{-\infty}^{\infty} R_{XY}(\tau)e^{-i\omega\tau}\,d\tau \tag{5.4.40}$$

We may think of the power density spectrum of a random process as the ensemble average of the power density spectra of all the sample functions.

$S_{XX}(\omega)$ is a positive, real and even function of ω. We can write (5.4.39) in the form

$$S_{XX}(\omega) = \int_{-\infty}^{\infty} R_{XX}(\tau)\cos\omega\tau\,d\tau - i\int_{-\infty}^{\infty} R_{XX}(\tau)\sin\omega\tau\,d\tau \tag{5.4.41}$$

Since $R_{XX}(\tau)$ is an even function of τ and $\sin\omega\tau$ is an odd function of τ, the second integral on the right-hand side is zero. Thus, we write (5.4.39) as

$$\begin{aligned} S_{XX}(\omega) &= \int_{-\infty}^{\infty} R_{XX}(\tau)\cos\omega\tau\,d\tau \\ &= 2\int_{0}^{\infty} R_{XX}(\tau)\cos\omega\tau\,d\tau \end{aligned} \tag{5.4.42}$$

EXAMPLE 5.13. Let us determine the relationship between the spectral densities of the input and output signals of a linear system. Substituting $R_{YY}(\tau)$ given by (5.4.39) into

$$S_{YY}(\omega) = \int_{-\infty}^{\infty} R_{YY}(\tau)e^{-i\omega\tau}\,d\tau$$

we obtain

$$\begin{aligned} S_{YY}(\omega) &= \int_{-\infty}^{\infty}\int_{-\infty}^{\infty}\int_{-\infty}^{\infty} w(\eta)w(\lambda)R_{XX}(\tau+\lambda-\eta)\,d\eta\,d\lambda\,d\tau \\ &= \int_{-\infty}^{\infty}\int_{-\infty}^{\infty}\int_{-\infty}^{\infty} e^{-i\omega(\tau+\lambda-\eta)}e^{i\omega\lambda}e^{-i\omega\eta}R_{XX}(\tau+\lambda-\eta)w(\lambda)w(\eta)\,d\eta\,d\lambda\,d\tau \\ &= \int_{-\infty}^{\infty} w(\eta)e^{-i\omega\eta}\,d\eta\int_{-\infty}^{\infty} w(\lambda)e^{i\omega\lambda}\,d\lambda\int_{-\infty}^{\infty} R_{XX}(\tau+\lambda-\eta)e^{-i\omega(\tau+\lambda-\eta)}\,d\tau \\ &= W(i\omega)W(-i\omega)S_{XX}(\omega) \\ &= |W(i\omega)|^2S_{XX}(\omega) \end{aligned}$$

Similarly, the relation between $S_{XY}(\omega)$ and $S_{XX}(\omega)$ is

$$S_{XY}(\omega) = W(i\omega)S_{XX}(\omega)$$

5.4.6 Discussion of the correlation functions

The autocorrelation function is the most significant characteristic in the analysis of random processes. This function is a measure of the rapidity of variation of a given random process. Consider a sample function $X(t)$, with $R(\tau) = \overline{X(t)X(t+\tau)}$. If the process varies slowly, it will not change significantly in τ seconds. Thus we expect some similarity between $X(t)$ and $X(t+\tau)$ such that it is possible to obtain information about $X(t+\tau)$ knowing $X(t)$. As τ increases, the interdependence of $X(t)$ and $X(t+\tau)$ will decrease. The value of τ at which $R(\tau)$ approaches zero will measure the speed of variation of $X(t)$. If $X(t)$ varies rapidly, $R(\tau)$ may approach zero for very small values of τ. If $X(t)$ varies rapidly or slowly, we say that $X(t)$ contains mainly high- or low-frequency components, respectively. In Figure 5.5 typical autocorrelation functions for a slowly and rapidly varying process are shown. For nonperiodic random processes, $R(\tau) \longrightarrow 0$ as $\tau \longrightarrow \infty$, as shown. From Figure 5.5

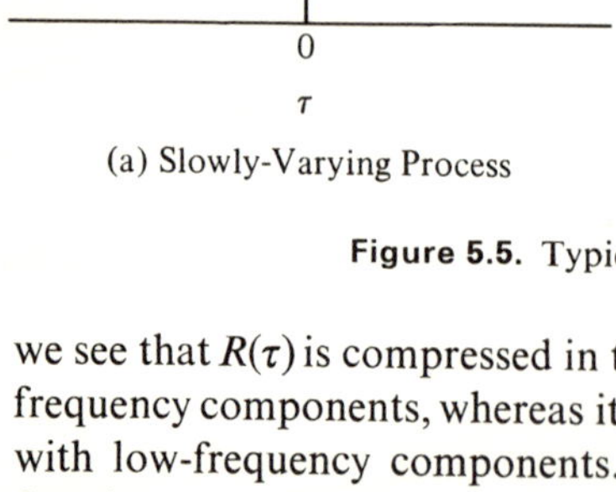

(a) Slowly-Varying Process

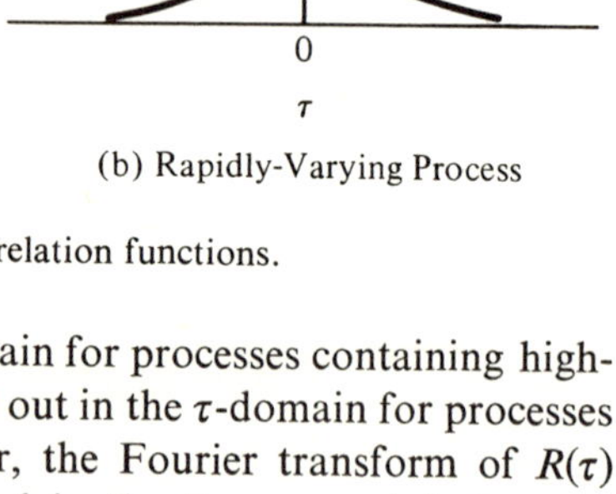

(b) Rapidly-Varying Process

Figure 5.5. Typical autocorrelation functions.

we see that $R(\tau)$ is compressed in the τ-domain for processes containing high-frequency components, whereas it is spread out in the τ-domain for processes with low-frequency components. However, the Fourier transform of $R(\tau)$ for slowly varying processes is compressed in the frequency (ω) domain, whereas it is expanded for high-frequency processes.

The autocorrelation function of a given stationary process can be computed from the second-order probability density function $p(x_1, x_2; \tau)$ by

$$R(\tau) = \int_{-\infty}^{\infty}\int_{-\infty}^{\infty} x_1 x_2 p(x_1, x_2; \tau)\, dx_1\, dx_2 \qquad (5.4.43)$$

The cross-correlation function of two stochastic processes $X(t)$ and $Y(t)$ is a characterization of the interdependence of the two variables. It is defined by

$$R_{XY}(t_1, t_2) = E[X(t_1)Y(t_2)] = \int_{-\infty}^{\infty}\int_{-\infty}^{\infty} xyp(x, y; t_1, t_2)\,dx\,dy \qquad (5.4.44)$$

If the processes are stationary, $R_{XY}(t_1, t_2) = R_{XY}(t_2 - t_1) = R_{YX}(t_1 - t_2)$. Cross-correlation functions are useful for (1) investigating the degree of statistical independence between two random processes, (2) estimating the transfer function of a system from statistical correlations (see Example 5.13), and (3) determining the time delay for errors to propagate through a system [$R_{XY}(\tau)$ will be a maximum at τ equal to the time delay]. In Figure 5.6 we show a cross-correlation function for two random processes. We note that $\tau = \tau_1$ is the delay time between errors in $X(t)$ producing responses in $Y(t)$.

We might also introduce two other functions of interest. The first is the *cross-covariance* for two random processes $X(t)$ and $Y(t)$, which is defined as

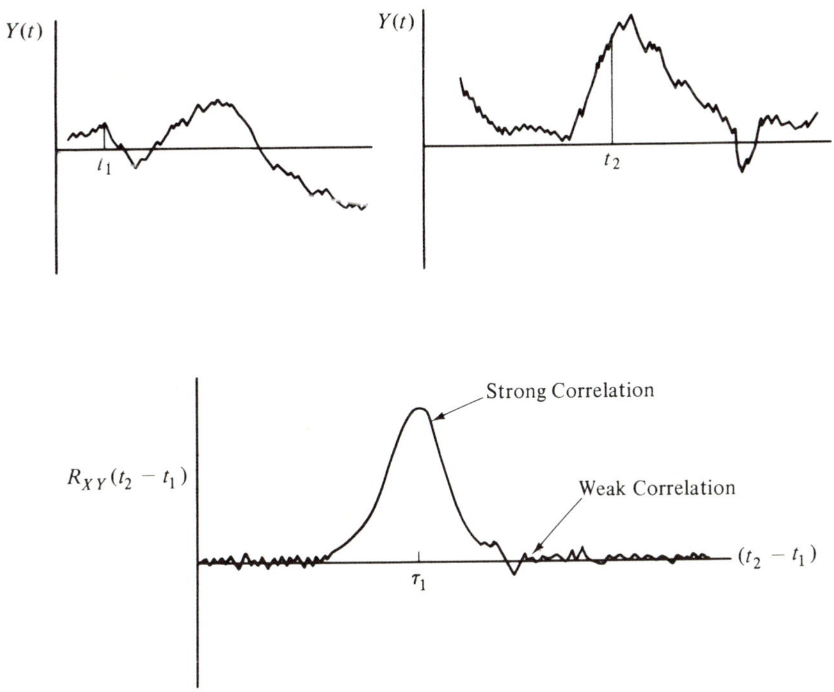

Figure 5.6. Typical cross-correlation function for two random processes $X(t)$ and $Y(t)$. (Reprinted from D. M. Himmelblau, *Process Analysis by Statistical Methods*, John Wiley & Sons, Inc., New York, 1970, p. 19. Reprinted by permission of the copyright owner.)

$\sigma_{XY}(t_1, t_2) = E\{[X(t_1) - \mu_X(t_1)][Y(t_2) - \mu_Y(t_2)]\}$. For stationary processes $\sigma_{XY}(t_1, t_2) = \sigma_{XY}(\tau) = R_{XY}(\tau) - \mu_X\mu_Y$. The second function is the *correlation coefficient* $\rho_{XY}(\tau) = \sigma_{XY}(\tau)/(\sigma_X(0)\sigma_Y(0))$. If X and Y are uncorrelated, $\sigma_{XY}(\tau)$ and $\rho_{XY}(\tau) = 0$. Also, if X and Y are independent, σ_{XY} and ρ_{XY} are zero, but the converse is not necessarily true.

5.4.7 Power density spectra of typical signals

Let us now consider the properties of some common random signals with respect to their power spectra. This is a problem central to communication theory. We shall be necessarily brief, although the interested reader can consult Davenport and Root [4] or Jenkins and Watts [10] for much deeper treatments.

By means of several examples, we shall illustrate the computation of power density spectra for some common random processes.

EXAMPLE 5.14. Determine the power density spectrum of the random process considered in Example 5.11, namely,

$$X(t) = a\cos(\omega_0 t + \Theta)$$

where Θ is a uniformly distributed random variable over $(0, 2\pi)$. In Example 5.12 we determined the autocorrelation function as

$$R(\tau) = \frac{a^2}{2}\cos\omega_0\tau$$

Then, $S(\omega)$ is the Fourier transform of $R(\tau)$, namely

$$S(\omega) = \frac{\pi a^2}{2}[\delta(\omega - \omega_0) + \delta(\omega + \omega_0)]$$

EXAMPLE 5.15. Let us determine the autocorrelation function and power density spectrum of a random process which consists of unit impulses randomly distributed along the time axis according to the Poisson distribution, as shown in Figure 5.7(a) [16]. First consider the process shown in Figure 5.7(b), consisting of pulses of height h and width ϵ such that $h\epsilon = 1$. Eventually, we shall let $h \longrightarrow \infty$ and $\epsilon \longrightarrow 0$. Recall that for the Poisson distribution the probability of observing an event in a time interval dt is $\lambda\,dt$, where λ is the average number of occurrences per second.

Since $X(t)$ is now discrete (it can be zero or h), the autocorrelation is given by

$$R(\tau) = \sum_{x_1}\sum_{x_2} x_1 x_2\, p(x_1, x_2)$$

For $\tau > \epsilon$, x_1 can be either zero or h, as can x_2. Thus $R(\tau)$ consists of four terms:

$$\begin{aligned} R(\tau) &= (h \times 0)p(h, 0) + (h \times h)p(h, h) + (0 \times h)p(0, h) + (0 \times 0)p(0, 0) \\ &= h^2 p(h, h) \end{aligned}$$

where $p(h, h)$ is the joint probability of $X_1 = h$ and $X_2 = h$. We can express this joint probability as

$$p(h, h) = p_{X_1}(h)p_{X_2}(h \mid X_1 = h)$$

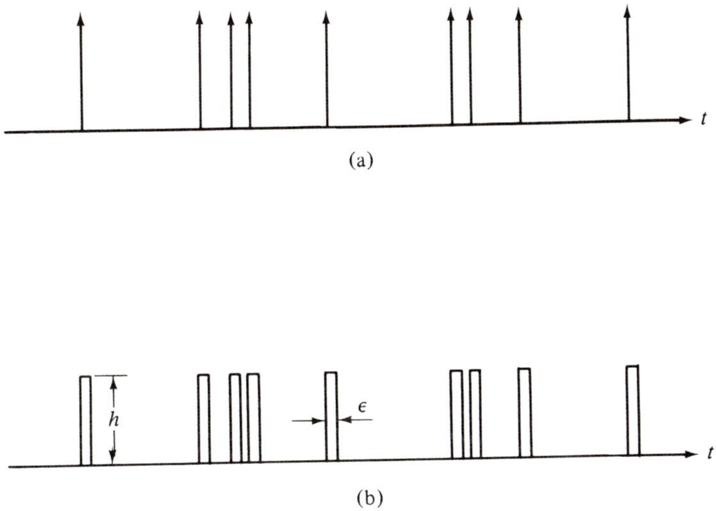

Figure 5.7. Poisson-distributed unit impulses.

If $\tau > \epsilon$, the two impulses must be different (since their widths are ϵ). The Poisson distribution is characterized by the fact that events are independent. Thus

$$p_{X_2}(h \mid X_1 = h) = p_{X_2}(h)$$

and

$$R(\tau) = h^2 p(h, h) = h^2 p_{X_1}(h) p_{X_2}(h)$$

To compute $p_{X_1}(h)$ we note that λ impulses per second occupy a total time of $\lambda\epsilon$ seconds. Therefore the probability of observing an impulse at any random time is simply $\lambda\epsilon$, the fraction of an average second occupied by pulses; i.e., $p_{X_1}(h) = p_{X_2}(h) = \lambda\epsilon$. Thus

$$R(\tau) = h^2 \lambda^2 \epsilon^2 = \lambda^2$$

Now consider $\tau < \epsilon$, for which $X_1 = h$ and $X_2 = h$ must represent the same pulse. If a pulse is observed at time t, the probability of observing the same pulse at $t + \tau$ is $1 - \tau/\epsilon$; i.e.,

$$P_{X_2}(h \mid X_1 = h) = 1 - \frac{\tau}{\epsilon}, \qquad \tau < \epsilon$$

Thus

$$p(h, h) = p_{X_1}(h) p_{X_2}(h \mid X_1 = h)$$
$$= \lambda\epsilon \left(1 - \frac{\tau}{\epsilon}\right)$$

and

$$R(\tau) = h^2 \lambda\epsilon \left(1 - \frac{\tau}{\epsilon}\right), \qquad \tau < \epsilon$$

Since $R(\tau)$ is an even function of τ,

$$R(\tau) = h^2\lambda\epsilon\left(1 - \frac{|\tau|}{\epsilon}\right), \qquad |\tau| < \epsilon$$

As $\epsilon \longrightarrow 0$, $R(\tau)$ approaches an impulse of strength λ:

$$R(\tau) = \lambda\,\delta(\tau)$$

Summing the contributions to $R(\tau)$ from $|\tau| > \epsilon$ and $|\tau| < \epsilon$, we obtain

$$R(t) = \lambda\delta(\tau) + \lambda^2$$

The power density spectrum is the Fourier transform of $R(\tau)$:

$$S(\omega) = \lambda + 2\pi\lambda^2\,\delta(\omega)$$

The random process in which the impulses can assume both positive and negative values is called *white noise*. It can be shown (Problem 5.15) that for such a process

$$R(\tau) = \lambda\,\delta(\tau)$$

$$S(\omega) = \lambda$$

The power density spectrum is a constant, independent of frequency ω. The process is called *white* noise, because of this fact, namely that the process contains all frequencies or "colors." The concept of white noise will be quite important for use in Chapter 10.

For additional examples of the computation of the autocorrelation and power density spectrum of specific random processes, we refer the reader to Davenport and Root [4] and Papoulis [18].

EXAMPLE 5.16. Let us consider the detection of a periodic signal in the presence of noise by means of the autocorrelation function. The problem is to determine whether a received signal, heavily corrupted with noise, contains a periodic component. Such problems are important in radar detection and brain-wave analysis.

First, we must show that if a random process has a periodic component, of period T, then the autocorrelation function also has a periodic component of period T. Say

$$X(t) = X(t + T) = X(t + nT)$$

then

$$\begin{aligned} R(\tau) &= \overline{X(t)X(t + \tau)} \\ &= \overline{X(t)X(t + nT + \tau)} \\ &= R(\tau + nT) \end{aligned}$$

Now suppose that a periodic random process $X(t)$ is corrupted by a random noise process $N(t)$. Assume that $X(t)$ and $N(t)$ are independent. The autocorrelation function of the process $X(t) + N(t)$ is

$$\begin{aligned} R(\tau) &= \overline{[X(t) + N(t)][X(t + \tau) + N(t + \tau)]} \\ &= R_{XX}(\tau) + R_{NN}(\tau) + R_{NX}(\tau) + R_{XN}(\tau) \end{aligned}$$

Since $X(t)$ and $N(t)$ are uncorrelated,

$$R_{NX}(\tau) = R_{XN}(\tau) = \overline{XN}$$

As $\tau \longrightarrow \infty$, from the properties of the autocorrelation function,

$$\lim_{\tau \to \infty} R_{NN}(\tau) \longrightarrow \bar{N}^2$$

Thus, since $R_{XX}(\tau)$ is periodic,

$$\lim_{\tau \to \infty} R(\tau) = R_{XX}(\tau) + 2\overline{XN} + \bar{N}^2$$

Since $\bar{X}$ and $\bar{N}$ are constants, $R(\tau)$, for large τ, exhibits periodicity of period T.

Now let us summarize the power spectral densities of some common stochastic processes. Table 5.4 shows both the autocorrelation function and spectral density of several processes.

TABLE 5.4
AUTOCORRELATIONS AND POWER DENSITY SPECTRA OF SOME COMMON STOCHASTIC PROCESSES

Stochastic process	Autocorrelation	Power density spectra
$a \cos(\omega_0 t + \Theta)$	$\frac{a^2}{2} \cos \omega_0 \tau$	$\frac{\pi a^2}{2}[\delta(\omega - \omega_0) + \delta(\omega + \omega_0)]$
Poisson-distributed positive unit impulses	$\lambda\, \delta(\tau) + \lambda^2$	$\lambda + 2\pi\lambda^2\, \delta(\omega)$
Poisson-distributed positive and negative unit impulses (white noise)	$\lambda\delta(\tau)$	λ
Random binary sequence (two possible states, ± 1, with switches occurring randomly at b-second intervals)	$1 - \frac{\lvert\tau\rvert}{b}$, $\lvert\tau\rvert < b$ 0, $\lvert\tau\rvert > b$	$\frac{4 \sin^2 (\omega b/2)}{b\omega^2}$
Thermal noise (abrupt changes in amplitude at random instants of time; the number of changes are Poisson-distributed with parameter α)	$\overline{X^2} e^{-\alpha\lvert\tau\rvert}$ $\overline{X^2}$ = mean square value of the process	$\frac{2\alpha\overline{X^2}}{\alpha^2 + \omega^2}$

5.4.8 Statistical analysis of a chemical reactor [1]

To illustrate the use of the correlation functions and power density spectra, let us consider the analysis of a continuous stirred tank reactor (CSTR) subject to random fluctuations in certain operating parameters. The object is to study the effect on the outputs (concentration and temperature) of random fluctuations in the inputs. We define

$X(t')$ = concentration of species being consumed in the reactor
$T(t')$ = temperature of the reactor fluid
q = volumetric flow rate through the reactor
V = reactor volume
X_0, T_0 = inlet concentration and temperature
q_c = rate of coolant flow through coil inside reactor
T_c = temperature of coolant
ρ = density of fluid in reactor
C_p = heat capacity of fluid in reactor
$r(X, T)$ = rate of consumption of species X
$U^*(T, T_c, q_c)$ = rate of heat removal by the internal coil
ΔH = heat of reaction
t' = time

The variables subject to fluctuation are q, X_0, T_0, q_c, and T_c. The mean values of these variables will be denoted by an overbar.

The dynamic mass and energy balance relations for the reactor are

$$V\frac{dX}{dt'} = q(X_0 - X) - Vr(X, T) \tag{5.4.45}$$

$$\rho VC_p\frac{dT}{dt'} = \rho C_p q(T_0 - T) + (-\Delta H)Vr(X, T) - VU^*(T, T_c, q_c) \tag{5.4.46}$$

Define the dimensionless variables:

$$t = \frac{qt'}{V} \qquad \zeta = \frac{X}{\bar{X}_0} \qquad \zeta_0 = \frac{X_0}{\bar{X}_0}$$

$$\eta = \frac{\rho C_p T}{(-\Delta H)\bar{X}_0} \qquad \kappa = \frac{q}{\bar{q}} \qquad \eta_0 = \frac{\rho C_p T_0}{(-\Delta H)\bar{X}_0}$$

$$P = \frac{Vr}{\bar{q}\bar{X}_0} \qquad \kappa_c = \frac{q_c}{\bar{q}} \qquad U = \frac{VU^*}{(-\Delta H)\bar{q}\bar{X}_0}$$

$$\eta_c = \frac{\rho C_p T_c}{(-\Delta H)\bar{X}_0}$$

where ζ_0 and κ have mean value 1, and η_0, η_c, and κ_c have mean values $\bar{\eta}_0$, $\bar{\eta}_c$, and $\bar{\kappa}_c$, respectively.

Then (5.4.45) and (5.4.46) become

$$\frac{d\zeta}{dt} = \kappa(\zeta_0 - \zeta) - P(\zeta, \eta) \tag{5.4.47}$$

$$\frac{d\eta}{dt} = \kappa(\eta_0 - \eta) + P(\zeta, \eta) - U(\eta, \eta_c, \kappa_c) \tag{5.4.48}$$

When the inputs q, X_0, T_0, q_c, and T_c are held constant at their mean values, the steady state of the reactor is

$$0 = 1 - \zeta_s - P(\zeta_s, \eta_s) \tag{5.4.49}$$

$$0 = \bar{\eta}_0 - \eta_s + P(\zeta_s, \eta_s) - U(\eta_s, \bar{\eta}_c, \bar{\kappa}_c) \tag{5.4.50}$$

Let us define the deviations or perturbations from steady state:

$$Z(t) = \zeta(t) - \zeta_s \tag{5.4.51}$$

$$Y(t) = \eta(t) - \eta_s \tag{5.4.52}$$

Subtracting the steady-state equations, (5.4.49) and (5.4.50), from (5.4.47) and (5.4.48) yields

$$\begin{aligned}\frac{dZ}{dt} &= -\kappa Z - [P(\zeta_s + Z, \eta_s + Y) - P(\zeta_s, \eta_s)] \\ &\quad + (\kappa\zeta_0 - 1) + (1 - \kappa)\zeta_s\end{aligned} \tag{5.4.53}$$

$$\begin{aligned}\frac{dY}{dt} &= -\kappa Y + [P(\zeta_s + Z, \eta_s + Y) - P(\zeta_s, \eta_s)] \\ &\quad + (\kappa\bar{\eta}_0 - \eta_0) + (1 - \kappa)\eta_s \\ &\quad - [U(\eta_s + Y, \eta_c, \kappa_c) - U(\eta_s, \bar{\eta}_c, \bar{\kappa}_c)]\end{aligned} \tag{5.4.54}$$

Now, assume that the fluctuations in the variables q, X_0, T_0, q_c, and T_c are small, so that perturbations away from the steady state are small. Then we can expand the bracketed terms in (5.4.53) and (5.4.54) in first-order Taylor series. Define

$$\alpha = \left(\frac{\partial P}{\partial \zeta}\right)_s, \qquad \beta = \left(\frac{\partial U}{\partial \eta}\right)_s, \qquad \gamma = \left(\frac{\partial P}{\partial \eta}\right)_s$$

$$\delta = -\left(\frac{\partial U}{\partial \eta_c}\right)_s, \qquad \epsilon = \left(\frac{\partial U}{\partial \kappa_c}\right)_s$$

and the new random variables λ, μ, ν, λ_c, and ν_c, all with zero means, by

$$\lambda = \kappa - 1, \qquad \mu = \zeta_0 - 1, \qquad \nu = \eta_0 - \bar{\eta}_0$$

$$\lambda_c = \kappa_c - \bar{\kappa}_c, \qquad \nu_c = \eta_c - \bar{\eta}_c$$

Neglecting second-order terms (products of perturbations), after expanding the bracketed terms in (5.4.53) and (5.4.54), we obtain

$$\frac{dZ}{dt} + (1 + \alpha)Z + \gamma Y = (1 - \zeta_s)\lambda + \mu \tag{5.4.55}$$

$$\frac{dY}{dt} - \alpha Z + (1 + \beta - \gamma)Y = (\bar{\eta}_0 - \eta_s)\lambda + \nu + \delta\nu_c - \epsilon\lambda_c \tag{5.4.56}$$

Let the right-hand sides of (5.4.55) and (5.4.56) be denoted by $E(t)$ and $F(t)$, respectively. Equations (5.4.55) and (5.4.56) are linear, constant coefficient, nonhomogeneous O.D.E.s. Taking the Fourier transform of these equations (letting lowercase letters denote the transformed variable), we can solve for $z(i\omega)$ and $y(i\omega)$:

$$z(i\omega) = \frac{i\omega + 1 + \beta - \gamma}{\Delta(\omega)} e(i\omega) + \frac{-\gamma}{\Delta(\omega)} f(i\omega) \tag{5.4.57}$$

$$y(i\omega) = \frac{\alpha}{\Delta(\omega)} e(i\omega) + \frac{i\omega + 1 + \alpha}{\Delta(\omega)} f(i\omega) \tag{5.4.58}$$

where

$$\Delta(\omega) = -\omega^2 + i\omega(1 + \alpha + 1 + \beta - \gamma) + (1 + \alpha)(1 + \beta) - \gamma \tag{5.4.59}$$

We write (5.4.57) and (5.4.58) in compact form as

$$\begin{bmatrix} z \\ y \end{bmatrix} = \begin{bmatrix} h_{11}(\omega) & h_{12}(\omega) \\ h_{21}(\omega) & h_{22}(\omega) \end{bmatrix} \begin{bmatrix} e \\ f \end{bmatrix} = \mathbf{h}(\omega) \begin{bmatrix} e \\ f \end{bmatrix} \tag{5.4.60}$$

Thus the states of the system are $Z(t)$ and $Y(t)$, with transforms $z(i\omega)$ and $y(i\omega)$ and the inputs are $E(t)$ and $F(t)$, with transforms $e(i\omega)$ and $f(i\omega)$. The system transfer function is the $\mathbf{h}(\omega)$ matrix in (5.4.60).

Now let $S_{ZZ}(\omega)$, $S_{YY}(\omega)$, and $S_{ZY}(\omega)$ be the power density spectra of the states $Z(t)$ and $Y(t)$, and let $S_{EE}(\omega)$, $S_{FF}(\omega)$, and $S_{EF}(\omega)$ be those of the inputs $E(t)$ and $F(t)$. We want to relate the power density spectra of the states to those of the inputs. It can be shown that for a linear system with two states and two inputs:

$$S_{ZZ}(\omega) = |h_{11}(\omega)|^2 S_{EE}(\omega) + 2Re[h_{11}^*(\omega)h_{12}(\omega)S_{EF}(\omega)] + |h_{12}(\omega)|^2 S_{FF}(\omega) \tag{5.4.61}$$

$$S_{YY}(\omega) = |h_{21}(\omega)|^2 S_{EE}(\omega) + 2Re[h_{21}^*(\omega)h_{22}(\omega)S_{EF}(\omega)] + |h_{22}(\omega)|^2 S_{FF}(\omega) \tag{5.4.62}$$

$$\begin{aligned} S_{ZY}(\omega) = {} & h_{11}^*(\omega)h_{21}(\omega)S_{EE}(\omega) + h_{11}^*(\omega)h_{22}(\omega)S_{EF}(\omega) \\ & + h_{12}^*(\omega)h_{21}(\omega)S_{FE}(\omega) + h_{12}^*(\omega)h_{22}(\omega)S_{FF}(\omega) \end{aligned} \tag{5.4.63}$$

Our strategy will be to compute the power density spectra of the states in terms of those of the inputs, and then to invert the power density spectra to obtain the auto- and cross-correlation functions among the states. This is in contrast to our prior procedure of first computing the correlation functions and then Fourier-transforming to obtain the power spectral densities.

Recall that

$$E(t) = (1 - \zeta_s)\lambda + \mu \tag{5.4.64}$$

$$F(t) = (\eta_0 - \eta_s)\lambda + \nu + \delta\nu_c - \epsilon\lambda_c \tag{5.4.65}$$

so that fluctuations in parameters λ, μ, ν, δ, and ϵ can be assumed and the resulting fluctuations in $E(t)$ and $F(t)$ calculated.

Aris and Amundson [1] considered the following parameter values:

$V = 100 \text{ ft}^3$ $\qquad -\Delta H = 20{,}000$ Btu/lb-mole

$q = 0.3 \text{ ft}^3/\text{sec}$ $\qquad \rho C_p = 60$ Btu/ft-°K

$\bar{X}_0 = 1 \text{ lb-mole/ft}^3$ $\qquad r(X, T) = X^2(2.7 \times 10^{11})\exp(-44{,}700/RT)$

$\bar{T}_0 = 650°\text{K}$

and the following separate disturbances:

1. $S_{\mu\mu}(\omega) = 1/\pi$ (white noise in inlet concentration).
2. $S_{\nu\nu}(\omega) = 1/\pi$ (white noise in inlet temperature).
3. $S_{\lambda\lambda}(\omega) = 1/\pi$ (white noise in reactor flow rate).

In the analysis, it is necessary to

1. Compute $S_{EE}(\omega)$, $S_{FF}(\omega)$, and $S_{EF}(\omega)$ from $S_{\mu\mu}(\omega)$, $S_{\nu\nu}(\omega)$, and $S_{\lambda\lambda}(\omega)$.
2. Compute $S_{ZZ}(\omega)$, $S_{YY}(\omega)$, and $S_{ZY}(\omega)$ from (5.4.61)–(5.4.63).
3. Invert to obtain $R_{ZZ}(\tau)$, $R_{YY}(\tau)$, and $R_{ZY}(\tau)$.

Figures 5.8–5.10 show $R_{ZZ}(\tau)$, $R_{YY}(\tau)$, and $R_{ZY}(\tau)$ for the three disturbance cases above (labeled 1, 2, and 3, correspondingly). We note that variations in the flow rate q (curves 3) cause the greatest variations in the output concentration and temperature, as reflected in R_{ZZ} and R_{YY}. $R_{ZY}(\tau) < 0$ because

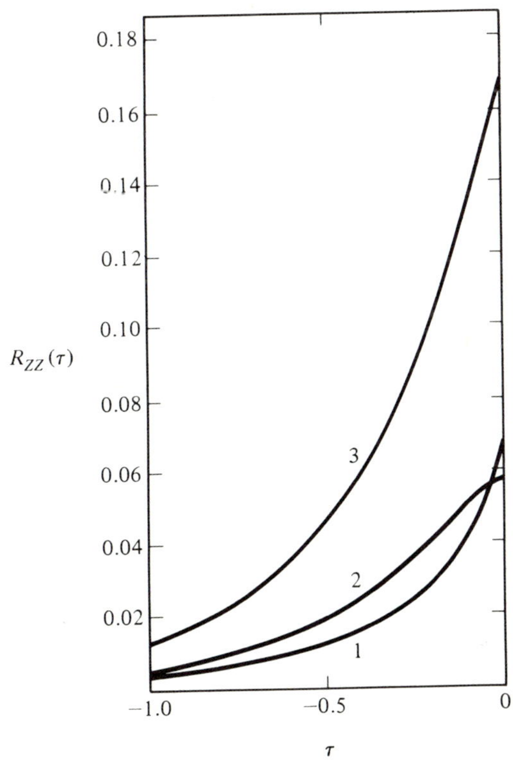

Figure 5.8. Autocorrelation function of output concentration of CSTR. [Reprinted from R. Aris and N. R. Amundson, "Statistical Analysis of a Reactor. Linear Theory," *Chem. Eng. Sci.* 9, 250 (1958). Reprinted by permission of the copyright owner.]

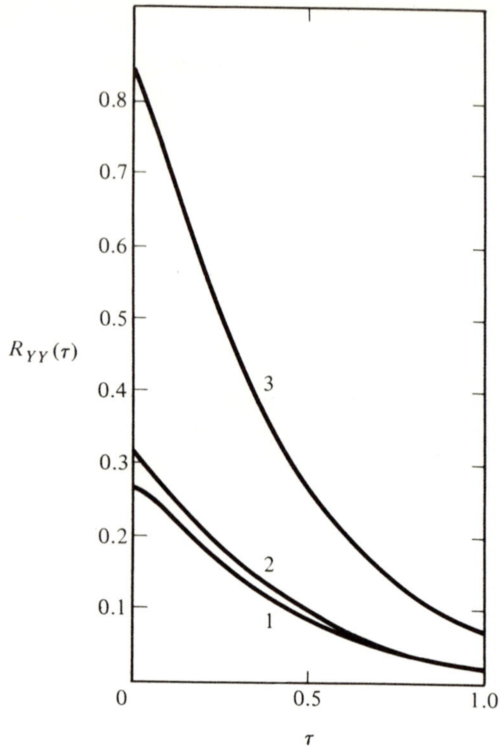

Figure 5.9. Autocorrelation function of output temperature of CSTR. [Reprinted from R. Aris and N. R. Amundson, "Statistical Analysis of a Reactor. Linear Theory," *Chem. Eng. Sci. 9*, 250 (1958). Reprinted by permission of the copyright owner.]

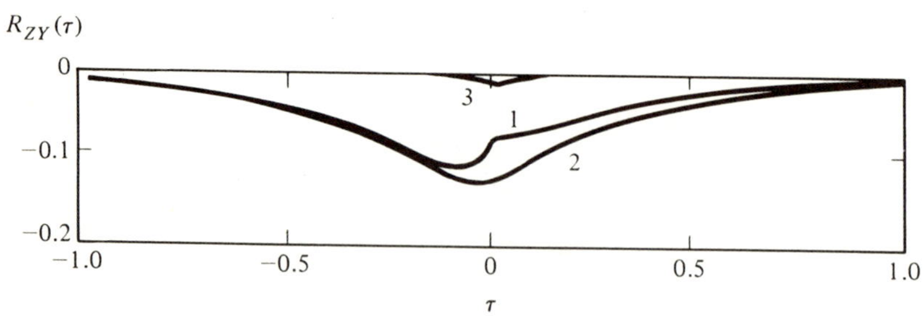

Figure 5.10. Cross-correlation of output concentration and temperature of CSTR. [Reprinted from R. Aris and N. R. Amundson, "Statistical Analysis of a Reactor. Linear Theory," *Chem. Eng. Sci. 9*, 250 (1958). Reprinted by permission of the copyright owner.]

an increase in temperature Y is generally accompanied by a decrease in concentration Z. $R_{ZY}(\tau)$ is largest in magnitude at about $\tau = -0.05$, and so $\overline{Z(t)Y(t - 0.05)}$ has the largest correlation. This indicates that the response of concentration lags a little behind that of the temperature. For curve 1, a sudden slight increase in inlet concentration increases the rate of reaction and increases the temperature, since the reaction is exothermic. However, the fall in concentration which accompanies an increase in the reaction rate is inhibited by the initial positive fluctuation of concentration. Thus the combined effect is to cause the decrease in concentration to occur at a time somewhat later than the initial fluctuation. As we see, for the variables chosen, this time is about 0.05.

We see also that a flow rate change has less effect on R_{ZY} than on R_{ZZ} and R_{YY} (curves 3 in all three figures), because an increase in q decreases the average residence time and thus the extent of reaction; at the same time serving to increase the temperature since the coil contacts the fluid for a shorter time and removes less heat. These two effects tend to balance each other and result in small R_{ZY}.

5.5 Mathematical Modeling of Stochastic Dynamic Systems

A large number of deterministic dynamic systems are modeled by a set of n-coupled O.D.E.s of the form (See Chapter 2)

$$\dot{\mathbf{x}} = \mathbf{f}(\mathbf{x}(t), t) \tag{5.5.1}$$

If we wish to modify (5.5.1) to include the effect of random disturbances on the system, one way to do this is to write

$$\dot{\mathbf{x}} = \mathbf{f}(\mathbf{x}(t), t) + \mathbf{G}(\mathbf{x}(t), t)\mathbf{v}(t) \tag{5.5.2}$$

where $\mathbf{v}(t)$ is an m-vector representing the random disturbance at time t, and $\mathbf{G}(\mathbf{x}(t), t)$ is an $n \times m$ matrix. We allow $\mathbf{G}$ to be a function of $\mathbf{x}$ and t, since the influence of the disturbance may depend on the state of the system. Since $\mathbf{v}(t)$ is a random process, $\mathbf{x}(t)$ is also a random process. We shall use lower-case letters here to conform with the usual state-space notation.

A commonly employed assumption is that $\mathbf{v}(t)$ is a Gaussian white-noise process. This means that for each fixed t, $\mathbf{v}(t)$ has a Gaussian distribution with zero mean and infinite variance. Also, for any two t_1 and t_2, $t_1 \neq t_2$, $\mathbf{v}(t_1)$ and $\mathbf{v}(t_2)$ are independent of each other. Thus

$$\begin{aligned} E[\mathbf{v}(t)] &= 0 \\ E[\mathbf{v}(t_1)\mathbf{v}(t_2)^T] &= \mathbf{C}(t_1)\delta(t_1 - t_2) \end{aligned} \tag{5.5.3}$$

where $\mathbf{C}(t_1)$ is an $m \times m$ covariance matrix, which expresses the degree of correlation among the components of the vector $\mathbf{v}(t_1)$. We can still speak of this correlation even though each component has infinite variance. If $\mathbf{C}$ is a constant, independent of time, the white noise is stationary. Strictly speaking, only in the stationary case can we justify the name *white* because only in this case can one define a power spectral density function. As we have seen (see Table 5.4), the power density spectrum is constant, independent of frequency. White noise plays a role in the theory of stochastic processes similar to that of the Dirac delta function in the theory of deterministic processes.

Essentially, what we are interested in doing is obtaining solutions to (5.5.2) if we are trying to model a system in this manner. Since $\mathbf{x}(t)$ is now a random vector, by solution of (5.5.2) we mean a solution for the statistical properties of $\mathbf{x}(t)$. It is the solution of this equation with which we shall be concerned in this section.

5.5.1 Langevin's equation and Brownian motion

Let us begin by considering the scalar equation

$$\dot{x}(t) + ax(t) = v(t) \tag{5.5.4}$$

where $v(t)$ is white noise with zero mean and power density spectrum λ. This equation, it turns out, is a classic equation of mathematical physics and is called Langevin's equation. This equation describes the motion of a free particle in a liquid, where $x(t)$ is the *velocity* of the particle (in any one of its coordinate directions). If the mass of the particle is m, the friction force on the particle is proportional to the velocity, and the random force due to collisions is $v(t)$, then Newton's law gives

$$m\dot{x}(t) + ax(t) = v(t)$$

On a microscopic scale $v(t)$ is very complicated, but, for all practical purposes, it can be represented as a process with independent, stationary increments with a flat spectrum. This differential equation is thus a model for Brownian motion.

Suppose that the particle begins at rest, $x(0) = 0$, and let us determine the autocorrelation of $x(t)$, governed by (5.5.4). It is easy to show that

$$R_{xx}(t_1, t_2) = \frac{\lambda}{2a}(1 - e^{-2at_2})e^{-a(t_1-t_2)}, \qquad t_1 > t_2 \tag{5.5.5}$$

For $t \gg 1/a$, $x(t)$ approaches stationarity. Since in the study of Brownian motion we are not too interested in the initial stages of the particle motion, we thus assume subsequently that the particle has been in the field long enough for its velocity to be a stationary process.

If $x(t)$ is the particle velocity in a particular coordinate direction, then its position $\xi(t)$ along that axis is governed by

$$\frac{d^2\xi(t)}{dt^2} + a\frac{d\xi}{dt} = v(t) \tag{5.5.6}$$

For $t \gg 1/a$, we can neglect the second derivative term, and so

$$a\frac{d\xi}{dt} = v(t) \tag{5.5.7}$$

and

$$\xi(t) = \frac{1}{a}\int_0^t v(\tau)\,d\tau \tag{5.5.8}$$

Therefore the autocorrelation of $\xi(t)$ is

$$R_{\xi\xi}(t_1, t_2) = \begin{cases} \lambda t_2, & t_1 > t_2 \\ \lambda t_1, & t_2 > t_1 \end{cases} \tag{5.5.9}$$

However, if we go back to Section 5.3.2 and compute the autocorrelation of the Wiener process, we would find that it is precisely (5.5.9). Thus we have uncovered an important relationship: The time integral of a white-noise process is a Wiener process. In addition, the Brownian motion process described by (5.5.4) and (5.5.7) is identically the Wiener process. What we have introduced here that we have not seen before is the use of an O.D.E. with a white-noise forcing term to describe a stochastic process.

The power density spectrum of $x(t)$ governed by (5.5.4) is quickly found to be

$$S_{xx}(\omega) = \frac{\lambda}{\omega^2 + a^2} \tag{5.5.10}$$

From (5.5.10) it follows that the autocorrelation of $x(t)$ is

$$R_{xx}(\tau) = \frac{\lambda}{2a}e^{-a|\tau|} \tag{5.5.11}$$

and that the density function for $x(t)$ is

$$p(x) = \sqrt{\frac{\lambda}{\pi a}}\exp\left(-\frac{\lambda x^2}{a}\right) \tag{5.5.12}$$

It turns out that $x(t)$ satisfying the more general equation

$$\dot{x}(t) = a(t)x(t) + g(t)v(t) \tag{5.5.13}$$

is a Gaussian random process. In fact, a Markov process can always be associated with the O.D.E.:

$$\dot{x}(t) = f(x(t), t) + g(t)v(t) \tag{5.5.14}$$

where $v(t)$ is such that $v(t_1), \ldots, v(t_n)$ are independent for any $n, t_1, \ldots, t_n$. For Gaussian white noise $E[v(t)] = 0$ and $E[v(t)v(s)] = \delta(t - s)$. Clearly, $x(t)$ is uniquely determined from $x(t_0)$ and the values of $v(t)$ in the interval (t_0, t). Thus, given any $x(t_0)$, any past conditions before t_0 will have no effect on

$x(t)$, $t > t_0$. The solution of (5.5.14) is therefore a Markov process. In (5.5.14) we shall assume that $|g(t)| > 0$ for all t, that $f(x, t)$ and $g(t)$ are at least piecewise continuous functions of t, and that f is at least once differentiable with respect to x. We note that in (5.5.14) the function $g(t)$ is a function of t only and not a function of x. We shall subsequently relax this assumption.

EXAMPLE 5.17. Vector Gaussian Random Process. Consider the special vector form of (5.5.14):

$$\dot{\mathbf{x}} = \mathbf{A}(t)\mathbf{x}(t) + \mathbf{G}(t)\mathbf{v}(t) \tag{A}$$

In this example we would like to examine the statistical properties of $\mathbf{x}(t)$ where (A) can be interpreted as the vector form of Langevin's equation. Let us assume that $E[\mathbf{x}(0)] = \boldsymbol{\mu}(0)$, that $E[(\mathbf{x}(0) - \bar{\mathbf{x}}_0)(\mathbf{x}(0) - \bar{\mathbf{x}}_0)^T] = \mathbf{P}(0)$, and that $E[\mathbf{v}(t)] = \bar{\mathbf{v}}(t)$. Note that it need not be necessary that the mean of $\mathbf{v}(t)$ be zero. However, we do require that $\mathbf{v}(t_1), \ldots, \mathbf{v}(t_n)$ are independent for any $t_1, \ldots, t_n$. The variance of $\mathbf{v}(t)$ is $E\{[\mathbf{v}(t) - \bar{\mathbf{v}}(t)][\mathbf{v}(\tau) - \mathbf{v}(\tau)]^T\} = \mathbf{C}(t)\,\delta(t-\tau)$. Our objective is to determine the behavior of the mean $\boldsymbol{\mu}(t)$ and covariance $\mathbf{P}(t)$ as a function of t.

Taking the expected value of (A), we obtain

$$\dot{\boldsymbol{\mu}} = \mathbf{A}(t)\boldsymbol{\mu}(t) + \mathbf{G}(t)\bar{\mathbf{v}}(t), \quad \boldsymbol{\mu}(0) \text{ given} \tag{B}$$

To determine the equation governing $\mathbf{P}(t)$, we subtract (B) from (A) and postmultiply by $[\mathbf{x}(t) - \boldsymbol{\mu}(t)]^T$:

$$\begin{aligned}[\dot{\mathbf{x}}(t) - \dot{\boldsymbol{\mu}}(t)][\mathbf{x}(t) - \boldsymbol{\mu}(t)]^T &= \mathbf{A}(t)[\mathbf{x}(t) - \boldsymbol{\mu}(t)][\mathbf{x}(t) - \boldsymbol{\mu}(t)]^T \\ &\quad + \mathbf{G}(t)[\mathbf{v}(t) - \bar{\mathbf{v}}(t)][\mathbf{x}(t) - \boldsymbol{\mu}(t)]^T\end{aligned} \tag{C}$$

We then add the transpose of (C) to (C) and take the expected value:

$$\begin{aligned}\dot{\mathbf{P}} &= \mathbf{A}(t)\mathbf{P}(t) + \mathbf{P}(t)\mathbf{A}(t)^T + \mathbf{G}(t)E\{[\mathbf{v}(t) - \bar{\mathbf{v}}(t)][\mathbf{x}(t) - \boldsymbol{\mu}(t)]^T\} \\ &\quad + E\{[\mathbf{x}(t) - \boldsymbol{\mu}(t)][\mathbf{v}(t) - \bar{\mathbf{v}}(t)]^T\}\mathbf{G}^T(t)\end{aligned} \tag{D}$$

We must determine the last two terms in (D). To do this we employ the fundamental relation (2.3.28), in the form

$$\mathbf{x}(t) - \boldsymbol{\mu}(t) = \boldsymbol{\Phi}(t, 0)[\mathbf{x}(0) - \boldsymbol{\mu}(0)] + \int_0^t \boldsymbol{\Phi}(t, \tau)\mathbf{G}(\tau)[\mathbf{v}(\tau) - \bar{\mathbf{v}}(\tau)]\,d\tau. \tag{E}$$

We postmultiply (E) by $[\mathbf{v}(t) - \bar{\mathbf{v}}(t)]^T$, take the expected value of the result, and assume that $E\{[\mathbf{x}(0) - \boldsymbol{\mu}(0)][\mathbf{v}(t) - \bar{\mathbf{v}}(t)]^T\} = 0$, and we obtain

$$\begin{aligned}E\{[\mathbf{x}(t) - \boldsymbol{\mu}(t)][\mathbf{v}(t) - \bar{\mathbf{v}}(t)]^T\} &= \int_0^t \boldsymbol{\Phi}(t, \tau)\mathbf{G}(\tau)E\{[\mathbf{v}(\tau) - \bar{\mathbf{v}}(t)][\mathbf{v}(t) - \bar{\mathbf{v}}(t)]^T\}\,d\tau \\ &= \mathbf{G}(t)\mathbf{C}(t)\end{aligned} \tag{F}$$

Thus (D) becomes

$$\dot{\mathbf{P}} = \mathbf{A}(t)\mathbf{P}(t) + \mathbf{P}(t)\mathbf{A}(t)^T + 2\mathbf{G}(t)\mathbf{C}(t)\mathbf{G}(t)^T \tag{G}$$

Equations (B) and (G) are differential equations for the mean and covariance of the random process $\mathbf{x}(t)$.

It is also of interest to compute the correlation matrix:

$$\mathbf{S}(t, \tau) = E\{[\mathbf{x}(t) - \boldsymbol{\mu}(t)][\mathbf{x}(\tau) - \boldsymbol{\mu}(\tau)]^T\} \tag{H}$$

First we write

$$\mathbf{x}(t+\tau) - \boldsymbol{\mu}(t+\tau) = \boldsymbol{\Phi}(t+\tau, t)[\mathbf{x}(t) - \boldsymbol{\mu}(t)] + \int_t^{t+\tau} \boldsymbol{\Phi}(t+\tau, \lambda)\mathbf{G}(\lambda)[\mathbf{v}(\lambda) - \bar{\mathbf{v}}(\lambda)]\, d\lambda \tag{I}$$

We then postmultiply (I) by $[\mathbf{x}(t) - \boldsymbol{\mu}(t)]^T$ and take the expected value:

$$\mathbf{S}(t+\tau, t) = \boldsymbol{\Phi}(t+\tau, t)\mathbf{P}(t) + \int_t^{t+\tau} \boldsymbol{\Phi}(t+\tau, \lambda)\mathbf{G}(\lambda)E\{\mathbf{v}(\lambda) - \bar{\mathbf{v}}(\lambda)] \cdot [\mathbf{x}(t) - \boldsymbol{\mu}(t)]^T\, d\lambda \tag{J}$$

The integrand in (J) equals zero for $\tau > 0$ since $\mathbf{v}(\lambda)$ is uncorrelated with $\mathbf{x}(t)$ for $\lambda > t$. Thus

$$\mathbf{S}(t+\tau, t) = \boldsymbol{\Phi}(t+\tau, t)\mathbf{P}(t), \qquad \tau \geq 0 \tag{K}$$

5.5.2 Stochastic differential equations

Our real objective here is to study the use of equations such as (5.5.2) to model stochastic dynamic systems. We have already noted that equations of the form (5.5.14) can be associated with a Markov process $x(t)$. In addition, in Section 5.3 we studied Markov processes in some detail, particularly their description by the Fokker-Planck equation. Therefore we would expect that there is a direct relation between the differential-equation representation of a Markov process, such as by (5.5.14), and the Fokker-Planck equations associated with that Markov process. In fact, the "solution" of (5.5.14) really implies the determination of the first-order density function for $x(t)$, and that in turn implies the solution of the associated Fokker-Planck equation.

Consider the process $x(t)$ satisfying the stochastic differential equation

$$\frac{dx(t)}{dt} = f(x(t), t) + g(t)\frac{dw(t)}{dt} \tag{5.5.15}$$

where $w(t)$ is a Wiener process [we have used (5.5.7), since the time derivative of a Wiener process is white noise]. We can write (5.5.15) in the form

$$dx(t) = f(x(t), t)\, dt + g(t)\, dw(t) \tag{5.5.16}$$

This may appear to be a rather curious way to express the stochastic differential equation, but it turns out that this equation is more useful than (5.5.15) because we would rather deal with the Wiener process than with white noise. The conditional mean and variance of dx are

$$\begin{aligned} E(dx \mid x) &= f(x(t), t)\, dt \\ E[(dx - f(x, t)\, dt)^2 \mid x] &= E[g^2(dw)^2 \mid x] = g^2(t)\, dt \end{aligned} \tag{5.5.17}$$

where $dw(t)$ is an increment of a Wiener process. From (5.3.14), however, we see that for the process of (5.5.16),

$$\alpha_1(x, t) = f(x, t)$$
$$\alpha_2(x, t) = g^2(t)$$

Thus the forward Fokker-Planck equation corresponding to (5.5.16) is

$$\frac{\partial p(x, t \mid x_0, t_0)}{\partial t} = -\frac{\partial}{\partial x}[f(x, t)p] + \frac{1}{2} g^2(t) \frac{\partial^2 p}{\partial x^2}$$

Actually, (5.5.16) is just a representation of the integral equation

$$x(t) = x(0) + \int_0^t f(x(\tau), \tau)\, d\tau + \int_0^t g(\tau)\, dw(\tau) \tag{5.5.18}$$

The last integral, being an integral with respect to a Wiener process, is a so-called *stochastic integral.* As long as g is a function of t only, and not $x(t)$, there are no difficulties in defining this integral. It can be defined uniquely in the customary manner for deterministic integrals as

$$\lim_{\delta \to 0} \sum_{i=0}^{n-1} g(t_i)[w(t_{i+1}) - w(t_i)], \qquad 0 < t_0 < t_1 < \cdots < t_n = t$$

where $\delta = \max(t_{i+1} - t_i)$.

However, when we want to allow $g(t)$ to be a function of the state x as well as of time t, we get into some problems. Mathematically, the trouble arises when we try to evaluate the stochastic integral when g is itself a function of a random process. Unfortunately, the proper interpretation of nonlinear stochastic differential equations of the type (5.5.2) involves mathematics well beyond our scope. Nevertheless, we shall attempt to give a nonrigorous analysis of the problem with, we hope, some acceptable engineering results. The material to follow is based on two excellent references: the review paper by Mortensen [17] and the lecture notes of King [15]. We might note that the classic treatise in this whole area is the book by Doob [5].

Let us begin by considering a little example of Mortensen's. Suppose that $w(t)$ is a Wiener process. Then for the case of $t_0 = 0$, $w_0 = 0$, we have already seen that

$$p(w, t) = \frac{1}{(2\pi t)^{1/2}} \exp\left(\frac{-w^2}{2t}\right) \tag{5.5.19}$$

In addition, the forward equation for a Wiener process is

$$\frac{\partial p}{\partial t} = \frac{1}{2}\frac{\partial^2 p}{\partial w^2} \qquad \lim_{t \to 0} p(w, t) = \delta(w) \tag{5.5.20}$$

Now suppose that the Wiener process is passed through a device to produce a new process

$$z(t) = \sinh[w(t)] \tag{5.5.21}$$

Define $p(\zeta, t) = \text{Prob}\{\zeta \leq z(t) < \zeta + d\zeta \mid z(0) = 0\}$. Then, to change variables, we employ (4.7.4):

$$\begin{aligned} p(\zeta, t) &= p(w, t)\frac{dw}{d\zeta}\bigg|_{w=\sinh^{-1}\zeta} \\ &= \left[\frac{1+\zeta^2}{2\pi t}\right]^{1/2} \exp\left[-\frac{(\sinh^{-1}\zeta)^2}{2t}\right] \end{aligned} \tag{5.5.22}$$

Either by making the change of independent variable $w = \sinh^{-1} \zeta$ in (5.5.20) or by direct differentiation of (5.5.22), we find that the Fokker-Planck equation satisfied by $p(\zeta, t)$ is

$$\frac{\partial p(\zeta, t)}{\partial t} = -\frac{\partial}{\partial \zeta}\left[\frac{\zeta}{2} p(\zeta, t)\right] + \frac{1}{2}\frac{\partial^2}{\partial \zeta^2}[(1 + \zeta^2)p(\zeta, t)] \tag{5.5.23}$$

If we want to find the stochastic differential equation for $z(t)$, we could follow the procedure just below (5.5.17) and write

$$dz(t) = \tfrac{1}{2} z(t)\, dt + [1 + z^2(t)]^{1/2}\, dw(t) \tag{5.5.24}$$

According to the theory given by Doob [5], this is the stochastic differential equation corresponding to (5.5.23). Even though (5.5.24) resembles (5.5.16) in form, here we see that the coefficient of $dw(t)$ is a function of $z(t)$, where we specifically said it should not be when considering (5.5.16).

Suppose that we were to compute $dz(t)$ from (5.5.21) simply using the chain rule of calculus:

$$\begin{aligned} dz(t) &= \frac{\partial}{\partial w} \sinh w \bigg|_{w(t) = \sinh^{-1} z(t)} dw(t) \\ &= \cosh w \bigg|_{w(t) = \sinh^{-1} z(t)} dw(t) \\ &= [1 + \sinh^2 w]^{1/2} \bigg|_{w(t) = \sinh^{-1} z(t)} dw(t) \\ &= [1 + z^2(t)]^{1/2}\, dw(t) \end{aligned} \tag{5.5.25}$$

However, this is not the same stochastic differential equation as (5.5.24) The question is, Which is the correct stochastic differential equation for generating the process $z(t)$ from a Wiener process?

Let us see where the trouble arises. Both the Wiener process and the process $z(t)$ are well defined, with (5.5.19) and (5.5.22) being correct. Also, (5.5.23) is the correct Fokker-Planck equation for the $z(t)$ process. The trouble arises when we try to integrate the differential equations (5.5.24) and (5.5.25) to get $z(t) = \sinh[w(t)]$. There are two ways in which we can integrate a stochastic differential equation such as (5.5.24) or (5.5.25), and these two ways constitute two approaches to the so-called *stochastic calculus*, the so-called *Ito* and *Stratonovich* approaches. By the Ito approach (5.5.24) is the correct equation, whereas by the Stratonovich approach (5.5.25) is the correct equation. Neither one is wrong, since as we shall see, both are internally consistent, and so let us try to pinpoint the issue of disagreement.

One unambiguous way to specify a Markov process mathematically is to specify its transition density or, equivalently, the Fokker-Planck equation obeyed by the transition density. The problem arises when we wish to represent the Markov process as a solution to a stochastic differential equation forced by the differential of a Wiener process. The problem comes down to two different ways of associating the coefficients in the Fokker-Planck equation with the coefficients in the stochastic differential equation and two ways of integrating this stochastic equation. Starting with the Fokker-Planck equation (5.5.23), the use of the Ito rules associates (5.5.24) with (5.5.23), whereas the use of the Stratonovich rules associates (5.5.25) with (5.5.23). Integrating (5.5.24) by the Ito rules yields $z(t) = \sinh[w(t)]$, and integrating (5.5.25) by the Stratonovich rules also yields $z(t) = \sinh[w(t)]$. Thus each set of rules is consistent within itself, and if the same set of rules is applied consistently throughout a problem, both methods yield the same result.

The mathematician usually begins with the transition density for a Markov process. He is then able to associate a Fokker-Planck equation unambiguously with this transition density. Upon finding that he has two different ways of representing this process as the solution of a stochastic differential equation, he will choose the approach with the most mathematical elegance. From this point of view, he would choose the Ito calculus.

However, the engineer faces a somewhat different problem. He does not start with the transition density, but rather he *starts* with a *differential equation* which is the basic deterministic representation of his dynamic system, for example, a continuous stirred tank reactor, a satellite, etc. He then adds a white-noise forcing term to the differential equation to account for the influence of random disturbances. If the coefficient of the noise is itself random, i.e., $g(x, t)$, then there are two possible ways of interpreting the equation, leading to two different Fokker-Planck equations and *two different Markov processes*. The engineer is then faced with the question, Which one is "right"?

In short, the paradox of obtaining two different stochastic processes from the same stochastic differential equation is a result of the pathological nature of white noise. Since white noise is only a mathematical idealization, we must view the stochastic differential equation as only an approximate model, and thus neither process is really "right." Once we accept this fact, we can employ either approach and obtain equally accurate results. Let us now explore the essential difference between the Ito and Stratonovich stochastic calculi.

5.5.3 The Ito and Stratonovich calculi

Following Mortensen [17], let F be any smooth, real-valued, nonlinear function of a real variable. Consider $F(w(t + \Delta t))$, where $w(t)$ is a Wiener process. By a Taylor series expansion

$$\begin{aligned} F(w(t + \Delta t)) &= F(w(t) + \Delta w(t)) \\ &= F(w(t)) + F'(w(t))\,\Delta w(t) + \tfrac{1}{2}F''(w(t))\,\Delta w(t)^2 + \cdots \end{aligned} \tag{5.5.26}$$

For a Wiener process

$$E\{(\Delta w)^k\} = \begin{cases} 0, & k \text{ odd} \\ 1\cdot3\cdot5 \cdots (k-1)(\Delta t)^{k/2}, & k \text{ even} \end{cases} \tag{5.5.27}$$

If we define

$$\dot{F}(w(t)) = \frac{d}{dt} F(w(t)) = \lim_{\Delta t \to 0} \frac{F(w(t + \Delta t)) - F(w(t))}{\Delta t} \tag{5.5.28}$$

then from (5.5.26) and (5.5.27) it follows that

$$E\{\dot{F}(w(t))\} = \tfrac{1}{2} E\{\ddot{F}(w(t))\} \tag{5.5.29}$$

Alternatively, if we compute the total differential $dF(w(t))$ using the chain rule of ordinary calculus,

$$dF(w(t)) = \dot{F}(w(t))\, dw(t) \tag{5.5.30}$$

and

$$\begin{aligned} E\{dF(w(t))\} &= E\{\dot{F}(w(t))\, dw(t)\} \\ &= E\{\dot{F}(w(t))\} E\{dw(t)\} \\ &= 0 \end{aligned} \tag{5.5.31}$$

which implies that

$$E\{\dot{F}(w(t))\} = 0 \tag{5.3.32}$$

This contradicts (5.5.29). Because $E\{(\Delta w)^2\} = \Delta t$, the definition (5.5.28) no longer leads to the usual rules of calculus. Ito corrected this deficiency by suggesting that the total differential (we now want to consider the differential dF because dw can be rigorously interpreted, whereas dw/dt cannot) be computed by

$$dF(w(t)) = \dot{F}(w(t))\, dw(t) + \tfrac{1}{2}\ddot{F}(w(t))\, dt \qquad \text{(Ito)} \tag{5.5.33}$$

If $F = \sinh[w(t)]$, for example,

$$dF = \cosh w(t)\, dw(t) + \tfrac{1}{2} \sinh w(t)\, dt$$

of if $z = \sinh w(t)$,

$$dz = \tfrac{1}{2} z(t)\, dt + \tfrac{1}{2}[1 + z^2(t)]^{1/2}\, dw(t)$$

which agrees with (5.5.24).

Ito has therefore changed the rule for computing a total differential from (5.5.30) to (5.5.33). Since the rule for computing total differentials has been changed, there will be a corresponding change in the rule for integration. First of all, we must require that

$$\int_{t_0}^{t_1} dF(w(t)) = F(w(t_1)) - F(w(t_0)) \tag{5.5.34}$$

Applying (5.5.34) to (5.5.33), we obtain

$$\int_{t_0}^{t_1} \dot{F}(w(t))\, dw(t) = F(w(t_1)) - F(w(t_0)) - \frac{1}{2}\int_{t_0}^{t_1} \ddot{F}(w(t))\, dt \tag{5.5.35}$$

Let $g(x)$ be any once-differentiable function. Define

$$G(x) = \int_0^x g(\xi) d\xi$$

Then, replacing F with G in (5.3.35) gives

$$\int_{t_0}^{t_1} g(w(t))\, dw(t) = \int_{w(t_0)}^{w(t_1)} g(\xi)\, d\xi - \frac{1}{2}\int_{t_0}^{t_1} \dot{g}(w(t))\, dt \qquad \text{(Ito)} \qquad (5.5.36)$$

On the other hand, the Stratonovich integral is defined by

$$\int_{t_0}^{t_1} g(w(t))\, dw(t) = \int_{w(t_0)}^{w(t_1)} g(\xi)\, d\xi \qquad \text{(Stratonovich)} \qquad (5.5.37)$$

as would arise if we define the total differential by (5.5.30).

Consider the evaluation of the two integrals if $g(w(t)) = w(t)$. We get

$$\int_{t_0}^{t_1} w(t)\, dw(t) = \frac{1}{2}[w^2(t_1) - w^2(t_0)] - \frac{1}{2}(t_1 - t_0) \quad \text{(Ito)} \qquad (5.5.38)$$

$$= \frac{1}{2}[w^2(t_1) - w^2(t_0)] \qquad \text{(Stratonovich)} \quad (5.5.39)$$

Taking expected values, we obtain, in the two cases,

$$E\left\{\int_{t_0}^{t_1} w(t)\, dw(t)\right\} = \frac{1}{2}(t_1 - t_0) - \frac{1}{2}(t_1 - t_0) = 0 \quad \text{(Ito)} \qquad (5.5.40)$$

$$= \frac{1}{2}(t_1 - t_0) \neq 0 \qquad \text{(Stratonovich)} \quad (5.5.41)$$

Recapitulating briefly, because Ito and Stratonovich propose different ways of defining the total derivative, the resulting integrals of these total derivatives differ. Crudely speaking, the Stratonovich integral always gives us the result we would expect from ordinary calculus, whereas the Ito definition leads to an unusual term, for example, $\frac{1}{2}(t_1 - t_0)$ in the case of $g(w(t)) = w(t)$.

We have seen how this discrepancy arises based on the definition of the total derivative. We can also define these integrals directly as the limit of a sequence of sums. The Ito definition is

$$\int_{t_0}^{t} g(w(t), t)\, dw(t) = \lim_{\delta \to 0} \sum_{i=0}^{n-1} g[w(t_i), t_i][w(t_{i+1}) - w(t_i)] \qquad \text{(Ito)} \qquad (5.5.42)$$

whereas the Stratonovich definition is

$$\int_{t_0}^{t} g(w(t), t)\, dw(t) = \lim_{\delta \to 0} \sum_{i=0}^{n-1} g\left[\frac{w(t_{i+1}) + w(t_i)}{2}\right][w(t_{i+1}) - w(t_i)]$$

$$\text{(Stratonovich)} \qquad (5.5.43)$$

Note that in the Ito definition the integrand is always evaluated at the beginning of the interval over which the increment $[w(t_{i+1}) - w(t_i)]$ is taken. Thus $g(w(t_i), t_i)$ and $[w(t_{i+1}) - w(t_i)]$ are always independent and the expected value of the integral equals zero, as in (5.5.40).

Let us now reconfirm (5.5.38) and (5.5.39) using the definitions (5.5.42) and (5.5.43). Following the Ito rule (5.5.42),

$$\begin{aligned}\int_{t_0}^{t} w(t)\,dw(t) &= \lim_{\delta\to 0}\sum_{i=0}^{n-1} w(t_i)[w(t_{i+1}) - w(t_i)] \\ &= \lim_{\delta\to 0}\sum_{i=0}^{n-1} [w(t_i)w(t_{i+1}) - w^2(t_i)] \\ &= \lim_{\delta\to 0}\sum_{i=0}^{n-1} \left\{\frac{1}{2}[w^2(t_{i+1}) - w^2(t_i)] - \frac{1}{2}[w^2(t_{i+1}) \right. \\ &\qquad \left. + w^2(t_i) + 2w(t_i)w(t_{i+1})]\right\} \\ &= \frac{1}{2}[w^2(t) - w^2(t_0)] - \frac{1}{2}\lim_{\delta\to 0}\sum_{i=0}^{n-1}[w(t_{i+1}) - w(t_i)]^2\end{aligned}$$

From the fundamental properties of the Wiener process, it can be shown that the sum on the right-hand side converges to $t - t_0$, verifying (5.5.38). The proof of this property is difficult and is not included here.

Next, following the Stratonovich rule (5.5.43),

$$\begin{aligned}\int_{t_0}^{t} w(t)\,dw(t) &= \lim_{\delta\to 0}\sum_{i=0}^{n-1}\left[\frac{1}{2}w(t_{i+1}) + \frac{1}{2}w(t_i)\right][w(t_{i+1}) - w(t_i)] \\ &= \lim_{\delta\to 0}\sum_{i=0}^{n-1}\frac{1}{2}[w^2(t_{i+1}) - w^2(t_i)] \\ &= \frac{1}{2}[w^2(t) - w^2(t_0)]\end{aligned}$$

confirming (5.5.39).

Actually, the Ito integral is more general than we have presented it, and the integrating process need not even be a Wiener process. It is beyond our scope to pursue this aspect. The Stratonovich integral, on the other hand, is not as fundamental and general as the Ito integral, although this point is also beyond our scope [20].

5.5.4 The associated Fokker-Planck equations

We shall return now to the real problem at hand, the determination of the Fokker-Planck equation associated with a given stochastic differential equation. As anticipated, we shall actually have *two* different Fokker-Planck equations associated with the differential equation

$$dx(t) = f(x(t), t)\,dt + g(x(t), t)\,dw(t) \tag{5.5.44}$$

depending on whether we employ Ito or Stratonovich rules. In this section we shall present the Fokker-Planck equations corresponding to the two rules.

In either case the forward Fokker-Planck equation will have the form

$$\frac{\partial p(x, t)}{\partial t} = -\frac{\partial}{\partial x}[\alpha_1(x, t)p] + \frac{1}{2}\frac{\partial^2}{\partial x^2}[\alpha_2(x, t)p]$$

in which

$$\alpha_1(x, t) = \lim_{\Delta t \to 0} \frac{1}{\Delta t} E[x(t + \Delta t) - x(t) \,|\, x(t) = x] \tag{5.5.45}$$

$$\alpha_2(x, t) = \lim_{\Delta t \to 0} \frac{1}{\Delta t} E[(x(t + \Delta t) - x(t))^2 \,|\, x(t) = x] \tag{5.5.46}$$

The connection between the Fokker-Planck equation and the stochastic differential equation (5.5.44) is through the moments α_1 and α_2. It is these two moments which will differ depending on whether we employ Ito or Stratonovich integration.

We write (5.5.44) as an integral equation:

$$x(t) = x(0) + \int_0^t f(x(\tau), \tau)\, d\tau + \int_0^t g(x(\tau), \tau)\, dw(\tau) \tag{5.5.47}$$

The solution of (5.5.47) will differ depending on which way we evaluate the last integral [even though $g = g(x, t)$ rather than $g(w, t)$, we can use both of the formerly developed Ito and Stratonovich rules].

We can write (5.5.47) as

$$x(t + \Delta t) - x(t) = \int_t^{t+\Delta t} f(x, \tau)\, d\tau + \int_t^{t+\Delta t} g(x, \tau)\, dw(\tau) \tag{5.5.48}$$

Using a Taylor series for f and g, (5.5.48) becomes

$$\begin{aligned} x(t + \Delta t) - x(t) &= f(x(t), t) \int_t^{t+\Delta t} d\tau + \frac{\partial f(x(t), t)}{\partial x} \int_t^{t+\Delta t} [x(\tau) - x(t)]\, d\tau \\ &\quad + g(x(t), t) \int_t^{t+\Delta t} dw(\tau) + \frac{\partial g(x(t), t)}{\partial x} \int_t^{t+\Delta t} [x(\tau) - x(t)]\, dw(\tau) \end{aligned} \tag{5.5.49}$$

However,

$$x(\tau) - x(t) = \int_t^\tau f(x(\theta), \theta)\, d\theta + \int_t^\tau g(x(\theta), \theta)\, dw(\theta) \tag{5.5.50}$$

Substituting (5.5.50) into (5.5.49), we obtain

$$\begin{aligned} x(t + \Delta t) - x(t) &= f(x(t), t)\, \Delta t + g(x(t), t)[w(t + \Delta t) - w(t)] \\ &\quad + \frac{\partial f(x(t), t)}{\partial x} \Bigg\{ \underbrace{\int_t^{t+\Delta t} \int_t^\tau f(x(\theta), \theta)\, d\theta\, d\tau}_{①} + \underbrace{\int_t^{t+\Delta t} \int_t^\tau g(x(\theta), \theta)\, dw(\theta)\, d\tau}_{②} \Bigg\} \\ &\quad + \frac{\partial g(x(t), t)}{\partial x} \Bigg\{ \underbrace{\int_t^{t+\Delta t} \int_t^\tau f(x(\theta), \theta)\, d\theta\, dw(\tau)}_{③} + \underbrace{\int_t^{t+\Delta t} \int_t^\tau g(x(\theta), \theta)\, dw(\theta)\, dw(\tau)}_{④} \Bigg\} \end{aligned} \tag{5.5.51}$$

Let us examine each of the four double integrals:

① $\simeq f(x(t), t)(\Delta t)^2$

$$② \simeq g(x(t), t) \int_t^{t+\Delta t} \int_t^{\tau} dw(\theta)\, d\tau = g(x(t), t) \int_t^{t+\Delta t} [w(\tau) - w(t)]\, d\tau$$

$$③ \simeq f(x(t), t) \int_t^{t+\Delta t} (\tau - t)\, dw(\tau)$$

$$④ \simeq g(x(t), t) \int_t^{t+\Delta t} \int_t^{\tau} dw(\theta)\, dw(\tau)$$

$$= g(x(t), t) \int_t^{t+\Delta t} [w(\tau) - w(t)]\, dw(\tau)$$

$$= g(x(t), t) \left\{ \int_t^{t+\Delta t} w(\tau)\, dw(\tau) - w(t)[w(t + \Delta t) - w(t)] \right\}$$

The ambiguity arises, as we have seen in (5.5.38) and (5.5.39), in the evaluation of the last integral. Using (5.5.38) and (5.5.39), we obtain

$$④ = g(x(t), t)\{\tfrac{1}{2}[w^2(t + \Delta t) - w^2(t)] - \tfrac{1}{2}\Delta t - w(t)[w(t + \Delta t) - w(t)]\} \qquad \text{(Ito)}$$

$$= g(x(t), t)\{\tfrac{1}{2}[w^2(t + \Delta t) - w^2(t)] - w(t)[w(t + \Delta t) - w(t)]\} \qquad \text{(Stratonovich)}$$

Neglecting integral ① as $o(\Delta t)$ and substituting the integrals into (5.5.51), we obtain

$$x(t + \Delta t) - x(t) = f(x, t)\, \Delta t + g(x, t)[w(t + \Delta t) - w(t)] + \frac{\partial f}{\partial x} g(x, t) \int_t^{t+\Delta t} [w(\tau) - w(t)]\, d\tau + \frac{\partial g}{\partial x} f(x, t) \int_t^{t+\Delta t} (\tau - t)\, dw(\tau) + \frac{\partial g}{\partial x} g(x, t) \begin{cases} \frac{1}{2}[w(t + \Delta t) - w(t)]^2 - \frac{1}{2}\Delta t & \text{(Ito)} \\ 0 & \text{(Stratonovich)} \end{cases} \tag{5.5.52}$$

We now take the expected value of (5.5.52) and using (5.5.45) we obtain

$$\alpha_1(x, t) = f(x, t) \qquad \text{(Ito)} \tag{5.5.53}$$

$$\alpha_1(x, t) = f(x, t) + \frac{1}{2} g \frac{\partial g}{\partial x} \qquad \text{(Stratonovich)} \tag{5.5.54}$$

In a similar manner we can show that

$$\alpha_2(x, t) = g^2(x, t) \qquad \text{(Ito)} \tag{5.5.55}$$

$$\alpha_2(x, t) = g^2(x, t) \qquad \text{(Stratonovich)} \tag{5.5.56}$$

The forward and backward Fokker-Planck equations for the Ito and Stratonovich interpretations of (5.5.44) are thus

$$\left.\begin{aligned} \frac{\partial p(\xi, t \mid \eta, s)}{\partial t} &= -\frac{\partial}{\partial \xi}[f(\xi, t)p] + \frac{1}{2} \frac{\partial^2}{\partial \xi^2}[g^2(\xi, t)p] \\ -\frac{\partial p(\xi, t \mid \eta, s)}{\partial s} &= f(\eta, s) \frac{\partial p}{\partial \eta} + \frac{1}{2} g^2(\eta, s) \frac{\partial^2 p}{\partial \eta^2} \end{aligned}\right\} \quad \text{(Ito)} \tag{5.5.57}$$

and

$$\left.\begin{aligned}\frac{\partial p(\xi, t \mid \eta, s)}{\partial t} &= -\frac{\partial}{\partial \xi}[f(\xi, t)p] + \frac{1}{2}\left[\frac{\partial}{\partial \xi} g(\xi, t)\frac{\partial}{\partial \xi}(g(\xi, t)p)\right] \\ -\frac{\partial p(\xi, t \mid \eta, s)}{\partial s} &= f(\eta, s)\frac{\partial p}{\partial \eta} + \frac{1}{2} g(\eta, s)\frac{\partial}{\partial \eta}\left[g(\eta, s)\frac{\partial p}{\partial \eta}\right]\end{aligned}\right\} \text{(Stratonovich)} \tag{5.5.58}$$

respectively. The case of a vector stochastic process is dealt with in Example 5.19.

We saw earlier that if (5.5.24) is interpreted in the Ito sense, then the appropriate forward equation is (5.5.23). If (5.5.25) is interpreted in the Stratonovich sense, then the associated forward equation is again (5.5.23), as it should be, since the Ito solution of (5.5.24) and the Stratonovich solution of (5.5.25) are both $z(t) = \sinh[w(t)]$. Therefore the differential equation (5.5.44) is ambiguous, and the ambiguity may be removed only by writing the equation in the form (5.5.47) with the type of integral (Ito or Stratonovich) prescribed.

Now let us return to the question at hand. Suppose that we have a real physical system and its deterministic model and that we want to include the influence of random disturbances by adding a random forcing term to the deterministic model. Which way should we interpret our model, Ito or Stratonovich?

As we shall see in later chapters, the mathematical modeling problem involves trying to make the output of the model agree with the observed output of the system. In the case at hand we are presumably trying to model the random output of a dynamic system as a Markov process. Even though the basic origin of the model (5.5.2) may have been based on sound principles, usually we really cannot account in detail for what is going on in the system. Therefore we must view (5.5.2) as a mathematical approximation and not necessarily an exact representation of the system. If we are willing to accept this concept, then the problem of modeling becomes one of matching model to output and the choice between the Ito or Stratonovich approach is a matter of preference. Mathematicians will prefer the Ito calculus because of its rigor and generality, while engineers will prefer the Stratonovich calculus because of their familiarity with its rules. In fact, in discussing the problem of converting a deterministic to a stochastic model, Mortensen has suggested that the engineer should throw away the deterministic model and remodel the whole problem, with the objective being to get the statistics of the output of a Monte Carlo computer simulation to agree with the statistics of the observed data.

We shall now present two examples illustrating some of the concepts in this section. Additional examples of the modeling of stochastic phenomena in chemical engineering systems are given by King [12–14] and Pell and Aris [19].

EXAMPLE 5.18. We desire to compute the stationary probability density and autocorrelation of the process described by (this example is due to King [15])

$$dx(t) = (\alpha + 1 - \tfrac{1}{2}D - \beta x)\,dt + (2Dx)^{1/2}\,dw(t) \tag{A}$$

Let us employ the Stratonovich integral, in which case the associated Fokker-Planck equation for $p(x, t \mid x_0, t_0)$ is

$$\frac{\partial p}{\partial t} = -\frac{\partial}{\partial x}(\alpha + 1 - \beta x)p + D\frac{\partial^2}{\partial x^2}(xp) \tag{B}$$

We seek the steady-state distribution, i.e., $\partial p/\partial t = 0$, and so we need to solve for $p(x)$ from

$$\frac{d}{dx}(\alpha + 1 - \beta x)p = D\frac{d^2(xp)}{dx^2} \tag{C}$$

The general solution of (C) is (using $p = dp/dx = 0$ at $x \longrightarrow \infty$)

$$p(x) = C_1 e^{-\beta x/D} x^{(\alpha+1-D)/D} \tag{D}$$

C_1 is evaluated by the condition

$$\int_0^\infty p(x)\,dx = 1$$

so that

$$p(x) = \frac{(\beta/D)^{(\alpha+1)/D} x^{(\alpha+1-D)/D} e^{-\beta x/D}}{\Gamma[(\alpha + 1)/D]} \tag{E}$$

We recognize (E) as the gamma density, and the process $x(t)$ is called a *gamma Markov process.*

Let us now compute the autocorrelation of this process:

$$R(\tau) = \int_0^\infty \int_0^\infty x_1 x_2 p(x_1, x_2; \tau)\,dx_1\,dx_2 \tag{F}$$

where $p(x_1, x_2; \tau)$, the second-order density function, can be expressed as

$$p(x_1, x_2; \tau) = p(x_2, t + \tau \mid x_1, t)p(x_1, t) \tag{G}$$

Substituting (G) into (F) and differentiating with respect to τ, we obtain

$$\frac{dR(\tau)}{d\tau} = \int_0^\infty \int_0^\infty x_1 x_2 p(x_1, t)\frac{\partial p(x_2, t + \tau \mid x_1, t)}{\partial \tau}\,dx_1\,dx_2 \tag{H}$$

Since $p(x_2, t + \tau \mid x_1, t)$ satisfies (B), we can substitute (B) into (H) and integrate by parts to obtain

$$\begin{aligned}\frac{dR(\tau)}{d\tau} &= \int_0^\infty \int_0^\infty x_1(\alpha + 1 - \beta x_2)p(x_1, t)p(x_2, t + \tau \mid x_1, t)\,dx_1\,dx_2 \\ &= \int_0^\infty \int_0^\infty x_1(\alpha + 1 - \beta x_2)p(x_1, x_2; \tau)\,dx_1\,dx_2 \\ &= \frac{(\alpha + 1)^2}{\beta} - \beta R(\tau) \qquad \text{(I)}\end{aligned}$$

This is simply an O.D.E. for $R(\tau)$, which can be solved subject to

$$R(0) = \frac{(\alpha + 1 + D)(\alpha + 1)}{\beta^2}$$

to yield

$$R(\tau) = \frac{D(\alpha + 1)}{\beta^2} e^{-\beta\tau} + \frac{(\alpha + 1)^2}{\beta^2}$$

EXAMPLE 5.19. *Control of the Liquid Level in a Tank* [15]. Consider the tank shown in Figure 5.11. It is desired to control the level $x_2(t)$ at a set value x_2^d in spite of input

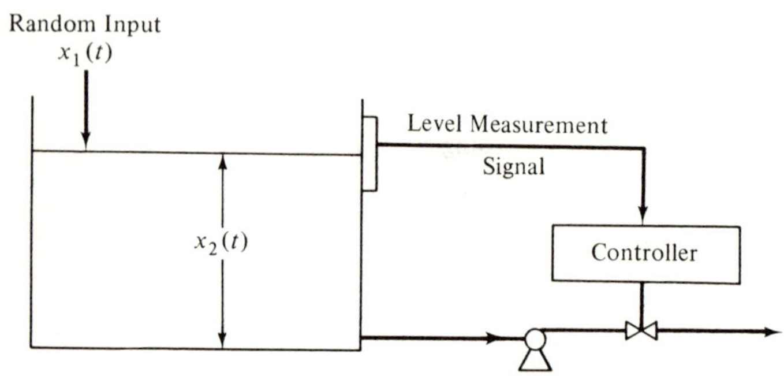

Figure 5.11 Control of the liquid level in a tank in the presence or random input of liquid [15].

flow rate fluctuations $x_1(t)$. Proportional control is employed; that is, the output flow rate through the control valve is made proportional to the difference between the current level and the desired level. A mass balance on the tank gives

$$\frac{dx_2}{dt} + K(x_2 - x_2^d) = x_1 \qquad \text{(A)}$$

where we see that if $x_2 = x_2^d$, no liquid is drained from the tank and K is a proportional constant. We note also that the output flow rate can never be negative, so that if $x_2 < x_2^d$, the second term on the left-hand side is equal to zero. Thus we just wait until enough liquid flows in to refill the tank up to x_2^d. $x_1(t)$ is the inlet flow rate. We shall assume that $x_1(t)$ is the gamma Markov process from the previous example; that is,

$$dx_1 = (\alpha + 1 - \tfrac{1}{2}D - \beta x_1)\, dt + (2Dx_1)^{1/2}\, dw(t) \qquad \text{(B)}$$

This is a reasonable choice, since we want $x_1 \geq 0$.

We rewrite (A) and (B) as

$$\begin{aligned} dx_1 &= (\alpha + 1 - \tfrac{1}{2}D - \beta x_1)\, dt + (2Dx_1)^{1/2}\, dw(t) \\ dx_2 &= (x_1 - Kx_2 + Kx_2^d)\, dt \end{aligned} \qquad \text{(C)}$$

We note that (C) are the equations for the two components of a vector stochastic

O.D.E. of the form (5.5.2). We can write (5.5.2) in component notation as

$$dx_i(t) = f_i(\mathbf{x}(t), t)\, dt + \sum_{j=1}^{m} g_{ij}(\mathbf{x}(t), t)\, dw_j(t), \qquad i = 1, 2, \ldots, n \tag{D}$$

The first-order probability density function for a vector stochastic process is $p(\mathbf{x}(t), t) = p(x_1(t), x_2(t), \ldots, x_n(t), t)$, where x_1, x_2, etc., are the components of the vector process $x(t)$ *at time t* and *not* values at other times. The second-order probability density function for a two-state variable stochastic process would be denoted, for example, by $p(x_1(t_\alpha), x_2(t_\alpha), t_\alpha; x_2(t_\beta), x_2(t_\beta), t_\beta)$.

The forward Fokker-Planck equation satisfied by a vector stochastic process can be written in component notation as

$$\frac{\partial p(\mathbf{x}(t), t)}{\partial t} = -\sum_{i=1}^{n} \frac{\partial}{\partial x_i}(\alpha_{1i} p) + \frac{1}{2} \sum_{i=1}^{n} \sum_{j=1}^{n} \frac{\partial^2}{\partial x_i\, \partial x_j}(\alpha_{2ij} p) \tag{E}$$

The moments are

$$\alpha_{1i} = f_i(\mathbf{x}(t), t) \qquad \text{(Ito)}$$

$$\alpha_{1i} = f_i(\mathbf{x}(t), t) + \frac{1}{2} \sum_{k=1}^{n} \sum_{l=1}^{m} g_{kl}(\mathbf{x}(t), t) \frac{\partial g_{kl}}{\partial x_k} \qquad \text{(Stratonovich)} \qquad i = 1, 2, \ldots, n \tag{F}$$

$$\alpha_{2ij} = \sum_{l=1}^{m} g_{il}(\mathbf{x}(t), t) g_{jl}(\mathbf{x}(t), t), \qquad i, j = 1, 2, \ldots, n \tag{G}$$

Returning to our example and employing the Stratonovich interpretation, we obtain the forward Fokker-Planck equation:

$$\frac{\partial p(x_1, x_2, t)}{\partial t} = -\frac{\partial}{\partial x_1}[\alpha_1(x_1) p] + \frac{1}{2} \frac{\partial^2}{\partial x_1^2}[\alpha_2(x_1) p] - \frac{\partial}{\partial x_2}[x_1 - K(x_2 - x_2^d)) p] \tag{H}$$

where

$$\begin{aligned} \alpha_1(x_1) &= \alpha + 1 - \beta x_1 \\ \alpha_2(x_1) &= 2D x_1 \end{aligned} \tag{I}$$

The performance of the control system is judged by its ability to reduce the variation in liquid level in the presence of random feed rates. It would be desirable to solve (H) for the first-order density p; however, we expect that (H), although linear, will be somewhat difficult to solve. A somewhat easier way to study the dynamics of this system is to consider the moments of the two state variables x_1 and x_2.

Define the mixed moment

$$m_{kj}(t) = \int_0^\infty \int_0^\infty x_1^k x_2^j p(x_1, x_2, t)\, dx_1\, dx_2 \tag{J}$$

By differentiation

$$\frac{dm_{kj}(t)}{dt} = \int_0^\infty \int_0^\infty x_1^k x_2^j \frac{\partial p(x_1, x_2, t)}{\partial t}\, dx_1\, dx_2 \tag{K}$$

Substituting (H) and (I) into (K) yields

$$\begin{aligned} \frac{dm_{kj}(t)}{dt} = k(\alpha + 1) m_{k-1,j} - k\beta m_{k,j} + k(k-1) D m_{k-1,j} + j m_{k+1,j-1} \\ - jK[m_{kj} - x_2^d m_{k,j-1}] \end{aligned} \tag{L}$$

Let us explore the stationary operation of the system, that is, when $\dot{m}_{kj} = 0$. In that case we find

$$m_{10} = \frac{\alpha + 1}{\beta}, \qquad m_{01} = \frac{\alpha + 1}{\beta K} + x_2^d$$

$$m_{20} = \frac{\alpha + 1 + D}{\beta} m_{10} = \frac{(\alpha + 1 + D)(\alpha + 1)}{\beta^2}$$

$$m_{11} = \frac{\alpha + 1}{\beta + K}\left[\frac{\alpha + 1}{\beta K} + x_2^d + \frac{\alpha + 1 + D}{\beta^2} + \frac{K x_2^d}{\beta}\right]$$

$$m_{02} = \frac{m_{11}}{K} + x_2^d m_{01}$$

5.6 Deterministic vs. Stochastic Models

The question now arises, When should one use a deterministic model and when should one use a stochastic model? Unfortunately, it is not possible to give a clear-cut answer, since it really depends on both the system being studied and the ultimate objective of the modeling effort.

Nevertheless, we can present some general rules of thumb. If the time scale for random fluctuations in the input to a process is the same order of magnitude as that for transients in the system itself, then it will be necessary to employ a stochastic description of the system. On the other hand, if the input fluctuations are either very much faster or very much slower than the time scale of the system, it is often possible to employ a deterministic description.

Randomness in a system can arise in three ways:

1. The system itself is "deterministic" but inputs to the system are random.
2. Inputs to the system are "deterministic" but the basic nature of the system is random.
3. Both the input and the system have random characteristics.

An example of the first situation would be a laminar flow reactor in which the inlet concentration of reactant fluctuates due to upstream disturbances. The second situation is exemplified by either a well-mixed vessel or a packed bed, each of which produces a random output from any deterministic input. A well-mixed vessel or a packed bed with a fluctuating input is an example of the third situation, as, for example, in Section 5.4.8.

An interesting comparison between deterministic and stochastic modeling approaches can be made with reference to a well-mixed vessel. We customarily describe a continuous stirred tank reactor deterministically, such as in Chapter 3. However, as we shall see in the next chapter, we can also model it from a stochastic viewpoint and obtain an *identical* description. We shall exploit this comparison in some detail in the next chapter.

To illustrate the effects of random disturbances on a particular system, consider the system of Section 5.4.8, a CSTR with an exothermic first-order reaction and heat removal by a coil or jacket. The dynamic mass and energy balances on the reactor are (5.4.45) and (5.4.46) (using a slightly changed nomenclature):

$$V\frac{dX}{dt'} = q(X_0 - X) - Vk_0X\exp\left(\frac{-E}{RT}\right) \tag{5.6.1}$$

$$\rho VC_p\frac{dT}{dt'} = \rho C_p q(T_0 - T) + (-\Delta H)Vk_0X\exp\left(\frac{-E}{RT}\right) - D(T - T_c) \tag{5.6.2}$$

Define the dimensionless variables:

$$t = \frac{qt'}{V}, \qquad \zeta = \frac{X}{X_0}, \qquad \eta = \frac{\rho C_p T}{(-\Delta H)X_0}$$

$$\beta = \ln\frac{Vk_0}{q}, \qquad \gamma = \frac{\rho C_p E}{(-\Delta H)X_0 R}, \qquad \theta = \frac{D}{\rho C_p q}$$

and (5.6.1) and (5.6.2) become

$$\frac{d\zeta}{dt} = 1 - \zeta - \exp\left(\beta - \frac{\gamma}{\eta}\right)\zeta \tag{5.6.3}$$

$$\frac{d\eta}{dt} = \eta_0 - \eta + \exp\left(\beta - \frac{\gamma}{\eta}\right)\zeta - \theta(\eta - \eta_c) \tag{5.6.4}$$

where η_c and η_0 have the obvious connections to η.

If feedback proportional control based on the measured temperature $\tilde{\eta}$ is used to manipulate the flow rate of coolant, the dimensionless heat transfer coefficient θ can be expressed as

$$\theta = \begin{cases} \theta_s + K\eta_{cr}, & \tilde{\eta} \geq \eta_s + \eta_{cr} \\ \theta_s + K(\tilde{\eta} - \eta_s), & |\tilde{\eta} - \eta_s| < \eta_{cr} \\ 0, & \tilde{\eta} \leq \eta_s + \eta_r \end{cases} \tag{5.6.5}$$

where K is the proportional gain, θ_s the steady-state heat transfer coefficient, η_s the desired reactor temperature, and $K\eta_{cr}$ a constant corresponding to the half range of the coolant flow valve. Thus $\theta = \theta_s + K\eta_{cr}$ and $\theta = 0$ correspond to the coolant flow valve full open and shut, respectively. We assume that there is a time lag of 0.1 in the controller. The object of the control is to keep η at η_s in the presence of disturbances.

As a low-frequency inlet disturbance we shall consider a step change in the inlet temperature η_0 at $t = 0$. High-frequency fluctuations in inlet concentration and temperature enter the right-hand sides of (5.6.3) and (5.6.4) additively. Thus we add the random variables $\xi_1(t)$ and $\xi_2(t)$ (these are not necessarily white noise):

$$\frac{d\zeta}{dt} = 1 - \zeta - \exp\left(\beta - \frac{\gamma}{\eta}\right)\zeta + \xi_1 \tag{5.6.6}$$

$$\frac{d\eta}{dt} = \eta_0 - \eta + \exp\left(\beta - \frac{\gamma}{\eta}\right)\zeta - \theta(\eta - \eta_c) + \xi_2 \tag{5.6.7}$$

The measured temperature $\tilde{\eta}(t)$ is not identical to $\eta(t)$ due to measurement errors:

$$\tilde{\eta}(t) = \eta(t) + \xi_3(t) \tag{5.6.8}$$

What we want to do is examine the performance of this system by computer simulation. At $t = 0$ we assume that the system is at η_s and θ_s and that a step disturbance arrives at $t = 0$. The following parameters are used; $\theta_s = 1$, $\eta_c = 1.75$, $K\eta_{cr} = 1$, $\beta = 25$, $\gamma = 50$, $\eta_0(t < 0) = 1.75$, $\eta_0(t \geq 0) = 1.85$, $\zeta_s = 0.5$, $\eta_s = 2.0$, and $K = 20$. The input disturbances are simulated by

$$\xi_i = A_i \cos \omega_i t, \qquad i = 1, 2$$

with $A_1 = 1.5$, $A_2 = 1.0$, $\omega_1 = 20\pi$, and $\omega_2 = 40\pi$; the measurement noise is simulated by

$$\xi_3 = G[0, 0.1]$$

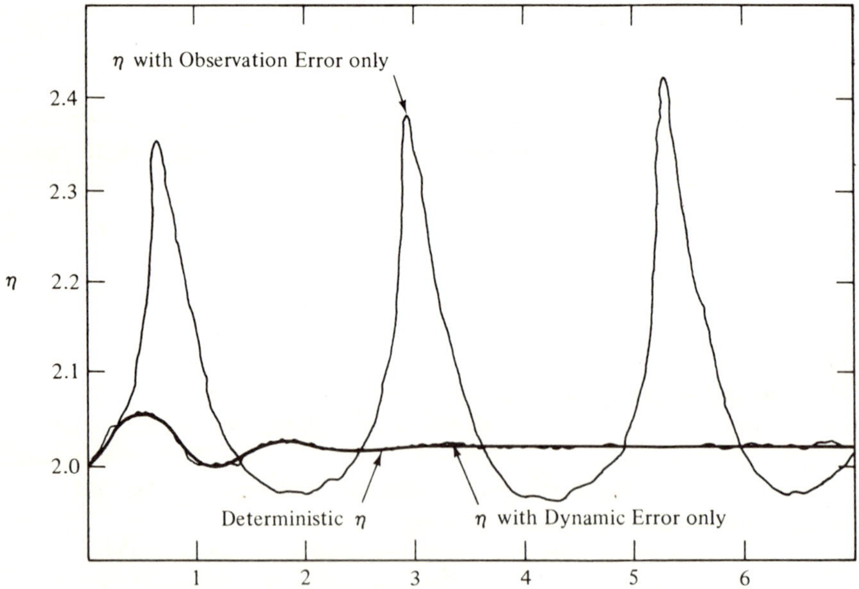

Figure 5.12 Effect of input disturbances and measurement errors on the feedback control of a CSTR.

a normally distributed random variable with mean zero and variance of 0.1.

Figure 5.12 shows the results of this simulation. The deterministic η is the response with no disturbances or measurement errors. We see a customary oscillation before reaching a steady state. The curve for η with input noise only (that is, $\xi_3 = 0$) corresponds very closely to the deterministic curve. The CSTR effectively damps out these input fluctuations, essentially because of the frequencies of the CSTR (the reciprocal of the time constant q/V). Thus in the absence of measurement errors the reactor response to high-frequency input noise is, for all practical purposes, identical to the deterministic response. In this case, a stochastic analysis is unnecessary. However, the measurement noise, which affects the control action (the coolant flow) through (5.6.5), has a devastating effect on the system (η curve with observation error only), driving it into a series of wild oscillations. If we expect errors of this type to be important, then a stochastic approach is warranted.

These simulated results confirm our rule of thumb stated at the beginning of this section concerning the comparison of the frequency of oscillations and the fundamental frequency of the system. We caution the reader, however, that each system must be approached individually with respect to the question of whether a stochastic representation is needed or whether a deterministic model will suffice.

REFERENCES

1. Aris, R., and N. R. Amundson, "Statistical Analysis of a Reactor Linear Theory," *Chem. Eng. Sci. 9*, 250 (1958).
2. Bartholomew, D. J., *Stochastic Models for Social Processes*, Wiley, New York, 1967.
3. Chandrasekhar, S., "Stochastic Problems in Physics and Astronomy," in *Selected Papers on Noise and Stochastic Processes* (N. Wax, ed.,) Dover, New York, 1954.
4. Davenport, W. B., and W. L. Root, *Random Signals and Noise*, McGraw-Hill, New York, 1958.
5. Doob, J. L., *Stochastic Processes*, Wiley, New York, 1953.
6. Feller, W., *An Introduction to Probability Theory and Its Applications*, vol. I, 3rd ed., Wiley, New York, 1968.
7. Fisz, M., *Probability Theory and Mathematical Statistics*, 3rd ed., Wiley, New York, 1963.
8. Gunn, D. J., "Theory of Axial and Radial Dispersion in Packed Beds," *Trans. Inst. Chem. Eng. 47*, T351 (1969).
9. Howard, R. A., *Dynamic Probabilistic Systems*, vol. I: Markov Models, Wiley, New York, 1971.

10. Jenkins, G. M., and D. G. Watts, *Spectral Analysis and Its Applications*, Holden-Day, Inc., San Francisco, 1968.

11. Karlin, S., *A First Course in Stochastic Processes*, Academic Press, New York, 1966.

12. King, R. P., "Continuous Flow Systems with Stochastic Transfer Functions," *Chem. Eng. Sci. 23*, 1035 (1968).

13. King, R. P., "Some Problems in Chemical Reactor Analysis with Stochastic Features," *Ind. Eng. Chem. Fund. 9*, 1, 190 (1970).

14. King, R. P., "Continuous Flow Systems with Stochastic Transfer Functions. II. Solution by the Associated Kolmogorov Equation," *Chem. Eng. Sci. 26*, 729 (1971).

15. King, R. P., *An Introduction to Stochastic Differential Equations—Abbreviated Lecture Notes*, Department of Chemical Engineering, University of Natal, Durban, South Africa, Mar. 19, 1971.

16. Lathi, B. P., *An Introduction to Random Signals and Communication Theory*, International Textbook Company, Scranton, Pa., 1968.

17. Mortensen, R. E., "Mathematical Problems of Modeling Stochastic Nonlinear Dynamic Systems," *J. Statist. Phys. 1*, 2, 271 (1969).

18. Papoulis, A., *Probability, Random Variables and Stochastic Processes*, McGraw-Hill, New York, 1965.

19. Pell, T. M., and R. Aris, "Some Problems in Chemical Reactor Analysis with Stochastic Features," *Ind. Eng. Chem. Fund. 8*, 339 (1969).

20. Stratonovich, R. L., "A New Representation for Stochastic Integrals and Equations," *SIAM J. Control 4*, 362 (1966).

PROBLEMS

5.1 Consider the Markov chain resulting from the random movement of a chess king around a vacant chessboard. The king moves to an adjacent square at each instant of time with equal probability of moving to any adjacent square. Recall that a chessboard (checkerboard) has 64 squares. Write the transition matrix **P** for this Markov chain. If once the king reaches any of the four corners, it is trapped there forever, what would be the form of **P**?

5.2 Consider a Markov chain with the transition matrix

$$\mathbf{P} = \begin{bmatrix} 0.3 & 0.2 & 0.5 \\ 0.1 & 0.8 & 0.1 \\ 0.4 & 0.4 & 0.2 \end{bmatrix}$$

a. Determine the n-step transition probability matrices for $n = 1, 2, 3$. What is the probability that if the process began in state 2 it is in state 3 after three transitions?

b. If the probabilities of initial state occupancy are [0.15 0.65 0.20], find the probability that after two transitions the process is in state 2.
c. Compute the asymptotic or ergodic probabilities for this process.

5.3 Consider a Markov chain with the transition matrix

$$\mathbf{P} = \begin{bmatrix} 0 & 0 & \frac{1}{2} & \frac{1}{2} \\ 0 & 0 & \frac{1}{2} & \frac{1}{2} \\ \frac{1}{2} & \frac{1}{2} & 0 & 0 \\ \frac{1}{2} & \frac{1}{2} & 0 & 0 \end{bmatrix}$$

a. Determine the n-step transition probability matrices for $n = 1, 2, 3$. What is the probability that if the process begins in state 3 it is in state 4 after three transitions?
b. If the probabilities of initial state occupancy are [0.1 0.1 0.1 0.7], find the probability that after two transitions the process is in state 3.
c. Discuss this chain with respect to reducibility, periodicity, and recurrence. What are the limiting probabilities?

5.4 Howard [9] presents a thorough analysis of the use of the z-transform in calculating $\mathbf{P}^n$. For transition matrices of dimension less than 5×5 this is probably the most expedient way to compute $\mathbf{P}^n$.
a. Beginning with (5.1.6),

$$p_{ij}(n+1) = \sum_{k=1}^{M} p_{ik}(n)p_{kj}, \qquad n = 0, 1, 2, \ldots \tag{A}$$

$$p_{ij}(0) = \delta_{ij}$$

Using the relation,

$$Z[f(n+k)] = z^{-k}[F(z) - f(0) - zf(1) - \cdots - z^{k-1}f(k-1)]$$

show that the matrix form of (A) is

$$Z[\mathbf{P}^n] = [\mathbf{I} - \mathbf{P}z]^{-1}$$

b. For the television-viewing example, Example 5.1, compute $\mathbf{P}^n$ by inverting the matrix $[\mathbf{I} - \mathbf{P}z]^{-1}$ and then by inverting from the z-domain to the time domain. The result you should obtain is

$$\mathbf{P}^n = \begin{bmatrix} 0.6 & 0.4 \\ 0.6 & 0.4 \end{bmatrix} + \left(\frac{1}{2}\right)^n \begin{bmatrix} 0.4 & -0.4 \\ -0.6 & 0.6 \end{bmatrix}$$

5.5 A drunk performs a random walk on the integers. An open manhole is located at zero. Find the probability that if the drunk begins at k, he will eventually fall into the manhole. Assume that the probabilities of his stepping to the right or left are p and q, respectively. *Note:* This is simply the gambler's ruin problem with an infinitely rich adversary.

5.6 A random walk with memory is performed along a line of length four units with reflecting ends. By *memory* we mean that there is a probability p of continuing in the same direction as the last step and a probability q of reversing direction. For the first step, let p and q be the probabilities of going to the right and left, respectively. Go as far as you can in exploring this process.

In particular, find the asymptotic form for the location probabilities after a very large number of steps.

5.7 Two tearful drunks (one-dimensional and symmetric $p = q = \frac{1}{2}$) bid each other farewell with the words "We'll probably never see each other again" and start off in opposite directions. Are they right?

5.8 Consider a one-dimensional random walk with a variable length X for the individual step. Let

$$\text{Prob}\,\{x \leq X \leq x + \Delta x\} = P_X(x)\,\Delta x$$

Find the probability density $P_S(x)$ of the distance from the origin S_n after n steps for the following cases:

a. Normal distribution of length of individual step:

$$P_X(x) = \frac{1}{\sqrt{2\pi}\,\sigma} e^{-(1/2)(x^2/\sigma^2)}, \qquad \text{zero mean, variance } \sigma^2$$

b. Each step X is a variable length depending on a number of little steps according to a Poisson distribution

$$\text{Prob}\{X = k\} = \frac{\lambda^k}{k!} e^{-\lambda}$$

5.9 Show that the solution of (5.2.42) and (5.2.43) is (5.2.44).

5.10 For stationary random sequences and processes, show that the autocorrelation functions R_i and $R(\tau)$ satisfy, respectively,

$$R_i \leq R_0$$

$$R(\tau) \leq R(0)$$

5.11 Determine the autocorrelation function $R(t_1, t_2)$ for

$$X(t) = A + Bt + Ct^2$$

if

a. A is the random variable uniformly distributed on $(-1, 1)$.
b. B is the random variable uniformly distributed on $(0, 1)$.
c. C is the random variable uniformly distributed on $(-1, 1)$.

5.12 Show that for the random process of Example 5.9 the time-averaged autocorrelation function

$$\bar{R}(\tau) = \overline{X(t)X(t+\tau)} = \frac{a^2}{2}\cos(\omega_0 \tau)$$

5.13 If $X(t)$ and $Y(t)$ are two independent random processes and two new processes are formed by

$$U(t) = \alpha X(t) + \beta Y(t)$$

$$V(t) = \alpha X(t) - \beta Y(t)$$

find R_{UU}, R_{VV}, R_{UV}, and R_{VU} in terms of R_{XX} and R_{YY}.

5.14 For the random process

$$X(t) = a\cos(\omega_0 t + \Theta)$$

where a and ω_0 are constants and Θ is a random variable with probability density $p(\theta)$, show that $X(t)$ is a wide-sense stationary process if the Fourier transform of $p(\theta)$ (the characteristic function of θ) has

$$\phi_\theta(1) = \phi_\theta(2) = 0$$

5.15 Show that for a random process that consists of unit impulses (both positive and negative) randomly distributed along the time axis according to a Poisson distribution with average frequency of occurrences λ

$$R(\tau) = \lambda\,\delta(\tau)$$
$$S(\omega) = \lambda$$

This process is one form of white noise.

5.16 A stochastic process $Y(t)$ is governed by

$$\dot{Y}(t) + Y(t) = X(t)$$
$$Y(0) = 0$$

where $X(t)$ is a stochastic process input. Determine the quantities $E[Y(t)]$, $R_{XY}(t_1, t_2)$, and $R_{YY}(t_1, t_2)$ for each of the following cases:
a. $X(t)$ is white noise, $R_{XX}(\tau) = \lambda\,\delta(\tau)$, $\tau = t_2 - t_1$.
b. $X(t)$ is thermal noise, $R_{XX}(\tau) = e^{-|\tau|}$.

5.17 The autocorrelation functions of three processes are

$$R(\tau) = ae^{-b|\tau|} + \cos(\omega_0\tau)$$
$$R(\tau) = ae^{-b|\tau|} + c$$
$$R(\tau) = ae^{-b|\tau|} + \cos(\omega_0\tau) + c$$

From this information alone, state all you can about these processes.

5.18 Consider the white-noise process of Problem 5.15. Assume that the impulses are not of uniform strength but can assume a random value A, where

$$p_A(a) = \frac{1}{\sqrt{2\pi}}e^{-a^2/2}$$

Find the autocorrelation function and power density spectrum for this process.

5.19 Derive (5.4.61)–(5.4.63).

5.20 In the Gaussian random process of Example 5.17, if **A**, **G**, and **C** are independent of time, it is possible that the variance of the process $\mathbf{P}(t)$ approaches a constant value $\mathbf{P}_\infty$ as $t \longrightarrow \infty$; that is, the process becomes stationary. Examine the conditions for stationarity of the second-order system

$$\ddot{x} + 2\zeta\omega\dot{x} + \omega^2 x = \omega^2 v(t)$$

where $v(t)$ is zero mean white noise with unit variance. *Hint:* Determine $\mathbf{P}(t)$ for this process and find the conditions under which $\mathbf{P}(t) \longrightarrow \mathbf{P}_\infty$ as $t \longrightarrow \infty$. Show that these conditions are identical to requiring negative real parts of the eigenvalues of **A**.

5.21 Pell and Aris [19] employed the differential equation

$$\frac{dx}{dt} = -(a + v(t))x + v_0(t)$$

as a model for a chemical reactor that is subject to random fluctuations in temperature, where $v(t)$ and $v_0(t)$ are both Gaussian white-noise processes. Derive the forward Fokker-Planck equation for this process using both Ito and Stratonovich rules. Solve the stationary form of this Fokker-Planck equation.

Residence Time Distribution Theory 6

The first five chapters have concentrated on the mathematical fundamentals of process modeling and analysis. The next five chapters will be devoted to the actual task of process modeling and analysis.

The analysis of a process can usually be divided into three parts. First, a model (often tentative) is selected for the process, based on any prior knowledge we may have on the internal nature of the system. Second, data are obtained from the process. Often these data are the result of a special test, for example, a tracer study, in which tracer is introduced in a known manner into the input stream to the system and measured as it leaves the system. Third, the model chosen and the available data are matched in some manner.

This chapter will be devoted to the first two phases: model selection and experimental test, whereas Chapters 7–10 will treat in detail the third aspect, the matching of model to data. We shall confine our attention in this chapter to linear systems. In addition, we shall consider only those types of process analysis experiments in which the modeling objective is to account for the behavior of fluid flow and mixing in the system. This is an important class of problems since to be able to predict the performance of equipment we must know the way in which fluid passes through the equipment, the amount of mixing that takes place, and the time that a given element of fluid can be expected to spend in the system. As noted, the experimental procedure usually employed to gain such knowledge is a tracer test. What we are interested in obtaining from a tracer test is the probability density for the *residence time* of an element of fluid in the system, and subsequently matching the experimental output curve to a model which accounts in some way for the behavior of the tracer. Such a procedure is commonly called a *residence time distribution* analysis.

Historically, residence time distribution analysis has played an important role in process modeling, as evidenced by the abundance of literature on the subject over the past 20 years. The basic concepts of residence time distribution theory are quite simple and, in most cases, follow directly from elementary probability theory. Nevertheless, the analysis of tracer studies via residence times is one of the most important tools available to the chemical engineer. In this chapter we shall present the theory and techniques of residence time distribution analysis.

6.1 Fundamental Definitions

Let us consider the flow of fluid through a vessel. Say at time $t = 0$ we inject instantaneously a quantity Q of tracer (a tracer pulse) and observe the output of the vessel for a long time. Let $c(t)$ represent the concentration of tracer in the effluent at time t, and let q be the volumetric flow rate of fluid through the vessel. Between times t and $t + dt$ the number of moles of tracer leaving is $qc(t)\,dt$. As $t \longrightarrow \infty$ we note that all the tracer will leave the system, and so

$$Q = q \int_0^\infty c(t)\,dt \tag{6.1.1}$$

We are interested in the fraction of tracer in the exit stream which has resided within the vessel between t and $t + dt$. This fraction is

$$\frac{qc(t)\,dt}{q \int_0^\infty c(t)\,dt} = \frac{c(t)\,dt}{\int_0^\infty c(t)\,dt} \tag{6.1.2}$$

particle in the system. We call $E(t)$ the *residence time density function*† for the system. Clearly,

$$\int_0^\infty E(t)\,dt = 1 \tag{6.1.3}$$

Also of interest is the fraction of tracer *in the system* of ages t to $t + dt$, as contrasted with $E(t)\,dt$, the fraction of tracer *leaving the system* of ages t to $t + dt$. We thus define $I(t')\,dt'$ as the fraction of tracer in the system with ages between t' and $t' + dt'$, where $I(t')$ is the *internal age density function* for the system. Note that we use t' because the random variable T' representing

†The term *residence time distribution function* is commonly employed for $E(t)$, even though $E(t)$ is a *density* function rather than a *distribution* function.

the internal age of a fluid particle is not the same as t, representing the existing age.

The average residence time of tracer before departure is $\int_0^\infty tE(t)\,dt$, the first noncentral moment of T, which we denote simply as μ. For a vessel in which there is no diffusion at the entrance and exit, only plug flow, μ is equal to the volume of the vessel available for flow divided by the volumetric flow rate of fluid through the vessel:

$$\mu = \frac{V}{q} \tag{6.1.4}$$

Experimentally, $E(t)$ and $I(t)$ are determined by a tracer test, in which tracer is injected into the inlet stream in a known manner, such as a pulse or step change, and the passage of tracer out of the system is subsequently monitored.

Let us determine the relationship between $E(t)$ and $I(t)$. Consider a vessel of volume V and volumetric flowrate q. Call the fluid entering after $t = 0$ *new fluid* and fluid that was present at $t = 0$ *old fluid*. At some later time t,

$$\text{new fluid in system} = V \int_0^t I(\tau)\,d\tau$$

$$\text{old fluid which has exited} = \int_0^t q\,dt' \int_{t'}^\infty E(\tau)\,d\tau$$

Clearly, these two quantities must be the same. Equating and differentiating with respect to t yields

$$\mu I(t) = \int_t^\infty E(\tau)\,d\tau = 1 - \int_0^t E(\tau)\,d\tau \tag{6.1.5}$$

Differentiating again,

$$E(t) = -\mu \frac{dI(t)}{dt} \tag{6.1.6}$$

Letting $\theta = t/\mu$ be a dimensionless time, (6.1.6) becomes

$$E(\theta) = -\frac{dI(\theta)}{d\theta} \tag{6.1.7}$$

We shall define two other common quantities in residence time analysis.

With no tracer initially present, introduce a step function of tracer in the entering fluid to the vessel. The concentration-time curve for tracer in the exit stream is called the F-curve. The F-curve, shown for a typical case in Figure 6.1, is normally plotted as $c(t)/c_0$ vs. t/μ, where c_0 is the tracer concentration in the entering stream.

The curve of exit tracer concentration vs. time corresponding to a pulse injection of tracer is called the C-curve. A typical C-curve is shown in Figure 6.2. The ordinate is usually plotted as $c(t)/c^0$, where c^0 is the concentration of tracer in the vessel if the injected tracer were evenly distributed through-

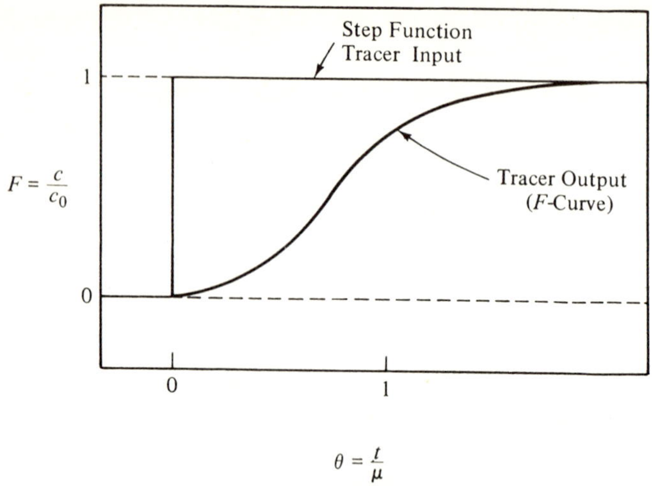

Figure 6.1. Typical output response to a step function of tracer as an input—the F-curve. (Reprinted from D. M. Himmelblau and K. B. Bischoff, *Process Analysis and Simulation-Deterministic Systems*, John Wiley & Sons, Inc., New York, 1968, p. 63. Reprinted by permission of the copyright owner.)

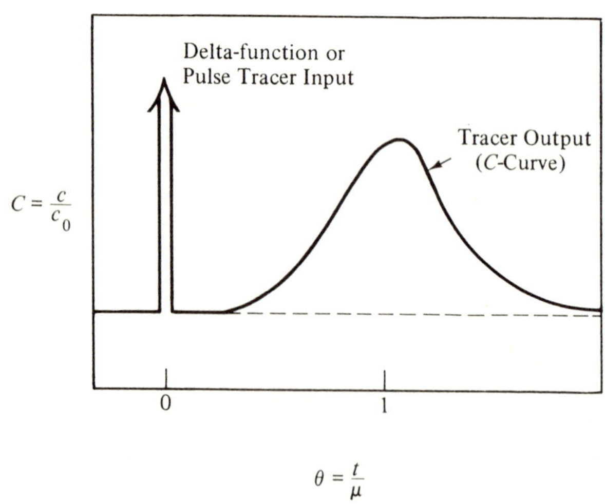

Figure 6.2. Typical output response to a step function of tracer as an input—the C-curve. (Reprinted from D. M. Himmelblau and K. B. Bischoff, *Process Analysis and Simulation-Deterministic Systems*, John Wiley & Sons, Inc., New York, 1968, p. 63. Reprinted by permission of the copyright owner.)

out the vessel. Using this dimensionless concentration, the area under the C-curve is always unity.

Given the experimental F- and C-curves, all that is needed is to relate these to the densities, $E(t)$ and $I(t)$. Consider first an impulse injection at $t = 0$. Since all the tracer is introduced at $t = 0$, the exit concentration, given by the C-curve, is identically the residence time density function, $E(t/\mu) = E(\theta)$:

$$C(\theta) = E(\theta) = \mu E(t) \tag{6.1.8}$$

Thus the residence time density can be determined directly from the response to a pulse input of tracer.

Now we shall determine the relation between $I(t)$ and $F(t)$. Let us assume that at $t = 0$ the inlet stream is changed entirely to tracer of concentration c_0. Let us also assume that the fluid in the system prior to $t = 0$ also has a concentration c_0. Call all fluid entering after $t = 0$ *new fluid*. The amount of new fluid in the system at time t is

$$Vc_0 \int_0^t I(\tau)\, d\tau$$

The fraction of fluid leaving at time t which is older than t is

$$\int_t^\infty E(\tau)\, d\tau = 1 - \int_0^t E(\tau)\, d\tau$$

The old fluid which has been purged from time zero to t is

$$qc_0 \int_0^t dt' \int_0^\infty E(\tau)\, d\tau$$

Since V is constant, the amount of new fluid which has entered must equal the old fluid which has left:

$$V \int_0^t I(\tau)\, d\tau = q \int_0^t dt' \int_t^\infty E(\tau)\, d\tau \tag{6.1.9}$$

Differentiating with respect to t,

$$VI(t) = q \int_t^\infty E(\tau)\, d\tau \tag{6.1.10}$$

However, from the definition of $F(\theta)$, we see that (6.1.10) becomes

$$\mu I(t) = 1 - F(\theta) \tag{6.1.11}$$

or, in dimensionless form,

$$I(\theta) = 1 - F(\theta) \tag{6.1.12}$$

Summarizing the relations among E, C, I, and F,

1. The C-curve gives directly the residence time distribution function E.
2. The F-curve can be subtracted from unity at every point to give the internal age distribution function I.

3. The C-curve is the slope of the F-curve for the same vessel:

$$C(\theta) = \frac{dF}{d\theta}$$

4. The F-curve represents the area under the corresponding C-curve up to that time:

$$F(\theta) = \int_0^\theta C(\theta')\, d\theta'$$

These relations provide the basis for pulse and step testing procedures to determine the age distribution of the contents and the residence time distribution for a vessel.

EXAMPLE 6.1. Compute the mean of $I(t)$ in terms of the mean and variance of $E(t)$. The mean of $I(t)$ is given by

$$\bar{T}_I = \int_0^\infty tI(t)\, dt$$

Using

$$\mu I(t) = 1 - \int_0^t E(\tau)\, d\tau$$

we have

$$\begin{aligned}\bar{T}_I &= \frac{1}{\mu}\int_0^\infty t\left[1 - \int_0^t E(\tau)\, d\tau\right] dt \\ &= \frac{1}{2\mu}\int_0^\infty t^2 E(t)\, dt \\ &= \frac{\sigma^2 + \mu}{2\mu}\end{aligned}$$

where σ^2 is the variance of $E(t)$.

6.2 Residence Time Distribution Studies

A residence time distribution tracer study can yield important information on the physical processes occurring within the system, specifically, the patterns of flow and degree of mixing among elements that entered the system at different times. Knowledge of the average residence time and degree of mixedness so obtained is then important in predicting the conversion of chemical reactions occurring in the system.

Comparison of residence time distributions in vessels of the same type but different size gives an indication of the effect of scale-up on flow properties. Also, in some heterogeneous systems the residence time distribution may be used directly for design. For example, in gas-solid catalytic reactors in which the catalyst is continuously replenished, the distribution of residence times of the catalyst particles is a primary design variable.

The single most important use for residence time distribution studies is to determine a mathematical model of the system. The approach usually taken is to perform one or more experimental tracer studies on the system. Having the tracer output curves, we then postulate models which reasonably approximate in behavior the output curves. The parameters of these models, usually correlated with fluid properties, vessel dimensions, flow rates, degree of mixing, etc., can then be determined so that the model output and the actual output match as closely as possible. The remainder of this chapter is devoted to the analysis of the models used in residence time distribution studies of flow systems.

Before proceeding to the actual models, we should discuss briefly the types of experimental tracer tests most often employed. The three most common tracer tests are pulse, step, and periodic injection. The advantage of a pulse test is that the output response is identically the residence time distribution function $E(t)$. The disadvantages of a pulse test are twofold. First, it is often difficult in practice to inject tracer rapidly enough so that it approximates a pulse. Second, accurate measurement of the low concentrations in the tail of the C-curve (see Figure 6.2) becomes quite difficult as time increases. As we shall see later, accurate measurement of the tail of the curve is important in model fitting.

The second common testing method consists of a step input of tracer. The advantage of step testing is principally experimental in that it is easy to generate a step change of concentration. Step testing has two disadvantages. First, as in the pulse testing case, after long times, in this case when $c/c_0 \cong 1$, measurement problems become significant. Second, since the residence time distribution function $E(t)$ must be obtained by numerical differentiation of the $F(t)$-curve, an inherently error-prone procedure, mathematical modeling based on the output to a step test is less accurate than with a pulse test.

The third testing method is that of a periodic input of tracer, that is, the frequency response testing of the system. Both the amplitude change and phase lag are used to characterize the system. The principal problems in this type of experiment are related to the generation of periodic concentration fluctuations.

Finally, we should mention a subtle point related to residence time distribution studies. We assume that the study is carried out in the following manner. The system itself is at steady state when we inject the tracer. If the amount of tracer is small and does not alter the flow behavior of the system, then, even though the tracer concentration varies with time in the system, the basic steady-state flow behavior of the system itself is not changed. Most systems we wish to study are turbulent and thus have changes occurring in flow patterns with time. It is assumed that these changes take place over time scales much shorter than those associated with the macroscopic dynamics of the system. Implicit in the assumption of steady state is that the residence time probability of a molecule in the system is independent of the time of introduction into the system. Also, the residence time distribution function is assumed the same regardless of whether it is determined by N particles introduced separately at different times or N particles entering at the same time. Thus we assume that systems at steady state exhibit both stationarity and ergodicity.

The tracer test on a system is useful in analyzing for the existence of dead space and bypassing. Dead space in a vessel is a region in the vessel in which fluid resides for times considerably *longer* than the overall mean residence time of the fluid actually flowing through the system. The existence of dead space can be discovered from $E(t)$. If $E(t)$ has a very long "tail," this indicates that fluid which enters the dead space leaves only very slowly. Clearly, if the experimental measurement of the outlet tracer concentration is carried out for a very long time, say θ, the calculated mean dimensionless residence time

$$\bar{\theta}_c = \int_0^{\theta} \theta' E(\theta')\, d\theta' \tag{6.2.1}$$

will be equal to the *true* mean dimensionless residence time, which is 1. However, if θ is only 3 or 4, the calculated $\bar{\theta}_c$ may be appreciably less than 1, since tracer is still left in the system.

Bypassing refers to a condition in which some fluid elements pass through the vessel much more rapidly than others; that is, a fraction of the fluid spends a time considerably *shorter* than the overall mean residence time of the fluid. If there is a large dead space or significant bypassing, there is no clear distinction between the two. An example of a system which often exhibits both dead spaces and bypassing is a fluidized bed in which some of the gas goes through rapidly in large bubbles while the rest is entrained in aggregates of solid particles for a much longer time. In simple terms, a system exhibits bypassing or dead space effects if there are two or more paths for flow with different residence times.

What we would like to have is another residence time function besides E and I which provides a measure of the dead space and/or bypassing of a system. Such a function is the *intensity function* $\Lambda(t)$, fiirst introduced by Naor and Shinnar [10]. We define

$\Lambda(t)\, dt$ = probability that a particle already in the system for a time t will exit in the next instant dt

$P(t)\, dt = \int_0^t E(\tau)\, d\tau$, the distribution function for T

We can write

$$E(t)\, dt = [1 - P(t)]\Lambda(t)\, dt \tag{6.2.2}$$

or

$$\Lambda(t) = \frac{E(t)}{1 - P(t)} = \frac{dP/dt}{1 - P(t)} = \frac{1}{\mu}\frac{E(t)}{I(t)} \tag{6.2.3}$$

The relation between $E(t)$ and $\Lambda(t)$ is thus

$$E(t) = \Lambda(t) \exp\left[-\int_0^t \Lambda(\tau)\, d\tau\right] \tag{6.2.4}$$

Let us consider what $\Lambda(t)$ might look like for a system with dead space and bypassing. For an ideally mixed vessel $\Lambda(t)$ is constant, since by definition every particle has an identical chance of escape. For a system with dead space or bypassing one path may move through the vessel fairly rapidly, whereas the other may be much longer due to dead regions, or possible absorption into a stagnant phase. Any particle which has resided in the system slightly shorter than the average residence time of the first path has a high escape probability. However, a particle which has remained in the system for a time longer than the first residence time has a high chance of being in the stagnant region and a lower probability of immediate escape. Thus stagnancy can be identified if $\Lambda(t)$ decreases with t over some interval. Figure 6.3 summarizes the characteristics of $E(\theta)$ and $\Lambda(\theta)$ when dead space and bypassing exist. The real utility of $\Lambda(\theta)$ is to determine the existence of dead space or bypassing from experimental data.

6.3 Typical Deterministic Flow System Models

Let us assume that we have carried out a tracer experiment and have computed $E(\theta)$, $I(\theta)$, and, perhaps, also $\Lambda(\theta)$ from the data. The question is now, What mathematical model do we use for representing the system? First, we have a choice between a deterministic or a stochastic model. The most common approach is to choose a deterministic model which contains one or more "adjustable" parameters to fit the model output to the experimental output.

Most systems exhibit behavior somewhere between the two extremes of plug flow and ideally mixed flow, the residence time characteristics of which are shown in Figure 6.4. An actual system will exhibit behavior more closely approaching the curves of Figure 6.3. In using deterministic flow system models, we have two approaches. First, we can use series and parallel arrangements of plug flow and ideally mixed flow systems to attempt to represent the actual system. Such models can be called *combined models*, since they

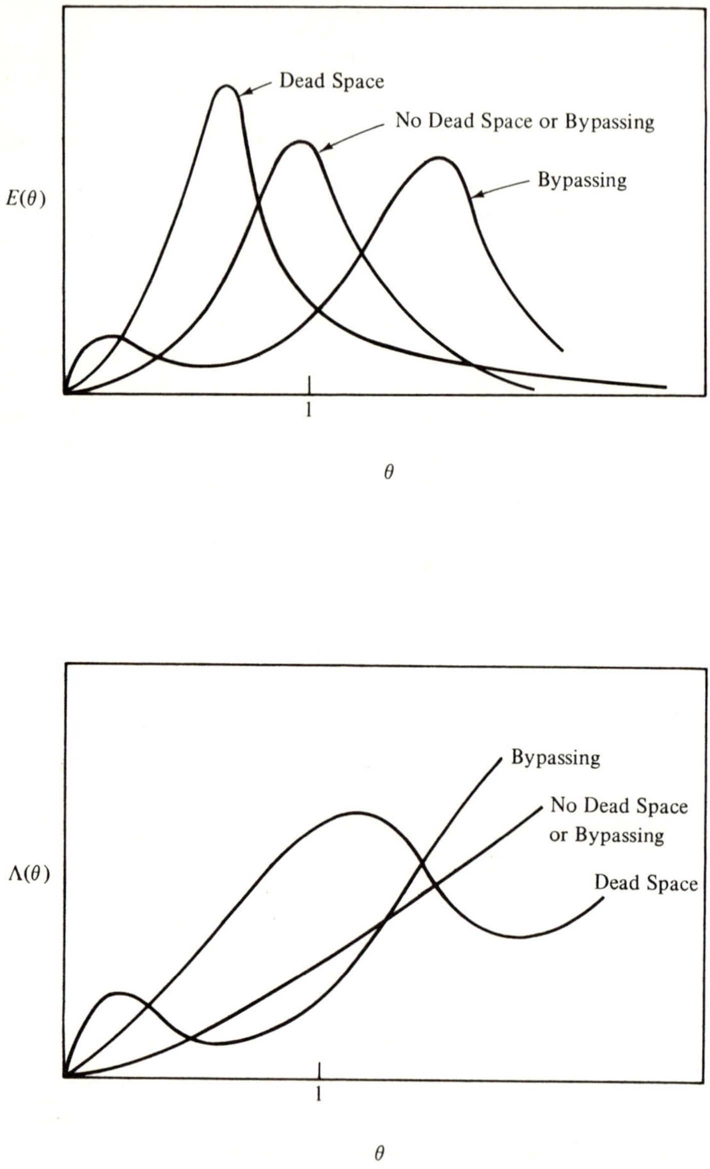

Figure 6.3. Characteristic residence time functions with dead space and bypassing. (Reprinted from D. M. Himmelblau and K. B. Bischoff, *Process Analysis and Simulation-Deterministic Systems*, John Wiley & Sons, Inc., New York, 1968, p. 73. Reprinted by permission of the copyright owner.)

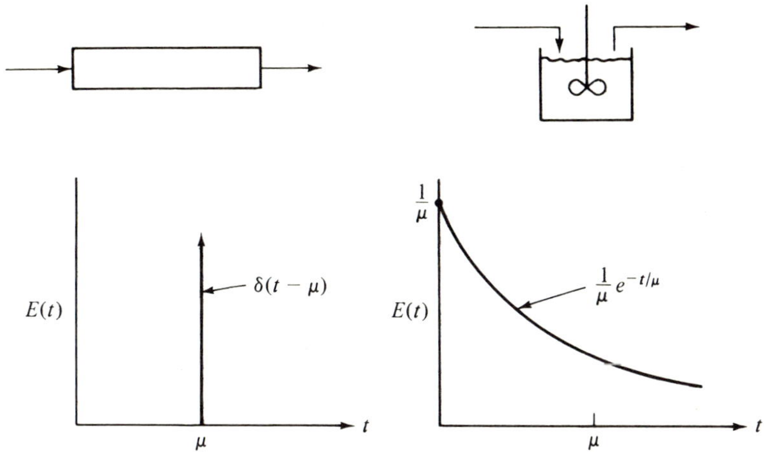

Plug Flow

$$E(t) = \delta(t - \mu)$$

$$I(t) = \begin{cases} 1 & t \leq \mu \\ 0 & t > \mu \end{cases}$$

$$\Lambda(t) = \delta(t - \mu)$$

Ideally Mixed Flow

$$E(t) = \frac{1}{\mu} e^{-t/\mu}$$

$$I(t) = e^{-t/\mu}$$

$$\Lambda(t) = \frac{1}{\mu}$$

Figure 6.4. Residence time characteristics of plug flow and ideally mixed flow.

represent combinations of the two ideal flow cases. Second, we can use a diffusional representation of mixing processes based on various forms of the parabolic diffusion equation. Such models are called *dispersion models.* Both types of models have been studied exhaustively in the chemical engineering literature, and an excellent review of the common models used is given by Levenspiel and Bischoff [9]. We shall now briefly summarize some common combined and dispersion models.

6.3.1 *N* ideally mixed vessels in series

Consider a sequence of N identical, ideally mixed vessels each of volume v. The volumetric flow rate through the system is q and the total volume of the system is $V = Nv$. A material balance on the nth vessel yields

$$v\frac{dc_n}{dt} = q(c_n - c_{n-1}), \qquad n = 1, 2, \ldots, N \tag{6.3.1}$$

The residence time distribution function $E(t)$ can be determined by solving (6.3.1) subject to the initial condition that all the vessels have zero tracer present and the input to vessel 1 has an impulse of tracer at $t = 0$:

$$c_n = 0, \qquad n = 1, 2, \ldots, N \tag{6.3.2}$$

$$c_0 = c^0 \frac{V}{q} \delta(t) \tag{6.3.3}$$

where c^0 is the average concentration of tracer if uniformly distributed throughout the entire system (all the tanks). The transfer function for the nth tank is (see Example 3.6)

$$\bar{c}_n(s) = \frac{c^0 V/q}{(1 + sv/q)^n} \tag{6.3.4}$$

and the solution is

$$c_N(t) = \frac{c^0 V}{q} \frac{1}{(N-1)!} \left(\frac{Nq}{V}\right)^N t^{N-1} \exp\left(\frac{-Nqt}{V}\right) \tag{6.3.5}$$

Letting $\theta = tq/V$, the $C(\theta)$-curve is [which is also $E(\theta)$]

$$C(\theta) = \frac{c}{c^0} = \frac{N^N}{N(-1)!} \theta^{N-1} e^{-N\theta} \tag{6.3.6}$$

which is shown for various values of N in Figure 6.5. Note that as N increases,

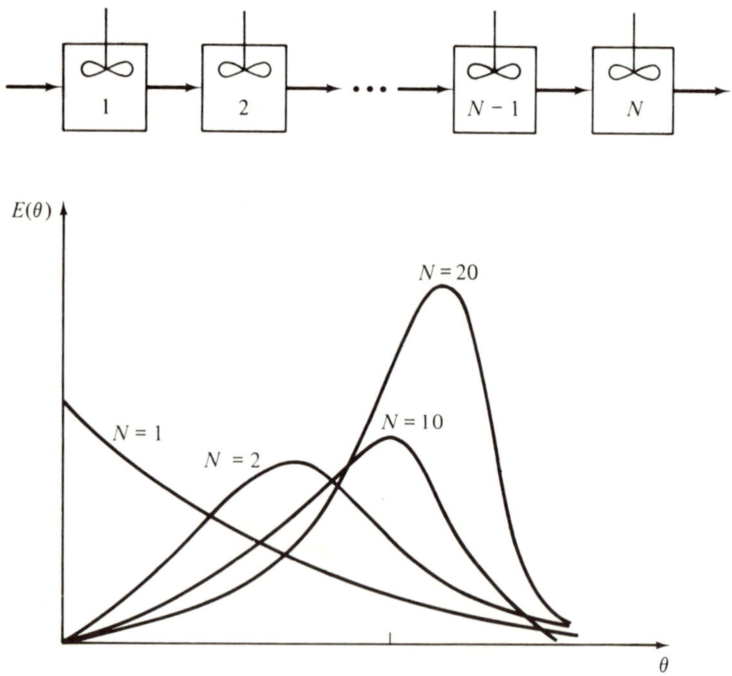

Figure 6.5. Residence time characteristics of N ideally mixed vessels in series.

the volume v of an individual tank decreases, such that the total volume V is held constant. Thus as N increases, the degree of mixing decreases since smaller stages are involved.

6.3.2 Two parallel vessels, one ideally mixed and one plug flow

Consider the situation of Figure 6.6, wherein a fraction α of the total flow passes through the plug flow vessel with $1 - \alpha$ through the ideally mixed

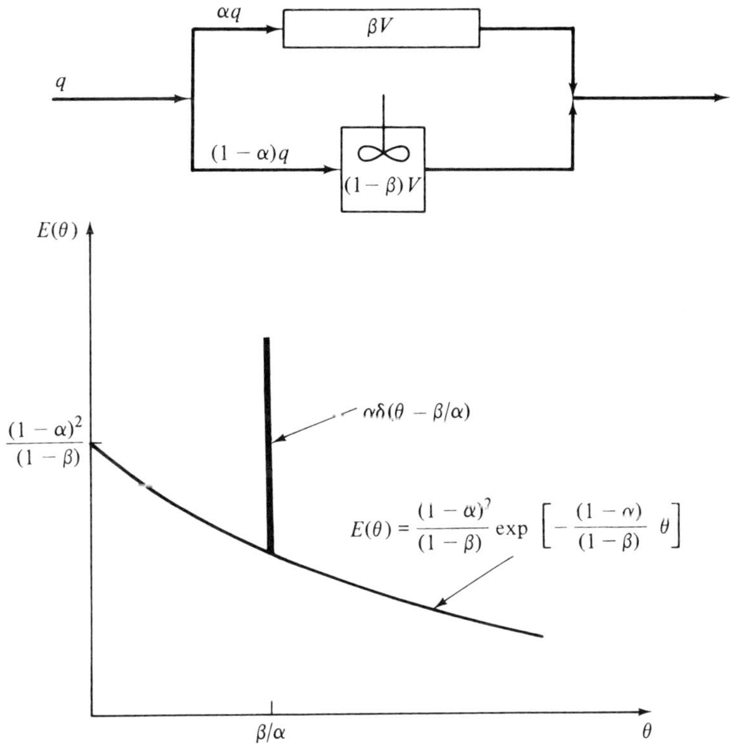

Figure 6.6. Residence time characteristics of two parallel vessels—one ideally mixed and one plug flow.

vessel. The residence time distribution $E(t)$ is also shown in Figure 6.6. Here, the reader will recall, we have an example of a mixed discrete and continuous probability density function (see Figure 4.2).

EXAMPLE 6.2. Let us explore the matching of several of the combined models to an experimental step response, i.e., the F-curve. In particular, we shall consider the following models: (1) one plug flow vessel, (2) one ideally mixed vessel, (3) a plug flow vessel in series with an ideally mixed vessel, (4) an ideally mixed vessel with dead space, and (5) an ideally mixed vessel with bypass flow. The F-functions for these models are

1. $F(t) = \begin{cases} 0, & t < \mu \\ 1, & t \geq \mu \end{cases}$
2. $F(t) = 1 - e^{-t/\mu}$
3. $F(t) = \begin{cases} 0, & t < \alpha \\ 1 - \exp\left[-\dfrac{1}{(1-\alpha)}\left(\dfrac{t-\alpha}{\mu}\right)\right] & t \geq \alpha \end{cases}$
4. $F(t) = 1 - e^{-t/\mu(1-\alpha)}$
5. $F(t) = 1 - (1-\alpha)e^{-(1-\alpha)t/\mu}$

It appears that it would be somewhat tedious to test each of these against an experimental F-curve. If we look again at each $F(t)$, we note many similarities. In fact, they are all special cases of a general $F(t)$ of the form

$$F(t) = 1 - \exp\left(-\eta\frac{t-\epsilon}{\mu}\right)$$

This form was proposed by Wolf and Resnick [14], with η and ϵ model parameters to be chosen to fit this $F(t)$ to the experimental curve. The parameter η is a measure of the efficiency of mixing, that is,

$$\begin{aligned} \eta &= 1 && \text{(ideally mixed flow)} \\ \eta &= \infty && \text{(plug flow)} \\ \eta &> 1 && \text{(dead space)} \\ \eta &< 1 && \text{(bypassing)} \end{aligned}$$

The parameter ϵ is a measure of the phase shift or response lag of the system:

$$\begin{aligned} \epsilon &= 0 && \text{(ideally mixed flow, dead space)} \\ \epsilon &> 0 && \text{(plug flow)} \\ \epsilon &< 0 && \text{(bypassing)} \end{aligned}$$

Figure 6.7 shows the fit of this function to an experimental F-curve for liquid in a fixed bed. The curve $\eta = 1$, $\epsilon = 0$ corresponds to an ideally mixed vessel; the dashed curve shows an almost perfect fit of the data for $\eta = 3.15$ and $\epsilon/\mu = 0.17$. It is important to realize that even though this $F(t)$ matches the data, it does not really give us much information on the true flow patterns in the packed bed.

EXAMPLE 6.3. Let us present the analysis of a combined model suggested by Chiang and Cholette [2]. The model consists of N vessels in series, the volume of each tank being V. Each tank has a dead space, characterized by a fraction $1 - m$ of the tank volume, or, conversely, a well-mixed region of fractional volume m. The fraction of the feed which enters the zone of ideal mixing is n, with $1 - n$ representing the fraction of the feed which is short-circuited directly to the outlet.

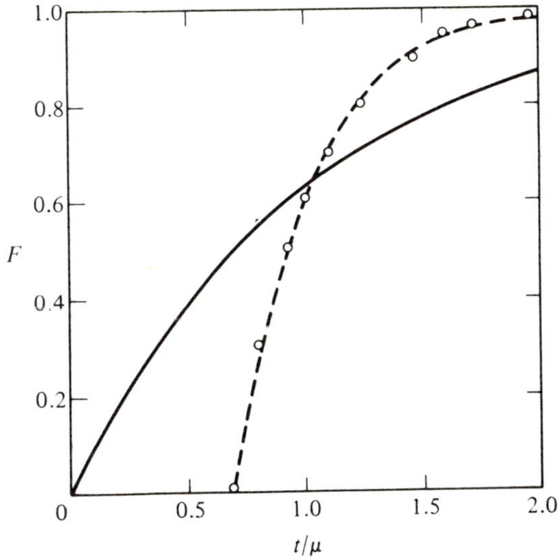

Figure 6.7. Experimental and theoretical F-curves for liquid flow in a packed bed. [Reprinted from D. Wolf and W. Resnick, "Residence Time Distribution in Real Systems," *Ind. Eng. Chem. Fund. 2*, 287 (1963). Reprinted by permission of the copyright owner.]

It is assumed that each of the N tanks has the same parameters m and n. Our objective is to compute the F-curves for such a model as a function of N, m, and n. The system is shown in Figure 6.8.

Assume that all the concentrations are normalized, so that at $t = 0$ the feed to the first vessel is changed from $C_0 = 0$ to $C_0 = 1$. As shown in Figure 6.8, we let

C'_j = concentration in the well-mixed zone of the jth tank
C_j = exit concentration from the jth tank $= (1 - n)C_{j-1} + nC'_j$

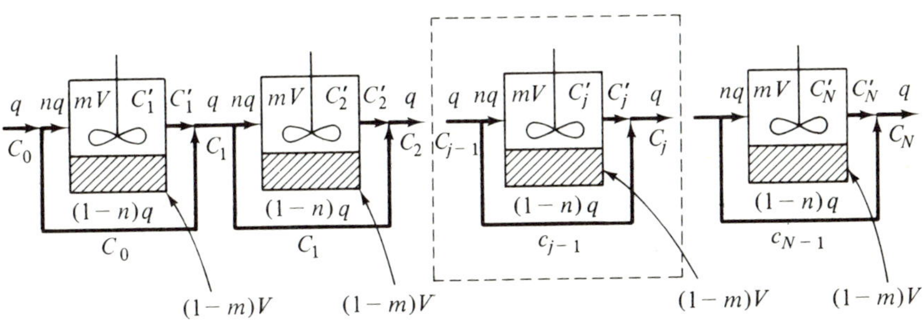

Figure 6.8. Cascade of ideally mixed vessels with dead space and bypassing. [Reprinted from D. Chiang and A. Cholette, "Performance of Tanks in Series of Non-ideal Mixing," *Can. J. Chem. Eng. 48*, 286 (1970). Reprinted by permission of the copyright owner.]

A material balance on the jth tank is

$$\frac{dC'_j}{dt} + \frac{nq}{mV}C'_j = \frac{nq}{mV}C_{j-1} \tag{A}$$

the solution of which is

$$C'_j = \frac{nq}{mV}e^{-nqt/mV}\int_0^t C_{j-1}e^{nqt/mV}\,dt \tag{B}$$

Letting

$$x = e^{nqt/mV} \tag{C}$$

(B) becomes

$$C'_j = \frac{1}{x}\int_1^x C_{j-1}\,dx \tag{D}$$

The feed to the $(j + 1)$st tank is then the total outlet of tank j, that is,

$$C_j = (1 - n)C_{j-1} + n\frac{1}{x}\int_1^x C_{j-1}\,dx \tag{E}$$

For a single tank, the outlet concentration C_1 from (E) is

$$\begin{aligned} C_1 &= 1 - \frac{n}{x} \\ &= 1 - ne^{-nqt/mV} \end{aligned} \tag{F}$$

For $N = 2$, application of (E) twice yields

$$\begin{aligned} C_2 &= (1 - n)^2 + 2n(1 - n)\frac{1}{x}\int_1^x dx + n^2\frac{1}{x}\int_1^x\frac{1}{x}\int_1^x (dx)^2 \\ &= 1 - n\left(2 - n + n\frac{nqt}{mV}\right)e^{-nqt/mV} \end{aligned} \tag{G}$$

For N tanks in series, the outlet concentration from the Nth tank is

$$\begin{aligned} C_N = (1 - n)^N &+ \binom{N}{1}n(1 - n)^{N-1}\frac{1}{x}\int_1^x dx + \cdots \\ &+ \binom{N}{j}n^j(1 - n)^{N-j}\frac{1}{x}\int_1^x\frac{1}{x}\int_1^x\cdots\frac{1}{x}\int_1^x (dx)^j + \cdots \\ &+ \binom{N}{N}n^N(1 - n)^{N-N}\frac{1}{x}\int_1^x\frac{1}{x}\int_1^x\cdots\frac{1}{x}\int_1^x (dx)^N \end{aligned} \tag{H}$$

Integrating and substituting for x,

$$\begin{aligned} C_N = (1 - n)^N &+ \binom{N}{1}n(1 - n)^{N-1}\{1 - e^{-nqt/mV}\} + \cdots \\ &+ \binom{N}{j}n^j(1 - n)^{N-j}\left\{1 - \left[\sum_{i=1}^{j}\frac{(nqt/mV)^{i-1}}{(i - 1)!}\right]e^{-nqt/mV}\right\} + \cdots \\ &+ \binom{N}{N}n^N(1 - n)^{N-N}\left\{1 - \left[\sum_{i=1}^{N}\frac{(nqt/mV)^{i-1}}{(i - 1)!}\right]e^{-nqt/mV}\right\} \end{aligned} \tag{I}$$

which simplifies to

$$C_N = 1 - \sum_{j=1}^{N} \binom{N}{j} n^j (1-n)^{N-j} \sum_{i=1}^{j} e^{-nqt/mV} \frac{(nqt/mV)^{i-1}}{(i-1)!} \tag{J}$$

Note that the term $\binom{N}{j} n^j(1-n)^{N-j}$ is simply the probability mass function for the binomial distribution (4.6.12). Also, the second summation in (J) is simply the probability integral of the chi-square distribution for which tabulated values are available.

Figure 6.9 shows C_N vs. qt/mV_{tot} where $V_{\text{tot}} = NV$ for $n = 0.6$ and N varying from 1 to 1000. We note the wide degree of flexibility in matching an experimental $C_N(t) = F(t)$-curve with this model.

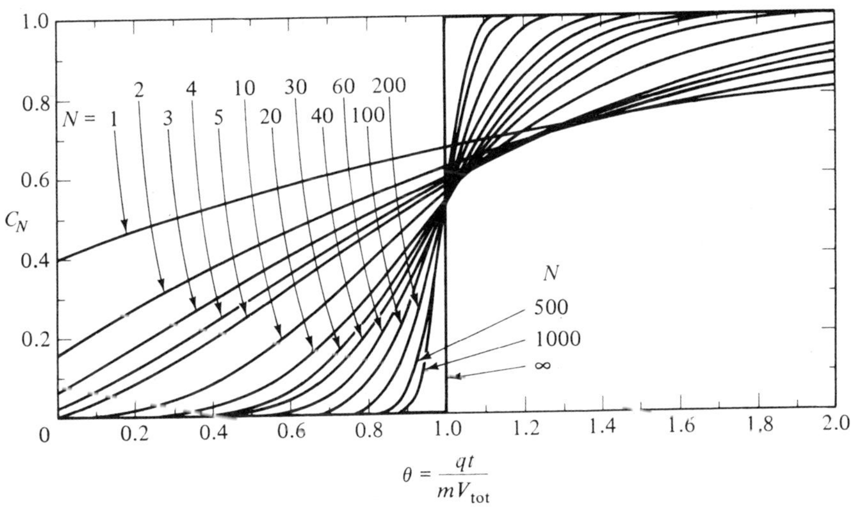

Figure 6.9. Variation of the outlet concentration C_N with dimensionless time for $n = 0.6$. [Reprinted from D. Chiang and A. Cholette, "Performance of Tanks in Series of Non-ideal Mixing," *Can. J. Chem. Eng. 48*, 286 (1970). Reprinted by permission of the copyright owner.]

6.3.3 Dispersion models

The second type of model used to represent the flow and mixing characteristics of complex chemical systems are dispersion models. In dispersion models, mixing processes are represented as diffusional in nature. As in the case of combined models, there exists an enormous literature over the past 15 years dealing with dispersion models.

Dispersion models have been used mainly to describe flow in empty tubes and packed beds, systems that are closer to plug flow than ideally mixed flow. These systems are characterized by turbulent flow and mixing both transverse to and in the direction of flow. The dispersion models take the form of a parabolic partial differential equation of the diffusion type. The most general form of a dispersion model is the equation of conservation of mass of species i (assume constant density):

$$\frac{\partial c_i}{\partial t} + \mathbf{u}\cdot\nabla c_i = \nabla\cdot(\mathbf{D}\nabla c_i) + R_i + S_i \tag{6.3.7}$$

where, if the system is in turbulent flow, each of the terms in (6.3.7) is time-averaged and $\mathbf{D}$ is the eddy diffusivity tensor, R_i the rate of generation of i by chemical reaction, S_i the rate of input of i from sources and $\mathbf{u}$ is a velocity vector.

Two special forms of (6.3.7) are often used for flow in a cylindrical tube. Both assume flow only in the x-direction; no radial variation of u, i.e., plug flow; and a constant value of $\mathbf{D}$. The first allows for both radial and axial diffusion, the special form of (6.3.7) being

$$\frac{\partial c_i}{\partial t} + u\frac{\partial c_i}{\partial x} = D_R\frac{1}{r}\frac{\partial}{\partial r}\left(r\frac{\partial c_i}{\partial r}\right) + D_L\frac{\partial^2 c_i}{\partial x^2} + R_i + S_i \tag{6.3.8}$$

This form is called the *dispersed plug flow model*. The second form neglects any radial concentration variations and is described by

$$\frac{\partial c_i}{\partial t} + u\frac{\partial c_i}{\partial x} = D_L\frac{\partial^2 c_i}{\partial x^2} + R_i + S_i \tag{6.3.9}$$

This form is called the *axial-dispersed plug flow model*.

Let us now consider some common situations that the axial-dispersed plug flow model has been used to describe.

This model is described by (for a single tracer in an inert fluid)

$$\frac{\partial c(x, t)}{\partial t} + u\frac{\partial c(x, t)}{\partial x} = D_L\frac{\partial^2 c(x, t)}{\partial x^2} + R + S \tag{6.3.10}$$

Let us henceforth neglect chemical reaction R. The boundary conditions of (6.3.10) depend on the particular system geometry and tracer injection. If the tracer is injected at $x = x_0$, then

$$S = Q'(t)\,\delta(x - x_0) \tag{6.3.11}$$

where $Q'(t)$ is the rate of tracer injection in units of concentration per time. If we let c_0 denote a reference concentration and l a reference length, we can place (6.3.10) in dimensionless form by defining

$$\theta = \frac{ut}{l} \qquad \text{Pe} = \frac{ul}{D_L} \quad \text{(Peclet number)}$$

$$z = \frac{x}{l} \qquad C = \frac{c}{c_0}$$

Thus (6.3.10) becomes

$$\frac{\partial C(z, \theta)}{\partial \theta} + \frac{\partial C(z, \theta)}{\partial z} = \frac{1}{\text{Pe}}\frac{\partial^2 C(z, \theta)}{\partial z^2} + Q'(\theta)\,\delta(z - z_0) \tag{6.3.12}$$

Table 6.1

Experimental Schemes for Axial-Dispersed Plug Flow Model*

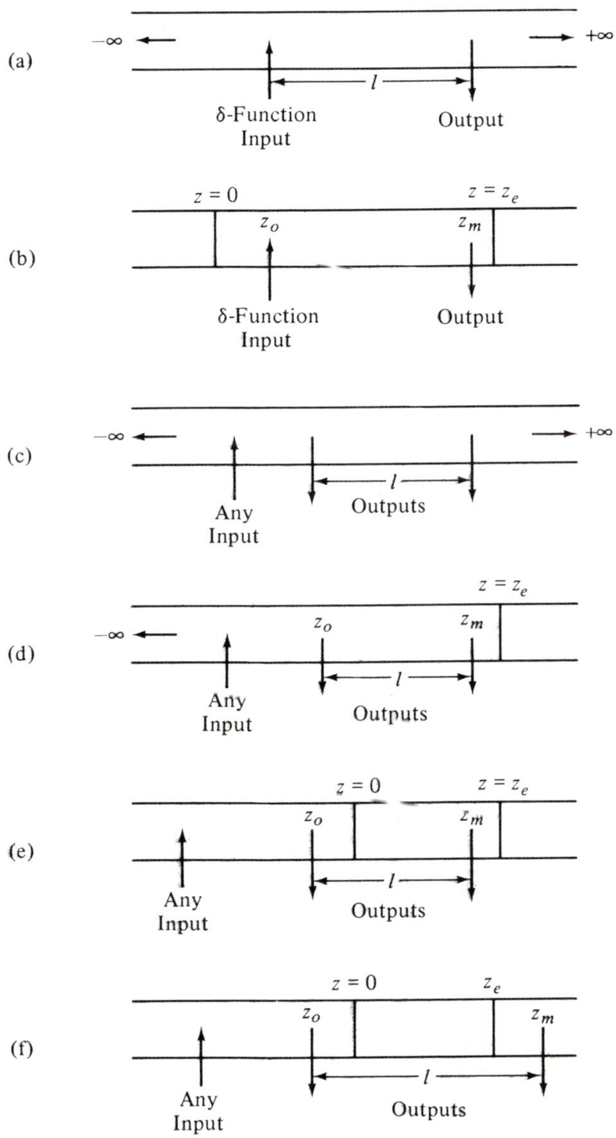

*Reprinted from O. Levenspiel and K. Bischoff, "Patterns of Flow in Chemical Process Vessels," *Advances in Chemical Engineering, Vol. 4,* John Wiley and Sons, Inc., New York, 1963. Reprinted by permission of the copyright owner.

Table 6.1, adapted from Levenspiel and Bischoff [9], shows the experimental schemes used in relation to the axial-dispersed plug flow model. Let us consider in each case a pulse injection, for which $Q'(\theta) = \delta(\theta)$. Scheme (a) is referred to as the doubly infinite reactor. The boundary conditions for this case are

$$C(z, 0) = 0 \tag{6.3.13}$$

$$C(+\infty, \theta) = \text{finite} \tag{6.3.14}$$

$$C(-\infty, \theta) = \text{finite} \tag{6.3.15}$$

The initial condition (6.3.13) states that just prior to tracer injection the concentration of tracer in the system is zero. Boundary conditions (6.3.14) and (6.3.15) are needed because only a finite amount of tracer is injected. The general solution of (6.3.12)–(6.3.15) is

$$C(z, \theta) = \frac{1}{2}\left(\frac{\text{Pe}}{\pi\theta}\right)^{1/2} \exp\left[\frac{-\text{Pe}(z-\theta)^2}{4\theta}\right] \tag{6.3.16}$$

and the concentration at the observation point, $z = 1$, is

$$C(1, \theta) = \frac{1}{2}\left(\frac{\text{Pe}}{\pi\theta}\right)^{1/2} \exp\left[\frac{-\text{Pe}(1-\theta)^2}{4\theta}\right] \tag{6.3.17}$$

Scheme (b) is a reactor of finite length. To develop the boundary conditions, let us assume that the sections $z \leq 0$ and $z \geq z_e$ are also packed. If we let C_a and C_b denote the concentrations in the sections $z \leq 0$ and $z \geq z_e$, the model for three sections is

$$\frac{\partial C_a}{\partial \theta} + \frac{\partial C_a}{\partial z} = \frac{1}{\text{Pe}_a}\frac{\partial^2 C_a}{\partial z^2}, \qquad z \leq 0 \tag{6.3.18}$$

$$\frac{\partial C}{\partial \theta} + \frac{\partial C}{\partial z} = \frac{1}{\text{Pe}}\frac{\partial^2 C}{\partial z^2} + \delta(z - z_0)\delta(\theta), \qquad 0 \leq z \leq z_e \tag{6.3.19}$$

$$\frac{\partial C_b}{\partial \theta} + \frac{\partial C_b}{\partial z} = \frac{1}{\text{Pe}_b}\frac{\partial^2 C_b}{\partial z^2} \qquad z \geq z_e \tag{6.3.20}$$

with boundary conditions

$$C_a(z, 0) = C(z, 0) = C_b(z, 0) = 0 \tag{6.3.21}$$

$$C_a(-\infty, \theta) = \text{finite} \tag{6.3.22}$$

$$C_a(0^-, \theta) = C_a(0^+, \theta) \tag{6.3.23}$$

$$C_a(0^-, \theta) - \frac{1}{\text{Pe}_a}\frac{\partial C_a}{\partial z}\bigg|_{z=0^-} = C(0^+, \theta) - \frac{1}{\text{Pe}}\frac{\partial C}{\partial z}\bigg|_{z=0^+} \tag{6.3.24}$$

$$C(z_e^-, \theta) - \frac{1}{\text{Pe}}\frac{\partial C}{\partial z}\bigg|_{z=z_e^-} = C_b(z_e^+, \theta) - \frac{1}{\text{Pe}_b}\frac{\partial C_b}{\partial z}\bigg|_{z=z_e^+} \tag{6.3.25}$$

$$C(z_e^-, \theta) = C_b(z_e^+, \theta) \tag{6.3.26}$$

$$C_b(+\infty, \theta) = \text{finite} \tag{6.3.27}$$

Conditions (6.3.23) and (6.3.26) insure continuity of concentration across the boundaries, while (6.3.24) and (6.3.25) state that the flux of tracer is the same on each side of the boundary.

Van der Laan [13] has obtained the transfer function of (6.3.18)–(6.3.27), that is, $\bar{C}(s, z_m)$, $z_0 < z_m \leq z_e$. The transform is

$$\bar{C}(s, z_m) = \frac{e^{[(1/2)-q]\mathrm{Pe}}}{2q} \frac{\{(q+q_a)(q+q_b)+(q-q_a)(q-q_b)e^{-2q\,\mathrm{Pe}\,(z_e-1)} + (q-q_a)(q+q_b)e^{-2q\,\mathrm{Pe}\,z_0} + (q+q_a)(q-q_b)e^{-2q\,\mathrm{Pe}(z_e-z_0)}\}}{(q+q_a)(q+q_b)-(q-q_a)(q-q_b)e^{-2q\,\mathrm{Pe}\,z_e}} \tag{6.3.28}$$

where the variable s is included in q as

$$q = \sqrt{\frac{1}{4} + \frac{s}{\mathrm{Pe}}} \tag{6.3.29}$$

Clearly, there is no chance of inverting (6.3.28) for the general case.

The experimental attainment of a perfect pulse of tracer is quite difficult. The necessity of a pulse input is circumvented if the concentration is measured at two points downstream of the injection point. Schemes (c)–(f) show the various alternatives in this case. We are then interested in the transfer function $\bar{C}(s, z_m)/\bar{C}(s, z_0)$. The general transfer function for cases (c)–(f) is presented by Levenspiel and Bischoff [9].

6.4 Moments of Deterministic Residence Time Distribution Models

We have now presented a number of deterministic models for flow systems. The purpose of these formulations is to provide a number of possible ways to represent the flow and mixing characteristics of real processes.† The question now arises, What properties of the residence time distribution functions from the model systems are useful to know in matching experimental curves?

†The residence time distribution analysis we have been presenting refers to the so-called *macromixing* characteristics of a system. Although the major features of nonideal flow patterns are reflected in the macromixing behavior of the system, so-called *micromixing* effects can have importance. If chemical reactions other than first order are taking place in the system, the residence time distribution alone does not yield a unique value of conversion. The key point is that the conversion is governed by the earliness or lateness of mixing of different fluid elements. Experimental studies of micromixing require measurements inside the system and are therefore often difficult to obtain. The extremes of micromixing are completely segregated flow (each fluid element is isolated from the others in the system) and maximum mixedness (fluid elements mix as soon as possible). In completely segregated flow all molecules in any one fluid particle have the same age, whereas in maximum mixedness the distribution of ages in any fluid particle is the same as in the whole system. We will not discuss micromixing in this book, however we refer the interested reader to the paper by Kattan and Adler (*Chem. Eng. Sci.*, *27*, 1013, 1972) for a complete analysis of micromixing in chemical reactors.

Since for these deterministic models the normalized concentration response to a pulse input is identically the residence time density E, those characterizations of a probability distribution will be important. The moments of E will therefore provide an important means of distinguishing among models.

6.4.1 Combined models

As we recall, combined models are series and parallel arrangements of plug flow, ideally mixed flow, and complete bypass. The moments of interest are the mean (μ'_1) and the variance ($\mu_2 = \sigma^2$) given by

$$\mu'_1 = \int_0^\infty tE(t)\,dt, \qquad \sigma^2 = \int_0^\infty t^2E(t)\,dt - \mu_1'^2 \tag{6.4.1}$$

1. *Plug flow:*

$$\mu'_1 = \mu, \qquad \mu'_2 = \mu, \qquad \mu'_3 = \mu$$
$$\mu_r = 0, \qquad r = 0, 1, 2, \ldots$$

2. *Ideally mixed flow*. Since the residence time density function $E(t)$ for a single ideally mixed vessel obeys the exponential distribution (4.6.26) with $a = 1/\mu$, the noncentral moments of $E(t)$ are given by (4.6.30):

$$\mu'_r = r!\mu^r \tag{6.4.2}$$

Thus $\mu'_1 = \mu$, $\mu'_2 = 2\mu^2$, and $\mu'_3 = 6\mu^3$. The central moments are

$$\mu_1 = 0$$
$$\mu_2 = \sigma^2 = \mu^2$$
$$\mu_3 = 2\mu^3$$
$$\mu_4 = 3\mu^4$$

3. *N ideally mixed vessels in series.* We have seen that the residence time distribution function $E(\theta)$ for N ideally mixed vessels in series (6.3.6) obeys the gamma distribution (4.6.33). The noncentral moments of $E(\theta)$ are given by

$$\mu'_r = \int_0^\infty \theta^r E(\theta)\,d\theta \tag{6.4.3}$$

The easiest way to compute μ'_r is by (4.5.17). The result is

$$\mu'_r = \frac{(r+N-1)(r+N-2)\cdots(N)}{N^r} \tag{6.4.4}$$

where $\mu'_1 = 1$, since θ has been made dimensionless. The variance of $E(\theta)$ is

$$\sigma^2 = \mu'_2 - \mu_1'^2 = \frac{1}{N} \tag{6.4.5}$$

We can express the moments in terms of the mean residence time μ of a single tank. Then

$$\mu'_1 = N\mu \qquad \mu'_2 = N(N+1)\mu^2$$
$$\mu_1 = 0 \qquad \mu_2 = \sigma^2 = N\mu^2 \tag{6.4.6}$$

EXAMPLE 6.4. Consider the cascade of N ideally mixed vessels shown in Figure 6.10. The average residence time of each vessel is μ, and there is a transport lag of

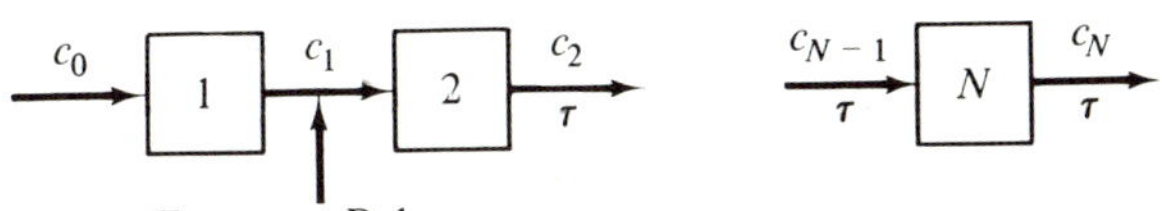

Figure 6.10. N ideally mixed vessels in series with a transport delay of time τ between each vessel.

time τ between each tank and after the last tank before measurement. Find the mean and variance of the residence time distribution function $E(t)$.

The concentration in tank n is governed by

$$\mu \frac{dc_n(t)}{dt} = c_{n-1}(t-\tau) - c_n(t)$$

$$c_n = 0, \qquad t \leq 0$$

Taking the L.T. and employing the shifting theorem (Section 3.1.4),

$$\frac{\bar{c}_n(s)}{\bar{c}_{n-1}(s)} = \frac{e^{-s\tau}}{1+\mu s}$$

For N vessels,

$$\bar{E}(s) = \frac{\bar{c}_N(s)}{\bar{c}_0(s)} = \left(\frac{e^{-s\tau}}{1+\mu s}\right)^N$$

To determine the moments we use (4.5.17). We note that

$$\frac{d\bar{E}(s)}{ds} = N\left(\frac{e^{-s\tau}}{1+\mu s}\right)^{N-1}\left[\frac{-\tau e^{-s\tau}}{1+\mu s} - \frac{\mu e^{-s\tau}}{(1+\mu s)^2}\right]$$

so that

$$\left.\frac{d\bar{E}(s)}{ds}\right|_{s=0} = -N(\tau+\mu)$$

Next,

$$\frac{d^2\bar{E}(s)}{ds^2} = N(N-1)\left(\frac{e^{-s\tau}}{1+\mu s}\right)^{N-2}\left[\frac{-\tau e^{-s\tau}}{1+\mu s} - \frac{\mu e^{-s\tau}}{(1+\mu s)^2}\right]^2$$
$$+ N\left(\frac{e^{-s\tau}}{1+\mu s}\right)^{N-1}\left[\frac{\tau[\tau(1+\mu s)+\mu]e^{-s\tau}}{(1+\mu s)^2} + \frac{\mu[\tau(1+\mu s)+2\mu]e^{-s\tau}}{(1+\mu s)^3}\right]$$

and

$$\left.\frac{d^2\bar{E}(s)}{ds^2}\right|_{s=0} = N^2(\tau+\mu)^2 + N\mu^2$$

Therefore

$$\mu_1' = N(\tau + \mu)$$
$$\mu_2 = \sigma^2 = N\mu^2$$

The transport delays only serve to shift the means by $N\tau$ but do not affect the variance, which is the same as for N vessels without any lags.

6.4.2 Dispersion models

We shall delay presentation of the moments of dispersion models until Section 7.2.

6.5 Stochastic Flow System Models

Deterministic flow system models account for the unpredictability of process flow systems through combined or dispersion models which contain parameters which can be chosen so that the model output matches the experimental output. Stochastic models attempt to account directly for the random nature of the system. In this section we shall consider the type of system which consists of a cascade of stages, such as a series of ideally mixed vessels.

Consider a series of N ideally mixed vessels. Instead of concerning ourselves with the concentration of tracer in each vessel as a function of time, let us instead concentrate on a single fluid element of tracer as it traverses the system. Physically, the fluid element enters the first stage at $t = 0$ and resides there for a random time. Then it moves to the next stage and stays for an additional random time, this process continuing until the particle leaves the Nth stage. Since the particle must at all times be in one of the stages (we assume that passage is instantaneous), the most logical description of the particle's traversal is as a discrete state, continuous time stochastic process. The *state* of the process is the vessel in which the particle resides at time t, whereas *time* runs continuously. Clearly, there are only N states, the N vessels, with the only possible transition being from state n to state $n + 1$.

This problem, the movement of a fluid particle through a sequence of ideally mixed vessels, is a Poisson process. Let $X(t)$ be the stage number that the fluid element is in at time t. Thus $X(t)$ can assume the values $1, 2, \ldots, N$ and is, in the terminology of Chapter 5, a discrete state continuous time Markov process. Then $V_n(t) = \text{Prob}\,\{X(t) = n\}$. For sufficiently small dt the probability that the element in vessel (state) n leaves vessel n is $\lambda_n\, dt + o(dt)$. The initial condition is $V_0 = 1$, since the fluid element begins in vessel 1.

Let q be the volumetric flow rate and V be the total volume of all the stages. The average residence time of the nth stage in general is $\bar{\mu}_n = V_n/q$, where V_n is the volume of the nth stage. Our objective is to compute the transfer function, the residence time distribution function, and the mean, variance, and skewness of the total system.

Let $V_n(t)$ be the probability that the fluid element is in state (stage) n at time t. The average frequency of passage through the nth stage is $\lambda_n = 1/\bar{\mu}_n$. Therefore the probability that the element is in state n at $t + dt$ is the sum of two probabilities: the probability that the element is in state $n - 1$ at t and transfers to state n during time dt, and the probability that the element is in state n at t and remains in state n during the time interval from t to $t + dt$:

$$V_n(t + dt) = (\lambda_{n-1}\, dt)V_{n-1}(t) + (1 - \lambda_n\, dt)V_n(t) \tag{6.5.1}$$

or

$$\begin{aligned} \dot{V}_1(t) &= -\lambda_1 V_1 \\ \dot{V}_n(t) &= -\lambda_n V_n + \lambda_{n-1} V_{n-1}, \qquad n = 2, 3, \ldots, N \end{aligned} \tag{6.5.2}$$

with initial conditions

$$V_n(0) = \begin{cases} 1, & n = 1 \\ 0, & n = 2, 3, \ldots, N \end{cases} \tag{6.5.3}$$

We can conveniently solve (6.5.2) and (6.5.3) using the L.T. Since

$$\frac{\bar{V}_n(s)}{\bar{V}_{n-1}(s)} = \frac{\lambda_{n-1}}{s + \lambda_n}, \qquad n = 2, 3, \ldots, N \tag{6.5.4}$$

we have that

$$\bar{V}_n(s) = \frac{\lambda_{n-1}\lambda_{n-2} \cdots \lambda_1}{(s + \lambda_n)(s + \lambda_{n-1}) \cdots (s + \lambda_1)}, \qquad n = 1, 2, \ldots, N \tag{6.5.5}$$

Inverting, using (3.1.42),

$$V_n(t) = \prod_{i=1}^{n-1} \lambda_i \left\{ \sum_{j=1}^{n} \frac{\exp(-\lambda_j t)}{\prod\limits_{\substack{k=1 \\ k \neq j}}^{n} (\lambda_k - \lambda_j)} \right\} \tag{6.5.6}$$

The probability that the particle exits from tank N between t and $t + dt$ is $V_N(t)\lambda_N\, dt$. Thus the residence time distribution function $E(t)$ is given by

$$\begin{aligned} E(t) &= V_N(t)\lambda_N \\ &= \prod_{i=1}^{N} \lambda_i \left\{ \sum_{j=1}^{N} \frac{\exp(-\lambda_j t)}{\prod\limits_{\substack{k=1 \\ k \neq j}}^{N} (\lambda_k - \lambda_j)} \right\} \end{aligned} \tag{6.5.7}$$

EXAMPLE 6.5. Fan et al. [4] have suggested that a convenient general model might be one consisting of a sequence of ideally mixed vessels the volumes of which obey either an arithmetic or a geometric progression. In this example we shall consider the case of an arithmetic progression; the geometric progression case is the subject of Problem 6.5.

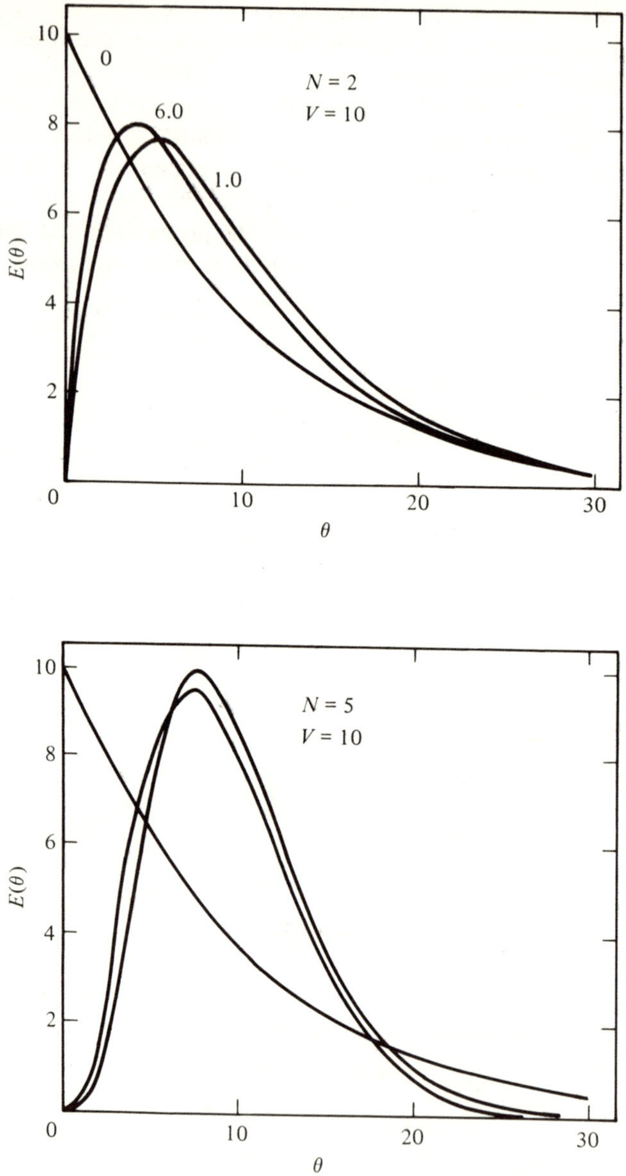

Figure 6.11. Residence time density $E(\theta)$ for a series of ideally mixed vessels with an arithmetic progression of volumes. [Reprinted from L. T. Fan, M. S. K. Chen, Y. K. Ahn, and C. Y. Wen, "Mixing Models with Varying Stage Size," *Can. J. Chem. Eng. 47*, 141 (1969). Reprinted by permission of the copyright owner.]

In an arithmetic progression of volumes

$$V_n = V_{n-1} + d \tag{A}$$

or, in terms of V_1,

$$V = NV_1 + \tfrac{1}{2}N(N-1)d \tag{B}$$

and

$$V_n = \frac{V}{N} - \frac{1}{2}(N+1)d + nd, \qquad n = 1, 2, \ldots, N \tag{C}$$

Substituting (C) into $\lambda_n = q/V_n$, and letting $\theta = t/\bar{\mu}(\bar{\mu} = V/q)$, $E(\theta)$ is

$$E(\theta) = \left\{\prod_{i=1}^{N}\left[\frac{1}{N} - \left(\frac{N+1}{2}\right)\left(\frac{d}{V}\right) + i\left(\frac{d}{V}\right)\right]\right\}^{-1}$$
$$\cdot \sum_{j=1}^{N} \frac{\exp\left[\dfrac{-\theta}{\{(1/N) - [(N+1)/2](d/V) + j(d/V)\}}\right]}{\prod\limits_{\substack{k=1\\k\neq j}}^{N}\left(\{(1/N) - [(N+1)/2](d/V) + k(d/V)\}^{-1} - \{(1/N) - [(N+1)/2](d/V) + j(d/V)\}^{-1}\right)}$$

$E(\theta)$ is shown in Figure 6.11 for $N = 2$ and 5, $V = 10$, and $d = 0$, 1.0, 6.0.

Let us now compute the mean, variance, and skewness of the residence time for the specific arithmetic progression of volumes. The transfer function for the system, the L.T. of $E(t)$, is

$$\bar{E}(s) = \lambda_N \bar{V}_N(s) = \prod_{i=1}^{N} \frac{\lambda_i}{(s + \lambda_i)} \tag{D}$$

For the arithmetic progression model, the dimensionless moments are

$$\mu'_1 = 1$$

$$\sigma^2 = \frac{1}{N} + \frac{N(N^2-1)}{12}\left(\frac{d}{V}\right)^2$$

$$\mu_3 = 2\sum_{i=1}^{N}\left[\frac{1}{N} - \left(\frac{N+1}{2}\right)\left(\frac{d}{V}\right) + i\left(\frac{d}{V}\right)\right]^3$$

6.6 Staged Models with Internal Backmixing

We shall now consider a class of residence time models which is in some ways intermediate between combined and dispersion models. This class of models is the series of ideally mixed vessels with backflow, or backmixing, between stages, as shown in Figure 6.12. This general model, consisting of a series of ideally mixed vessels with both forward and backward flow, has great flexibility. Not only can it be used to describe staged systems with inten-

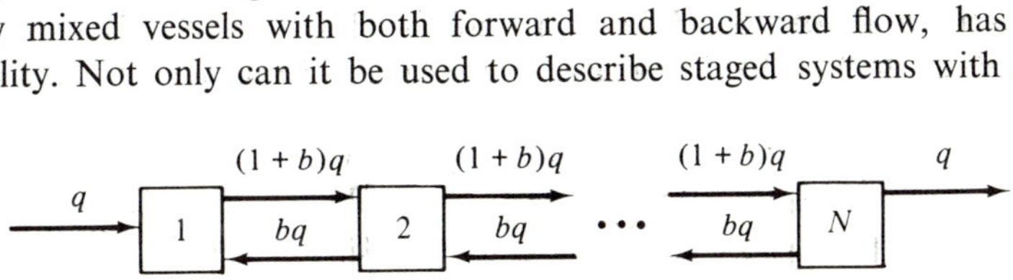

Figure 6.12. Series of ideally mixed vessels with backmixing.

tionally induced reflux, such as distillation, but it can also be used as a model for complex diffusional systems, such as packed beds.

The analysis of this class of models follows the three steps we have been using:

1. Derivation of the transient differential equations for the tracer concentrations in each stage (or, equivalently, for the discrete state, continuous time Markov probabilities).
2. Solution of the differential equations to determine the residence time distribution function $E(t)$.
3. Evaluation of the moments of $E(t)$ by direct computation or successive differentiation of the L.T. (the generating function) of $E(t)$.

Consider the case in which each of the N vessels in Figure 6.12 are of identical volume V. The net forward flow rate through the cascade is q. We define $\bar{\mu}$ as V/q, the volume divided by the net forward flow. Note that $\bar{\mu}$ is *not* the average residence time of a fluid element in each vessel, since the flow rate through each vessel is greater than q because of the backflow. At each stage except the first a constant fraction b of the net forward flow rate q is returned to the previous vessel.

The differential equations describing this system can be derived in two ways:

1. Writing a transient mass balance for tracer on each stage.
2. Considering the location at time t of a fluid element which entered the cascade at $t = 0$ to be a discrete state, continuous time Markov process.

Let us consider these two approaches.

1. *Transient mass balance.* A transient mass balance for tracer on each stage readily yields

$$\bar{\mu}\frac{dc_1}{dt} = c_0 - (1 + b)c_1 + bc_2 \tag{6.6.1}$$

$$\bar{\mu}\frac{dc_n}{dt} = (1 + b)c_{n-1} - (1 + 2b)c_n + bc_{n+1} \qquad N > n > 1 \tag{6.6.2}$$

$$\bar{\mu}\frac{dc_N}{dt} = (1 + b)c_{N-1} - (1 + b)c_N \tag{6.6.3}$$

where $c_n(t)$ is the concentration of tracer in stage n at time t. For a pulse injection of tracer in the inlet to stage 1 at time zero,

$$\begin{aligned} c_0(t) &= \delta(t) \\ c_n(0) &= 0, \qquad n = 1, 2, \ldots, N \end{aligned} \tag{6.6.4}$$

2. *Discrete state, continuous time Markov process.* The differential equations can also be derived from a probabilistic interpretation. In fact, the movement of a fluid particle through the cascade can be viewed as a birth and death process. A particle begins its residence time in the system when it enters the first vessel. It stays in the first vessel for a random time and then moves on to the second vessel. After staying in the second vessel for a random time, the fluid particle may either move forward to the third vessel or backward to the first vessel. The particle may, however, move only to neighboring states. If we envision the movement forward as a birth process and the movement backward as a death process, the entire process can be described in terms of the birth and death process.

The length of life T of a fluid particle in any ideally mixed stage is a random variable which obeys the exponential distribution:

$$f(t) = \begin{cases} 0, & t \leq 0 \\ \dfrac{1}{\xi} e^{-t/\xi}, & t > 0 \end{cases} \tag{6.6.5}$$

where ξ is the average residence time in each vessel. This distribution implies that for small Δt the probability that the particle will exit from a stage during the time interval t to $t + \Delta t$, if it has resided in the stage for t units of time, is $(1/\xi)\,\Delta t + o(\Delta t)$ and is independent of t. When the particle exits, the probability of moving forward and backward is

$$\text{prob. of moving forward} = \begin{cases} 1, & n = 1 \\ \dfrac{1+b}{1+2b}, & N > n > 1 \\ \dfrac{1}{1+b}, & n = N \end{cases}$$

$$\text{prob. of moving backward} = \begin{cases} 0, & n = 1 \\ \dfrac{b}{1+2b}, & N > n > 1 \\ \dfrac{b}{1+b}, & n = N \end{cases}$$

Let $X(t)$ be the state of the fluid element (the stage in which it resides) at time t, and let $c_j(t)$ be the probability that the particle will be in stage j at time t, $c_j(t) \doteq \text{Prob}\{X(t) = j\}$, $j = 1, 2, \ldots, N$. For stage 1, the probability that the fluid particle is in stage 1 at time $t + \Delta t$, if it was introduced in stage 1 at $t = 0$, is the sum of two probabilities: the probability that the particle was in stage 1 at t and stays for an additional Δt, and the probability that the particle was in stage 2 at t and flowed back to stage 1 during Δt. Since the total flow through vessel 1 is $(1 + b)q$, the average residence time in vessel is $V/[(1+b)q] = \bar{\mu}/(1+b)$, and so $\xi = \bar{\mu}/(1+b)$ for vessel 1. Thus

$$\begin{aligned} c_1(t+\Delta t) = c_1(t)&\left[1 - \frac{(1+b)}{\bar{\mu}}\Delta t + o(\Delta t)\right] \\ &+ bc_2(t)\left[\frac{1}{\bar{\mu}}\Delta t + o(\Delta t)\right] \end{aligned} \tag{6.6.6}$$

Rearranging,

$$\frac{c_1(t+\Delta t) - c_1(t)}{\Delta t} = -\frac{(1+b)}{\bar{\mu}}\left(c_1(t) + \frac{o(\Delta t)}{\Delta t}\right) + \frac{b}{\bar{\mu}}\left(c_2(t) + \frac{o(\Delta t)}{\Delta t}\right) \tag{6.6.7}$$

Taking the limit as $\Delta t \longrightarrow 0$,

$$\bar{\mu}\frac{dc_1}{dt} = -(1+b)c_1(t) + bc_2(t) \tag{6.6.8}$$

which agreees with (6.6.1), assuming that $c_0 = 0$ for all $t > 0$.

The probability that the fluid element is in stage n at time t is the sum of three probabilities:

1. The probability that the particle was in stage $n-1$ at t and moved forward to n during Δt:

$$c_{n-1}(t)\left[\frac{(1+b)/(1+2b)}{\bar{\mu}/(1+2b)}\Delta t + o(\Delta t)\right]$$

2. The probability that the particle was in stage n at t and remained in stage n during Δt:

$$c_n(t)\left[1 - \frac{1}{\bar{\mu}/(1+2b)}\Delta t + o(\Delta t)\right]$$

3. The probability that the particle was in stage $n+1$ at t and moved backward to stage n during Δt:

$$c_{n+1}(t)\left[\frac{b/(1+2b)}{\bar{\mu}/(1+2b)}\Delta t + o(\Delta t)\right]$$

where the average residence time ξ in an internal stage is $\bar{\mu}/(1+2b)$. Thus

$$\begin{aligned} c_n(t+\Delta t) = c_{n-1}(t)&\left[\frac{(1+b)/(1+2b)}{\bar{\mu}/(1+2b)}\Delta t + o(\Delta t)\right] \\ &+ c_n(t)\left[1 - \frac{1}{\bar{\mu}/(1+2b)}\Delta t + o(\Delta t)\right] \\ &+ c_{n+1}(t)\left[\frac{b/(1+2b)}{\bar{\mu}/(1+2b)}\Delta t + o(\Delta t)\right] \end{aligned} \tag{6.6.9}$$

Rearranging,

$$\begin{aligned} \frac{c_n(t+\Delta t) - c_n(t)}{\Delta t} = c_{n-1}(t)&\left[\frac{1+b}{\bar{\mu}} + \frac{o(\Delta t)}{\Delta t}\right] - c_n(t)\left[\frac{1+2b}{\bar{\mu}} + \frac{o(\Delta t)}{\Delta t}\right] \\ &+ c_{n+1}(t)\left[\frac{b}{\bar{\mu}} + \frac{o(\Delta t)}{\Delta t}\right] \end{aligned}$$

Taking the limit as $\Delta t \longrightarrow 0$,

$$\bar{\mu}\frac{dc_n}{dt} = (1 + b)c_{n-1}(t) - (1 + 2b)c_n(t) + bc_{n+1}(t) \tag{6.6.10}$$

which is identical to (6.6.2). A similar procedure for stage N yields (6.6.3) for $c_N(t)$:

$$\bar{\mu}\frac{dc_N}{dt} = (1 + b)c_{N-1}(t) - (1 + b)c_N(t) \tag{6.6.11}$$

This model has been used to represent the behavior of systems wherein mixing occurs as a result of turbulent eddying such as in a packed bed. As the actual mixing phenomena approach finely detailed dispersion, the diffusion equation becomes the theoretically desirable model. However, as we shall see in Section 6.7, as the number of stages N in the backflow model increases, this model converges to the diffusion model. To be able to compare the backflow cell model to the diffusion model, we shall introduce a slight modification into the scheme shown in Figure 6.12. At the two ends of the cascade we shall replace the inlet and outlet streams by the reflux-type arrangements shown in Figure 6.13. We shall do this to provide a means for movement of tracer upstream from stage 1 and back into stage N from the exit stream.

Figure 6.13. Series of ideally mixed vessels with backmixing with end reflux.

This modification changes the differential equations for $c_n(t)$ slightly. In particular, for a pulse injection of tracer, we have

$$\delta(t) = (1 + b)c_0 - bc_1 \tag{6.6.12}$$

$$\bar{\mu}\frac{dc_n}{dt} = (1 + b)c_{n-1} - (1 + 2b)c_n + bc_{n+1}, \qquad 1 \leq n \leq N \tag{6.6.13}$$

$$c_{N+1} - c_N = 0 \tag{6.6.14}$$

where c_0 is still the concentration entering stage 1, and c_{N+1} is the concentration finally leaving the cascade.

The diffusion model to which we shall compare the backflow cell model is the axial-dispersed plug flow model:

$$\frac{\partial C}{\partial \theta} + \frac{\partial C}{\partial z} = \frac{1}{\text{Pe}}\frac{\partial^2 C}{\partial z^2} \tag{6.6.15}$$

$$C_0(\theta) = C(0, \theta) - \frac{1}{\text{Pe}}\frac{\partial C}{\partial z}\bigg|_{z=0}, \qquad z = 0 \tag{6.6.16}$$

$$\frac{\partial C}{\partial z} = 0, \qquad z = 1 \qquad (6.6.17)$$

Boundary condition (6.6.16) allows for diffusion upstream of the inlet $z = 0$ and is the reason for the above modification of the flow at the ends of the backflow cell cascade. For a pulse injection $c_0(\theta) = \delta(\theta)$.

6.7 Moments of the Backflow Cell Model and Comparison to Continuous Models for Systems with Axial Mixing

In this chapter we have introduced the basic concepts of residence time distribution theory and illustrated these concepts with a variety of models for flow systems, which we divided into two categories, deterministic and stochastic models. We have computed the residence time distribution functions and the moments of most of these models. The reason for introducing all these models is, of course, to have a number of possible ways to describe flow systems. Combined models can be used for situations ranging from plug flow to ideally mixed flow, whereas dispersion models are normally used for situations close to plug flow where diffusion or a velocity profile causes longitudinal or axial mixing. Both combined models consisting of ideally mixed stages in series and dispersion models are used to represent packed bed operations. In a packed bed process where the flow is characterized by fluid jumping from void space to void space, a discrete cell model may actually be a more realistic description than a dispersion model. As the number of stages N increases, under certain conditions staged models converge to the dispersion models, a fact which we shall demonstrate in this section.

We shall consider, as an illustration of the correspondence of the staged and dispersion models as the number of stages get large, the relation of the backflow cell model to the axially dispersed plug flow model. This relationship will be developed by comparison of the moments of the residence time distribution functions of the two models.

Consider first the computation of the moments of the backflow cell model, for which there are several techniques available:

1. Solving for $c_N(t)$ and using $\int_0^\infty t^r c_N(t)\, dt$. This is the most tedious approach of all and is seldom employed.
2. Successive differentiation of the L.T., $\bar{c}_N(s)$. This is faster than approach 1 since $\bar{c}_N$ need not be inverted; nevertheless $\bar{c}_N(s)$ still must be computed. This approach has been used by Bell and Babb [1], Roemer and Durbin [11], and Shinnar and Noar [12].

3. Multiplying the original differential equations for the c_n by t^r and integrating from zero to ∞ to obtain partial difference equations with stage number and order of moment as independent variables. This can be the most expedient of the three approaches and was developed by Klinkenberg [5, 6].

We shall illustrate approaches 2 and 3.

Let us now develop the comparison between the backflow cell model and the axial-dispersed plug flow model for the doubly infinite systems shown in Table 6.1.

A mass balance on tracer in any stage n of the backflow cell model yields

$$V\frac{dc_n}{dt} = (1 + b)qc_{n-1} - (1 + 2b)qc_n + bqc_{n+1} \tag{6.7.1}$$

or, with $\bar{\mu} = V/q$,

$$\bar{\mu}\frac{dc_n}{dt} = (1 + b)c_{n-1} - (1 + 2b)c_n + bc_{n+1} \tag{6.7.2}$$

The initial conditions for (6.7.2) depend on where the tracer is introduced. We usually assume it is introduced into tank zero at $t = 0$. Since tank 0 has finite volume, it is most convenient to assume that tank 0 has unit concentration at $t = 0$, although it could be assumed that a delta function was introduced into the stream to cell 0 at $t = 0$. We shall not need to select one of these initial conditions since only the moments are desired.

The boundary conditions are

$$c_n(t) \longrightarrow 0, \qquad n \longrightarrow \pm\infty \tag{6.7.3}$$

We carry out the solution by means of L.T.s. The L.T. of (6.7.2) is

$$\bar{\mu}s\bar{c}_n = (1 + b)\bar{c}_{n-1} - (1 + 2b)\bar{c}_n + b\bar{c}_{n+1} \tag{6.7.4}$$

or, rearranging,

$$b\bar{c}_{n+1} - (1 + 2b + \bar{\mu}s)\bar{c}_n + (1 + b)\bar{c}_{n-1} = 0 \tag{6.7.5}$$

This is a linear, constant coefficient, second-order difference equation. It has a general solution of the form

$$\bar{c}_n = K_1\lambda_1^n + K_2\lambda_2^n \tag{6.7.6}$$

where K_1 and K_2 are constants determined by the boundary conditions and λ_1 and λ_2 are the roots of the characteristic equation

$$b\lambda^2 - (1 + 2b + \bar{\mu}s)\lambda + (1 + b) = 0 \tag{6.7.7}$$

Thus the solution $\bar{c}_n(s)$ is

$$\begin{aligned}\bar{c}_n(s) = K_1&\left\{\frac{1}{2b}[1 + 2b + \bar{\mu}s + \sqrt{1 + 2\bar{\mu}s(1 + 2b) + \bar{\mu}^2s^2}]\right\}^n \\ &+ K_2\left\{\frac{1}{2b}[1 + 2b + \bar{\mu}s - \sqrt{1 + 2\bar{\mu}s(1 + 2b) + \bar{\mu}^2s^2}]\right\}^n\end{aligned} \tag{6.7.8}$$

To satisfy (6.7.3) we see that the first term on the right-hand side of (6.7.8) applies for $n < 0$ and the second term for $n > 0$.

Consider now the axial-dispersed plug flow model (6.3.28):

$$\frac{\partial c}{\partial t} + u\frac{\partial c}{\partial x} = D_L \frac{\partial^2 c}{\partial x^2} \tag{6.7.9}$$

The L.T. of (6.7.9) for $c(x, 0) = 0$ is

$$s\bar{c}(x, s) + u\frac{d\bar{c}(x, s)}{dx} = D_L \frac{d^2\bar{c}(x, s)}{dx^2} \tag{6.7.10}$$

the general solution of which is

$$\begin{aligned}\bar{c}(x, s) = A_1 \exp\left[\frac{x}{2D_L}(u + \sqrt{u^2 + 4sD_L})\right] \\ + A_2 \exp\left[\frac{x}{2D_L}(u - \sqrt{u^2 + 4sD_L})\right]\end{aligned} \tag{6.7.11}$$

Again, we use only the first term on the right-hand side of (6.7.11) upstream of the source, $x < 0$, and the second term downstream of the source, $x > 0$.

We now have relations for the transformed concentration in stage n, given by (6.7.8), in the backflow cell model and for the transformed concentration at location x, given by (6.7.11), in the axially dispersed plug flow model. To compare (6.7.8) and (6.7.11) we let $x = nd$, where d is the length of the continuous model associated with one stage of the discrete model. Thus

$$d = \bar{\mu} u \tag{6.7.12}$$

Since (6.7.8) and (6.7.11) are L.T.s of residence time distribution functions, and thus generating functions, they can easily be characterized by their moments. Performing the differentiations with respect to s, evaluating at $s = 0$, and equating like noncentral moments of (6.7.8) and (6.7.11), we find that (moment order r)

$$r = 0 \qquad \frac{ud}{D_L} = \ln\frac{1 + b}{b} \tag{6.7.13}$$

$$r = 1 \qquad \bar{\mu} = \bar{\mu} \tag{6.7.14}$$

$$r = 2 \qquad \frac{ud}{D_L} = \frac{2}{1 + 2b} \tag{6.7.15}$$

The group ud/D_L is the Peclet number based on the mixing length d, which we can denote Pe_d to distinguish it from the prior used Peclet number Pe_L based on the reactor characteristic length L.

We can obtain (6.7.15) in a different way, which is somewhat faster than that of comparing moments. Let us relate the discrete and continuous models by

$$c_{n+1}(t) \longrightarrow c(x + d, t)$$
$$c_n(t) \longrightarrow c(x, t)$$
$$c_{n-1}(t) \longrightarrow c(x - d, t)$$

Expanding $c(x+d,t)$ and $c(x-d,t)$ in Taylor series,

$$c(x+d,t) = c(x,t) + d\frac{\partial c}{\partial x} + \frac{d^2}{2!}\frac{\partial^2 c}{\partial x^2} + \cdots \tag{6.7.16}$$

$$c(x-d,t) = c(x,t) - d\frac{\partial c}{\partial x} + \frac{d^2}{2!}\frac{\partial^2 c}{\partial x^2} + \cdots \tag{6.7.17}$$

Now let us substitute (6.7.16) and (6.7.17) into (6.7.9), and we obtain

$$\begin{aligned}\left(\frac{d}{u}\right)\frac{\partial c(x,t)}{\partial x} &= (1+b)\left[c(x,t) - d\frac{\partial c}{\partial x} + \frac{d^2}{2!}\frac{\partial^2 c}{\partial x^2}\right] - (1+2b)c(x,t)\\ &\quad + b\left[c(x,t) + d\frac{\partial c}{\partial x} + \frac{d^2}{2!}\frac{\partial^2 c}{\partial x^2}\right]\\ &= -d\frac{\partial c}{\partial x} + (1+2b)d^2\frac{\partial^2 c}{\partial x^2}\end{aligned} \tag{6.7.18}$$

or

$$\frac{\partial c}{\partial t} + u\frac{\partial c}{\partial x} = \frac{(1+2b)ud}{2}\left(\frac{\partial^2 c}{\partial x^2}\right) \tag{6.7.19}$$

Thus, for the two approaches to be compatible,

$$D_L = \frac{(1+2b)ud}{2} \tag{6.7.20}$$

or

$$\frac{ud}{D_L} = \frac{2}{1+2b} \tag{6.7.21}$$

We have obtained two relations between Pe_d and the backflow parameter b, (6.7.13) and (6.7.15). When b is large, these two relations are about the same. When $b \geq 1$, we have

$$\ln\frac{1+b}{b} = \frac{1}{b} - \frac{1}{2b^2} + \frac{1}{3b^3} - \cdots \tag{6.7.22}$$

while for $b > \frac{1}{2}$,

$$\frac{2}{1+2b} = \frac{1}{b} - \frac{1}{2b^2} + \frac{1}{4b^3} - \cdots \tag{6.7.23}$$

The discrepancy arising between (6.7.13) and (6.7.15) for small values of b ($b \ll 1$) can be explained on the basis of the origin of these two relations. The relation (6.7.13) arises from equating of the zeroth moments, that is, to have each model predict the correct total amount of material transported upstream against the flow. When the backmixing goes to zero, an infinite Peclet number is predicted by (6.7.13), i.e., a zero diffusivity, since no material is diffusing upstream. The relation (6.7.15) is based on equating of the second moments and describes the spreading of an impulse as it travels downstream. Thus (6.7.13) and (6.7.15) define Pe_d upstream and downstream of the source, respectively. As $b \rightarrow 0$, (6.7.15) predicts a Peclet number of 2. Thus the axial-dispersed plug flow model with $\text{Pe}_d = 2$ is equivalent to a model of ideally mixed vessels in series with no backflow.

We compare below the two expressions for the Peclet number as a function of b:

b	$\ln \frac{1+b}{b}$	$\frac{2}{1+2b}$
0	∞	2.00
0.1	2.40	1.67
0.2	1.79	1.43
0.3	1.47	1.25
0.4	1.25	1.11
0.5	1.10	1.00
1.0	0.693	0.67

In summary, when the backmixing is large, the two estimates for the Peclet number are equivalent. Thus, if backmixing is the predominant effect, we may use an effective diffusivity and the axial-dispersed plug flow model or the backflow cell model with very similar results. Probably the easier choice is the axial-dispersed plug flow model. However, when the amount of backflow is small and the predominant diffusive effect is by stagewise mixing ($\text{Pe}_d \cong 2$), such as is often the case in very short reactors, the proper model to use is the backflow cell model with a small value of b.

We can now proceed to the finite length systems in our comparison of the backflow cell model and the axial-dispersed plug flow model. A complete comparison of the two models has been presented by Roemer and Durbin [11]. Roemer and Durbin compared the transient responses of the backflow cell model and the axial-dispersed plug flow model when the Peclet number in the diffusion model, $\text{Pe}_L = uL/D_L$, based on the characteristic reactor length L, is equated to $2N/(1 + 2b)$, the backflow parameter in the cell model. Thus

$$\text{Pe}_L = \frac{uL}{D_L} = \frac{2N}{1 + 2b} \tag{6.7.24}$$

which is equivalent to (6.7.15) if $d = L/N$. Comparisons of the peaks of the impulse response curves indicated that the cell model responses converge to the continuous model responses on the order of N^{-2}. Also, the variances and central skew moments μ_3 of the outlet responses converge as N^{-2}. At interior points of the reactor the convergence of the moments goes as N^{-1}.

A discrete staged model to represent time-dependent packed bed behavior with both radial and axial gradients was formulated by Deans and Lapidus [3]. This model has been shown to be quite effective in simulating complex packed bed dynamics, in which both chemical reaction and heat transfer effects take place.

EXAMPLE 6.6. Determine the mean and variance of the finite backflow cell model described by (6.7.1)–(6.7.3) using the third technique mentioned in the beginning of this section. Let

$$\mu'_{r,k} = \int_0^\infty t^r c_k(t)dt \tag{A}$$

Carrying out the procedure described, we obtain from (6.7.1)–(6.7.3)

$$(1-b)\mu'_{r,1} - b\mu'_{r,2} = r\bar{\mu}\mu'_{r-1,1} + \delta(r) \tag{B}$$

$$[(1+b)\mu'_{r,k} - b\mu'_{r,k+1}] - [(1+b)\mu'_{r,k-1} - b\mu'_{r,k}] = r\bar{\mu}\mu'_{r-1,k} \tag{C}$$

$$(1+b)\mu'_{r,N} - (1+b)\mu'_{r,N-1} = r\bar{\mu}\mu'_{r-1,N} \tag{D}$$

For $r = 0$, the $\mu'_{r-1,k}$ are meaningless but have a zero coefficient. Summing these equations over the N stages, we note that

$$\mu'_{0,N} = 1 \tag{E}$$

$$\mu'_{r,N} = r\bar{\mu}\sum_{k=1}^{N} \mu_{r-1,k} \tag{F}$$

Also, for all k, all the zeroth moments are unity:

$$\mu'_{0,k} = 1 \tag{G}$$

For $r = 1$, the mean of the residence time distribution for the cascade is

$$\mu'_{1,N} = \bar{\mu}\sum_{k=1}^{N} \mu'_{0,k} = N\bar{\mu} \tag{H}$$

From (B)–(D) we obtain relations for the means of the other cells:

$$\mu'_{1,k} = \left[\frac{1-\beta^{N-k+1}}{1-\beta} + k - 1\right]\bar{\mu} \tag{I}$$

where $\beta = b/(1+b)$. The second noncentral moment for the cascade is

$$\begin{aligned}\mu'_{2,N} &= 2\bar{\mu}\sum_{k=1}^{N} \mu'_{1,k} \\ &= N^2\bar{\mu}^2 + [N(1-\beta^2) - 2\beta(1-\beta^N)]\frac{\bar{\mu}^2}{(1-\beta)^2}\end{aligned} \tag{J}$$

In terms of b, the variance of the overall residence time distribution is

$$\sigma^2 = (1+2b)N\bar{\mu}^2 - 2b(1+b)\left[1 - \left(\frac{b}{1+b}\right)^N\right]\bar{\mu}^2 \tag{K}$$

For $N \gg 1$, σ^2 approaches

$$\sigma^2 \simeq (1+2b)N\bar{\mu}^2 \tag{L}$$

Often the procedure in matching an experimental curve is to fix N and adjust b so that the variances coincide as closely as possible. For reasonably large values of N, (L) can be used in place of (K), with the calculated value of b too low, but only by a fraction $(1+b)/N$ of its exact value from (K) [5, 6].

6.8 Stochastic Mixing Models for Chemical Reactors

Up to now we have assumed that the residence time distribution of a fluid particle is independent of the time the particle entered the system. In such a case, combined and dispersion models are very successful in representing the behavior of tracer particles injected into the system and thus of the flow patterns in the system itself. However, many systems of importance exhibit pronounced nonstationarity, in that the flow behavior is unsteady enough to cause successive tracer experiments to yield significantly different results. It is of interest to try to represent the unsteady nature of the flow for two reasons. First, even in systems in which the fluctuations average out sufficiently to produce stationarity of tracer results, the unsteady nature of the flow might have strong effects on nonlinear reaction systems. Second, we must be able to describe systems in which the turbulent fluctuations are strong enough to cause different tracer experiments to yield different results. An important step toward the modeling of turbulent flow reactor systems has been taken by Krambeck et al. [7, 8] in introducing the concept of stochastic mixing models for chemical reactors. Their work will be the subject of this section.

The basic model suggested is that of a network of ideally mixed vessels with interconnecting flows that are random functions of time. It is hoped with a large number of tanks and interconnections to describe both the steady and unsteady behavior of a turbulent flow system to any desired degree of approximation. The interconnecting flows are assumed to be stationary random processes; in particular each interconnecting flow is taken to be a discrete state, continuous time Markov process. Thus the intercell flows can assume only a finite number of values. Although certainly an idealization, this assumption simplifies the mathematics involved. The cellular concentrations, on the other hand, are not restricted to a finite number of values but vary continuously over the range of physically possible values.

Consider an array of N ideally mixed vessels connected arbitrarily by interstage flows. The volumes of the vessels are constant and denoted by $v_1, v_2, \ldots, v_N$, whereas the volumetric flow rate from tank i to tank j is denoted by ω_{ij}, $i = 0, 1, \ldots, N$, $j = 1, 2, \ldots, N + 1$. The indices $i = 0$ and $j = N + 1$ refer to inlet and outlet streams, respectively. We direct the reader to Figure 6.14 for a typical flow network involving three tanks. The flow $\omega_{0,N+1}$ bypasses the array of tanks completely.

Even though the flows ω_{ij} will be permitted to vary with time, at any instant the total volumetric flow into any tank must equal the total volumetric flow out:

$$\sum_{\substack{i=0\\ i\neq j}}^{N} \omega_{ij} = \sum_{\substack{k=1\\ k\neq j}}^{N+1} \omega_{jk}, \qquad j = 1, 2, \ldots, N \tag{6.8.1}$$

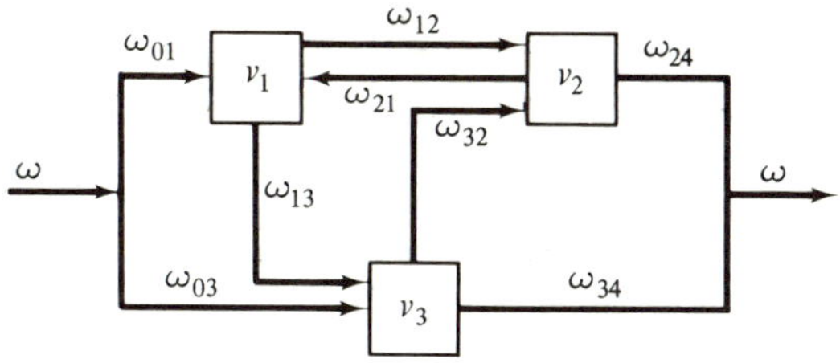

Figure 6.14. Typical flow network of ideally mixed vessels. [Reprinted from F. J. Krambeck, R. Shinnar, and S. Katz, "Stochastic Mixing Models for Chemical Reactors," *Ind. Eng. Chem. Fund.* 6, 276 (1967). Reprinted by permission of the copyright owner.]

To simplify our notation we shall let the common value of the two sides of (6.8.1) be $-\omega_{jj}$, so that at any instant we have an $(N+1) \times (N+1)$ matrix of intercell flows for which

$$\sum_{i=0}^{N} \omega_{ij} = \sum_{k=1}^{N+1} \omega_{jk} = 0, \qquad j = 1, 2, \ldots, N \tag{6.8.2}$$

The total flow through the network, including bypass, will be denoted by ω.

We want to allow the flows to have a random character. We therefore assume that the ω_{ij} are each a discrete state, continuous time homogeneous Markov process. We let the transition probability function

$$p_{\alpha\beta}(\tau) = \text{Prob}\{\omega_{ij}(t+\tau) = \beta \,|\, \omega_{ij}(t) = \alpha\}$$

We assume that in a small time τ the probability of a transition of an intercell flow from state α to state β is given by

$$p_{\alpha\beta}(\tau) = \lambda_{\alpha\beta}\tau + o(\tau), \qquad \alpha \neq \beta \tag{6.8.3}$$

where $\lambda_{\alpha\beta}$ are the elements of a given matrix of switching rates. The probability that the state of a flow remains unchanged in time τ is clearly

$$p_{\alpha\alpha}(\tau) = 1 - (\sum_{\beta \neq \alpha} \lambda_{\alpha\beta})\tau + o(\tau) \tag{6.8.4}$$

Thus we have

$$p_{\alpha\beta}(\tau) = \delta_{\alpha\beta} + \lambda_{\alpha\beta}\tau + o(\tau) \tag{6.8.5}$$

as long as $\lambda_{\alpha\alpha}$ is defined such that $\sum_{\beta} \lambda_{\alpha\beta} = 0$.

We now define

$$p_{\alpha}(t) = \text{Prob}\{\omega_{ij}(t) = \alpha\}$$

Thus we can write

$$p_{\beta}(t+\tau) = \sum_{\alpha} p_{\alpha}(t) p_{\alpha\beta}(\tau) \tag{6.8.6}$$

in which the probability of a flow rate being in state β at time $t+\tau$ is the product of the probability that the flow rate was in state α at time t and the probability that in time τ a transition from state α to β occurred, summed over all possible states α. Substituting (6.8.5) into (6.8.6), dividing by τ, and taking the limit as $\tau \longrightarrow 0$, we obtain

$$\frac{dp_\beta(t)}{dt} = \sum_\alpha \lambda_{\alpha\beta} p_\alpha(t) \tag{6.8.7}$$

We shall introduce one final simplification. We shall assume that the flows have attained a stationary condition in which the p_β no longer vary with time. Thus the p_β become the ergodic probabilities $\bar{p}_\beta$ and are found from

$$\sum_\alpha \lambda_{\alpha\beta} \bar{p}_\alpha = 0 \tag{6.8.8}$$

It is important to note that even though the *probabilities* of the flow rates occupying various states do not change with time, the *flows themselves* continue to exhibit random fluctuations. This is, of course, the concept of ergodic probabilities detailed in Chapter 5. We assume that the $\bar{p}_\alpha$ are unique.

As has been our custom, let us describe the passage of a fluid particle through the system. The location $X(t)$ of the fluid particle as a function of time is another discrete state, continuous time stochastic process. However, the location of the fluid particle as a function of time is not, by itself, a Markov process, since the transition probabilities from one tank to another depend on the state of the flow. Since the tanks are ideally mixed, the residence time of a particle in any tank is exponentially distributed; that is, the probability that the particle will exit to tank j in a short time τ is $(\omega_{ij}/v_i)\tau$.

Now a key assumption is made. Even though the particle location cannot be a Markov process, the two random variables, the state of the intercell flows and the particle's location, together form a *joint* Markov process. We shall define the joint transition probability by

$$\begin{aligned} p[(\alpha, i) \longrightarrow (\beta, j); \tau] = \text{Prob}\{&\omega(t+\tau) = \beta, \\ &X(t+\tau) = j \mid \omega(t) = \alpha, X(t) = i\} \end{aligned}$$

where we employ the argument notation shown with p to avoid the clumsiness of four subscripts. Analogous to (6.8.3) we let

$$p[(\alpha, i) \longrightarrow (\beta, j); \tau] = [\delta_{\alpha\beta} + \lambda_{\alpha\beta}\tau + o(\tau)]\left[\delta_{ij} + \frac{\omega_{ij\alpha}}{v_i}\tau + o(\tau)\right] \tag{6.8.9}$$

and

$$p[(\alpha, N+1) \longrightarrow (\beta, j); \tau] = [\delta_{\alpha\beta} + \lambda_{\alpha\beta}\tau + o(\tau)]\delta_{N+1,j} \tag{6.8.10}$$

which simply implies that once the particle reaches state $N + 1$, it cannot leave that state.

We now define

$$p_{\alpha i}(t) = \text{Prob}\{\omega(t) = \alpha, X(t) = i\}$$

Thus we can write

$$p_{\beta j}(t+\tau) = \sum_\alpha \sum_{i=0}^{N+1} p_{\alpha i}(t) p[(\alpha, i) \longrightarrow (\beta, j); \tau] \tag{6.8.11}$$

which states that the probability that the flow rate is in state β and the particle is in tank j at time $t + \tau$ is the product of the probability that the flow rate and particle location were in states α and i, respectively, at time t and the probability of transition from states (α, i) to (β, j) in time τ, summed over all α and i. Substituting (6.8.9) and (6.8.10) into (6.8.11), dividing by τ, and taking the limit as $\tau \longrightarrow 0$ yields

$$\frac{dp_{\beta j}}{dt} = \sum_{i=1}^{N} \frac{\omega_{ij\beta}}{v_i} p_{\beta i}(t) + \sum_{\alpha} \lambda_{\alpha\beta} p_{\beta j}(t), \qquad j = 1, 2, \ldots, N+1 \tag{6.8.12}$$

At $t = 0$ the particle is in the input line, and so its initial location probabilities are set in proportion to the feed rates to each of the N tanks:

$$p_{\beta j}(0) = \bar{p}_\beta \frac{\omega_{0j\beta}}{\omega_\beta}, \qquad j = 1, 2, \ldots, N+1 \tag{6.8.13}$$

This means that we do not know in which tank the particle will begin its journey, but only the probabilities of its being in each tank. Note that saying that the particle is in the input line means that at $t = 0$ it is in one of the tanks fed by the input line, since feed to the tanks and intercell transitions are assumed to be instantaneous.

The differential equation (6.8.12) subject to initial condition (6.8.13) describes the whole development in time of the joint probability history of flow state and particle location.

We note that if we sum $p_{\beta j}$ over all possible locations j, we obtain the probability distribution of flow states $p_\beta(t)$, which is now the marginal probability for the flow states. Similarly, summing (6.8.12) and (6.8.13) over j, we recover (6.8.7). Now, $\sum_\beta p_{\beta j}(t)$ is just the probability that a particle, initially in the inlet, is at location j at time t. Then the normalized age distribution

$$g_i(t) = \frac{\sum_\beta p_{\beta j}(t)}{\int_0^\infty \sum_\beta p_{\beta j}(t)\, dt} \tag{6.8.14}$$

can be defined. Also, $\sum_\beta p_{\beta, N+1}(t)$ is the probability that the particle has exited by time t, the cumulative residence time distribution.

6.8.1 The equations for the tracer concentrations

Now let us consider the injection of tracer into the input lines of the system at a rate $\phi(t)$ moles per unit time. If the concentration of tracer in tank j is $c_j(t)$, the mass balance on tank j is

$$v_j \frac{dc_j}{dt} = \frac{\omega_{0j\alpha}}{\omega_\alpha} \phi(t) + \sum_{\substack{i=1 \\ i \neq j}}^{N} \omega_{ij\alpha} c_i - c_j \sum_{\substack{k=1 \\ k \neq j}}^{N+1} \omega_{jk\alpha} \tag{6.8.15}$$

The terms on the right-hand side account, respectively, for the input of tracer directly from the feed lines, the input of tracer from all other tanks, and the outflow of tracer from tank j to all other tanks. A more compact notation is possible if we use the pseudoflows $\omega_{jj\alpha}$:

$$v_j \frac{dc_j}{dt} = \frac{\omega_{0j\alpha}}{\omega_\alpha}\phi(t) + \sum_{i=1}^{N} \omega_{ij\alpha} c_i \tag{6.8.16}$$

where (6.8.2) has been used for each flow state α. If there is no tracer present in any of the tanks at $t = 0$, the initial condition for (6.8.16) is

$$c_j(0) = 0, \qquad j = 1, 2, \ldots, N \tag{6.8.17}$$

Just as the tracer particle location $X(t)$ was not itself a Markov process, the tank concentrations $c_j(t)$ alone are not Markov processes. However, the joint process of flow states and tank concentrations is Markovian. The concentrations $c_j(t)$ form continuous state, continuous time stochastic processes. We define the transition probability density as

$$\begin{aligned} p[(\alpha, \mathbf{x}) \longrightarrow (\beta, \mathbf{y}); t, \tau]\, dy_1\, dy_2 \cdots dy_N &= \text{Prob}\{\omega(t+\tau) = \beta, \\ &\quad \mathbf{y} < \mathbf{c}(t+\tau) \le \mathbf{y} + d\mathbf{y} \,|\, \omega(t) = \alpha, \mathbf{x} < \mathbf{c}(t) \le \mathbf{x} + d\mathbf{x}\} \end{aligned}$$

We need the explicit dependence on t because of the time dependence of the tracer feed $\phi(t)$. Analogous to (6.8.9) the relation for the transition probability density is

$$\begin{aligned} p[(\alpha, \mathbf{x}) &\longrightarrow (\beta, \mathbf{y}); t, \tau] \\ &= [\delta_{\alpha\beta} + \lambda_{\alpha\beta}\tau + o(\tau)] \\ &\quad \cdot \delta\left[y_1 - x_1 - \left(\frac{\omega_{01\alpha}}{\omega_\alpha}\frac{\phi(t)}{v_1} + \sum_{i=1}^{N}\frac{\omega_{i1\alpha}}{v_i}x_i\right)\tau + o(\tau)\right] \\ &\quad \cdots \delta\left[y_N - x_N - \left(\frac{\omega_{0N\alpha}}{\omega_\alpha}\frac{\phi(t)}{v_N} + \sum_{i=1}^{N}\frac{\omega_{iN\alpha}}{v_N}x_i\right)\tau + o(\tau)\right] \end{aligned} \tag{6.8.18}$$

Finally, we define the probability density,

$$\begin{aligned} p_\alpha(\mathbf{x}, t)\, dx_1 \cdots dx_N \\ = \text{Prob}\{\omega(t) = \alpha, x_1 < c_1 \le x_1 + dx_1, \cdots, x_N < c_N \le x_N + dx_N\} \end{aligned}$$

for which the following relation holds:

$$p_\beta(y, t+\tau) = \sum_\alpha \int \cdots \int p_\alpha(\mathbf{x}, t) p[(\alpha, \mathbf{x}) \longrightarrow (\beta, \mathbf{y}); t, \tau]\, dx_1 \cdots dx_N \tag{6.8.19}$$

where the integrals are overall values of the x_j. Combining (6.8.18) and (6.8.19), dividing by τ, and taking the limit as $\tau \longrightarrow 0$ yields

$$\frac{\partial p_\beta(\mathbf{y}, t)}{\partial t} + \sum_{j=1}^{N} \frac{\partial}{\partial y_j}\left[\left\{\frac{\omega_{0j\beta}}{\omega_\beta}\frac{\phi(t)}{v_j} + \sum_{i=1}^{N}\frac{\omega_{ij\beta}}{v_j}y_i\right\} p_\beta(\mathbf{y}, t)\right] = \sum_\alpha \lambda_{\alpha\beta} p_\beta(\mathbf{y}, t) \tag{6.8.20}$$

The initial condition for (6.8.20) is obtained from (6.8.17) as

$$p_\beta(\mathbf{y}, 0) = \bar{p}_\beta \delta(y_1) \cdots \delta(y_N) \tag{6.8.21}$$

The solution of (6.8.20), subject to (6.8.21), yields the full probability information on the tank tracer concentrations and flow states. However, for any reasonable number of flow states and tanks, solution of (6.8.20) will be nearly an impossible task. Thus the authors consider only the first two moments of the concentrations.

6.8.2 Moments of the tracer concentrations

The mean concentration in tank j for flow state p is

$$\mu_{\beta j}(t) = \int \cdots \int y_j p_\beta(\mathbf{y}, t)\, dy_1 \cdots dy_N, \qquad j = 1, 2, \ldots, N \tag{6.8.22}$$

Noting that

$$\int \cdots \int p_\beta(\mathbf{y}, t)\, dy_1 \cdots dy_N = p_\beta(t) = \bar{p}_\beta \tag{6.8.23}$$

we integrate (6.8.20) and (6.8.21) over the entire space of $\mathbf{y}$-values to obtain the differential equations governing the $\mu_{\beta j}(t)$:

$$\frac{d\mu_{\beta j}(t)}{dt} = \frac{\bar{p}_\beta}{v_j} \frac{\omega_{0j\beta}}{\omega_\beta} \phi(t) + \sum_{i=1}^{N} \frac{\omega_{ij\beta}}{v_j} \mu_{\beta i}(t) + \sum_{\alpha} \lambda_{\alpha\beta} \mu_{\alpha j}(t),$$
$$\mu_{\beta j}(0) = 0, \qquad j = 1, 2, \ldots, N \tag{6.8.24}$$

The mean rate at which tracer leaves the system, including bypass, in moles per unit time is

$$\mu(t) = \sum_\beta \int \cdots \int \left[\frac{\omega_{0,N+1,\beta}}{\omega_\beta} \phi(t) + \sum_{j=1}^{N} \omega_{j,N+1,\beta} y_j\right] p_\beta(\mathbf{y}, t)\, dy_1 \cdots dy_N$$
$$= \sum_\beta \left[\bar{p}_\beta \frac{\omega_{0,N+1,\beta}}{\omega_\beta} \phi(t) + \sum_{j=1}^{N} \omega_{j,N+1,\beta} \mu_{\beta j}(t)\right] \tag{6.8.25}$$

where the solution of (6.8.24) is required.

We notice a very interesting facet of (6.8.25), namely that both $\mu_{\beta j}(t)$ and $\mu(t)$ are related linearly and homogeneously to the tracer feed rate $\phi(t)$. We can thus write

$$\mu(t) = \int_0^t f(t - \tau) \phi(\tau)\, d\tau \tag{6.8.26}$$

Thus the mean tracer outlet response is simply a convolution of $\phi(t)$ with the impulse response $f(t)$. If we let $\phi(t) = \delta(t)$, a unit impulse tracer, in (6.8.24), we find that the average molar amounts of tracer $v_j \mu_{\beta j}(t)$ are identically the $p_{\beta j}(t)$, the probabilities that the single tracer particle is in tank j at time t. Also, when $\phi(t) = \delta(t)$, (6.8.25) becomes

$$\mu(t) = f(t) = \sum_\beta \left[\bar{p}_\beta \frac{\omega_{0,N+1,\beta}}{\beta} \delta(t) + \sum_{j=1}^{N} \frac{\omega_{j,N+1,\beta}}{v_j} p_{\beta j}(t)\right] \tag{6.8.27}$$

Therefore the mean impulse response $f(t) = \mu(t)$ is just the probability density function of residence times in the entire system, that is, by our earlier notation, $E(t)$. The first moment, then, gives the same result as the deterministic residence time density.

Next we shall proceed to the second noncentral moments of the concentrations. Although the manipulations become quite messy, the principles of the calculation are the same as for the means. Define the second noncentral moment of the concentrations in tanks j and k for flow state β by

$$S_{\beta jk}(t) = \int \cdots \int y_j y_k p_\beta(\mathbf{y}, t)\, dy_1 \cdots dy_N, \qquad j, k = 1, 2, \ldots, N \tag{6.8.28}$$

We can multiply (6.8.20) and (6.8.21) by $y_j y_k$ and integrate over the entire space of y-values to obtain the differential equations governing the $S_{\beta jk}(t)$:

$$\begin{aligned}\frac{dS_{\beta jk}}{dt} &= \left(\frac{\omega_{0j\beta}}{\omega_\beta}\frac{\mu_{\beta k}(t)}{v_j} + \frac{\omega_{0k\beta}}{\omega_\beta}\frac{\mu_{\beta j}(t)}{v_k}\right)\phi(t) \\ &\quad + \sum_{i=1}^{N}\left(\frac{\omega_{ij\beta}}{v_j}S_{\beta ik}(t) + \frac{\omega_{ik\beta}}{v_k}S_{\beta ij}(t)\right) + \sum_\alpha \lambda_{\alpha\beta}S_{\alpha jk}(t), \qquad (6.8.29) \\ S_{\beta jk}(0) &= 0, \qquad j, k = 1, 2, \ldots, N\end{aligned}$$

The mean square rate that tracer leaves the system is, from (6.8.25),

$$S^2(t) = \sum_\beta \int \cdots \int \left[\frac{\omega_{0,N+1,\beta}}{\omega_\beta}\phi(t) + \sum_{j=1}^{N} \omega_{j,N+1,\beta} y_j\right]^2 p_\beta(\mathbf{y}, t)\, dy, \ldots, dy_N \tag{6.8.30}$$

which we may write as

$$\begin{aligned}S^2(t) = \sum_\beta \Bigg[\bar{p}_\beta\left(\frac{\omega_{0,N+1,\beta}}{\omega_\beta}\right)^2 \phi^2(t) &+ 2\frac{\omega_{0,N+1,\beta}}{\omega_\beta}\phi(t)\sum_{j=1}^{N} \omega_{j,N+1,\beta}\mu_{\beta j}(t) \\ &+ \sum_{j=1}^{N}\sum_{k=1}^{N} \omega_{j,N+1,\beta}\omega_{k,N+1,\beta}S_{\beta jk}\Bigg]\end{aligned} \tag{6.8.31}$$

Finally, the variance of the tracer outlet response is

$$\sigma^2(t) = S^2(t) - \mu(t)^2 \tag{6.8.32}$$

6.8.3 Tracer calculations

Krambeck et al., in their original paper [8], applied the above analysis to two simple systems, two ideally mixed vessels in parallel and two ideally mixed vessels in series. In each case the flow rates were allowed to assume only two values. Then, since $\lambda_{12} = -\lambda_{11} = \lambda_1$ and $\lambda_{21} = -\lambda_{22} = \lambda_2$, we have

$$\begin{aligned} p_{12}(\tau) &= \lambda_1\tau + o(\tau) \\ p_{11}(\tau) &= 1 - \lambda_1\tau + o(\tau) \\ p_{21}(\tau) &= \lambda_2\tau + o(\tau) \\ p_{22}(\tau) &= 1 - \lambda_2\tau + o(\tau) \end{aligned} \tag{6.8.33}$$

Solving (6.8.8) for the stationary probabilities of the flow states, we get

$$\bar{p}_1 = \frac{\lambda_2}{\lambda_1 + \lambda_2}, \qquad \bar{p}_2 = \frac{\lambda_1}{\lambda_1 + \lambda_2} \tag{6.8.34}$$

The mean rate of switching $\bar{\lambda}$ is

$$\bar{\lambda} = \bar{p}_1\lambda_1 + \bar{p}_2\lambda_2 = \frac{2\lambda_1\lambda_2}{\lambda_1 + \lambda_2} \tag{6.8.35}$$

Let us consider two ideally mixed vessels in parallel, as shown in Figure 6.15, in which the total flow ω is constant but the division of the flow between the two tanks varies. Let the flow state be characterized by $r_\alpha = \omega_{01\alpha}/\omega$, the fraction of the total flow entering tank 1.

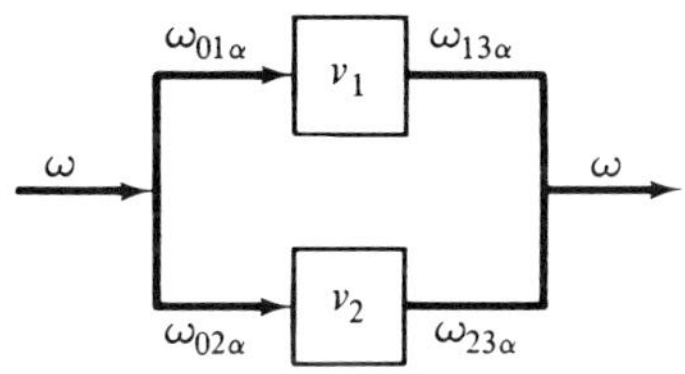

Figure 6.15. Model consisting of two ideally mixed vessels in parallel with stochastic flows. [Reprinted from F. J. Krambeck, R. Shinnar, and S. Katz, "Stochastic Mixing Models for Chemical Reactors," *Ind. Eng. Chem. Fund.* 6, 276 (1967). Reprinted with permission of the copyright owner.]

The material balance on the two tanks is

$$\begin{aligned} v_1 \frac{dc_1}{dt} &= r_\alpha \phi(t) - \omega r_\alpha c_1 \\ v_2 \frac{dc_2}{dt} &= (1 - r_\alpha)\phi(t) - \omega(1 - r_\alpha)c_2 \end{aligned} \tag{6.8.36}$$

We also have equations for μ_{11}, μ_{12}, μ_{21}, and μ_{22} from (6.8.24) and for S_{111}, S_{112}, S_{121}, S_{122}, S_{211}, S_{212}, S_{221}, and S_{222} from (6.8.29), a system of 10 linear O.D.E.s. From their solutions, $\mu(t)$, $S^2(t)$, and $\sigma^2(t)$ can be computed.

The mean step response for various fluctuation magnitudes ϵ, where $\epsilon = r_1 - r_2$, is shown in Figure 6.16, and in Figure 6.17, $1 - F(t)$ is plotted. The negative of the slope of this curve is the intensity function $\Lambda(t)$; that is, if a particle has been in the system for a time t, its probability of escaping in the next dt is $\Lambda(t)\,dt$. As the magnitude of the flow disturbance increases, $\Lambda(t)$ for large t decreases. As we noted in introducing $\Lambda(t)$, a decrease in Λ over some interval is characteristic of systems exhibiting bypassing, and so the effect of flow fluctuations on the residence time distribution of this simple model is similar to these effects. The coefficient of variation $\sigma(t)/\mu(t)$ is shown in Figures 6.18 and 6.19 for various fluctuation magnitudes and switching rates, respectively. We note that $\sigma(t)/\mu(t)$ increases as ϵ increases, verifying that the spreading of the output curve is increased as the fluctuations increase

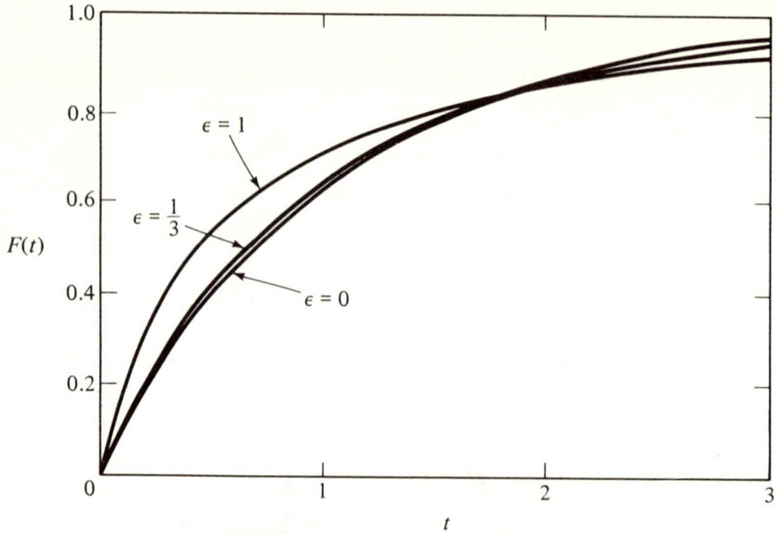

Figure 6.16. Effect of fluctuation magnitude on mean response to step input for parallel ideally mixed vessels. [Reprinted from F. J. Krambeck, R. Shinnar, and S. Katz, "Stochastic Mixing Models for Chemical Reactors," *Ind. Eng. Chem. Fund. 6*, 276 (1967). Reprinted by permission of the copyright owner.]

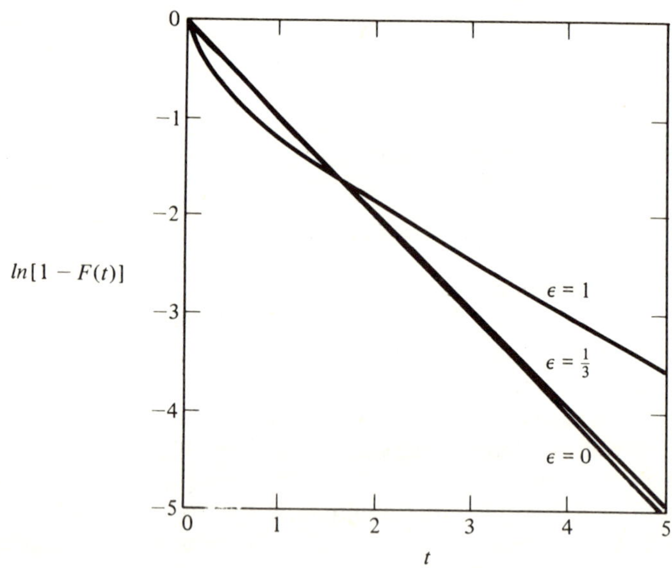

Figure 6.17. Effect of fluctuation magnitude on mean response to step input for parallel ideally mixed vessels. [Reprinted from F. J. Krambeck, R. Shinnar, and S. Katz, "Stochastic Mixing Models for Chemical Reactors," *Ind. Eng. Chem. Fund. 6*, 276 (1967). Reprinted by permission of the copyright owner.]

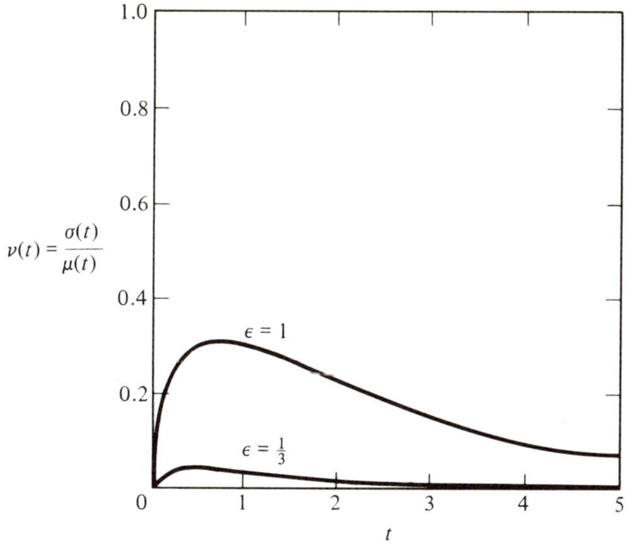

Figure 6.18. Effect of fluctuation magnitude on $\nu(t) = \sigma(t)/\mu(t)$, the coefficient of variation, for parallel ideally mixed vessels. [Reprinted from F. J. Krambeck, R. Shinnar, and S. Katz, "Stochastic Mixing Models for Chemical Reactors," *Ind. Eng. Chem. Fund. 6*, 276 (1967). Reprinted by permission of the copyright owner.]

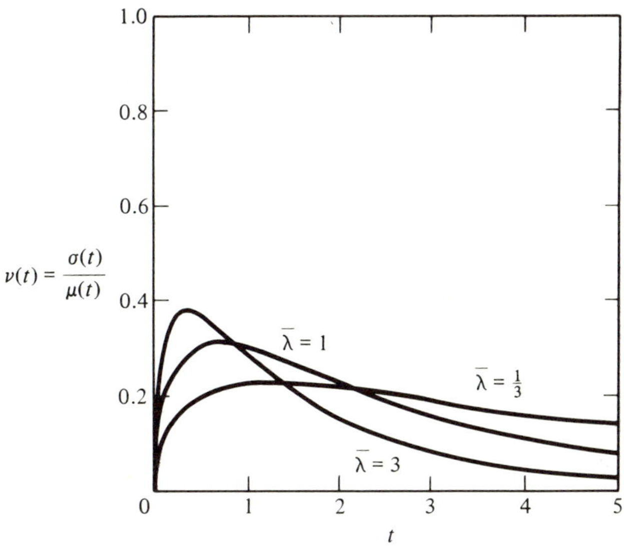

Figure 6.19. Effect of switching rate $\bar{\lambda}$ on $\nu(t) = \sigma(t)/\mu(t)$, the coefficient of variation, for parallel ideally mixed vessels. [Reprinted from F. J. Krambeck, R. Shinnar, and S. Katz, "Stochastic Mixing Models for Chemical Reactors," *Ind. Eng. Chem. Fund. 6*, 276 (1967). Reprinted by permission of the copyright owner.]

in size. As the switching rate $\bar{\lambda}$ increases, $\sigma(t)/\mu(t)$ peaks earlier and tends to zero more quickly.

REFERENCES

1. Bell, R. L., and A. L. Babb, "On the Extension of the Method of Moments to a Cascade of Well-Mixed Discrete Stages with Back-Flow Between Stages," *Chem. Eng. Sci. 20*, 1001 (1965).
2. Chiang, D. and A. Cholette, "Performance of Tanks in Series for Non-Ideal Mixing," *Can. J. Chem. Eng. 48*, 286 (1970).
3. Deans, H. and L. Lapidus, "A Computational Model for Predicting and Correlating the Behavior of Fixed-Bed Reactors," *AIChE J. 6*, 656 (1961).
4. Fan, L. T., M. S. K. Chen, Y. K. Ahn, and W. Y. Wen, "Mixing Models with Varying Stage Size," *Can. J. Chem. Eng. 47*, 141 (1969).
5. Klinkenberg, A., "Distribution of Residence Times in a Cascade of Mixed Vessels with Backmixing," *Ind. Eng. Chem. Fund. 5*, 283 (1966).
6. Klinkenberg, A., "Moments of Residence Time Distributions for Cascades of Stirred Vessels with Backmixing," *Chem. Eng. Sci. 23*, 1975 (1968).
7. Krambeck, F. J., S. Katz, and R. Shinnar, "Interpretation of Tracer Experiments in Systems with Fluctuating Throughput," *Ind. Eng. Chem. Fund. 8*, 431 (1969).
8. Krambeck, F. J., S. Katz, and R. Shinnar, "Stochastic Mixing Models for Chemical Reactors," *Ind. Eng. Chem. Fund. 6*, 276 (1967).
9. Levenspiel, O., and K. B. Bischoff, "Patterns of Flow in Chemical Process Vessels," *Advan. Chem. Eng. 4*, 95 (1963).
10. Naor, P., and R. Shinnar, "Representation and Evaluation of Residence Time Distributions," *Ind. Eng. Chem. Fund. 2*, 278 (1963).
11. Roemer, M. H., and L. D. Durbin, "Transient Response and Moments Analysis of Backflow Cell Model for Flow Systems with Longitudinal Mixing," *Ind. Eng. Chem. Fund. 6*, 121 (1967).
12. Shinnar, R., and P. Naor, "Residence Time Distributions in Systems with Internal Reflux," *Chem. Eng. Sci. 22*, 1369 (1967).
13. van der Laan, E. T., "Note on the Diffusion-Type Model for the Longitudinal Mixing in Flow," *Chem. Eng. Sci. 7*, 187 (1957).
14. Wolf, D., and W. Resnick, "Residence Time Distribution in Real Systems," *Ind. Eng. Chem. Fund. 2*, 287 (1963).

PROBLEMS

6.1 Determine $E(t)$, $I(t)$, and $\Lambda(t)$ for two ideally mixed vessels in series, one with average residence time μ_1 and the other with average residence time μ_2.

6.2 Determine $E(t)$ and the mean and variance of $E(t)$ as a function of β for the three systems in Figure P6.2.

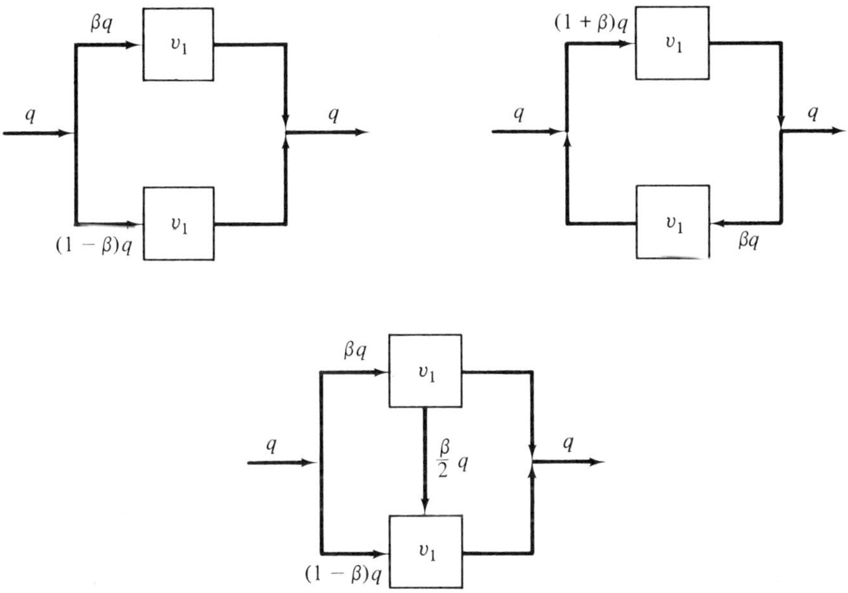

Figure P6.2. Parallel arrangements of ideally mixed vessels.

6.3 Determine the mean and variance of $E(t)$ for the system shown in Figure 6.6.

6.4 Solve (6.3.12)—(6.3.15) to obtain (6.3.16). What are the mean and variance of this distribution?

6.5 Extend Example 6.5 to the case of a geometric progression of volumes.

6.6 Present as complete an analysis as you can of a finite series of ideally mixed vessels with backflow with a recycle stream, as shown in Figure P6.6.

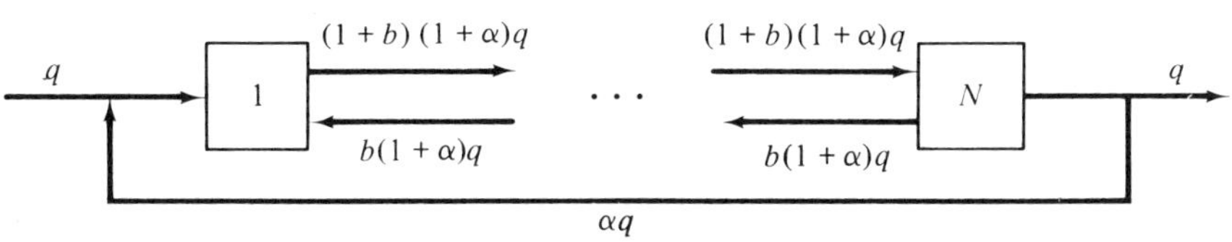

Figure P6.6. N ideally mixed vessels with backflow and recycle.

6.7 In Chapter 2 we discussed the way in which individual transfer functions are utilized to construct an overall transfer function relating some arrangement of series and parallel individual units. Try to show how analogous conclusions would result in terms of the moments of individual units.

6.8 Analyze the recirculation reactor system shown in Figure P6.8. Entry and exit flow rates q are constant. Show that the natural limits of the overall system

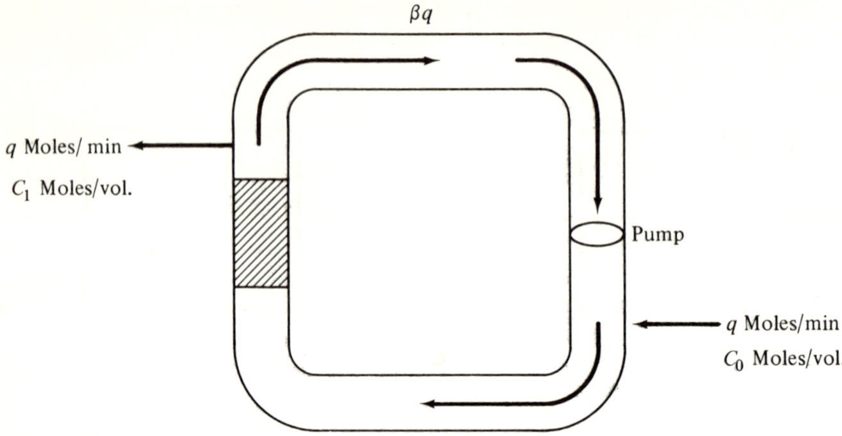

Figure P6.8. Recycle reactor.

operation, as the recycle ratio β is changed widely, is a plug flow reactor at one limit and an ideally mixed vessel at the other limit. The packed bed catalytic reaction is first order with no volume change; plug flow and isothermal conditions also hold.

6.9 Consider a source of tracer introduced centrally into a tube packed with spherical particles in random array. There are many particles in the cross section of the bed, and the average fluid velocity may be considered independent of radial location. Formulate this problem with regard to radial spread of tracer. Sketch the solution in terms of the appropriate dimensionless variables for the regions close to the source, where the tracer reaches the wall, and in an infinitely long bed.

Parameter Estimation 7

The remainder of this book is devoted to parameter estimation and process identification. To provide motivation for this material we shall indicate the general steps of process analysis, as depicted in Figure 7.1. The first step is to obtain experimental data from the system, either from an observed response to a test input or from the normal operating data from the system. The second part of the analysis is to form a mathematical model of the system based on both the physical nature of the system and the experimental output at hand. The formation of a model can proceed along several avenues, depending on the amount of structural detail we are able to ascribe to the model, namely on the amount of a priori information we have on the functional form of the model. We shall distinguish the two cases of full information on the structure of the model and no information on the structure:

1. *Full information on model structure.* In this case the form of the model is specified, usually from conservation equations and transport phenomena principles, and the only step required is the estimation of parameters in the model so that the model output can be matched as closely as possible to the experimental output. This problem is called the *parameter estimation problem*, and it will be the subject of Chapters 7 and 8.
2. *No information on model structure.* In this case the system is so complicated that we are forced to view it as a black box. We have no a priori information on the form of the model or any parameters that might be important. The problem of constructing a model purely from input/output

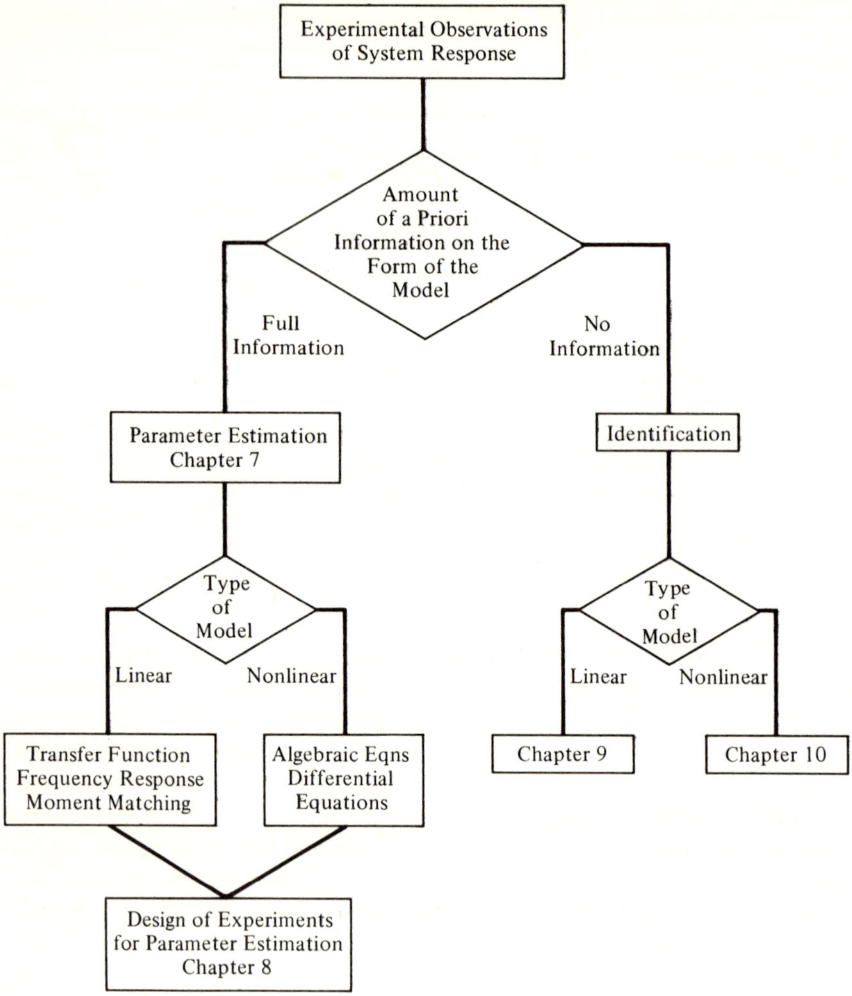

Figure 7.1. Process identification and parameter estimation.

data is called the *identification problem*. This will be the subject of Chapters 9 and 10.

When formulating and solving an identification or parameter estimation problem it is important to have in mind the ultimate purpose of the study. Two common objectives of identification and parameter estimation are

1. *System diagnosis*. The need to obtain more knowledge about a system. For a mechanical or chemical system this may involve estimation of parameters or more elaborate identification. For a highly complex system,

such as the human body or an economic system, this may take the form of a black box approach.

2. *System control.* The need to develop a model of the system that can be used in conjunction with a controller. Depending on the type of controller, the complexity of the model required may vary. For example, to design a stable regulator (to keep the system at a certain desired point) it is often sufficient to have a fairly crude model. On the other hand, to design a control program for optimal transition of the process from one state to another generally requires an accurate model of the system. Finally, if we want to design a regulator which minimizes the variations in process variables due to external disturbances, we may need a model of the environment of the system.

Since experiments will usually have to be conducted on the system, there are several questions which must be faced before these are designed, namely:

1. How should the experiment be planned? Should all the data be accumulated at once or should a sequential design be employed? That is, perform an experiment, evaluate the results, and, on that basis, plan the next experiment.
2. What type of perturbing input should be used to test the system?
3. What type of analysis should be applied to the results of the experiment?

In spite of the large amount of work on process identification and parameter estimation there are still no general answers to the problems above. These must be approached freshly for each new identification study. We can, however, make some general observations about identification problems. Zadeh [53] has stated, "Identification is the determination, on the basis of input and output, of a system within a specified class of systems, to which the system under test is equivalent." In our terminology, the system under test is the *process* and the system within a specified class of systems is the *model.* Equivalence is usually defined in terms of an error criterion which quantifies the deviation between the model and the data. Thus, if the error criteria of two models for the same process are equal, the two models must be called equivalent. In general, identification problems can be divided according to

1. The specified class of models from which one is to be chosen.
2. The types of input and output used.
3. The error criterion employed.

All identification problems are simply optimization problems in which we desire to minimize the error criterion by choice of a model, where "choice of a model" implies anything from simply estimating unknown parameters to determining the form of the model from scratch. We can call the process *identifiable* if the optimization problem has a unique solution.

In this and the next chapter we shall consider parameter estimation problems, whereas the subject of Chapters 9 and 10 will be identification problems.

7.1 Model Forms for Parameter Estimation

Parameter estimation problems result when we attempt to match a model of known form to experimental data by optimal determination of unknown model parameters. The exact nature of the parameter estimation problem will depend on the type of mathematical model. We must make an important distinction at this point. A model will contain both state variables (concentrations, temperatures, pressures) and parameters (rate constants, dispersion coefficients, activation energies, etc.). When we refer to a model as linear we emphasize that the linearity to which we refer is with respect to the *state variables* and *not* the parameters. Even a system which is linear with respect to state variables is almost never linear with respect to parameters. Therefore we shall always assume in general that the parameters enter the model in a nonlinear fashion.

The linearity of a model with respect to state variables may, however, influence the ultimate manner in which the parameters occurring in that model are determined. For example, if the model for the process is linear, the transfer function will normally be available. Then the model parameters, which all appear in the transfer function, can, in principle, be determined by matching the transfer function to the data, which has been likewise transformed into the s-domain. However, if the model is nonlinear, the actual parameter estimation must be carried out by matching the solution of the model directly to the experimental data in the time domain. This normally requires the numerical solution of the model equations by means of a computer.

Additional considerations include [36]

1. *Dimension of the system.* Some techniques will handle systems in which the state variable is only a scalar, whereas others can treat systems of n state variables.
2. *Parameters constant or varying.* In some model formulations the parameters are assumed to be functions of time, while in others they are constant. It should be noted that most methods which handle time-varying parameters can handle constant parameters but the calculational effort is often significantly greater than for techniques designed specifically for constant parameter estimation. In this and the following chapter we shall consider only those estimation techniques based on constant parameters.

3. *Type of input forcing*. In some cases the choice of the estimation technique defines the input forcing function that must be used to excite the system. For example, frequency response analysis necessitates the use of pure sinusoidal variations in the input. In general, techniques with the fewest restrictions on choice of the forcing function have the greatest computational requirements.
4. *Number of measured states*. Some methods require only measurement of the output, whereas other more restrictive methods require measurement of all the states.
5. *Sensitivity to disturbances*. In practical applications the test signals are often corrupted with noise. Estimation methods can be divided according to their ability to tolerate randomness in the inputs to the system. Clearly, the methods which depend on a particular deterministic input are least sensitive to disturbances. Some techniques are more tolerant of noise than others, but, in general, an increase in noise level decreases the reliability of the parameter estimates. In some cases, we intentionally use stochastic inputs.
6. *Processing of data*. In some estimation techniques all the data must be collected before the analysis is undertaken and the result is a single estimate of the parameters. This case is called *nonsequential* estimation. However, other techniques, called *sequential* estimation techniques, can be employed in real time as the data are being collected; that is, each time a new data point is collected it is fed into the algorithm to produce an updated estimate of the parameters. In this book we shall concentrate on nonsequential estimation.
7. *Error criterion*. The error criterion of a parameter estimation method is a measure of the discrepancy between the model predictions and the data. The type of error criterion used will very often dictate the type of computations required in the parameter estimation algorithm.

7.1.1 Algebraic models

We shall begin this chapter by specifying in general form the classes of models we shall be considering. First, we shall consider algebraic models, defining the following variables:

$\mathbf{k}$ = p-dimensional column vector of parameters whose numerical values are unknown, $[k_1, k_2, \ldots, k_p]^T$

$\mathbf{x}$ = n-dimensional column vector of states; these are the independent variables which are either fixed arbitrarily for each experiment or which are known precisely for each experimental observation, $[x_1, x_2, \ldots, x_n]^T$

$\boldsymbol{\eta}$ = m-dimensional column vector of observed variables; these are the model variables which are actually measured in the experiments, $[\eta_1, \eta_2, \ldots, \eta_m]^T$; note that $\boldsymbol{\eta}$ is *not* the actual data but only the dependent or observed variables

A single experiment consists of the measurement of each of the m observed variables for a given set of values of the independent variables. The models which we shall consider in this subsection are those in which the observed variables at each experiment are related to the states and parameters by an algebraic equation of the form

$$\boldsymbol{\eta} = \mathbf{f}(\mathbf{x}, \mathbf{k}) \tag{7.1.1}$$

where $\mathbf{f}$ is a vector function of known form. In component form, (7.1.1) is

$$\begin{bmatrix} \eta_1 \\ \eta_2 \\ \cdot \\ \cdot \\ \cdot \\ \eta_m \end{bmatrix} = \begin{bmatrix} f_1(x_1, \ldots, x_n, k_1, \ldots, k_p) \\ f_2(x_1, \ldots, x_n, k_1, \ldots, k_p) \\ \cdot \\ \cdot \\ \cdot \\ f_m(x_1, \ldots, x_n, k_1, \ldots, k_p) \end{bmatrix}$$

Since we want to determine p parameters, we must perform at least p different experiments. Thus for each individual experiment we have a set of ηs and xs. For the jth experiment, (7.1.1) is

$$\boldsymbol{\eta}_j = \mathbf{f}(\mathbf{x}_j, \mathbf{k}), \qquad j = 1, 2, \ldots, M \tag{7.1.2}$$

where we assume that M separate experiments have been performed.

In a study of a chemical reaction, for example, η might be the conversion and the variables might be time of reaction x_1, temperature x_2, and pressure x_3. The experimenter may make a series of M runs using various experimental conditions. Equation (7.1.2) expresses the fact that the basic model is assumed valid for each one of these M runs. In the jth run the independent variables are kept at $x_{1j}, \ldots, x_{nj}$. A table showing the values of the variables x employed in a series of runs is called the experimental design matrix. For example, suppose that time x_1, temperature x_2, and pressure x_3 are the variables and four runs are performed. The design matrix below shows that in run 1 ($j = 1$) a reaction time of 10 minutes, a temperature of 25°C, and a pressure of 1 atmosphere are employed.

Run no., j	Time (min), x_1	Temp. (°C), x_2	Pressure (atm), x_3
1	10	25	1
2	20	30	0.5
3	30	30	0.2
4	40	50	0.2

In addition to quantitative variables we could also include qualitative variables such as the type of catalyst, which would then be x_4.

Let us show how some familiar problems fit into the framework of (7.1.2). Perhaps the most well-known class of problems requiring parameter estimation using algebraic models are those involving the estimation of kinetic parameters from concentration measurements in reaction systems.

Suppose that we have a unidirectional, unimolecular homogeneous reaction

$$A \xrightarrow{k}$$

Under isothermal conditions the rate equation for this reaction is

$$\frac{dc_A}{dt} = -kc_A$$

where c_A is the concentration or partial pressure of A, t is reaction time, and k is the rate constant. Integrating this equation with a given initial c_A at $t = 0$, c_{A_0}, we obtain

$$c_A = c_{A_0}e^{-kt} \tag{7.1.3}$$

If we now let $\eta = c_A$, $\mathbf{x} = [t, c_{A_0}]^T$, and $k = k$, (7.1.3) is equivalent to (7.1.1). If we repeat this experiment for M values of t and c_{A_0}, we have (7.1.2.)

Now consider the isothermal rate equations corresponding to the reactions

$$A \xrightarrow{k_1} B \xrightarrow{k_2} C$$

They are

$$\frac{dc_A}{dt} = -k_1c_A$$

$$\frac{dc_B}{dt} = k_1c_A - k_2c_B$$

Note that if the initial conditions are such that the total number of initial moles are known, say

$$c_{A_0} = 1, \qquad c_{B_0} = 0, \qquad c_{C_0} = 0, \qquad t = 0$$

we do not need a separate rate equation for C since $c_A + c_B + c_C = 1$. The integrated forms are

$$\begin{aligned} c_A &= e^{-k_1t} \\ c_B &= \frac{k_1}{k_2 - k_1}(e^{-k_1t} - e^{-k_2t}) \end{aligned} \tag{7.1.4}$$

Say that we measure both A and B. Then, letting $\boldsymbol{\eta} = [c_A, c_B]^T$, $x = t$, and $\mathbf{k} = [k_1, k_2]^T$, we see that we once again have the form of (7.1.1). If we measure c_A and c_B at M values of t, then we have (7.1.2). Say that we measure B only. Then $\eta = c_B$ only, and the second equation of (7.1.4) becomes

$$c_{B_j} = \frac{k_1}{k_2 - k_1}(e^{-k_1t_j} - e^{-k_2t_j}), \qquad j = 1, 2, \ldots, M$$

the scalar form of (7.1.2).

As a final example, let us consider a heterogeneous-type kinetic system,

$$A + R \underset{k_2}{\overset{k_1}{\rightleftharpoons}} B + C$$

carried out in a differential flow reactor.

If the controlling mechanism is that of surface reaction on dual sites without dissociation, the Langmuir-Hinshelwood expression for the rate of disappearance of A, r_A, becomes

$$r_A = \frac{k_1 c_A c_R - k_2 c_B c_C}{(1 + K_A c_A + K_R c_R + K_B c_B + K_C c_C)^2} \tag{7.1.5}$$

where K_A, K_R, K_B, and K_C are adsorption equilibrium coefficients. If we let $\mathbf{x} = [c_A, c_R, c_B, c_C]^T$, and $\mathbf{k} = [k_1, k_2, K_A, K_R, K_B, K_C]^T$, $\eta = r_A$, we obtain (7.1.1). It is important to note in (7.1.5) that it is assumed that the initial *rate* r_A is the measured quantity rather than the concentration directly.

Having seen how several common problems in reaction kinetics modeling fall into the framework of (7.1.2), let us now consider the question of the relation of the model to the actual experimental data. Recall that $\boldsymbol{\eta}$ is the vector of dependent variables, those variables that we measure. Now, let us denote by $\mathbf{y}_j = [y_{1_j}, y_{2_j}, \ldots, y_{m_j}]^T, j = 1, 2, \ldots, M$, the actual *data points*, that is, experimentally measured values of the $\boldsymbol{\eta}_j$. We do this because we really do not know the exact values of the $\boldsymbol{\eta}_j$, but, rather, we only have the experimental measurements of the $\boldsymbol{\eta}_j$.

Because of the errors in the measurements and the inaccuracies in the model, it is impossible to hope for an exact fit such that

$$\mathbf{y}_j = \mathbf{f}(\mathbf{x}_j, \mathbf{k})$$

Instead we shall try to find values of $\mathbf{k}$ which minimize some appropriate measure of the errors involved. Let us, therefore, rewrite our basic equation in the form

$$\mathbf{y}_j = \mathbf{f}(\mathbf{x}_j, \mathbf{k}) + \boldsymbol{\epsilon}_j, \qquad j = 1, 2, \ldots, M \tag{7.1.6}$$

where $\boldsymbol{\epsilon}_j = [\epsilon_{1_j}, \epsilon_{2_j}, \ldots, \epsilon_{m_j}]^T$ is the vector of errors, or *residuals*, between the observations and the predicted dependent variables. Parameter estimation then tries to find a set of parameters $\mathbf{k}$ such that some scalar function of the errors is minimized. In general, we shall write this function as $S(\mathbf{k})$ to indicate the dependence on the chosen parameters.

1. *Least squares estimation.* In least squares estimation the function $S(\mathbf{k})$ is defined as the sum of squares of the errors:

$$\begin{aligned} S(\mathbf{k}) &= \sum_{j=1}^{M} [\mathbf{y}_j - \mathbf{f}(\mathbf{x}_j, \mathbf{k})]^T \mathbf{Q}_j [\mathbf{y}_j - \mathbf{f}(\mathbf{x}_j, \mathbf{k})] \\ &= \sum_{j=1}^{M} \sum_{l=1}^{m} \sum_{r=1}^{m} [y_{l_j} - f_l(\mathbf{x}_j, \mathbf{k})] q_{lr_j} [y_{r_j} - f_r(\mathbf{x}_j, \mathbf{k})] \end{aligned} \tag{7.1.7}$$

where $\mathbf{Q}_j$ is an $m \times m$ weighting matrix with elements q_{lr_j} which is chosen to reflect knowledge of the relative precision of the various measurements. We shall discuss the selection of $\mathbf{Q}_j$ shortly. Generally, $\mathbf{Q}_j$ is taken to be diagonal. If $\mathbf{Q}_j = \mathbf{I}$, then (7.1.7) is an *unweighted* least squares criterion.

We can distinguish two cases associated with (7.1.7). In the first case the parameters $\mathbf{k}$ enter $\mathbf{f}(\mathbf{x}_j, \mathbf{k})$ in a linear fashion (the result of the minimization is referred to as linear least squares); in the second case no such linearity occurs and we have nonlinear least squares. For the moment we shall consider the unweighted form of (7.1.7). The value of $\mathbf{k}$ which minimizes $S(\mathbf{k})$ is denoted by $\hat{\mathbf{k}}$.

Linear least squares is also known as multiple linear regression. In such a case the model need only be substituted into (7.1.7), the derivative of $S(\mathbf{k})$ taken with respect to $\mathbf{k}$, and the result set equal to zero. This yields p linear algebraic equations corresponding to the p parameters (called the normal equations). By means of a single matrix inversion it is then possible to solve for the best $\hat{\mathbf{k}}$.

If the $\boldsymbol{\epsilon}_j$ are independently and identically distributed errors with zero means, then by this procedure the $\hat{\mathbf{k}}$ are efficient linear unbiased estimates of $\mathbf{k}$. In other words, the parameter estimates which minimize the sum of squares of the residuals will, on the average, equal the true values of the parameters (unbiased) and will be estimated with maximum precision (minimum variance).

Unfortunately this method has a number of decided defects. Foremost among these is the fact that few models of any complexity occur in the desired linear form. As a result the model equations must be linearized or rearranged to be handled by this method. Thus for the familiar expression (7.1.3)

$$\eta = e^{-kt}$$

a conventional logarithmic transformation leads to

$$\eta' = \ln \eta = -kt$$

which is linear in the parameter k and is therefore amenable to treatment by linear least squares. However, if the corresponding error equation for the original expression is given by

$$y = \exp\{-kt\} + \epsilon_j$$

it is apparent that the logarithmic transformation does not preserve the original distribution of errors which might have been judged appropriate.

This situation is further aggravated when a temperature-dependency condition is introduced through the Arrhenius relation. Then a two-step estimation process with logarithmic transformation for each step is required:

1. From $\log \eta = -kt$, estimate k.
2. From $\log k = \ln k_0 - (E/R)(1/T)$, estimate k_0 and E.

A similar difficulty ensues when heterogeneous catalytic reactions represented by the Langmuir-Hinshelwood relations are studied by linear least squares. Thus suppose that the model of a heterogeneous reaction is taken as

$$r_A = \frac{k'(p_A - p_R p_S / K_{\text{equil}})}{(1 + K_A p_A + K_R p_R + K_S p_S)^2} \tag{7.1.8}$$

where the $p_A, \ldots,$ are partial pressures and K_{equil} is a known constant. Obviously the parameters k', K_A, K_R, and K_S do not appear linearly in the equation. If, however, the equation is rearranged to the form

$$\left(\frac{p_A - p_R p_S / K_{\text{equil}}}{r_A}\right)^{1/2} = \frac{1}{k'} + \frac{K_A p_A}{k'} + \frac{K_R p_R}{k'} + \frac{K_S p_S}{k'} \tag{7.1.9}$$

or

$$\eta_A = k_1 + k_2 p_A + k_3 p_R + k_4 p_S$$

then combinations of parameters occur linearly. However, now the partial pressures appear on both sides of the equation and therefore play the role of both dependent and independent variables. Furthermore, the errors of (7.1.9) are minimized and not the original errors of (7.1.8).

As a result of such transformations the method of linear least squares will give parameter estimates which guarantee neither a best prediction of reaction rates [the left-hand side of (7.1.8)] nor reasonably satisfactory extrapolations. Furthermore, in models involving several dependent variables, linearization is usually quite impossible.

In summary, we see that linear least squares is applicable for parameter estimation in only very special cases and that it is a viable method only under special circumstances.

Since the linear least squares approach has these obvious defects, it seems natural to turn to a method which does not require that the model equations be linear in the unknown parameters. We refer to this method as nonlinear least squares since the parameters may occur in any fashion, linear or nonlinear.

In a broad-gauged description of this method, one attempts to minimize $S(\mathbf{k})$ in an iterative fashion rather than in a single step. As such, and because of the extensive calculations which are then required, the method almost certainly requires implementation via a computer. Methods for performing this minimization are described in a subsequent section.

It is apparent that nonlinear least squares, in which the dependency of the dependent variables on the parameters is essentially unrestricted, obviates many of the difficulties inherent in linear least squares. The algebraic model is used in the form originally proposed without imposing arbitrary transformations; the minimization of the sum of squares of errors is appropriate in terms of untransformed variables.

The principal problem with nonlinear least squares is the choice of the weight functions $\mathbf{Q}$. Unfortunately, one usually does not know what numerical values to use. On theoretical grounds, if the residuals have zero mean with a known covariance matrix, $\mathbf{Q}$ should be the inverse of this covariance, if it is known. We can overcome this problem by using maximum likelihood estimation.

2. *Maximum likelihood estimation.* The maximum likelihood principle offers a powerful and versatile tool to the parameter estimator. A price must be paid, however, in that explicit assumptions concerning the form of the probability distribution of the errors must be made. Once the form has been assumed, any parameters appearing in the distribution function may be estimated along with the unknown parameters in the model. This is really as it should be; the errors in the observations are as much a part of physical reality as is the process being observed. Meaningful parameter estimation can proceed only if we provide a mathematical model not only of the system but also of the errors, and a probability distribution is an appropriate mathematical model for the errors.

Suppose that we assume $p(\boldsymbol{\epsilon}, \boldsymbol{\psi})$ to be the joint probability density function of all the errors $\boldsymbol{\epsilon}$ in the observed variables. Here $\boldsymbol{\psi}$ represents a set of unknown parameters (e.g., means, variances) which appear in the formulas for the distribution. Given any values for the model parameters $\mathbf{k}$, we substitute the residuals $\mathbf{y} - \mathbf{f}(\mathbf{x}, \mathbf{k})$ for the errors $\boldsymbol{\epsilon}$ in the expression for the probability density. This yields a function depending on $\mathbf{k}$ and $\boldsymbol{\psi}$ (the $\mathbf{x}$ and $\mathbf{y}$ being given by the observations):

$$L(\mathbf{k}, \boldsymbol{\psi}) = p[\mathbf{y} - \mathbf{f}(\mathbf{x}, \mathbf{k}), \boldsymbol{\psi}] \tag{7.1.10}$$

We refer to $L(\mathbf{k}, \boldsymbol{\psi})$ as the likelihood function. The maximum likelihood method simply consists of finding those values of $\mathbf{k}$ and $\boldsymbol{\psi}$ which maximize L. For reasons of convenience one usually maximizes $S_{ML} = \log L$. This is clearly equivalent to maximizing L. Thus we have

$$S_{ML}(\mathbf{k}, \boldsymbol{\psi}) = \log p[\mathbf{y} - \mathbf{f}(\mathbf{x}, \mathbf{k}), \boldsymbol{\psi}] \tag{7.1.11}$$

If, as is frequently assumed, the observation errors in different experiments are uncorrelated, the joint probability density p is the product of the individual experiment probability densities p, and

$$S_{ML}(\mathbf{k}, \boldsymbol{\psi}) = \sum_{j=1}^{M} \log p[\mathbf{y}_j - \mathbf{f}(\mathbf{x}_j, \mathbf{k}), \boldsymbol{\psi}] \tag{7.1.12}$$

The most frequently used distribution for the residuals is the normal distribution. If the errors of each experiment, $\boldsymbol{\epsilon}_j$, are independent and normally distributed with zero means and covariance matrix $\mathbf{V}_j$, then (see Table 4.2)

$$p(\boldsymbol{\epsilon}_j, \mathbf{V}_j) = \frac{1}{(2\pi)^{m/2}|\mathbf{V}_j|^{1/2}} \exp\left(-\frac{1}{2}\boldsymbol{\epsilon}_j^T\mathbf{V}_j^{-1}\boldsymbol{\epsilon}_j\right) \qquad j = 1, 2, \ldots, M \tag{7.1.13}$$

Substituting (7.1.13) in (7.1.12), we obtain

$$S_{ML}(\mathbf{k}, \mathbf{V}_j) = \frac{mM}{2}\log 2\pi - \frac{1}{2}\sum_{j=1}^{M}\log(|\mathbf{V}_j|) - \frac{1}{2}\sum_{j=1}^{M}\boldsymbol{\epsilon}_j^T\mathbf{V}_j^{-1}\boldsymbol{\epsilon}_j \tag{7.1.14}$$

where we use $\boldsymbol{\epsilon}_j$ for the residual $\mathbf{y}_j - \mathbf{f}(\mathbf{x}_j, \mathbf{k})$.

Let us now consider some special cases of (7.1.14). Suppose, first, that all the $\mathbf{V}_j$, $j = 1, 2, \ldots, M$, are known. The only nonconstant term in (7.1.14) is the last one on the R.H.S. Maximizing $S_{ML}(\mathbf{k})$ is equivalent to minimizing

$$S(\mathbf{k}) = \sum_{j=1}^{M}\sum_{l=1}^{m}\sum_{r=1}^{m}[y_{l_j} - f_l(\mathbf{x}_j, \mathbf{k})][y_{r_j} - f_r(\mathbf{x}_j, \mathbf{k})][\mathbf{V}_j^{-1}]_{lr} \tag{7.1.15}$$

where $[\mathbf{V}_j^{-1}]_{lr}$ is the l, r-element of $\mathbf{V}_j^{-1}$. However, by comparing (7.1.15) to (7.1.7) we see that if $q_{lr_j} = [\mathbf{V}_j^{-1}]_{lr}$, then the least squares estimates are also the maximum likelihood estimates. If the covariance matrices $\mathbf{V}_j$ are identical for all experiments and equal to $\mathbf{V}$, then (7.1.14) becomes

$$S_{ML}(\mathbf{k}, \mathbf{V}) = -\frac{mM}{2}\log 2\pi - \frac{M}{2}\log|\mathbf{V}| - \frac{1}{2}\sum_{j=1}^{M}\boldsymbol{\epsilon}_j^T\mathbf{V}^{-1}\boldsymbol{\epsilon}_j \tag{7.1.16}$$

Consider first the case where all errors are independent. This is equivalent to saying that the matrix $\mathbf{V}$ is diagonal. Let v_i be the ith diagonal element of $\mathbf{V}$. Then (7.1.16) becomes

$$S_{ML}(\mathbf{k}, \mathbf{V}) = -\frac{mM}{2}\log 2\pi - \frac{M}{2}\sum_{i=1}^{m}\log v_i - \frac{1}{2}\sum_{j=1}^{M}\sum_{i=1}^{m}\epsilon_{ji}^2 v_i^{-1} \tag{7.1.17}$$

If the v_i are known, we minimize

$$\sum_{j=1}^{M}\sum_{i=1}^{m}\epsilon_{ji}^2 v_i^{-1}$$

If the v_i are unknown, we maximize (7.1.17) first with respect to the v_i by setting $(\partial S/\partial v_i) = 0$. In this case we find

$$v_i = \frac{1}{M}\sum_{j=1}^{M}\epsilon_{ji}^2 \tag{7.1.18}$$

Substituting (7.1.18) into (7.1.17),

$$S_{ML}(\mathbf{k}) = -\frac{mM}{2}\left(1 + \log\frac{2\pi}{M}\right) - \frac{M}{2}\sum_{i=1}^{m}\log\sum_{j=1}^{M}\epsilon_{ji}^2 \tag{7.1.19}$$

Then maximizing (7.1.19) is equivalent to minimizing

$$S(\mathbf{k}) = \sum_{i=1}^{m}\log\sum_{j=1}^{M}\epsilon_{ji}^2 \tag{7.1.20}$$

This may be regarded as a weighted least squares problem with unknown weights. Once the maximizing values of $\mathbf{k}$ have been found, the unknown weights are determined from (7.1.18).

3. *Bayesian estimation.* It occurs frequently that even before we start estimating the parameters from current data, some information concerning the values of the parameters is available from previous experiments or from general physical considerations. It is frequently possible to summarize this a priori information in the form of a relative probability density function $p_0(\mathbf{k})$. If, e.g., $p_0(\mathbf{k}_1)/p_0(\mathbf{k}_2) = 10$, this is interpreted to mean that we think $\mathbf{k} = \mathbf{k}_1$ to be 10 times more likely than $\mathbf{k} = \mathbf{k}_2$. We refer to $p_0(\mathbf{k})$ as the prior distribution. The following are typical examples:

a. A reaction rate constant must be positive. Hence $p_0(k) = 0$ for $k \leq 0$. If we have no further information on k, we set $p_0(k) = c$ (a positive constant whose value is immaterial) for $k > 0$. Computational convenience often requires that an upper bound α be prescribed for k. In this case $p_0(k) = 0$ also for $k > \alpha$.
b. The ratio K of forward and reverse reaction rate constants may have been estimated as being $K_0 \pm \sigma$ from measurements of equilibrium concentrations. It is then reasonable to use the normal prior distribution:

$$p_0(k) = \frac{1}{\sigma\sqrt{2\pi}} \exp\left[-\frac{(K - K_0)^2}{2\sigma^2}\right]$$

When we come to estimate the parameters on the basis of the new data, we take the prior information into account by multiplying the likelihood function by the prior distribution. In terms of the logarithms, we maximize

$$S_B(\mathbf{k}, \boldsymbol{\psi}) = \log p[\mathbf{y} - \mathbf{f}(\mathbf{x}, \mathbf{k}), \boldsymbol{\psi}] + \log p_0(\mathbf{k}) \tag{7.1.21}$$

In case (a.) above, $\log p_0$ is negatively infinite outside the prescribed bounds. Maximizing (7.1.21) then consists of finding the maximum of $\log p$ subject to the constraints; e.g., $0 \leq k \leq \alpha$.

Sequential estimation, i.e., the reestimation of the parameters after the results of each in a series of experiments become available, is an important application of the Bayesian technique. Suppose that after N experiments the parameters are estimated to be $\mathbf{k} = \mathbf{k}_N$, with covariance matrix $\mathbf{V}_N$. The relevant information contained in the data from these experiments may be summarized in the posterior distribution $p_N(\mathbf{k}) \approx c \exp[-\frac{1}{2}(\mathbf{k} - \mathbf{k}_N)^T \mathbf{V}_N^{-1}(\mathbf{k} - \mathbf{k}_N)]$, where c is an irrelevant constant. When results of subsequent experiments become available, the likelihood function is constructed from these alone, and $p_N(\mathbf{k})$ is used as the prior distribution.

One cannot overstate the importance of using all available prior information, in the form of either constraints on the parameters or of prior densities. Use of such information frequently spells the difference between convergence and nonconvergence of the estimation procedure.

7.1.2 Differential equation models

1. *Ordinary differential equations.* Let us consider the problem of estimating a set of p constant parameters, $k_1, \ldots, k_p$, in a set of n-coupled O.D.E.s of the form

$$\frac{dx_i}{dt} = f_i(t, \mathbf{x}(t), \mathbf{k}), \qquad x_i(0) = x_{i0}, \qquad i = 1, 2, \ldots, n \tag{7.1.22}$$

with the observations given by the m-dimensional ($m \leq n$) vector $\mathbf{y}$ at R discrete values of t,

$$\mathbf{y}_r = \mathbf{h}(t_r, \mathbf{x}(t_r)) + \boldsymbol{\epsilon}_r, \qquad r = 1, 2, \ldots, R \tag{7.1.23}$$

or continuously in t,

$$\mathbf{y}(t) = \mathbf{h}(t, \mathbf{x}(t)) + \boldsymbol{\epsilon}(t) \tag{7.1.24}$$

For example, consider a batch reactor in which the reaction $A \xrightarrow{k_1} B \xrightarrow{k_2} C$ is being carried out. Let x_1 and x_2 denote the concentrations of A and B, respectively, so that

$$\frac{dx_1}{dt} = -k_1 x_1$$

$$\frac{dx_2}{dt} = k_1 x_1 - k_2 x_2$$

Say that we measure the concentration of B only. These measurements are $y_1, y_2, \ldots, y_R$ and they are clearly related to the states by

$$y_r = x_2(t_r) + \text{measurement errors}$$

so that $\mathbf{h}(t_r, \mathbf{x}(t_r)) = x_2(t_r)$.

The simplest case of this parameter estimation problem results if all the state variables $x_i(t)$ and their derivatives $\dot{x}_i(t)$ are measured directly. Then we need only consider (7.1.22) and the estimation problem involves only n algebraic equations. For example, if the system (7.1.22) represents a set of reaction rate equations, then measurement of the derivatives $\dot{x}_i$ of all the concentrations might be possible in a differential flow reactor. If the derivatives are not available by direct measurement, they could in theory be obtained by numerical differentiation of the x vs. t data. However, this procedure is highly inaccurate and error-prone even with the smoothest of data. Thus unless the derivatives of the state variables are measured *directly*, the left-hand side of (7.1.22) must *not* be approximated from the x vs. t data.

The estimation of parameters in O.D.E.s is important, for example, in the determination of kinetic parameters from batch and integral reactor data, the determination of parameters associated with missile and spacecraft trajectories from tracking data, and the determination of parameters characterizing transfer coefficients in the human body. In short, whenever we must analyze experimental data in terms of an O.D.E. model, parameter estimation will be important.

In this chapter we shall assume that the parameters **k** are time-invariant and that the initial conditions in (7.1.22) are known exactly. Our interest is again in estimating **k** such that the model output $\mathbf{h}(t, \mathbf{x})$ matches the data **y** in some optimal way. Since we assume that no disturbances enter into (7.1.22) and that the x_{i0} are known precisely, we shall not be concerned with *state* estimation. This is a vast subject which would take us somewhat afield. In addition, we shall consider only initial value problems of the form (7.1.22).

2. *Partial differential equations*. We shall now turn to the estimation of constant parameters in P.D.E.s. Two factors tend to complicate the P.D.E. parameter estimation problem as opposed to the O.D.E. problem. First, parameters may enter into the boundary conditions as well as the equations themselves. Such a case occurs in estimating heat and mass transfer coefficients on the boundary of a distributed medium. Second, the measurements of the process are functions of both space and time. Thus the experimental data can be values of the state at a certain location or integrated values of the state over some or all of the spatial domain of the process. The first case is exemplified by measurements of the temperature of a system at various points in the system, while the second case might be represented by a measurement of the average temperature over some region of the system.

Consider a distributed parameter system whose state at any time t can be specified by a set of n functions $u_i(t, x)$, where the scalar variable x assumes values from a spatial domain Ω. We assume that if there exist manipulatable input variables to the system (e.g., controls), distributed either wholly or partially on Ω, they are completely known.

We desire to obtain a mathematical model of the system. We assume that the model consists of a set of partial differential equations of known form but containing a number of unknown parameters. The method we shall use to determine those parameters is based on using the difference between the solution of the model equations and the actual measured system response to continually improve on an initial guess of the unknown parameters.

Consider the class of parabolic or hyperbolic systems whose behavior can be described by the set of nonlinear partial differential equations and initial conditions

$$\frac{\partial u_i(t, x)}{\partial t} = f_i\left(t, x, \mathbf{u}(t, x), \frac{\partial \mathbf{u}(t, x)}{\partial x}, \frac{\partial^2 \mathbf{u}(t, x)}{\partial x^2}, \mathbf{k}\right) \qquad (7.1.25)$$

$$u_i(0, x) = u_{i0}(x), \qquad i = 1, 2, \ldots, n$$

where **k** is a p-dimensional vector of unknown parameters. The boundary conditions can be written in the general form

$$g_i\left(t, \mathbf{u}(t, x), \frac{\partial \mathbf{u}}{\partial x}, \mathbf{k}\right) = 0, \qquad x \in \partial\Omega, \qquad i = 1, 2, \ldots, n \qquad (7.1.26)$$

where $\partial\Omega$ denotes the boundary of the region Ω.

In actual situations the state variables $u_i(t, x)$ may not be directly measurable; instead, only certain prescribed functions of $u_i(t, x)$ are obtained. We shall call these measured variables the output of the system and denote them by $y_j, j = 1, 2, \ldots, m \leq n$. The observed output variables can be divided into four classes, dependent on the nature of the transformation from the state $u(t, x)$ to the output y_j:

a. Time- and spatially dependent output transformation. The observed variables y_j depend on both t and x; for example,

$$y_j(t, x) = h_j(t, x, \mathbf{u}(t, x)) + \text{errors}, \qquad j = 1, 2, \ldots, m_1 \tag{7.1.27}$$

b. Time-dependent output transformation. The observed variables y_j depend on t only; for example,

$$y_j(t) = \int_\Omega h_j(t, x, \mathbf{u}(t, x))\, d\Omega + \text{errors}, \qquad j = m_1 + 1, \ldots, m_2 \tag{7.1.28}$$

c. Spatially dependent output transformation. The observed variables y_j depend on x only; for example,

$$y_j(x) = \int_0^T h_j(t, x, \mathbf{u}(t, x))\, dt + \text{errors}, \qquad j = m_2 + 1, \ldots, m_3 \tag{7.1.29}$$

d. Time- and spatially independent output transformation. The observed variables y_j are independent of both t and x; for example,

$$y_j = \int_0^T \int_\Omega h_j(t, x, \mathbf{u}(t, x))\, d\Omega\, dt + \text{errors}, \qquad j = m_3 + 1, \ldots, m \tag{7.1.30}$$

Corresponding to the output forms a–d, we can define the corresponding errors, measuring the deviation between prediction and data:

$$e_j(t, x) = y_j(t, x) - h_j(t, x, \mathbf{u}(t, x)), \qquad j = 1, \ldots, m_1 \tag{7.1.31}$$

$$e_j(t) = y_j(t) - \int_\Omega h_j(t, x, \mathbf{u}(t, x))\, d\Omega, \qquad j = m_1 + 1, \ldots, m_2 \tag{7.1.32}$$

$$e_j(x) = y_j(x) - \int_0^T h_j(t, x, \mathbf{u}(t, x))\, dt, \qquad j = m_2 + 1, \ldots, m_3 \tag{7.1.33}$$

$$e_j = y_j - \int_0^T \int_\Omega h_j(t, x, \mathbf{u}(t, x))\, d\Omega\, dt, \qquad j = m_3 + 1, \ldots, m \tag{7.1.34}$$

where y_j are the observed output variables and $\mathbf{u}(t, x)$ is the solution of (7.1.25) and (7.1.26) corresponding to the value of $\mathbf{k}$ we have assigned.

Two forms of error functional are possible, an instantaneous error criterion defined at a definite spatial location and/or time and an integral average error criterion defined over all or part of $\mathbf{\Omega}, \times [0, T]$. To be specific, an instantaneous error criterion evaluated at $t = T$ and $x = 0$, for example, takes the forms corresponding to (7.1.31)–(7.1.33), respectively:

$$S = \sum_{i=1}^{m_1} \sum_{j=1}^{m_1} e_i(T, 0) q_{ij} e_j(T, 0) \tag{7.1.31a}$$

$$S = \sum_{i=m_1+1}^{m_2} \sum_{j=m_1+1}^{m_2} e_i(T) q_{ij} e_j(T) \tag{7.1.32a}$$

$$S = \sum_{i=m_2+1}^{m_3} \sum_{j=m_2+1}^{m_3} e_i(0) q_{ij}(0) e_j(0) \tag{7.1.33a}$$

where the q_{ij} are the elements of a weighting matrix $\mathbf{Q}$. Integral average criteria based on (7.1.31)–(7.1.33) take the form

$$S = \int_0^T \int_\Omega \sum_{i=1}^{m_1} \sum_{j=1}^{m_1} e_i(t, x) q_{ij} e_j(t, x)\, d\Omega\, dt \tag{7.1.31b}$$

$$S = \int_0^T \sum_{i=m_1+1}^{m_2} \sum_{j=m_1+1}^{m_2} e_i(t) q_{ij} e_j(t)\, dt \tag{7.1.32b}$$

$$S = \int_\Omega \sum_{i=m_2+1}^{m_3} \sum_{j=m_2+1}^{m_3} e_i(x) q_{ij} e_j(x)\, d\Omega \tag{7.1.33b}$$

If the measurements are of the type in (7.1.31), they may be considered to be carried out at R values of t, $t_1, t_2, \ldots, t_R$, and at S' spatial locations, $x_1, x_2, \ldots, x_{S'}$. Then the measurements are related to the state by the discrete form of (7.1.31):

$$y_{r,s} = h(t_r, x_s, u(t_r, x_s)) + \text{errors}, \qquad r = 1, 2, \ldots, R, \qquad s = 1, 2, \ldots, S' \tag{7.1.35}$$

In this case the least squares criterion is

$$S = \sum_{r=1}^{R} \sum_{s=1}^{S'} [\mathbf{y}_{r,s} - \mathbf{h}(t_r, x_s, \mathbf{u}(t_r, x_s))]^T \mathbf{Q}_{r,s} \cdot [\mathbf{y}_{r,s} - \mathbf{h}(t_r, x_s, \mathbf{u}(t_r, x_s))] \tag{7.1.36}$$

The problem is to find the global minimum of $S(\mathbf{k})$ in the parameter space $\mathbf{k}$.

A rather special situation arises when the differential equations (either O.D.E.s or P.D.E.s) are linear. Then either the analytical solution or the transfer function is available. If the analytical solution is available, then the parameter estimation problem reduces to one involving an algebraic model. If the transfer function is available, we can estimate the parameters using the transfer function and the data which have been transformed into the s-domain. As a final point in this section, we should consider the correspondence between time and frequency domain parameter estimation.

7.1.3 Time domain vs. frequency domain estimation for linear models

The general form of an error criterion in the time domain (assuming that we start collecting data at $t = 0$) is

$$S = \int_0^\infty G[y(t) - h(t, \mathbf{x}(t; \mathbf{k}))] q(t)\, dt \tag{7.1.37}$$

If the data are available only at R discrete times, then S can be written as

$$S = \sum_{i=1}^{R} G[y(t_i) - h(t_i, \mathbf{x}(t_i; \mathbf{k}))]q_i \tag{7.1.38}$$

For example, an error criterion based on matching the rth noncentral moments of a model and data takes the form (for continuous data)

$$S = \int_0^\infty [y(t) - x(t; \mathbf{k})]t^r \, dt \tag{7.1.39}$$

where, in this case, $q(t) = t^r$. This particular error criterion gives substantial influence to large times, which results in fitting the tail of the response at the expense of other regions of the output. Since the tail of an impulse response often presents problems in accurate concentration measurement, this is a drawback in the use of moment matching for parameter estimation.

The two error criteria given above involve the time domain model output and data. We may, however, want to carry out the matching in the transform domain, particularly if

1. The transfer function is too complicated to be inverted to the time domain pulse response, or
2. The experimental data are from frequency response testing, that is, gain and phase angle vs. frequency.

We have two options in formulating a transform domain error criterion. First, we can take a time domain error function of the form (7.1.37) and convert it to the transform domain. Second, we can formulate a transform criterion directly. Let us consider these two options.

Let us determine the frequency domain form of what is perhaps the most useful and popular form of (7.1.37), the unweighted least squares criterion:

$$S = \int_0^\infty [y(t) - x(t; \mathbf{k})]^2 \, dt \tag{7.1.40}$$

This can easily be carried out with the aid of Parseval's theorem (5.4.26). If we let $F(t) = y(t) - x(t; \mathbf{k})$ in (5.4.23), then (5.4.26) is

$$S = \frac{1}{2\pi} \int_0^\infty [\bar{y}(\omega) - \bar{x}(\omega; \mathbf{k})]^2 \, d\omega \tag{7.1.41}$$

It is convenient to write (7.1.41) in the form

$$S = \frac{1}{2\pi} \int_0^\infty \{(\Delta R)^2 + (\Delta I)^2\} \, d\omega \tag{7.1.42}$$

where

$$\begin{aligned} \Delta R &= \mathrm{Re}[\bar{y}(\omega)] - \mathrm{Re}[\bar{x}(\omega; \mathbf{k})] \\ \Delta I &= \mathrm{Im}[\bar{y}(\omega)] - \mathrm{Im}[\bar{x}(\omega; \mathbf{k})] \end{aligned} \tag{7.1.43}$$

Unfortunately, only the simple unweighted least squares criterion has a readily determinable frequency domain counterpart. Thus, minimization of

S with respect to $\mathbf{k}$ should yield identical results using either (7.1.40) or (7.1.41).

The second option is specification of a transform domain error criterion. Consider the integral absolute error in the s-domain:

$$S = \int_{s_1}^{s_2} |\bar{y}(s) - \bar{x}(s; \mathbf{k})| \, ds \tag{7.1.44}$$

We now want to determine the time domain error criterion corresponding to (7.1.44). This can be done only approximately. Let $\epsilon(t) = y(t) - x(t; \mathbf{k})$. We can write (7.1.44) as [1]

$$S = \int_{s_1}^{s_2} \left| \int_0^\infty e^{-st}\epsilon(t) \, dt \right| ds \tag{7.1.45}$$

Assuming that $\epsilon(t) = 0$ for $t > T$ and approximating the inner integral by the sum of M terms,

$$S = \int_{s_1}^{s_2} \left| \sum_{i=1}^{M} e^{-st_i}\epsilon(t_i) \frac{T}{M} \right| ds \tag{7.1.46}$$

Since $\epsilon(t)$ may be positive or negative,

$$\begin{aligned} S &\leq \int_{s_1}^{s_2} \frac{T}{M} \sum_{i=1}^{M} e^{-st_i} |\epsilon(t_i)| \, ds \\ &\leq \frac{T}{M} \sum_{i=1}^{M} \int_{s_1}^{s_2} e^{-st_i} \, ds \, |\epsilon(t_i)| \\ &\leq \frac{T}{M} \sum_{i=1}^{M} \frac{|\epsilon(t_i)|}{t_i} (e^{-s_2t_i} - e^{-s_1t_i}) \\ &\leq \int_0^T |\epsilon(t)| \, q(t) \, dt \end{aligned} \tag{7.1.47}$$

where

$$q(t) = \frac{e^{-s_1t} - e^{-s_2t}}{t} \tag{7.1.48}$$

Therefore we see that if the criterion (7.1.44) is used for parameter estimation, the process corresponds to fitting in the time domain using the weighting function (7.1.48) in which errors at small times are given the greatest weight.

Consider the integral square error criterion in the s-domain:

$$S = \int_0^\infty [\bar{y}(s) - \bar{x}(s; \mathbf{k})]^2 \, ds \tag{7.1.49}$$

Note that (7.1.49) is more general than (7.1.41) since s may assume complex rather than only purely imaginary values. We write (7.1.49) as

$$\begin{aligned} S &= \int_0^\infty \left\{ \int_0^\infty \epsilon(t) e^{-st} \, dt \right\}^2 ds \\ &= \int_0^\infty \left\{ \frac{T}{M} \sum_{i=1}^{M} \epsilon(t_i) e^{-st_i} \right\}^2 ds \\ &= \left(\frac{T}{M} \right)^2 \sum_{i=1}^{M} \sum_{j=1}^{M} \frac{\epsilon(t_i)\epsilon(t_j)}{t_i + t_j} \end{aligned} \tag{7.1.50}$$

Again we see that the weighting of errors in the time domain is heavily biased toward small values of t. These two results are in contrast to the moment criterion (7.1.39), wherein the errors incurred at large values of t are the most heavily weighted.

We can summarize these time and frequency domain error criteria for continuous measurements as follows:

1. Matching of moments:

$$S = \int_0^\infty [y(t) - x(t;\mathbf{k})]t^r\,dt$$

2. Unweighted least squares in the time domain:

$$S = \int_0^\infty [y(t) - x(t;\mathbf{k})]^2\,dt$$

3. Frequency response (matching of transfer function and transformed data

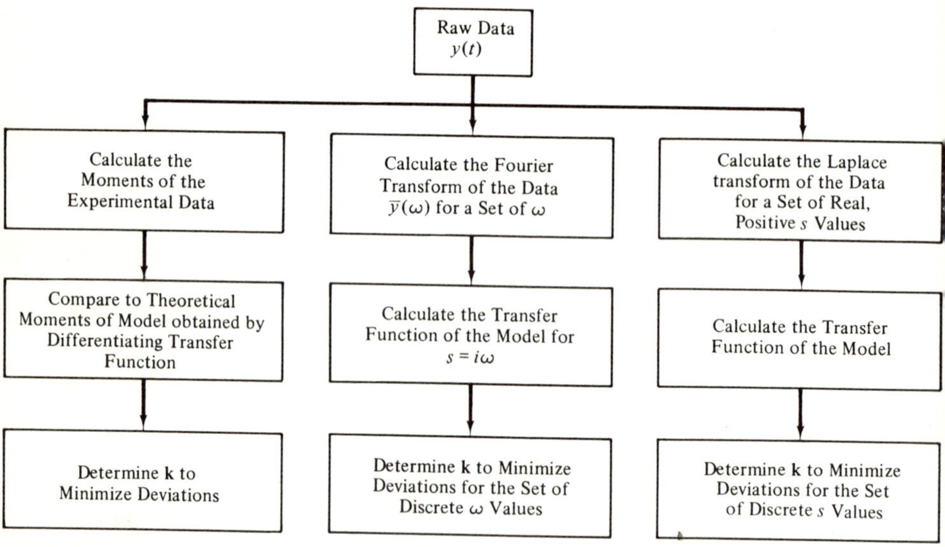

Figure 7.2. Comparison of moment matching, frequency response, and s-plane parameter estimation procedures. [Reprinted from J. L. Johnson, L. T. Fan, and Y. S. Wu, "Comparison of Moments, s-Plane, and Frequency Response Methods for Analyzing Pulse Testing Data from Flow Systems," *Ind. Eng. Chem. Fund. 10*, 425 (1971). Reprinted by permission of the copyright owner.]

for $s = i\omega$):

$$S = \frac{1}{2\pi}\int_0^\infty \{[\mathrm{Re}(\bar{y}(\omega)) - \mathrm{Re}(\bar{x}(\omega; \mathbf{k}))]^2 + [\mathrm{Im}(\bar{y}(\omega)) - \mathrm{Im}(\bar{x}(\omega; \mathbf{k}))]^2\}\, d\omega$$

4. Integral absolute error in the transform domain:

$$S = \int_{s_1}^{s_2} |\bar{y}(s) - \bar{x}(s; \mathbf{k})|\, ds$$

5. Integral square error in the transform domain:

$$S = \int_0^\infty [\bar{y}(s) - \bar{x}(s; \mathbf{k})]^2\, ds$$

Most commonly used for linear models are criteria 1, 2, 3, and 5. When carrying out estimation in the s-domain, it is important to remember that s is a complex number. Very seldom, however, do we use values of s containing both real and imaginary components. We usually restrict s either to purely imaginary values (frequency response) or to real, positive values. Thus criteria 4 and 5 are almost always employed with s equal to real, positive values. Actually in practice we use only a discrete set of s-values, and so the integrations above are replaced by summations.

The procedures involved in moment matching, frequency response, and s-plane fitting are shown in Figure 7.2. We note finally that since $\bar{x}(\omega; \mathbf{k})$ or $\bar{x}(s; \mathbf{k})$ is just an algebraic function of $\mathbf{k}$, the actual estimation procedure will be one from the general class of those for algebraic models.

7.2 Parameter Estimation Using Moments

As our first technique for dealing with parameter estimation in linear models we shall consider the matching of the moments of the model output to the like moments of the experimental data. In most cases the process model is deterministic, although, as we have already seen, we can define the moments of a deterministic variable in precisely the same way as a stochastic variable. Parameter estimation using moments has two desirable features:

1. Often a full, time domain analytical solution of the model cannot be obtained when a transfer function can. The transfer function readily yields the moments of the solution since the moments can be computed without the necessity of inverting the transform.
2. Even when an analytical, time domain solution can be obtained, the moments are much simpler functions of the parameters than the solution.

To illustrate parameter estimation using moments we shall concentrate on a problem that has received extensive attention, the use of dispersion models for representing packed bed operations.

We shall begin by presenting the moments for several forms of the axial-dispersed plug flow model (See Table 6.1). Consider a pulse injection of tracer in a packed bed. The tracer pulse spreads out as it is carried downstream from the source, the degree of spreading dependent on the nature of the bed. The parameter which characterizes the axial dispersion properties of the bed is the Peclet number, Pe, appearing in the model (6.3.12). Therefore we often want to estimate Pe from measurements of the amount of spreading of a tracer pulse at a fixed location downstream of the injection point. This estimation can be carried out very neatly using moments.

We define the rth noncentral moment of the dimensionless tracer concentration $C(z, \theta)$ at any location z by

$$\mu_r' = \int_0^\infty \theta^r C(z, \theta)\, d\theta \tag{7.2.1}$$

Given $C(z, \theta)$ from the analytical solution of (6.3.12) subject to appropriate boundary conditions, we can compute μ_r' directly from (7.2.1). However, because it is often far easier to compute simply the L.T. of $C(z, \theta)$, the most efficient way of computing these moments is by

$$\mu_r' = (-1)^r \frac{\partial^r \bar{C}(z, s)}{\partial s^r}\bigg|_{s=0} \tag{7.2.2}$$

When two measurements are made downstream from the tracer injection, we are interested in the differences between the moments at the two locations:

$$\begin{aligned} \Delta\mu_r' &= \mu_{r_2}' - \mu_{r_1}' \\ \Delta\mu_r &= \mu_{r_2} - \mu_{r_1} \end{aligned} \tag{7.2.3}$$

The moments corresponding to some of the experimental schemes in Table 6.1 are presented in Table 7.1. For more detailed treatments of the moments of dispersion models corresponding to different experimental configurations, we refer the reader to Levenspiel and Bischoff [34]. We notice that the moments differ depending on the experimental setup, the particular details of which enter as boundary conditions to (6.3.12). We also notice that if it is possible to make measurements far enough away from the ends of the bed so that an infinite system can be assumed, the moments simplify immensely. Quantitative assessments of the distance from the end that one may safely neglect end effects are presented by Levenspiel and Bischoff [34].

Let us now proceed to parameter estimation using moments in several systems. Our line of attack will be as follows. First, we shall compute the moments from the transfer function. Then we shall compute the moments from the data by means of numerical quadrature. Finally, we shall compare the two to estimate parameters in the theoretical expression.

TABLE 7.1
MOMENTS OF EXPERIMENTAL SCHEMES USED IN RELATION TO THE AXIAL-DISPERSED PLUG FLOW MODEL [34]

Experimental scheme	Mean and variance
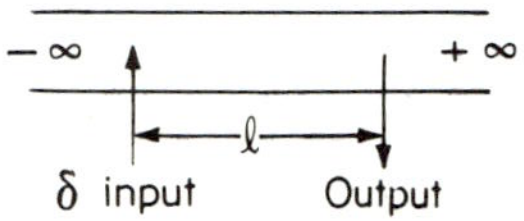	$\mu'_1 = 1 + \frac{2}{\mathrm{Pe}}$ $\sigma^2 = \frac{2}{\mathrm{Pe}} + \frac{8}{\mathrm{Pe}^2}$
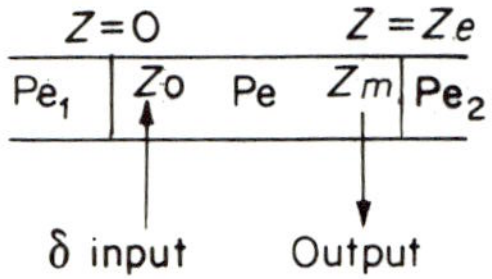	$\mu'_1 = 1 + \frac{1}{\mathrm{Pe}}\left[2 - \left(1 - \frac{\mathrm{Pe}}{\mathrm{Pe}_1}\right)e^{-\mathrm{Pe}Z_0} - \left(1 - \frac{\mathrm{Pe}}{\mathrm{Pe}_2}\right)e^{-\mathrm{Pe}(Z_e - Z_m)}\right]$ $\sigma^2 = \frac{2}{\mathrm{Pe}} + \frac{1}{\mathrm{Pe}^2}\left\{8 + 2\left(1 - \frac{\mathrm{Pe}}{\mathrm{Pe}_1}\right)\left(1 - \frac{\mathrm{Pe}}{\mathrm{Pe}_2}\right)e^{-\mathrm{Pe}Z_e} - \left(1 - \frac{\mathrm{Pe}}{\mathrm{Pe}_1}\right)e^{-\mathrm{Pe}Z_0}\left[4Z_0\mathrm{Pe} + 4\left(1 + \frac{\mathrm{Pe}}{\mathrm{Pe}_1}\right)\right.\right.$ $\left. + \left(1 - \frac{\mathrm{Pe}}{\mathrm{Pe}_1}\right)e^{-\mathrm{Pe}Z_0}\right] - \left(1 - \frac{\mathrm{Pe}}{\mathrm{Pe}_2}\right)e^{-\mathrm{Pe}(Z_e - Z_m)}\left[4(Z_e - Z_m)\mathrm{Pe} + 4\left(1 + \frac{\mathrm{Pe}}{\mathrm{Pe}_2}\right)\right.$ $\left.\left. + \left(1 - \frac{\mathrm{Pe}}{\mathrm{Pe}_2}\right)e^{-\mathrm{Pe})Z_e - Z_m)}\right]\right\}$
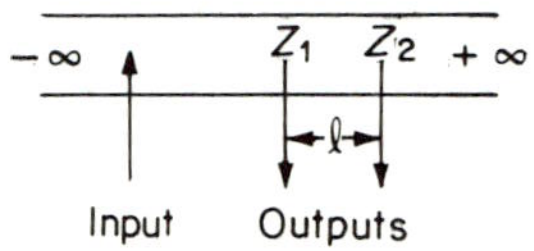	$\Delta\mu'_1 = \mu'_{1_2} - \mu'_{1_1} = 1$ $\Delta\sigma^2 = \sigma_2^2 - \sigma_1^2 = \frac{2}{\mathrm{Pe}}$
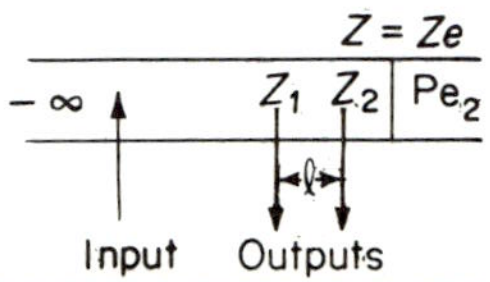	$\Delta\mu'_1 = 1 - \frac{1}{\mathrm{Pe}}\left(1 - \frac{\mathrm{Pe}}{\mathrm{Pe}_2}\right)(1 - e^{-\mathrm{Pe}})e^{\mathrm{Pe}(Z_2 - Z_e)}$ $\Delta\sigma^2 = \frac{2}{\mathrm{Pe}} + \frac{1}{\mathrm{Pe}}\left(1 - \frac{\mathrm{Pe}}{\mathrm{Pe}_2}\right)\exp[\mathrm{Pe}(Z_2 - Z_e)]\left\{4\left(1 + \frac{\mathrm{Pe}}{\mathrm{Pe}_2}\right)(e^{-\mathrm{Pe}} - 1) + 4\mathrm{Pe}(Z_2 - Z_e)\right.$ $\left. + \left(1 - \frac{\mathrm{Pe}}{\mathrm{Pe}_2}\right)(e^{-2\mathrm{Pe}} - 1)e^{\mathrm{Pe}(Z_2 - Z_e)} + 4\mathrm{Pe}(Z_e - Z_1)e^{-\mathrm{Pe}}\right\}$

7.2.1 Axial dispersion and adsorption in a packed bed

Let us investigate the exchange of mass or energy between a moving fluid stream and the solid particles in a packed bed. The following processes can take place in a packed bed:

1. Axial mixing or dispersion in the flowing fluid.
2. Radial mixing or dispersion in the flowing fluid.
3. Mass or energy transport from the flowing fluid to the packing.
4. Mass or energy transport from the flowing fluid to stagnant fluid pockets.
5. Mass or energy transport in the interior of the packing.
6. Chemical reactions in the fluid phase or in the packing.

Depending on the nature of the bed, the fluid, and the packing, any or all of these processes may be important.

In this section we shall consider the process of adsorption of a species onto the particles in a packed bed. Note that this system has been treated in Section 3.6. Of the above processes we shall neglect any radial gradients of concentration or fluid velocity, any gradients of concentration of the adsorbed species in the packing, and chemical reactions. Also, we shall assume an isothermal system, so that mass transfer is the only process of interest. Let

$c(x, t)$ = concentration of the adsorbing species in the fluid at axial location x and time t

$n(x, t)$ = concentration of the adsorbing species in the packing at axial location x and time t (uniform throughout a particle)

$c^*(x, t)$ = the fluid phase adsorbent concentration at the fluid-solid interface which is in equilibrium with the adsorbent concentration $n(x, t)$ on the solid

v = fluid velocity through the bed

D_L = effective axial turbulent eddy diffusivity of the adsorbing species (which plays the same role as a tracer) in the inert carrier fluid

k_f = mass transfer coefficient for the adsorbent between the fluid phase and the surface of the solid particles, based on the difference between $c(x, t)$ and $c^*(x, t)$

ϵ = void fraction of the packing

Assuming that the adsorbent concentration $n(x, t)$ is uniform throughout the packing, that is, the diffusivity of adsorbent within the particle is very large, the differential mass balance for the adsorbing species is

$$\frac{\partial c(x, t)}{\partial t} = D_L \frac{\partial^2 c(x, t)}{\partial x^2} - v\frac{\partial c(x, t)}{\partial x} - \frac{1}{\epsilon}\frac{\partial n(x, t)}{\partial t} \tag{7.2.4}$$

$$\frac{1}{\epsilon}\frac{\partial n(x, t)}{\partial t} = k_f[c(x, t) - c^*(x, t)] \tag{7.2.5}$$

The physical situation to which (7.2.4) and (7.2.5) correspond is shown in Figure 7.3. We note in (7.2.4) that the accumulation of adsorbent in the packing represents a sink for adsorbent in the fluid. The driving force for mass transfer of adsorbent from the fluid to the solid is the difference between the actual fluid concentration $c(x, t)$ and the equilibrium concentration $c^*(x, t)$ corresponding to the adsorbent concentration in the packing $n(x, t)$.

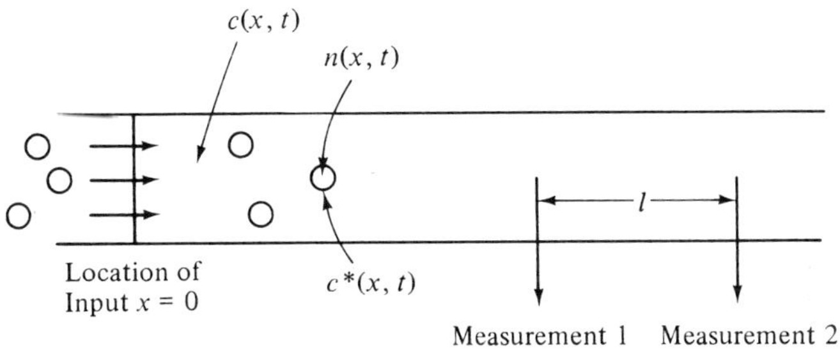

Figure 7.3. Packed bed adsorption.

Next we shall make a key assumption, namely that the equilibrium concentration c^* is *linearly* related to n:

$$n(x, t) = k_1 c^*(x, t) \tag{7.2.6}$$

The substitution of (7.2.6) into (7.2.5) makes (7.2.4) and (7.2.5) a closed set of equations for c and n.

As an initial condition, we shall assume that there is no adsorbent in the bed at $t = 0$:

$$c(x, 0) = n(x, 0) = 0 \tag{7.2.7}$$

We assume that the bed extends to infinity in both directions with measurements of $c(x, t)$ made at two locations separated by a distance l.

Neglecting diffusion across the bed entrance, we simply specify a general time-varying input at $z = 0$:

$$c(0, t) = f(t) \tag{7.2.8}$$

Defining reference concentrations c_0 and n_0, we can let

$$\begin{aligned} C &= \frac{c}{c_0} & N &= \frac{n}{n_0} \\ z &= \frac{x}{l} & \theta &= \frac{vt}{l} \\ Pe &= \frac{vl}{D_L} & K &= \frac{v}{k_f l} \\ \beta &= \frac{n_0}{c_0} \end{aligned} \tag{7.2.9}$$

Note that the Peclet number is defined on the basis of the length between measurements l rather than the packing size. This is done because if the concentration of adsorbent is uniform throughout each solid particle, the particle size is not important, only the void fraction ϵ.

Using (7.2.9), (7.2.4)–(7.2.7) become

$$\frac{\partial C(z,\theta)}{\partial \theta} = \frac{1}{Pe}\frac{\partial^2 C(z,\theta)}{\partial z^2} - \frac{\partial C(z,\theta)}{\partial z} - \frac{\beta}{\epsilon}\frac{\partial N(z,\theta)}{\partial \theta} \tag{7.2.10}$$

$$\frac{\beta}{\epsilon}\frac{\partial N(z,\theta)}{\partial \theta} = \frac{1}{K}\left[C(z,\theta) - \frac{\beta}{k_1}N(z,\theta)\right] \tag{7.2.11}$$

$$C(z,0) = N(z,0) = 0 \tag{7.2.12}$$

$$C(0,\theta) = F(\theta) \tag{7.2.13}$$

which are the dimensionless equations for adsorption of a single solute in a packed bed.

Let us apply the L.T. using the variable θ. We obtain

$$\frac{1}{Pe}\frac{d^2\bar{C}}{dz^2} - \frac{d\bar{C}}{dz} - s\bar{C} - \frac{\beta}{\epsilon}s\bar{N} = 0 \tag{7.2.14}$$

$$\frac{\beta}{\epsilon}s\bar{N} = \frac{1}{K}\left[\bar{C} - \frac{\beta}{k_1}\bar{N}\right] \tag{7.2.15}$$

Solving (7.2.15) for $\bar{N}$ and substituting into (7.2.14), we obtain

$$\frac{1}{Pe}\frac{d^2\bar{C}}{dz^2} - \frac{d\bar{C}}{dy} - h\bar{C} = 0 \tag{7.2.16}$$

where

$$h = s\left[\frac{sK + (\epsilon/k_1) + 1}{sK + (\epsilon/k_1)}\right] \tag{7.2.17}$$

Let us solve (7.2.16) for an impulse input, $F(\theta) = \delta(\theta)$. The solution then is

$$\bar{C}(z,s) = e^{Pe/2}e^{-[(Pe^2/4)+hPe]^{1/2}z} \tag{7.2.18}$$

Since we have employed an impulse input, $\bar{C}(z, s)$ is simply the transfer function of the system. Thus we can compute the moments of $C(z,\theta)$ by differentiating $\bar{C}(z, s)$. This yields for the mean, variance, and skewness

$$\begin{aligned}
\mu_1' &= 1 + \frac{k_1}{\epsilon} \\
\sigma^2 &= \frac{2}{\mathrm{Pe}}\left(1 + \frac{k_1}{\epsilon}\right)^2 + 2\left(\frac{k_1}{\epsilon}\right)^2 K \\
\mu_3 &= \frac{12}{\mathrm{Pe}^2}\left(1 + \frac{k_1}{\epsilon}\right)^3 + \frac{12}{\mathrm{Pe}}\left(\frac{k_1}{\epsilon}\right)^2\left(1 + \frac{k_1}{\epsilon}\right)K + 6\left(\frac{k_1}{\epsilon}\right)^3 K^2
\end{aligned} \tag{7.2.19}$$

Simplified versions of these moments may be obtained. If, for example, there is no adsorption on the packing and thus $k_1 = 0$, these reduce to the values for the axial-dispersed plug flow model:

$$\mu_1' = 1, \qquad \sigma^2 = \frac{2}{\text{Pe}}, \qquad \mu_3 = \frac{12}{\text{Pe}^2} \tag{7.2.20}$$

If, on the other hand, no axial dispersion occurs, $D_L = 0$ or $\text{Pe} = \infty$, the moments become

$$\begin{aligned} \mu_1' &= 1 + \frac{k_1}{\epsilon} \\ \sigma^2 &= 2\left(\frac{k_1}{\epsilon}\right)^2 K \\ \mu_3 &= 6\left(\frac{k_1}{\epsilon}\right)^3 K^2 \end{aligned} \tag{7.2.21}$$

which are those for conventional chromatography, in which axial dispersion is often neglected.

Note that with adsorption occurring, the mean residence time μ_1' increases over that when no adsorption takes place. In fact, the increase is k_1/ϵ, regardless of the amount of axial dispersion. Also, the variance and skewness are larger when adsorption occurs. These effects can, of course, be attributed to the retention of solute on the solid.

The problem of heat transfer from a fluid of a single component to the solid packing of a bed can be shown to be mathematically identical to the above problem of adsorption with a linear equilibrium relation. For an analysis of this problem, we refer the reader to Jeffreson [25].

EXAMPLE 7.1. *Transport to a Stagnant Fluid.* Consider the flow of fluid through a packed bed in which stagnant pockets of fluid act as dead spaces. We want to develop the transient mass balance equations and moments for a tracer material injected into the fluid. We neglect any adsorption of tracer onto the packing, and thus the packing itself does not enter into consideration. However, we must take into account transfer of tracer between the flowing and stagnant fluid. A situation of this type often occurs in trickle flow packed bed systems.

In particular, Van Swaaij et al. [49], Villermaux and Van Swaaij [50], and Bennett and Goodridge [6] have investigated this system. Using the notation that γ is the ratio of the volume of the moving stream to the total volume of the moving and stagnant streams and that $C(z, \theta)$ and $N(z, \theta)$ are the tracer concentrations in the moving and stagnant phases, respectively, the differential mass balances are

$$\gamma \frac{\partial C}{\partial \theta} = \frac{1}{Pe}\frac{\partial^2 C}{\partial z^2} - \frac{\partial C}{\partial z} - (1-\gamma)\frac{\partial N}{\partial \theta} \tag{A}$$

$$(1-\gamma)\frac{\partial N}{\partial \theta} = \nu(C - N) \tag{B}$$

where ν is a dimensionless mass transfer coefficient.

Let us consider a semi-infinite bed. The initial conditions are

$$C(z, 0) = N(z, 0) = 0 \tag{C}$$

with boundary conditions

$$C - \frac{1}{\text{Pe}}\frac{\partial C}{\partial z} = \delta(\theta), \qquad z = 0 \tag{D}$$

$$C \longrightarrow 0 \qquad z \longrightarrow \infty \tag{E}$$

The transfer function $\bar{C}(z, s)$ can be obtained in a straightforward manner, from which the mean and variance of $C(z, \theta)$ are found as

$$\begin{aligned} \mu'_1 &= 1 + \frac{1}{\text{Pe}} \\ \sigma^2 &= \frac{2}{\text{Pe}} + \frac{3}{\text{Pe}^2} + \frac{2(1-\gamma)^2}{\nu}\left[1 + \frac{1}{\text{Pe}}\right] \end{aligned} \tag{F}$$

As $\gamma \longrightarrow 1$, the stagnant phase disappears, leaving

$$\begin{aligned} \mu'_1 &= 1 + \frac{1}{\text{Pe}} \\ \sigma^2 &= \frac{2}{\text{Pe}} + \frac{3}{\text{Pe}^2} \end{aligned} \tag{G}$$

We shall now cite some reported experimental packed bed modeling studies. Chao and Hoelscher [9] reported experiments in which an impulse of hydrocarbon gas was passed through a bed of activated carbon. Two measuring probes were inserted within the bed, far enough away from the ends of the bed so that it could be considered as a doubly infinite system. The hydrocarbon was adsorbed on the packing, and so the appropriate model was (7.2.10) and (7.2.11). The object of the experiment was to determine the three unknown parameters, Pe, K, and k_1. Measurement of the mean residence time μ'_1 yielded k_1 from (7.2.19), assuming that ϵ is known. From the data, σ^2 and μ_3 were computed, and the second two equations of (7.2.19) were solved simultaneously for Pe and K.

The behavior of an air-water trickle bed reactor system was analyzed by Van Swaaij et al. [49]. A salt tracer was injected into the water stream at the bed entrance and the salt concentration in the bed was measured downstream by an electrical conductivity probe. Since the first measurement was at the bed entrance and the bed extended effectively to $+\infty$, the appropriate dispersion model was that of Example 7.1. The authors determined μ'_1 and σ^2 from the experimental salt concentration and determined Pe and ν from (F). It was assumed that γ was available from independent measurements.

Finally, we mention the work of Bennett and Goodridge [6], in which salt tracer was also used in a packed bed. However, in their experiments the downstream measuring device was placed in the exiting stream from the bed rather than within the column itself. The appropriate dispersion model is now the finite bed model rather than the infinite bed model.

7.2.2 Intraphase transport in a packed bed

In the previous section we considered external (fluid-to-particle) transport in a packed bed, where we assumed that the diffusivity of the particles for mass or heat transfer was infinite, thereby eliminating the need to consider intraparticle gradients. We shall now relax this assumption and include the possibility of intraparticle gradients.

We shall let

$c(x, t)$ = solute concentration in the flowing fluid at axial location x and time t
$q(r, x, t)$ = solute concentration in the fluid in the particle pores at radial location r, axial location x, and time t
$n(r, x, t)$ = solute concentration adsorbed on the solid at radial location r, axial location x, and time t
R_1 = radius of the packing, assumed to be identical, porous spheres; thus $0 \leq r \leq R_1$
β = internal porosity of the particles
ϵ = void fraction of the bed
D_L = effective axial turbulent eddy diffusivity of the solute
D_s = effective diffusivity of solute in the pores of the particles
v = fluid velocity through the bed

The transient material balances are

$$\frac{\partial c(x, t)}{\partial t} = D_L \frac{\partial^2 c(x, t)}{\partial x^2} - v \frac{\partial c(x, t)}{\partial x} + \frac{3D_s(1 - \epsilon)\beta}{\epsilon R_1} \left.\frac{\partial q}{\partial r}\right|_{r=R_1} \tag{7.2.22}$$

$$\frac{\partial q(r, x, t)}{\partial t} + \frac{\partial n(r, x, t)}{\partial t} = D_s\left[\frac{\partial^2 q(r, x, t)}{\partial r^2} + \frac{2}{r}\frac{\partial q(r, x, t)}{\partial r}\right] \tag{7.2.23}$$

$$\frac{\partial n(r, x, t)}{\partial t} = k_s(K_s q(r, x, t) - n(r, x, t)) \tag{7.2.24}$$

The first two relations, (7.2.22) and (7.2.23), represent the differential mass balances for the solute in the external and internal fluid, respectively. The last term in (7.2.22) represents the flux of solute between the external and internal fluids. Equation (7.2.24) states that the rate of accumulation of solute on the solid surfaces is equal to the difference between the actual adsorbed concentration n and the linear equilibrium adsorbed concentration, $K_s q$, multiplied by a mass transfer coefficient, k_s.

Thus the model consists of (7.2.22)–(7.2.24), with dependent variables $c(x, t)$, $q(r, x, t)$, and $n(r, x, t)$. We assume that the bed is doubly infinite, and so the ranges of the independent variables are

$$-\infty < x < \infty$$

$$t \geq 0$$

$$0 \leq r \leq R_1$$

We need initial conditions for $c(x, t)$, $q(r, x, t)$, and $n(r, x, t)$. We shall specify that the bed is initially free of solute:

$$c(x, 0) = 0 \tag{7.2.25}$$

$$q(r, x, 0) = n(r, x, 0) = 0 \tag{7.2.26}$$

We now require two boundary conditions on x for $c(x, t)$ and two boundary conditions on r for $q(r, x, t)$. Since the bed is doubly infinite,

$$c(x, t) \longrightarrow 0, \qquad x \longrightarrow \pm\infty \tag{7.2.27}$$

The symmetry of the spherical particle implies that

$$\left.\frac{\partial q}{\partial r}\right|_{r=0} = 0 \tag{7.2.28}$$

Finally, we shall specify that the flux of solute between the external and internal fluids is linearly related to the driving force, $c(x, t) - q(R_1, x, t)$, through a mass transfer coefficient k_f:

$$\frac{3D_s(1-\epsilon)\beta}{\epsilon R_1}\left.\frac{\partial q}{\partial r}\right|_{r=R_1} = k_f[c(x, t) - q(R_1, x, t)] \tag{7.2.29}$$

The only remaining point is the injection of tracer at $x = 0$. For the usual pulse injection at $x = 0$,

$$c(0, t) = \delta(t) \tag{7.2.30}$$

The full model is thus (7.2.22)–(7.2.30).

Even though the model is quite complicated, it is still linear and can be handled with the L.T. Therefore, taking the L.T. of (7.2.22)–(7.2.24), we obtain

$$D_L\frac{d^2\bar{c}}{dx^2} - v\frac{d\bar{c}}{dx} - s\bar{c} = -\frac{3D_s(1-\epsilon)\beta}{\epsilon R_1}\left.\frac{\partial \bar{q}}{\partial r}\right|_{r=R_1} \tag{7.2.31}$$

$$D_s\left[\frac{\partial^2\bar{q}}{\partial r^2} + \frac{2}{r}\frac{\partial \bar{q}}{\partial r}\right] = s\bar{q} + s\bar{n} \tag{7.2.32}$$

$$s\bar{n} = k_s[K_s\bar{q} - \bar{n}] \tag{7.2.33}$$

The L.T. of (7.2.27)–(7.2.29) follows directly.

Obtaining the inverse of $\bar{c}$, $\bar{q}$, and $\bar{n}$ would be an enormous task if possible at all. However, we can compute the moments of the response $c(l, t)$ after solving for $\bar{c}(x, s)$ and differentiating. Several studies have been concerned with this packed bed model, namely, Kucera [29], Grubner [17], Grubner and Underhill [18], and Smith and coworkers [40, 42, 43]. The distinctions among these studies lie in the particular physical configuration of the bed. The first three studies consider a doubly infinite system, while Smith and coworkers employ a singly infinite bed. From our previous experiences we know that these differences in experimental configuration lead to slight differences in the moments. The mean and variance $c(l, t)$ for the doubly infinite system have been obtained by Kucera and are

$$\mu_1' = \left[\frac{l}{v} + \frac{2D_L}{v^2}\right][1 + \alpha + \alpha K_2] \qquad \alpha = \frac{(1-\epsilon)\beta}{\epsilon} \tag{7.2.34}$$

$$\sigma^2 = \left[\frac{2D_L l}{v^3} + \frac{8D_L^2}{v^4}\right][1 + \alpha + \alpha K_s]^2 + \left[\frac{2l}{v} + \frac{4D_L}{v^2}\right]\alpha\left[\frac{R_1^2}{D_s}\frac{(1+K_s)^2}{15} + \frac{\alpha(1+K_s)}{k_f} + \frac{K_s}{k_s}\right] \tag{7.2.35}$$

We can also examine certain simplified cases of (7.2.34) and (7.2.35). For example, if $k_f = k_s = \infty$, $K_s = 0$, and $R_1^2/D_s = 0$, that is, there is no resistance to mass transport outside or inside the particle and infinite diffusivity within the particle,

$$\mu_1' = \left[\frac{l}{v} + \frac{2D_L}{v^2}\right](1 + \alpha) \tag{7.2.36}$$

$$\sigma^2 = \left[\frac{2D_L l}{v^3} + \frac{8D_L^2}{v^4}\right](1 + \alpha)^2 \tag{7.2.37}$$

If, in addition, the packing is nonporous, $\beta = 0$, we have

$$\mu_1' = \left[\frac{l}{v} + \frac{2D_L}{v^2}\right] \tag{7.2.38}$$

$$\sigma^2 = \left[\frac{2D_L}{v^3} + \frac{8D_L^2}{v^4}\right] \tag{7.2.39}$$

In dimensionless forms, (7.2.38) and (7.2.39), become the moments shown in the first entry of Table 7.1.

There are a number of competing processes in the above formulation: axial dispersion of solute in the external fluid, mass transfer of solute from the external fluid to the internal fluid of the particles, diffusion of solute in the internal fluid, mass transfer of solute from the internal fluid to the solid surface, and adsorption on the surface. The parameters characterizing these processes are D_L, k_f, D_s, k_s, and K_s. Thus an obvious approach to the evaluation of these five parameters is to use a response curve from which five moments can be evaluated. Such a procedure is, unfortunately, not feasible because of the inaccuracies associated with computation of the higher moments from the experimental tracer curve. Thus it is necessary to turn to the use of cleverly designed experiments, in which different processes are important, so that only one or two parameters need to be calculated from one experimental run. This procedure is described for several cases by Grubner and Underhill [18], Cerro and Smith [8], and Sagara et al. [40].

It is of interest to mention briefly other work which has been carried out in the modeling of packed beds in which adsorption occurs. Suzuki and Smith [48] have included, in addition to the processes above, a first-order surface reaction after adsorption in the pore structure of the packing. Gunn [16] has analyzed the transfer functions and moments for a variety of inputs in packed beds. Jeffreson [25] has used an external discrete staged model with an intraparticle balance of the form (7.2.23) to generate moments of the response. The transfer function between the external and surface solute concentrations is presented. Jeffreson compares the moments for the dispersion and discrete staged models and points out a close correspondence. Finally, Buffham [7] used a discrete staged model for the external fluid, as Jeffreson, to account for the dispersion of solute in the external fluid. However, instead of considering unsteady diffusion in the particle, the dynamics of the interphase exchange are described in terms of a distribution of the duration of residence times in the solid phase. Of importance is that this approach easily accommodates different particle shapes and has a small number of parameters, relatively easy to evaluate.

EXAMPLE 7.2. *Dispersion in a Gas-Liquid Fluidized Bed.* Ostergaard and Michelsen [37] present data for a gas-liquid fluidized bed, shown in Figure 7.4, from which

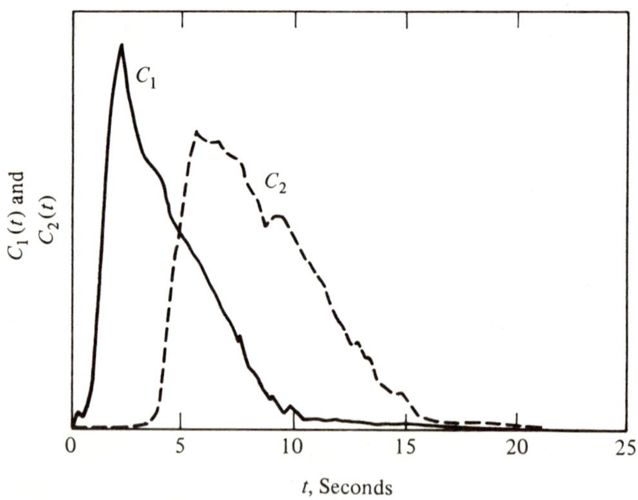

Figure 7.4. Experimental measurements of tracer concentrations at two locations in a gas-liquid fluidized bed. [Reprinted from K. Ostergaard and M. L. Michelsen, "On the Use of the Imperfect Tracer Pulse Method for Determination of Hold-up and Axial Mixing," *Can. J. Chem. Eng. 47*, 107 (1969). Reprinted by permission of the copyright owner.]

they desired to compute the liquid phase Peclet number assuming the axial-dispersed plug flow model. Given the tracer concentration measurements at two points downstream of the injection, the authors matched the difference in the moments of the two curves in Figure 7.4 to the two relations for the axial dispersed plug flow model:

$$\Delta\mu_1' = \mu, \qquad \Delta\sigma^2 = \frac{2\mu^2}{\text{Pe}} \tag{A}$$

where $\mu = l/v$, the average flow time between the two measurements. At each location the mean and variance of the data were computed by the simple summations:

$$\mu = \frac{\sum_{i=1}^{N} C_i t_i \,\Delta t_i}{\sum_{i=1}^{N} C_i \,\Delta t_i}$$

$$\sigma = \frac{\sum_{i=1}^{N} C_i t_i^2 \,\Delta t_i}{\sum_{i=1}^{N} C_i \,\Delta t_i} - \mu^2 \tag{B}$$

Using this method for several experiments, they obtained values of Pe ranging from 335 to -38, illustrating the difficulties one often encounters in parameter estimation using moments when the tails of the experimental data are inaccurate.

To circumvent this problem, the authors proposed another approach using moments. Say that $C_1(t)$ and $C_2(t)$ are experimentally determined, normalized transient responses to an arbitrary forcing function measured at two locations in the system downstream of the point of injection. We define the transfer function $F(s)$ by

$$F(s) = \frac{\bar{C}_2(s)}{\bar{C}_1(s)} = \frac{\int_0^\infty C_2(t)e^{-st}\,dt \Big/ \int_0^\infty C_2(t)\,dt}{\int_0^\infty C_1(t)e^{-st}\,dt \Big/ \int_0^\infty C_1(t)\,dt} \tag{C}$$

Numerical values of $F(s)$ may be computed for a number of values of s using the experimental C_1, C_2-curves in (C). If $F(s)$ contains p parameters, then we need to compute at least p values of $F(s)$ corresponding to p arbitrary values of s, although it is usually advantageous to compute more than p values of $F(s)$ and evaluate the parameters statistically.

For the doubly infinite axial-dispersed plug flow model,

$$F(s) = \exp\left\{\frac{\text{Pe}}{2}\left[1 - \left(1 + \frac{4\mu s}{\text{Pe}}\right)^{1/2}\right]\right\} \tag{D}$$

which may be rearranged as

$$\left(\ln\frac{1}{F(s)}\right)^{-1} = \mu s\left(\ln\left(\frac{1}{F(s)}\right)\right)^{-2} - \frac{1}{\text{Pe}} \tag{E}$$

If the left-hand side of (E) is plotted vs. $s\{\ln[1/F(s)]\}^{-2}$, the result should be a straight line of slope μ and intercept $-\text{Pe}^{-1}$ if the axial-dispersed plug flow model is applicable. The procedure is to compute a number of $F(s)$ values from (C) corresponding to several arbitrary real values of s. These values are then plotted in the manner described, from which μ and Pe are obtained. The data of Figure 7.4 were analyzed in this manner, the result of which is shown in Figure 7.5. The values of $\mu = 4.31 \pm 00$ seconds and Pe $= 25.66 \pm 0.54$ were obtained.

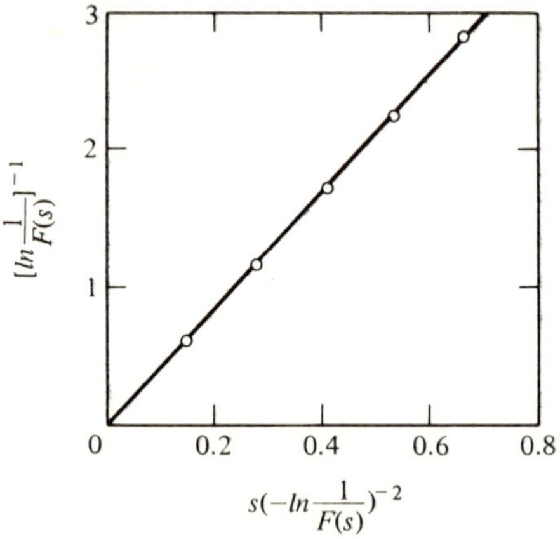

Figure 7.5. Evaluation of transfer functions by the method of Ostergaard and Michelsen. [Reprinted from K. Ostergaard and M. L. Michelsen, "On the Use of the Imperfect Tracer Pulse Method for Determination of Hold-up and Axial Mixing," *Can. J. Chem. Eng. 47*, 107 (1969). Reprinted by permission of the copyright owner.]

This example puts us a little ahead of ourselves, since we have not yet considered the use of transfer functions for estimating parameters. Nevertheless, the technique finally arrived at by Ostergaard and Michelsen is simple and illustrates an important point, namely, that one will usually have to try several approaches to analyzing data from a tracer test on a complex system, anywhere from a direct matching of moments to time domain estimation procedures. The particular technique which will produce the most consistent parameter estimates will depend on the quality of the data and the form of the model.

7.2.3 Determination of moments from experimental data

Say that we inject a quantity of tracer in a flowing system and that we desire to determine parameters from moment matching. The dimensionless concentration $C(z, \theta)$ is referred to a reference concentration c_0. Normally, c_0 would be Q/V, the mass of tracer Q divided by the volume of the system V. The mean residence time $\bar{t}$ would be V/q, where q is the volumetric flow rate of fluid.

The rth noncentral moment μ'_r defined by (7.2.1) is automatically normalized since C is normalized. The nonnormalized experimental output can be represented by $c(x, t)$. It is necessary to compute the three integrals

$$I_r = \int_0^\infty t^r c(x, t)\, dt, \qquad r = 0, 1, 2$$

The normalized mean and variance of the experimental curve are then computed by

$$\mu_1'^{\,\text{exp}} = \frac{1}{\bar{t}}\frac{I_1}{I_0} = \frac{q^2}{QV} I_1 \tag{7.2.40}$$

$$\begin{aligned} \mu_2^{\text{exp}} &= \frac{1}{\bar{t}^2}\left[\frac{I_2}{I_0} - \left(\frac{I_1}{I_0}\right)^2\right] \\ &= \frac{q}{Q}\left(\frac{q}{V}\right)^2\left[I_2 - 2\frac{q}{Q}I_1^2 + \left(\frac{q}{Q}\right)^2 I_1^2 I_0\right] \end{aligned} \tag{7.2.41}$$

The integrals I_0, I_1, and I_2 can be evaluated numerically from the experimental data using the trapezoidal rule, Simpson's rule, or any other convenient quadrature method. An example illustrating the use of the trapezoidal rule to evaluate these integrals from actual data is given by Himmelblau [19].

7.3 Parameter Estimation Using Transfer Functions with Deterministic Inputs

It is often desirable, or necessary, to estimate model parameters in the transfer function of a system rather than in the time domain solution. The desirability rests with the fact that the transfer function is an algebraic relationship as opposed to the time domain differential equation relationship. The necessity arises because we often cannot obtain the time domain solution for complex linear models, leaving only the transfer function available, aside from numerical solution of the differential equations. In this section we shall explore the estimation of parameters using transfer functions with deterministic inputs (pulse, step, and sinusoidal). We shall assume that the form of the transfer function is known from specification of the model and that we desire to estimate parameters in the transfer function from experimental data. We shall begin by considering the relationship of moments to the transfer function.

7.3.1 Relationship of moments to the transfer function

Let $F(t)$, $F(t) = 0$, $t < 0$, be the impulse response function of a stable linear system, i.e., the residence time density function of a linear flow system. The characteristic function of $F(t)$ is simply the Fourier transform of $F(t)$ (or the L.T. with $s = i\omega$):

$$\bar{F}(i\omega) = \int_0^\infty F(t)\, e^{-i\omega t}\, dt \tag{7.3.1}$$

The frequency response of the system is given by $\bar{F}(i\omega)$ as a function of ω and is usually represented as $|\bar{F}(i\omega)|$ and $\angle\bar{F}(i\omega)$ vs. ω. We call $\bar{F}(i\omega)$ the frequency response transfer function.

We can write (7.3.1) as

$$\bar{F}(i\omega) = \int_0^\infty F(t)\cos\omega t\,dt - i\int_0^\infty F(t)\sin\omega t\,dt \tag{7.3.2}$$

If we substitute the series expansions of $\cos\omega t$ and $\sin\omega t$ into (7.3.2) and use the definitions of the noncentral moments of $F(t)$, we obtain

$$\bar{F}(i\omega) = \mu'_0 - \mu'_2\frac{\omega^2}{2!} + \mu'_4\frac{\omega^4}{4!} - \cdots + i\left[-\mu'_1\omega + \mu'_3\frac{\omega^2}{3!} - \mu'_5\frac{\omega^5}{5!} + \cdots\right] \tag{7.3.3}$$

Therefore the Fourier transform of the impulse response function $F(t)$ of a linear system effectively involves evaluation of all the noncentral moments of $F(t)$.

Since the frequency response transfer function $\bar{F}(i\omega)$ is often used in the form

$$\ln\bar{F}(i\omega) = \ln|\bar{F}(i\omega)| + i\angle\bar{F}(i\omega) \tag{7.3.4}$$

it is convenient to consider the relation between $\ln\bar{F}(i\omega)$ and the moments of $F(t)$. We recall (4.5.38) wherein the cumulants of a probability distribution were defined. Thus we can write

$$\ln\bar{F}(i\omega) = \sum_{n=0}^{\infty}\kappa_n\frac{(-i\omega)^n}{n!} \tag{7.3.5}$$

Comparing (7.3.4) and (7.3.5), we find

$$\ln|\bar{F}(i\omega)| = \kappa_0 - \kappa_2\frac{\omega^2}{2!} + \kappa_4\frac{\omega^4}{4!} - \kappa_6\frac{\omega^6}{6!} + \cdots \tag{7.3.6}$$

$$\angle\bar{F}(i\omega) = -\kappa_1\omega + \kappa_3\frac{\omega^3}{3!} - \kappa_5\frac{\omega^5}{5!} + \cdots \tag{7.3.7}$$

Therefore the frequency response transfer function magnitude and phase angle are directly related to the moments of the residence time distribution function $F(t)$ of the system.

Now consider the determination of the moments from a pulse test. The fundamental time and frequency domain relationships for a linear system with input $X(t)$ and output $Y(t)$ are

$$Y(t) = \int_0^t X(\lambda)F(t-\lambda)\,d\lambda \tag{7.3.8}$$

$$\bar{Y}(i\omega) = \bar{X}(i\omega)\bar{F}(i\omega) \tag{7.3.9}$$

Taking the logarithms of each side of (7.3.9),

$$\ln|\bar{Y}(i\omega)| = \ln|\bar{X}(i\omega)| + \ln|\bar{F}(i\omega)| \tag{7.3.10}$$

$$\angle \bar{Y}(i\omega) = \angle \bar{X}(i\omega) + \angle \bar{F}(i\omega) \tag{7.3.11}$$

If $X(t)$ and $Y(t)$ are impulse input and response, respectively, then relations of the form (7.3.6) and (7.3.7) can be written for $\bar{X}(i\omega)$ and $\bar{Y}(i\omega)$:

$$\ln|\bar{X}(i\omega)| = \kappa_{X_0} - \kappa_{X_2}\frac{\omega^2}{2!} + \kappa_{X_4}\frac{\omega^4}{4!} - \kappa_{X_6}\frac{\omega^6}{6!} + \cdots \tag{7.3.12}$$

$$\ln|\bar{Y}(i\omega)| = -\kappa_{Y_1}\omega + \kappa_{Y_3}\frac{\omega^3}{3!} - \kappa_{Y_5}\frac{\omega^5}{5!} + \cdots \tag{7.3.13}$$

Substituting (7.3.6) and (7.3.13) into (7.3.10) and equating like powers of ω, we obtain the following relation among the cumulants of $X(t)$, $Y(t)$, and $F(t)$:

$$\kappa_{Y_n} = \kappa_{X_n} + \kappa_{F_n} \tag{7.3.14}$$

A specific case of (7.3.14) was encountered earlier in using the difference between response curves measured at two locations in a packed bed.

Frequency response testing is one of the classic techniques for analysis of linear systems, particularly in the designing of control systems. The normal parameter estimation procedure is to determine the frequency response from sinusoidal forcing. Then $|\bar{F}(i\omega)|$ and $\angle \bar{F}(i\omega)$ of an assumed transfer function are fit to the data by the adjustment of parameters in $\bar{F}(i\omega)$. When a pure sine wave is used as an input the ratio of the amplitudes and relative phase of the input and output signals gives one point on a frequency response plot of $|\bar{F}|$ and $\angle \bar{F}$ vs. ω. Since a separate test must be carried out for each frequency, the procedure can be time-consuming, particularly with slow-responding industrial processes. Nevertheless, the method is excellent for determining the system response to specific frequencies and is relatively insensitive to noise.

7.3.2 Determination of moments from frequency response testing

Let us assume that we have experimentally determined an accurate value of $\bar{F}(i\omega)$ either from a pulse test or direct sinusoidal testing. The modular part of the transfer function, $|\bar{F}(i\omega)|$, is expressed by (7.3.6) in terms of the even cumulants, whereas the angular part of $\bar{F}(i\omega)$, $\angle \bar{F}(i\omega)$, representing the phase shift between output and input sine waves in cyclic testing, is expressed in terms of the odd cumulants, the first two being μ'_1 and μ_3.

When the system has the characteristics of a pure time delay, i.e., plug flow, the phase shift is

$$\angle \bar{F}(i\omega) = -\mu'_1\omega \tag{7.3.15}$$

In many flow systems of interest, e.g., turbulent flow in empty tubes and flow through packed and fluidized beds, the impulse response $F(t)$ has the form of a Gaussian distribution with negligible skewness. The phase shift is then

$$\begin{aligned}\angle \bar{F}(i\omega) &= -\mu'_1\omega + \mu_3\frac{\omega^3}{3!} \\ &\cong -\mu'_1\omega \end{aligned} \tag{7.3.16}$$

since $\kappa_3 = \mu_3$. Thus it is usually quite difficult to determine μ_3 accurately from phase shift measurements, since μ_3 will generally be quite small.

Least squares polynomial curve fitting in powers of ω can be used to determine the moments of the impulse response function if experimental values of $|\bar{F}(i\omega)|$ and $\angle \bar{F}(i\omega)$ are available. Clearly, if $|\bar{F}(i\omega)|$ and $\angle \bar{F}(i\omega)$ have been obtained by pulse testing and conversion to the frequency domain, the moments estimated by the curve fitting to $|\bar{F}(i\omega)|$ and $\angle \bar{F}(i\omega)$ can be no more accurate than those that would be obtained by a direct moment analysis of the time domain data [26].

EXAMPLE 7.3. Jeffreson [26] has considered the estimation of the Peclet number in the doubly infinite axially dispersed plug flow model by frequency response testing. The impulse response and frequency response transfer function between any two points separated by a mean residence time μ are

$$F(t) = \sqrt{\frac{\text{Pe}}{4\pi t\mu}}\exp\left[\frac{-\text{Pe}(\mu - t)^2}{4t\mu}\right] \tag{A}$$

$$\bar{F}(i\omega) = \exp\left\{\frac{\text{Pe}}{2}\left[1 - \sqrt{1 + \frac{4i\omega\mu}{\text{Pe}}}\right]\right\} \tag{B}$$

Let us determine the first four cumulants of $F(t)$ from $\bar{F}(i\omega)$. The magnitude and phase angle of $\bar{F}(i\omega)$ are

$$|\bar{F}(i\omega)| = (1 + a^2)^{-1/4}\exp\left\{\frac{\text{Pe}}{2}[1 - (1 + a^2)^{1/4}\cos b]\right\} \tag{C}$$

$$\angle \bar{F}(i\omega) = -\frac{\text{Pe}}{2}(1 + a^2)^{1/4}\sin b - b \tag{D}$$

where

$$a = \frac{4\omega\mu}{\text{Pe}}, \qquad b = \tan^{-1} a \tag{E}$$

Let $z = 4i\omega\mu/\text{Pe}$, and so

$$\bar{F}(i\omega) = \exp\left\{\frac{\text{Pe}}{2}[1 - (1 + z)^{1/2}]\right\} \tag{F}$$

The series expansion for $(1 + z)^{1/2}$ is

$$(1 + z)^{1/2} = 1 + \frac{1}{2}z + \frac{1}{2}\left(\frac{1}{2} - 1\right)\frac{z^2}{2!} + \frac{1}{2}\left(\frac{1}{2} - 1\right)\left(\frac{1}{2} - 2\right)\frac{z^3}{3!} + \cdots \tag{G}$$

provided $|z| \leq 1$; i.e., $\omega\mu < \text{Pe}/4$. Substitute (G) into (F) and take logarithms:

$$\begin{aligned}\ln \bar{F}(i\omega) &= \frac{\text{Pe}}{2}\left[-\frac{1}{2}z - \frac{1}{2}\left(\frac{1}{2} - 1\right)\frac{z^2}{2!} - \cdots\right] \\ &= \text{Pe}\left[-\frac{1}{2^2}z + \frac{1}{2^3}\frac{z^2}{2!} - \frac{3}{2^4}\frac{z^3}{3!} + \cdots\right]\end{aligned} \tag{H}$$

Comparing (H) term by term with (7.3.5), we obtain

$$\kappa_0 = 0, \qquad \kappa_1 = \mu, \qquad \kappa_2 = \frac{2\mu^2}{\text{Pe}}$$

$$\kappa_3 = \frac{12\mu^3}{\text{Pe}^2}, \qquad \kappa_4 = \frac{120\mu^4}{\text{Pe}^3}$$

Thus

$$\ln|\bar{F}(i\omega)| = \frac{(\mu\omega)^2}{\text{Pe}} + \frac{5(\mu\omega)^4}{\text{Pe}^3} \tag{I}$$

Consider an experiment for which Pe $\simeq$ 20. Then

$$\begin{aligned} \ln|\bar{F}(i\omega)| &= -0.05(\mu\omega)^2 + 0.000625(\mu\omega)^4 - \cdots \\ \angle\bar{F}(i\omega) &= -\mu\omega + 0.01(\mu\omega)^2 \end{aligned} \tag{J}$$

In the normalized frequency range $0 < \mu\omega < 1$ the second terms are more than 100 times smaller than the first. Therefore in this frequency range fitting of data to (J) will be almost identical to a direct moment analysis of the time domain data since the higher moments will not be used. At higher frequencies the higher power terms in ω become larger.

7.3.3 Estimation in the transform domain using transformed data

In Section 7.1.3 we developed the relations between time and frequency domain error criteria. We saw, for example, that the time domain unweighted least squares criterion (7.1.40) is identical to the frequency domain form (7.1.41). The subject of this subsection is those methods in which a frequency domain error criterion is used with transformed data.

The general technique is straightforward. We shall assume that the transfer function is of the form $\bar{F}(s, k_1, \ldots, k_p)$, where $k_1, \ldots, k_p$ are p parameters to be determined. The experimental data, $y(t)$, is transformed into the Laplace domain by numerical evaluation of

$$\bar{y}(s) = \int_0^\infty y(t)e^{-st}\,dt \tag{7.3.17}$$

We then select an appropriate error criterion, several of which were presented in Section 7.1. When it is possible to compute both the real and imaginary parts of $\bar{F}$, probably the best criterion to use is (7.1.42), since it corresponds to time domain least squares fitting. In this case, we would want to minimize $S(k_1, \ldots, k_p)$ from the summation approximation to (7.1.42):

$$S(k_1, \ldots, k_p) = \sum_{k=1}^{N} \{\Delta R(\omega_k)^2 - \Delta I(\omega_k)^2\} \tag{7.3.18}$$

If both real and imaginary components of $\bar{F}$ cannot be computed, we would probably want to use (7.1.49), the integral square error in the s-domain, in the following form:

$$S(k_1, \ldots, k_p) = \sum_{i=1}^{N} [\bar{F}(s_i, k_1, \ldots, k_p) - \bar{y}(s_i)]^2 \tag{7.3.19}$$

The minimization of S with respect to $\mathbf{k}$ is carried out by an appropriate minimization procedure for a nonlinear algebraic function of p parameters.

There have been several studies of parameter estimation in the transform domain reported in the literature. Hopkins et al. [22] determined values of Pe in the axial-dispersed plug flow model. They used a logarithmic transformation of the transfer function (similar to that in Example 7.2) linearizing the model with respect to the two parameters, μ and Pe. In their study, only real, positive values of s were used. Williams et al. [51] determined parameters in a dispersion model of gas mixing in a fluidized bed using error criterion (7.3.19), in which integer values of s from 1 to 10 were used. An interesting comparison of several techniques used to estimate μ and Pe in the axial-dispersed plug flow model is given by Anderssen and White [1]. The results indicate that the error criteria in decreasing order of preference are

1. Time domain least squares (7.1.40).
2. Minimum square amplitude ratio of difference in Fourier transforms:

$$\sum_{j=1}^{N} \{|\bar{F}(\omega_j) - \bar{y}(\omega_j)|\}^2$$

3. Minimum square error in amplitude ratios:

$$\sum_{j=1}^{N} \{|\bar{F}(\omega_j)| - |\bar{y}(\omega_j)|\}^2$$

4. Integral square error in the s-domain (7.3.19).
5. Matching of moments.

7.4 Parameter Estimation Using Transfer Functions with Stochastic Inputs

In each of the techniques discussed up to this point, we have assumed that we induce a transient in a process by means of a change in input condition. In addition, this change in input condition has always been assumed to be deterministic (pulse, step, sinusoid), except in the case in which we make two measurements downstream of an arbitrary input in a moment analysis. Often it may be inconvenient or even inadvisable to intentionally upset an operating process for the purpose of carrying out modeling studies. What would be desired is a technique for process modeling that can use, if necessary, the normal operating data for a system without the need for a formal experimental testing program. Such a technique does indeed exist, based on the inherently random nature of the inputs and outputs of a process during normal operating conditions. The general problem is thus to use the response of a system to stochastic inputs as a means of determining the transfer function of the system.

Consider a linear system with input $X(t)$, output $Y(t)$, and system weighting function $F(t)$. We shall use capital letters because these are random variables. If all initial conditions are zero at $t = 0$, the fundamental time domain relation is

$$Y(t) = \int_0^t F(\lambda)X(t - \lambda)\, d\lambda \tag{7.4.1}$$

In theory, $F(t)$ can be obtained directly from (7.4.1) or from the equivalent Laplace domain relationship $\bar{Y}(s) = \bar{F}(s)\bar{X}(s)$ from actual operating data. Unfortunately, the inevitable presence of noise in both inputs and outputs associated with normal operation renders these approaches practically useless. To evaluate $F(t)$ or $\bar{F}(s)$ we thus turn to statistical correlation techniques.

Let us recall the relationship between the cross-correlation function $R_{XY}(\tau)$ between the input and output of a linear system and the autocorrelation function of the input $R_{XX}(\tau)$ as given by (5.4.21):

$$R_{XY}(\tau) = \int_0^\infty F(t)R_{XX}(\tau - t)\, dt \tag{7.4.2}$$

where we assume all inputs equal zero for $t < 0$.

Of key interest to us will be the results of Example 5.13, namely the relations between the power density spectra of the input and output of a linear system:

$$S_{YY}(\omega) = |\bar{F}(\omega)|^2 S_{XX}(\omega) \tag{7.4.3}$$

$$S_{XY}(\omega) = \bar{F}(\omega)S_{XX}(\omega) \tag{7.4.4}$$

Let us concentrate on (7.4.4) which is more useful than (7.4.3) because (7.4.3) will not give any information on the phase angle of the transfer function $\bar{F}(s)$ since only $|\bar{F}(s)|$ is involved in the equation. We shall want to generalize (7.4.4) to the case of multiple inputs and outputs (actually this has already been demonstrated in Section 5.4.8 and Problem 5.19). Let the vectors $\mathbf{X}(t)$ and $\mathbf{Y}(t)$ be the n- and m-dimensional input and output vectors, respectively. The transfer function then becomes a transfer matrix with m rows and n columns, $\bar{\mathbf{F}}(s)$. We define the correlation matrices $\mathbf{R}_{XX}(\tau)$ and $\mathbf{R}_{XY}(\tau)$ accordingly, such that the ijth element of $\mathbf{R}_{XX}(\tau)$ is the cross-correlation between the ith and jth inputs, etc. We also define the following spectral density matrices:

$\mathbf{S}_{XX}(\omega)$ = $(n \times n)$ spectral density matrix of inputs; for example, diagonal element $S_{XX_{ii}}(\omega)$ is the power density spectrum of input $X_i(t)$, whereas element $S_{XX_{ij}}(\omega)$ is the cross power density spectrum of inputs i and j

$\mathbf{S}_{YY}(\omega)$ = $(m \times m)$ spectral density matrix of outputs; for example, diagonal element $S_{YY_{ii}}(\omega)$ is the power density spectrum of output $Y_i(t)$, whereas element $S_{YY_{ij}}(\omega)$ is the cross power density spectrum of outputs i and j

$\mathbf{S}_{XY}(\omega)$ = $(m \times n)$ cross spectral density matrix of inputs and outputs; for example, diagonal element $S_{XY_{ii}}(\omega)$ is the cross power density spectrum of input i and output i, whereas element $S_{XY_{ij}}(\omega)$ is the cross power density spectrum of input i and output j

The matrix forms of (7.4.3) and (7.4.4) are then

$$\mathbf{S}_{YY}(\omega) = \bar{\mathbf{F}}^*(\omega)\mathbf{S}_{XY}(\omega) = \bar{\mathbf{F}}^*(\omega)\mathbf{S}_{XX}(\omega)\bar{\mathbf{F}}^T(\omega) \tag{7.4.5}$$

and

$$\mathbf{S}_{XY}(\omega) = \mathbf{S}_{XX}(\omega)\bar{\mathbf{F}}^T(\omega) \tag{7.4.6}$$

We can then compute $\bar{\mathbf{F}}(\omega)$ from

$$\bar{\mathbf{F}}(\omega) = \{\mathbf{S}_{XX}(\omega)^{-1}\mathbf{S}_{XY}(\omega)\}^T \tag{7.4.7}$$

To use (7.4.4) or (7.4.7) it is necessary that both correlation functions $R_{XX}(\tau)$ and $R_{XY}(\tau)$ approach zero as $\tau \longrightarrow \infty$. This can be achieved by defining new correlation functions by

$$\begin{aligned} R'_{XX}(\tau) &= R_{XX}(\tau) - \{E[X(t)]\}^2 \\ R'_{XY}(\tau) &= R_{XY}(\tau) - E[X(t)]E[Y(t)] \end{aligned} \tag{7.4.8}$$

Since both R_{XX} and R_{XY} are defined as the differences from the mean values, they will approach zero as $\tau \longrightarrow \infty$. From now on let us assume that R_{XX} and R_{XY} approach zero as $\tau \longrightarrow \infty$; if not, we can always redefine them as such by (7.4.8).

Let us now discuss the actual use of these equations in determining the transfer function of a process. We have basically two types of stochastic inputs available for use:

1. The normal process inputs.
2. Intentional, externally generated stochastic inputs.

Normal process inputs possess the advantage of always being present. However, their use presents several serious drawbacks. First, normal process inputs are probably not stationary, as is assumed in our theoretical development. Second, if the process input noise is of small magnitude relative to the deterministic portion of the input, its accurate measurement may be quite difficult. Finally, we may need to wait a long time for correlations between the noise and the deterministic part of the input to become negligible. If the system includes a feedback controller, then noise is returned from the output back through the controller, which may further complicate the inputs.

For these reasons, we most often use externally generated stochastic inputs. Aside from the disadvantage of often requiring exotic equipment to generate the inputs, this approach has the very definite advantage that the precise form of the input spectral density is known, thereby removing a possible source of error in the problem.

Assuming that we know $S_{XX}(\omega)$ from a given input function, we face the question of the computation of $S_{XY}(\omega)$. Ideally we would like to determine these from the ensemble average correlation function $R_{XY}(\tau)$. However, what is usually available is the time average correlation function $\bar{R}_{XY}(\tau)$. For a process record of length T_f we define $\bar{R}_{XY}(\tau)$ by

$$\bar{R}_{XY}(\tau) = \frac{1}{T_f}\int_0^{T_f} X(t)Y(t+\tau)\,dt \tag{7.4.9}$$

We note that since $\bar{R}_{XY}$ is essentially a sum of products of random variables, it itself is a random variable. Of course, if the process is ergodic (a property which we can never verify), this is equivalent to the ensemble average, which is not a random variable since it is the sum of the products of given numbers from the ensemble of experiments. Ergodicity requires not only that the time average equal the ensemble average but also that the variance of the time average approaches zero as $T_f \longrightarrow \infty$. Only in this case will the ensemble and time averages coincide. For any stationary process the expected value of the correlation functions equal the ensemble correlation functions. Thus the time average correlation functions are unbiased estimates of the ensemble averages. Unfortunately, however, the spectral density $S_{XY}(\omega)$ computed on the basis of Fourier transformation of $\bar{R}_{XY}(\tau)$ is a *biased* estimate of the ensemble average spectral density. Also, the variance of the spectral density estimates does not decrease as $T_f \longrightarrow \infty$.

To obtain unbiased estimates of the ensemble average spectral densities it is necessary to modify either the correlation function estimates before Fourier transformation or the spectral density estimates after Fourier transformation. This modification takes the form of multiplying $\bar{R}_{XY}(\tau)$ by some weighting function $W(\tau)$ or, alternatively, multiplying $S_{XY}(\omega)$ by some weighting function $\bar{W}(\omega)$. The weighting function in the time domain $W(\tau)$ is called the *lag window*, whereas the Fourier transform of $W(\tau)$, $\bar{W}(\omega)$, is called the *spectral window*. Lag windows tend to concentrate the values near $\tau = 0$ at the total exclusion of values at far values of τ and have the form $W(0) = 1$, $W(\tau) = 0$ for $|\tau| > \tau_m$. The simplest pair is the so-called Bartlett window:

$$W(\tau) = \begin{cases} 1 - \dfrac{|\tau|}{\tau_m}, & |\tau| < \tau_m \\ 0, & |\tau| > \tau_m \end{cases} \tag{7.4.10}$$

$$\bar{W}(\omega) = \tau_m\left(\frac{\sin \frac{1}{2}\omega\tau_m}{\frac{1}{2}\omega\tau_m}\right)^2$$

The formulation and comparison of different window pairs is beyond our scope and intentions, although the subject is of significant importance in the spectral analysis of data. For a thorough treatment, we refer the reader to the book by Jenkins and Watts [27].

This technique has been applied to an oil refinery depropanizer column by Hutchinson and Shelton [24], to heat exchangers by Florentin et al. [13], to reactor systems by Homan and Tierney [20] and Angus and Lapidus [2], and to packed towers by Woodburn et al. [52].

7.5 Algorithms for Parameter Estimation in Algebraic Models

In Section 7.1 the parameter estimation problem was formulated as that of finding those values of the parameters (possibly subject to constraints) which minimize (or maximize) a certain objective function, the explicit form of which was demonstrated for a number of cases. In the next three sections we shall describe a number of numerical algorithms which can be used to find $\hat{\mathbf{k}}$. We divide these algorithms according to the type of model: algebraic and ordinary and partial differential equations. It might be noted that this distinction need not really be made, in that $\mathbf{f}(\mathbf{x}, \mathbf{k})$ in (7.1.1) could be viewed either as an explicit function (the algebraic case) or as an implicit relationship merely expressing the functional dependence of the solution of an O.D.E. or P.D.E. We choose not to follow the latter path, since we wish to develop specific aspects of algorithms for these three classes of systems. We shall devote this section to methods particularly amenable to parameter estimation in algebraic models. Much of this section has been adapted from Bard and Lapidus [4].

7.5.1 Direct search methods

Methods for finding the minimum of $S(\mathbf{k})$ that do not require the computation of the derivatives $\partial S/\partial \mathbf{k}$ are known as direct search methods, in contrast to gradient methods, which require derivative evaluations. This distinction is not always clear-cut since gradient methods in which derivatives are computed by finite difference approximations may be regarded as direct search methods. Among proper direct search methods one may mention those due to Hooke and Jeeves [21], Rosenbrock [39], and Powell [38], the last reference being specific to least squares problems.

At present there exists no conclusive evidence for preferring one method or set of methods over another. Direct search methods have the obvious advantage of not requiring differentiations, and they seem to perform well on well-conditioned problems. In difficult problems, however, precise knowledge of the derivatives can be crucial, and gradient methods tend to be more reliable. For this reason we shall omit any detailed description of the direct search methods.

7.5.2 Gradient methods

The problem under investigation here is to find the minimum of $S(\mathbf{k})$. It may help the reader to visualize a set of mountains and the need to locate the lowest valley within these mountains. To solve the problem we proceed in an iterative sequence: Given an initial value $\mathbf{k}^{(0)}$ of the parameters, we seek a new value of $\mathbf{k}^{(1)}$ which is nearer the minimum, in the sense that $S(\mathbf{k}^{(1)}) < S(\mathbf{k}^{(0)})$. Once $\mathbf{k}^{(1)}$ has been obtained, we proceed to find $\mathbf{k}^{(2)}, \mathbf{k}^{(3)}, \ldots,$ each, in turn, having the property of being closer to the minimum. In the class of methods which have proved successful for parameter estimation, the formula used for finding the new value is

$$\mathbf{k}^{(1)} = \mathbf{k}^{(0)} - \lambda \mathbf{R}\mathbf{g} \tag{7.5.1}$$

where λ is a scalar, $\mathbf{R}$ a matrix, and $\mathbf{g}$ the gradient vector of S; i.e., $g_i = \partial S/\partial k_i$. Gradient methods differ from each other in the choice of $\mathbf{R}$ and of λ and we shall now examine both items. Before doing so, however, we see that $\mathbf{R}$ when it premultiplies the vector $\mathbf{g}$ twists $\mathbf{g}$ in vector space to produce a new vector; $\mathbf{R}$ thus determines the direction to go from $\mathbf{k}^{(0)}$. λ, being a scalar, merely defines how far along this direction to go and determines the length of the step.

The choice $\mathbf{R} = \mathbf{I}$ constitutes the *method of steepest descent*. It converges slowly in most practical problems. The choice of $\mathbf{R}$ as the inverse of the Hessian matrix, $\mathbf{G}$,

$$\mathbf{R} = \left[\frac{\partial^2 S}{\partial \mathbf{k}^2}\right]^{-1} = \mathbf{G}^{-1}$$

is the *Newton-Raphson* method. It performs beautifully when one is near the minimum but suffers from two major difficulties: (1) Except near the minimum, a step taken along the Newton-Raphson direction is not guaranteed to reduce S, no matter what value is chosen for λ; and (2) the method requires computation of the second derivatives of S, usually a laborious procedure.

These difficulties may be overcome in the following ways:

1. If $\mathbf{R}$ is positive-definite and $\mathbf{g} = 0$, then it can be shown that $S(\mathbf{k}^{(1)}) < S(\mathbf{k}^{(0)})$ for sufficiently small λ. Unfortunately, $\mathbf{G}$ (and therefore $\mathbf{G}^{-1}$) is not necessarily positive-definite away from the minimum. However, let $\rho_1, \rho_2, \ldots, \rho_p$ be the eigenvalues of $\mathbf{G}$ and $\mathbf{e}_1, \mathbf{e}_2, \ldots, \mathbf{e}_p$ be the corresponding eigenvectors. Then the ij-element of the inverse of $\mathbf{G}$ is found from

$$[\mathbf{G}^{-1}]_{ij} = \sum_{k=1}^{p} \rho_k^{-1} e_{ki} e_{kj}$$

where e_{ki} is the ith component of the p-dimensional column vector $\mathbf{e}_k$. Greenstadt [15] has recommended that one compute $\mathbf{R}$ by

$$[\mathbf{R}]_{ij} = \sum_{k=1}^{p} |\rho_k^{-1}| e_{ki} e_{kj}$$

If some $\rho_k = 0$, we replace it by a small positive number. $\mathbf{R}$ is then positive-definite and coincides with $\mathbf{G}^{-1}$, where the latter is also positive-definite. The advantages of the Newton-Raphson method are thus retained at the cost of having to compute the eigenvalues and eigenvectors of $\mathbf{G}$.

An alternative solution, suggested by Levenberg [33], Marquardt [35], and Goldfeld et al. [14] is to set

$$\mathbf{R} = [\mathbf{G} + \nu \mathbf{I}]^{-1}$$

If ν is greater in magnitude than the largest negative eigenvalue of $\mathbf{G}$, $\mathbf{R}$ will be positive-definite. The required value of the scalar ν can be determined by trial and error.

2. It is fortunately possible in most parameter estimation problems to obtain a reasonable approximation to the second derivative matrix $\mathbf{G}$ without actually computing any second derivatives of the model equations. We shall illustrate this point by means of the least squares criterion (scalar measurements, $m = 1$)

$$S(\mathbf{k}) = \sum_{j=1}^{M} [y_j - f(\mathbf{x}_j, \mathbf{k})]^2$$

where

$$g_i = \frac{\partial S}{\partial k_i} = -2\sum_{j=1}^{M} \epsilon_j \frac{\partial f_j}{\partial k_i}, \qquad f_j \equiv f(\mathbf{x}_j, k) \tag{7.5.2}$$

$$G_{il} = \frac{\partial^2 S}{\partial k_i \partial k_l} = -2\sum_{j=1}^{M} \epsilon_j \frac{\partial^2 f_j}{\partial k_i \partial k_l} + 2\sum_{j=1}^{M} \frac{\partial f_j}{\partial k_i}\frac{\partial f_j}{\partial k_l} \tag{7.5.3}$$

If the fit of the model to the experimental data is at all good, the ϵ_j will be small, and the first term of (7.5.3) will be negligible compared to the second. Thus we replace $\mathbf{G}$ by the approximation

$$G_{il} \cong 2\sum_{j=1}^{M} \frac{\partial f_j}{\partial k_i}\frac{\partial f_j}{\partial k_l} \tag{7.5.4}$$

Using $\mathbf{R}$ as the inverse of $\mathbf{G}$ calculated by (7.5.4) is called the *Gauss-Newton method.* The Gauss-Newton method can be interpreted as replacing the model by their tangents and solving the resultant linear regression problem at each iteration. The same treatment can be applied to weighted least squares and maximum likelihood estimation. In all cases it can be shown that in the expression for $(\partial^2 S/\partial k_i \, \partial k_l)$ the second derivatives of f are always multiplied by some residual ϵ and may therefore be neglected.

Alternative approaches, such as the Davidon-Fletcher-Powell method [11, 12], exist, but we shall not detail these here. In a large number of test problems, Bard [3] found the Gauss-Newton method to be somewhat more efficient than its competitors.

The selection of the step length λ varies with each method. In the Newton-Raphson method, $\lambda = 1$. This is usually not a satisfactory value, except very near the minimum. In Marquardt's method also, $\lambda = 1$ is used. If $S(\mathbf{k}^{(1)}) \geq S(\mathbf{k}^{(0)})$, the value of ν is increased and $\mathbf{k}^{(1)}$ recomputed until $S(\mathbf{k}^{(1)}) < S(\mathbf{k}^{(0)})$. In the Gauss-Newton method we initially set $\lambda = 1$. If $S(\mathbf{k}^{(1)}) > S(\mathbf{k}^{(0)})$, we choose a smaller value, perhaps $\lambda/2$. In any method, an extensive search for an optimum value of λ rarely is justified. It is usually sufficient to test one or two values of λ in any given problem.

All algorithms then reduce to repetitive use of (7.5.1). To start the procedure we need initial guesses, $\mathbf{k}^{(0)}$. The success of the minimization of S often depends on these guesses, and it is clearly advantageous to make these guesses as close to the true values (the minimum) as possible. Several factors to be considered are

1. Use all available prior information.
2. Use the results of linear least squares estimates based on linearized or rearranged equations.
3. Fix the values of some of the parameters (e.g., those on which most prior information is available) and estimate only the other parameters. Use these results as initial guesses for estimating all parameters simultaneously.
4. Compute the objective function on a sparse grid of parameter values. Use the grid point with optimal objective function value as the initial guess.

If the number of unknown parameters is too large to make a grid search feasible, it is possible to conduct a random search instead.

None of these methods is infallible, and which one will work best in any given situation can be determined only by experience.

Once the iterative procedure has been started, it will continue to run on the computer until some termination criterion is satisfied. At the minimum of $S(\mathbf{k})$, the gradient $\mathbf{g}$ should vanish. Because of roundoff errors, however, the condition $\mathbf{g} = 0$ can never be attained precisely and cannot be used as a termination criterion. In practice it seems preferable to terminate whenever the iterative procedure ceases to cause significant changes in the value of S.

Failure to converge to a stationary point is usually due to one of the following reasons:

1. A constraint has been reached. If the constraint is one about which there is no doubt (e.g., the condition that a rate constant must be positive), it is likely that the mechanism chosen is inappropriate.
2. The matrix $\mathbf{R}$ is not, or is insufficiently, positive-definite. We have described above how this problem may be overcome.

3. The termination criterion is not stringent enough.
4. The gradient is computed with insufficient precision. Whenever possible, gradients should be computed by analytic differentiation, rather than by means of finite difference approximations.

To handle constraints on the k_i, it is necessary to adjust the objective function in such a way that one pays a penalty whenever one violates the constraint. Say that **k** must satisfy the set of constraints $\beta_i(\mathbf{k}) \geq 0$, $i = 1, 2, \ldots$. One technique is to introduce a new objective function

$$S^*(\mathbf{k}) = S(\mathbf{k}) + \sum_i \frac{\alpha_i}{\beta_i(\mathbf{k})}$$

where the α_i are suitably chosen small positive constants. As the ith constraint is approached, β_i approaches zero and $S^*(\mathbf{k})$ increases without bound. At a point far from any constraint, the function $S^*(\mathbf{k})$ differs but little from $S(\mathbf{k})$, and the minima of the two should nearly coincide. After finding the minimum of $S^*(\mathbf{k})$, one may reduce the α_i by, e.g., a factor of 10, find the minimum of the new $S^*(\mathbf{k})$, and repeat the procedure until the contribution of the *penalty function*

$$\sum_i \frac{\alpha_i}{\beta_i(\mathbf{k})}$$

becomes negligible. If the minimum of $S(\mathbf{k})$ is actually on a constraint, this procedure may approach the true minimum as closely as one wishes.

7.6 Algorithms for Parameter Estimation in Ordinary Differential Equations

We shall now proceed to the case (7.1.22) in which the model consists of a set of ordinary differential equations. We shall confine our attention to least squares estimation, since it is the most widely applicable and computationally most efficient. In general, each of the methods in Section 7.5 can be adapted to the present situation if we consider **f** to represent the solution of the O.D.E. model rather than an explicit function.

In the cases of discrete and continuous observations, the least squares criteria are

$$S = \sum_{r=1}^{R} [\mathbf{y}_r - \mathbf{h}(t_r, \mathbf{x}(t_r; \mathbf{k}))]^T \mathbf{Q}_r [\mathbf{y}_r - \mathbf{h}(t_r, \mathbf{x}(t_r; \mathbf{k}))] \tag{7.6.1}$$

and

$$S = \int_0^{T_f} [\mathbf{y}(t) - \mathbf{h}(t, \mathbf{x}(t; \mathbf{k}))]^T \mathbf{Q}(t) [\mathbf{y} - \mathbf{h}(t, \mathbf{x}(t; \mathbf{k}))] \, dt \tag{7.6.2}$$

where $\mathbf{Q}_r$ and $\mathbf{Q}(t)$ are weighting matrices, which can be chosen arbitrarily or related to the inverse of the error covariances.

The key aspect of least squares estimation procedures is to locate the minimum of S with respect to $\mathbf{k}$. Thus S can be envisioned as a surface in p-dimensional space, the global minimum of which we seek. Various methods differ in the manner in which they find the value of $\mathbf{k}$ which minimizes S. We shall now present two methods which employ the least squares criterion.

7.6.1 Method of steepest descent

The method of steepest descent is based on the successive correction of an initial guess, $\mathbf{k}^{(0)}$, by means of $\mathbf{k}^{(i+1)} = \mathbf{k}^{(i)} + \Delta\mathbf{k}^{(i)}$, where $\Delta\mathbf{k}^{(i)}$ is a correction to the ith iterate computed so that $S(\mathbf{k}^{(i+1)}) < S(\mathbf{k}^{(i)})$. We apply this formula repeatedly until $|S(\mathbf{k}^{(i+1)}) - S(\mathbf{k}^{(i)})|$ is less than some prescribed tolerance. Let us now derive the expression for $\Delta\mathbf{k}^{(i)}$.

The objective is to move toward the minimum of $S(k_1, \ldots, k_p)$ from any location. We let the distance in parameter space be given by the usual relation,

$$ds^2 = \sum_{i=1}^{p} dk_i^2$$

The gradient of S with respect to s is

$$\frac{dS}{ds} = \sum_{i=1}^{p} \frac{\partial S}{\partial k_i} \frac{dk_i}{ds}$$

We want to find the values of dk_i/ds that produce the largest negative value of dS/ds. To do this we minimize dS/ds with respect to dk_i/ds subject to

$$\sum_{i=1}^{p} \left(\frac{dk_i}{ds}\right)^2 = 1$$

This is simply the minimization of a linear function subject to an equality constraint and can be conveniently carried out using a Lagrange multiplier λ. Let

$$L = \sum_{i=1}^{p} \frac{\partial S}{\partial k_i} \frac{dk_i}{ds} + \lambda\left[1 - \sum_{i=1}^{p} \left(\frac{dk_i}{ds}\right)^2\right]$$

and compute the values of dk_i/ds so that

$$\frac{\partial L}{\partial(dk_i/ds)} = 0$$

These are found to be

$$\frac{dk_i}{ds} = \frac{1}{2\lambda} \frac{\partial S}{\partial k_i}$$

Use of the constraint gives

$$\lambda = \pm\frac{1}{2}\left[\sum_{i=1}^{p} \left(\frac{\partial S}{\partial k_i}\right)^2\right]^{1/2}$$

and thus

$$\frac{dk_i}{ds} = \pm\left[\sum_{j=1}^{p}\left(\frac{\partial S}{\partial k_j}\right)^2\right]^{-1/2}\left(\frac{\partial S}{\partial k_i}\right) \tag{7.6.3}$$

where we would use the negative sign to decrease S.

To improve the k_i with time we need to compute dk_i/dt. We see that

$$\frac{dk_i}{dt} = \frac{dk_i}{ds}\frac{ds}{dt}$$

If we now choose ds/dt proportional to the gradient of S,

$$\frac{ds}{dt} = \gamma'\left[\sum_{j=1}^{p}\left(\frac{dS}{dk_j}\right)^2\right]^{1/2}, \qquad \gamma' > 0$$

we find that

$$\frac{dk_i}{dt} = -\gamma'\frac{\partial S}{dk_i}, \qquad i = 1, 2, \ldots, p \tag{7.6.4}$$

This equation gives the time evolution of the values of k_i such that S is minimized as much as possible at every instant. For the purposes of actual implementation we rewrite (7.6.4) in the discrete form:

$$\frac{k_i^{(l+1)} - k_i^{(l)}}{\Delta t} = -\gamma'\frac{\partial S}{\partial k_i}\bigg|_{\mathbf{k}=\mathbf{k}^{(l)}}$$

or

$$k_i^{(l+1)} = k_i^{(l)} - \gamma\frac{\partial S}{\partial k_i}\bigg|_{\mathbf{k}=\mathbf{k}^{(l)}} \tag{7.6.5}$$

Thus we have determined that the correction $\Delta\mathbf{k}^{(l)}$ is given by $-\gamma(\partial S/\partial\mathbf{k})$. Now all that we need to do is compute this quantity.

Consider for illustration the case of discrete measurements (7.6.1). Differentiating S with respect to k_i gives

$$\frac{\partial S}{\partial k_i} = 2\sum_{r=1}^{R}[\mathbf{y}_r - \mathbf{h}(t_r, \mathbf{x}(t_r))]^T\mathbf{Q}_r\left(\frac{\partial\mathbf{h}}{\partial\mathbf{x}}\right)\left(\frac{\partial\mathbf{x}}{\partial k_i}\right) \tag{7.6.6}$$

where $(\partial\mathbf{h}/\partial\mathbf{x})$ is an $m \times n$ matrix with elements $(\partial h_j/\partial x_l)$. We need, finally, to compute the elements $(\partial x_i/\partial k_j)$ of the $n \times p$ matrix $(\partial\mathbf{x}/\partial\mathbf{k})$. These elements are called the *sensitivity coefficients* of the state $\mathbf{x}$ with respect to the parameters $\mathbf{k}$. To obtain the equations governing the sensitivity coefficients we differentiate both sides of (7.1.22) with respect to k_j. Assuming that the order of differentiation can be reversed,

$$\frac{\partial}{\partial k_j}\left(\frac{dx_i}{dt}\right) = \frac{d}{dt}\left(\frac{\partial x_i}{\partial k_j}\right)$$

we obtain

$$\frac{d}{dt}\left(\frac{\partial x_i}{\partial k_j}\right) = \sum_{l=1}^{n}\left(\frac{\partial f_i}{\partial x_l}\right)\left(\frac{\partial x_l}{\partial k_j}\right) + \left(\frac{\partial f_i}{\partial k_j}\right), \qquad i = 1, 2, \ldots, n, j = 1, 2, \ldots, p$$

Denoting $\partial x_i/\partial k_j$ by $\lambda_{ij}(t)$, we rewrite this equation and introduce the conditions shown

$$\dot{\lambda}_{ij} = \sum_{l=1}^{n}\left(\frac{\partial f_i}{\partial x_l}\right)\lambda_{lj} + \left(\frac{\partial f_i}{\partial k_j}\right) \tag{7.6.7}$$
$$\lambda_{ij}(0) = 0$$

where $\lambda_{ij}(0) = 0$ follows since the x_{i0} are independent of the **k** values.

Summarizing the steepest descent algorithm,

1. Guess the p initial values $k_1^{(0)}, \ldots, k_p^{(0)}$.
2. Integrate (7.1.22) and evaluate S by (7.6.1) or (7.6.2) depending on whether the data are discrete or continuous.
3. Integrate the sensitivity equations (7.6.7) from zero to t_R.
4. Evaluate $\partial S/\partial k_i$ from (7.6.6), $i = 1, 2, \ldots, p$.
5. Form updated estimates of the k_i from (7.6.5).
6. Continue until $|S(\mathbf{k}^{(l+1)}) - S(\mathbf{k}^{(l)})|/|S(\mathbf{k}^{(l)})|$ is less than some preset criterion.

The steepest descent algorithm possesses the advantage that no assumptions need be made concerning the distribution of the errors ϵ. It is easy to program for a computer and can handle both discrete and continuous data. However, the method has two possible disadvantages. The first is that for each iteration, $n(p + 1)$ O.D.E.s must be integrated, and for large values of n or p the computing time required for convergence might be excessive. Second, the method is based on (7.6.4), and thus for (7.6.5) to be valid, γ should be small. For this reason, convergence with the method of steepest descent is often slow. We shall now introduce another method for estimation of parameters in O.D.E.s based on least squares, which may increase significantly the speed of convergence over steepest descent.

7.6.2 Quasilinearization

The procedure we shall outline is popularly called quasilinearization. We begin by defining the $(n + p)$-dimensional vectors

$$\mathbf{z} = \begin{bmatrix} \mathbf{x} \\ \mathbf{k} \end{bmatrix}, \qquad \mathbf{g} = \begin{bmatrix} \mathbf{f} \\ 0 \end{bmatrix} \tag{7.6.8}$$

Since **k** is constant,

$$\frac{d\mathbf{k}}{dt} = 0 \tag{7.6.9}$$

If we adjoin (7.6.9) to (7.1.22) and use (7.6.8), we have

$$\begin{aligned} \dot{\mathbf{z}} &= \mathbf{g}(t, \mathbf{z}) \\ z_i(0) &= x_{i_0}, \qquad i = 1, 2, \ldots, n \\ z_i(0) &= ?, \qquad i = n + 1, \ldots, n + p \end{aligned} \tag{7.6.10}$$

The problem is now to find the initial conditions $z_i(0)$, $i = n + 1, \ldots, n + p$, to minimize the least squares criterion

$$S = \sum_{r=1}^{R} [\mathbf{y}_r - \mathbf{h}(t_r, \mathbf{z}(t_r; \mathbf{z}(0)))]^T \mathbf{Q}_r [\mathbf{y}_r - \mathbf{h}(t_r, \mathbf{z}(t_r; \mathbf{z}(0)))] \tag{7.6.11}$$

where $\mathbf{z}(t; \mathbf{z}(0))$ is the solution of (7.1.22) corresponding to a particular set of $z_i(0)$, $i = n + 1, \ldots, n + p$.

First, make an initial guess of the unknown parameters, $k_i = z_{n+i}(0)$, $i = 1, 2, \ldots, p$. Then integrate (7.6.10) numerically from $t = 0$ to $t = t_R$. This solution, corresponding to the first guess of $\mathbf{k}$, may be indicated by $\mathbf{z}^{(0)}(t)$. Next, we can linearize (7.6.10) to first order about the nominal trajectory $\mathbf{z}^{(0)}(t)$, the result of which is

$$\dot{\mathbf{z}}^{(1)}(t) = \mathbf{g}[\mathbf{z}^{(0)}, t] + \left[\frac{\partial \mathbf{g}}{\partial \mathbf{z}}\right]_{\mathbf{z}^{(0)}} [\mathbf{z}^{(1)}(t) - \mathbf{z}^{(0)}(t)] \tag{7.6.12}$$

where $[\partial \mathbf{g}/\partial \mathbf{z}]_{\mathbf{z}^{(0)}}$ is the $(n + p) \times (n + p)$ matrix, the elements of which are $(\partial g_i/\partial z_j)$ evaluated at $\mathbf{z} = \mathbf{z}^{(0)}$. This equation is just a set of $(n + p)$-coupled linear O.D.E.s for which we can obtain an analytical solution as the sum of a particular solution and $n + p$ homogeneous solutions. If the particular solution of (7.6.12) is denoted $\mathbf{p}^{(1)}(t)$, then the overall solution of (7.6.12) can be written as (see Section 2.3)

$$\mathbf{z}^{(1)}(t) = \mathbf{M}^{(1)}(t)\boldsymbol{\alpha}^{(1)} + \mathbf{p}^{(1)}(t) \tag{7.6.13}$$

where the $(n + p) \times (n + p)$ matrix $\mathbf{M}^{(1)}(t)$ is the solution of the matrix homogeneous equation,

$$\dot{\mathbf{M}}^{(1)}(t) = \left[\frac{\partial \mathbf{g}}{\partial \mathbf{z}}\right]_{\mathbf{z}^{(0)}} \mathbf{M}^{(1)}(t) \tag{7.6.14}$$

$$\mathbf{M}^{(1)}(0) = \mathbf{I} \tag{7.6.15}$$

and $\mathbf{p}^{(1)}(t)$ is the solution of

$$\dot{\mathbf{p}}^{(1)}(t) = \mathbf{g}[\mathbf{z}^{(0)}, t] + \left[\frac{\partial \mathbf{g}}{\partial \mathbf{z}}\right]_{\mathbf{z}^{(0)}} [\mathbf{p}^{(1)}(t) - \mathbf{z}^{(0)}(t)] \tag{7.6.16}$$

The $(n + p)$-dimensional vector $\boldsymbol{\alpha}^{(1)}$ is, at this point, simply a set of arbitrary coefficients for the $n + p$ homogeneous solutions of (7.6.12), which are the columns of $\mathbf{M}^{(1)}(t)$.

The key step is the assignment of boundary conditions to (7.6.16). We choose these as follows:

$$p_i(0) = \begin{cases} x_{i_0}, & i = 1, 2, \ldots, n \\ 0, & i = n + 1, \ldots, n + p \end{cases} \tag{7.6.17}$$

With this choice of $p_i(0)$ we find that $\alpha_1^{(1)} = \alpha_2^{(1)} = \cdots = \alpha_n^{(1)} = 0$, and so (7.6.13) can be written as

$$\mathbf{z}^{(1)}(t) = \mathbf{p}^{(1)}(t) + \sum_{i=n+1}^{n+p} \alpha_i \mathbf{m}_i(t) \tag{7.6.18}$$

where the $(n + p)$-dimensional vectors $\mathbf{m}_i(t)$ satisfy

$$\dot{\mathbf{m}}_i(t) = \left[\frac{\partial \mathbf{g}}{\partial \mathbf{z}}\right]_{\mathbf{z}^{(0)}} \mathbf{m}_i(t), \qquad i = n + 1, \ldots, n + p \tag{7.6.19}$$

subject to the initial conditions

$$m_{ij}(0) = \delta_{ij}, \qquad i = n + 1, \ldots, n + p, j = 1, 2, \ldots, n + p \tag{7.6.20}$$

Evaluating (7.6.18) at $t = 0$, we have

$$\mathbf{z}^{(1)}(0) = \mathbf{p}^{(1)}(0) + \sum_{i=n+1}^{n+p} \alpha_i \mathbf{m}_i(0)$$

or

$$\begin{aligned} z_i^{(1)}(0) &= x_{i_0} + \alpha_{n+1} m_{n+1,i}(0) + \alpha_{n+2} m_{n+2,i}(0) + \cdots \\ &\quad + \alpha_{n+p} m_{n+p,i}(0), \qquad i = 1, 2, \ldots, n \\ z_j^{(1)}(0) &= \alpha_{n+1} m_{n+1,j}(0) + \alpha_{n+2} m_{n+2,j}(0) + \cdots \\ &\quad + \alpha_{n+p} m_{n+p,j}(0), \qquad j = n + 1, \ldots, n + p \end{aligned}$$

However, from (7.6.20),

$$z_i^{(1)}(0) = x_{i_0}, \qquad i = 1, 2, \ldots, n \tag{7.6.21}$$

$$z_j^{(1)}(0) = \alpha_j, \qquad j = n + 1, \ldots, n + p \tag{7.6.22}$$

Therefore the $\alpha_{n+1}^{(1)}, \ldots, \alpha_{n+p}^{(1)}$ *represent the new estimates of* $k_1, \ldots, k_p$. These are still arbitrary, and so we use the final piece of information, namely, minimizing S by choice of these p values of $\boldsymbol{\alpha}^{(1)}$:

$$\frac{\partial S}{\partial \alpha_i^{(1)}} = 0, \qquad i = n + 1, \ldots, n + p \tag{7.6.23}$$

If $\mathbf{h}(t, \mathbf{x})$ is a nonlinear function of $\mathbf{x}$, (7.6.23) will lead to p nonlinear algebraic equations for the $\alpha_i^{(1)}$. If $\mathbf{h}(t, \mathbf{x}) = \mathbf{H}(t)\mathbf{x}(t)$, then (7.6.23) gives p linear algebraic equations for the $\alpha_i^{(1)}$. Once the $\alpha_i^{(1)}$ are obtained, these values are used as the next guess of the initial conditions $z_i(0)$, $i = n + 1, \ldots, n + p$, and the procedure is repeated.

Summarizing, the algorithm is used as follows:

1. Select a first guess $z_i(0)$, $i = n + 1, \ldots, n + p$, and integrate (7.6.10) to $t = t_R$. Denote this solution $\mathbf{z}^{(0)}(t)$.
2. Solve (7.6.16) with initial condition (7.6.17).
3. Solve the p equations (7.6.19) with initial condition (7.6.20).
4. Solve the algebraic equations resulting from (7.6.23) to determine $\alpha_{n+1}^{(1)}, \ldots, \alpha_{n+p}^{(1)}$. Determine $\mathbf{z}^{(1)}(t)$ from (7.6.18).
5. Return to step 2 with $\mathbf{z}^{(1)}(t)$ in place of $\mathbf{z}^{(0)}(t)$.
6. Continue until each $|\alpha_i^{(l+1)} - \alpha_i^{(l)}|/|\alpha_i^{(l)}|$, $i = n + 1, \ldots, n + p$, is less than some preset criterion.

This algorithm, if it converges, will do so rapidly. As with most iterative estimation techniques, convergence depends on a reasonable starting guess.

There are a large number of examples of parameter estimation in O.D.E.s using quasilinearization in the literature, for example, [5]. A more complete account of quasilinearization can be found in the book by Lee [32]. The following example illustrates the application of the method to the determination of reaction rate constants.

EXAMPLE 7.4. *Analysis of Kinetic Parameters from Batch and Integral Reaction Experiments* [47]. One of the most important applications of parameter estimation in O.D.E.s is the estimation of reaction rate constants from experimental kinetic data. If differential reactor data are available, the rate of reaction can be determined directly, and the estimation of rate constants involves only algebraic equations. If the study is conducted in a batch or integral flow reactor, the experimental data usually take the form of concentrations at a series of times (or reciprocal space velocities). The problem of estimating kinetic parameters is then one of estimating parameters in O.D.E.s. We shall now illustrate this procedure for the pyrolytic dehydrogenation of benzene to diphenyl and triphenyl:

$$2C_6H_6 \rightleftharpoons C_{12}H_{10} + H_2$$

$$C_6H_6 + C_{12}H_{10} \rightleftharpoons C_{10}H_{14} + H_2$$

Hougen and Watson [23] proposed the following kinetic model for these two reactions:

$$\frac{dx_1}{dt} = -r_1 - r_2, \qquad x_1(0) = 1$$

$$\frac{dx_2}{dt} = \frac{r_1}{2} - r_2, \qquad x_2(0) = 0$$

$$r_1 = k_1[x_1^2 - x_2(2 - 2x_1 - x_2)/3K_1]$$

$$r_2 = k_2[x_1x_2 - (1 - x_1 - 2x_2)(2 - 2x_1 - x_2)/9K_2]$$

where x_1 = pound-moles of benzene per pound-mole of pure benzene feed, x_2 = pound-moles of diphenyl per pound-mole of pure benzene feed, k_1 and k_2 are reaction constants and K_1 and K_2 are equilibrium constants.

Hougen and Watson report experimental measurements of x_1 and x_2 in a flow reactor at eight values of the reciprocal space velocity t. The feed to the reactor was pure benzene. The data are

$t \times 10^4$	5.63	11.32	16.97	22.62	34.0	39.7	45.2	169.7
y_1	0.828	0.704	0.622	0.565	0.499	0.482	0.470	0.443
y_2	0.0737	0.113	0.1322	0.1400	0.1468	0.1477	0.1477	0.1476

These data are actually not the raw experimental data but represent the average of a number of similar runs. Thus they can be expected to be somewhat smoother than customary data from a single experiment. We shall compensate for this later by intentionally corrupting this data with additional error.

The measurements y_1 and y_2 are related to the states by

$$y_1 = x_1 + \text{errors}, \qquad y_2 = x_2 + \text{errors}$$

The equilibrium constants K_1 and K_2 were determined from the run at lowest space velocity to be 0.242 and 0.428, respectively.

The problem we wish to solve is to estimate k_1 and k_2 from the y_1 and y_2 experimental data. We use the quasilinearization algorithm derived in this section. The least squares criterion (7.6.7) becomes

$$S = \sum_{r=1}^{8} [\mathbf{y}(t_r) - \mathbf{x}(t_r; \mathbf{k})]^T \mathbf{Q}_r [\mathbf{y}(t_r) - \mathbf{x}(t_r; \mathbf{k})]$$

where $\mathbf{y} = [y_1, y_2]^T$, $\mathbf{x} = [x_1, x_2]^T$, and $\mathbf{k} = [k_1, k_2]^T$. We take $\mathbf{Q}_r$ to be diagonal with elements q_1 and q_2, independent of r. Thus in this example $n = p = 2$.

Let $z_1 = x_1$, $z_2 = x_2$, $z_3 = k_1$, and $z_4 = k_2$ in accordance with (7.6.8)–(7.6.10). Also, let $f_1 = -r_1 - r_2$ and $f_2 = (r_1/2) - r_2$. We need to compute $[\partial \mathbf{g}/\partial \mathbf{z}]$, which in this case is

$$\left[\frac{\partial \mathbf{g}}{\partial \mathbf{z}}\right] = \begin{bmatrix} \dfrac{\partial f_1}{\partial z_1} & \dfrac{\partial f_1}{\partial z_2} & \dfrac{\partial f_1}{\partial z_3} & \dfrac{\partial f_1}{\partial z_4} \\ \dfrac{\partial f_2}{\partial z_1} & \dfrac{\partial f_2}{\partial z_2} & \dfrac{\partial f_2}{\partial z_3} & \dfrac{\partial f_2}{\partial z_4} \\ 0 & 0 & 0 & 0 \\ 0 & 0 & 0 & 0 \end{bmatrix}$$

The particular solution satisfies (7.6.16). The homogeneous equations (7.6.19) are

$$\dot{\mathbf{m}}_i(t) = \left[\frac{\partial \mathbf{g}}{\partial \mathbf{z}}\right]_{\mathbf{z}^{(0)}} \mathbf{m}_i(t), \qquad i = 3, 4$$

where

$$m_3(t) = \begin{bmatrix} m_{31}(t) \\ m_{32}(t) \\ m_{33}(t) \\ m_{34}(t) \end{bmatrix}, \qquad m_4(t) = \begin{bmatrix} m_{41}(t) \\ m_{42}(t) \\ m_{43}(t) \\ m_{44}(t) \end{bmatrix}$$

In this case the initial conditions become

$$m_3(0) = \begin{bmatrix} 0 \\ 0 \\ 1 \\ 0 \end{bmatrix}, \qquad m_4(0) = \begin{bmatrix} 0 \\ 0 \\ 0 \\ 1 \end{bmatrix}$$

The total solution $\mathbf{z}^{(1)}(t)$, after the first iterate, is

$$\begin{bmatrix} z_1^{(1)}(t) \\ z_2^{(1)}(t) \\ z_3^{(1)}(t) \\ z_4^{(1)}(t) \end{bmatrix} = \begin{bmatrix} p_1^{(1)}(t) \\ p_2^{(1)}(t) \\ 0 \\ 0 \end{bmatrix} + \alpha_3^{(1)} \begin{bmatrix} m_{31}(t) \\ m_{32}(t) \\ 1 \\ 0 \end{bmatrix} + \alpha_4^{(1)} \begin{bmatrix} m_{41}(t) \\ m_{42}(t) \\ 0 \\ 1 \end{bmatrix}$$

so that $z_3^{(1)}(t) = k_1 = \alpha_3^{(1)}$, and $z_4^{(1)}(t) = k_2 = \alpha_4^{(1)}$.

The final step per iteration is to determine $\alpha_3^{(1)}$ and $\alpha_4^{(1)}$ to minimize S, which is

$$S = \sum_{r=1}^{8} [q_1(y_1(t_r) - z_1(t_r))^2 + q_2(y_2(t_r) - z_2(t_r))^2]$$

Substituting for $z_1^{(1)}(t)$ and $z_2^{(1)}(t)$ we obtain, for the first iteration,

$$S = \sum_{r=1}^{8} [q_1(y_1(t_r) - p_1^{(1)}(t_r) - \alpha_3^{(1)} m_{31}(t_r) - \alpha_4^{(1)} m_{41}(t_r))^2 + q_2(y_2(t_r) - p_2^{(1)}(t_r) - \alpha_3^{(1)} m_{32}(t_r) - \alpha_4^{(1)} m_{42}(t_r))^2]$$

We want to find $\alpha_3^{(1)}$ and $\alpha_4^{(1)}$ so that

$$\frac{\partial S}{\partial \alpha_3^{(1)}} = \frac{\partial S}{\partial \alpha_4^{(1)}} = 0$$

The result is two linear algebraic equations for $\alpha_3^{(1)}$ and $\alpha_4^{(1)}$, the solution of which is

$$\alpha_3 = \frac{\sum_{j=1}^{8} \Gamma_3(t_j) \sum_{j=1}^{8} \theta_4(t_j) - \sum_{j=1}^{8} \Gamma_4(t_j) \sum_{j=1}^{8} \Omega(t_j)}{\sum_{j=1}^{8} \theta_3(t_j) \sum_{j=1}^{8} \theta_4(t_j) - \left[\sum_{j=1}^{8} \Omega(t_j)\right]^2}$$

$$\alpha_4 = \frac{\sum_{j=1}^{8} \theta_3(t_j) \sum_{j=1}^{8} \Gamma_4(t_j) - \sum_{j=1}^{8} \Omega(t_j) \sum_{j=1}^{8} \Gamma_3(t_j)}{\sum_{j=1}^{8} \theta_3(t_j) \sum_{j=1}^{8} \theta_4(t_j) - \left[\sum_{j=1}^{8} \Omega(t_j)\right]^2}$$

where

$$\Gamma_3(t_j) = q_1(y_1(t_j) - p_1(t_j))m_{31}(t_j) + q_2(y_2(t_j) - p_2(t_j))m_{32}(t_j)$$
$$\Gamma_4(t_j) = q_1(y_1(t_j) - p_1(t_j))m_{41}(t_j) + q_2(y_2(t_j) - p_2(t_j))m_{42}(t_j)$$
$$\theta_3(t_j) = q_1 m_{31}^2(t_j) + q_2 m_{32}^2(t_j)$$
$$\theta_4(t_j) = q_1 m_{41}^2(t_j) + q_2 m_{42}^2(t_j)$$
$$\Omega(t_j) = q_1 m_{31}(t_j) m_{41}(t_j) + q_2 m_{32}(t_j) m_{42}(t_j)$$

The algorithm can now be programmed in a straightforward way to estimate k_1 and k_2. Several questions can be studied:

1. The range of initial guesses required for convergence of the algorithm.
2. The effect of different weighting of the observations, as reflected in q_1 and q_2, in particular the number of components to be measured.
3. The effect of using fewer than the eight data points on the estimates.
4. The effect of the level of experimental error on the estimates.

1. The range of initial guesses required for convergence of the algorithm. The algorithm was used with $q_1 = 1$ and $q_2 = 1$ with a convergence criterion of

$$\frac{|k_1^{(i+1)} - k_1^{(i)}|}{k_1^{(i)}} + \frac{|k_2^{(i+1)} - k_2^{(i)}|}{k_2^{(i)}} \le \epsilon = 0.01$$

Results are shown in Table 7.2 for four different initial guesses. Figure 7.6 has been prepared from Table 7.2 to illustrate the approximate region of convergence of initial guesses in the $(k_1 - k_2)$-plane. Obviously, many further trials with other sets of initial k_1, k_2-values would more closely define the two-dimensional region in the

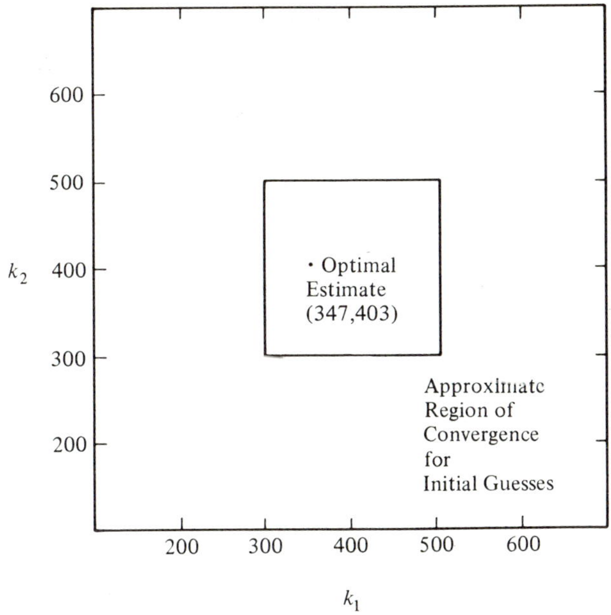

Figure 7.6. Approximate region of convergence for initial guesses of k_1 and k_2 in the quasilinearization algorithm. [Reprinted from J. H. Seinfeld and G. R. Gavalas, "Analysis of Kinetic Parameters from Batch and Integral Reaction Experiments," *AIChE J. 16*, 644 (1970). Reprinted by permission of the copyright owner.]

TABLE 7.2
EFFECT OF INITIAL GUESS ON CONVERGENCE

Case	Initial guess k_1	Initial guess k_2	Final values k_1	Final values k_2	Iterations
1	100	100	Diverged		—
2	300	300	347.443	403.132	4
3	500	500	347.402	402.751	4
4	1000	1000	Diverged		—

$(k_1 - k_2)$-plane within which all initial guesses will produce convergence of the algorithm. In any event, the convergence of the algorithm depends on the goodness of the initial guesses. Although Figure 7.6 is not to be construed as an exact region of convergence, it is evident that the initial guesses should be the same order of magnitude as the convergence values. Often, experimenters have no idea about the order of magnitude of the parameter values. Thus in such a case it is necessary to affix to the quasilinearization algorithm a procedure for generating initial guesses from the raw experimental data. Such a procedure would be the following.

One could easily fit a polynomial to the first several data points and compute the value of the derivative of the polynomial at some point in this range, perhaps at $t = 0$. This derivative is then an estimate of the rate $r(x, \mathbf{k})$. The values of the

k_i can then be determined by using the kinetic model. Although the estimates obtained in this manner are crude at best, they should be within the order of magnitude of the true values. If this procedure cannot be trusted alone, then the method of steepest descent can be used to improve these initial estimates. The steepest descent procedure can be stopped when the percentage change in the k_i is less than some predetermined criterion. These values are then used as initial guesses for the quasilinearization algorithm.

2. The effect on the estimates of different weighting of the observations. In the previous runs, each observation y_1 and y_2 is given equal weighting ($q_1 = 1, q_2 = 1$). To examine the effect of the weighting of the observations, $q_1 = 1$ and $q_2 = 5$ were used with $\epsilon = 0.01$ (case 5). See Table 7.3.

TABLE 7.3
EFFECT OF WEIGHTING ON k_1 AND k_2

Case	Initial guess k_1	k_2	q_1	q_2	Final values k_1	k_2	Iterations
5	300	300	1	5	347.386	404.196	5
6	300	300	1	0	351.124	366.366	4
7	300	300	0	1	346.826	401.076	7

By comparison of cases 2 and 5 it is seen that the final values of k_1 and k_2 are not affected appreciably by different weighting of the states. The limit of different weighting factors is represented by cases 6 and 7, wherein only one output is observed. It appears that better estimates are obtained by observing y_2 than y_1. This point will be discussed in more detail in the next chapter.

3. The effect on the estimates of using fewer than the eight data points. In all preceding cases the estimating procedure utilized each of the eight data points. It is interesting to examine the effect on the estimates of k_1 and k_2 of using experimental measurements at fewer than eight values of t. Insight should be provided into the problem of choosing a priori the number of values of the independent variable at which to make measurements. Values of $R = 6$, 4, and 3 corresponding to $t = 5.63$, 16.97, 22.62, 34.0, 42.2, and 169.7; $t = 5.63$, 16.97, 34.0, and 169.7; and $t = 5.63$, 34.0, and 169.7 ($\times 10^{-4}$), respectively, were used in cases 8, 9, and 10 with $q_1 = 1$, $q_2 = 1$, and $\epsilon = 0.01$ (Table 7.4).

TABLE 7.4
EFFECT OF FEWER DATA POINTS ON k_1 AND k_2

Case	Initial guess k_1	k_2	Data points, M	Final values k_1	k_2	Iterations
8	300	300	6	346.421	402.129	4
9	300	300	4	346.890	398.742	4
10	300	300	3	346.853	406.032	4

There is some variation in the estimates of k_2 as the number of data points is reduced, but the estimates of k_1 remain generally constant. The results from cases 8–10 seem to indicate that for this example reasonable estimates are obtained with observations at as few as three values of t.

The next question one might raise is the following. If we know that a number of data points, for example, four, are enough to yield acceptable (again, we shall define *acceptable* later) estimates of k_1 and k_2, how should the four points be distributed over the range of the observed conversions? In particular, should the points be evenly distributed, or is a clustering near the low or high conversion more desirable? Cases 8–10 represent an even distribution of the points. Consider now cases 11 and 12, wherein four points are used at values of $t = 5.63$, 11.32, 16.97, and 22.62, and $t = 34.0$, 39.7, 45.2, and 169.7, respectively (Table 7.5).

TABLE 7.5
EFFECT OF DISTRIBUTION OF DATA POINTS ON k_1 AND k_2

Case	Initial guess k_1	Initial guess k_2	Data points, M	Final values k_1	Final values k_2	Iterations
11	300	300	4	347.722	403.329	3
12	300	300	4	345.318	403.260	4

At this point it is difficult to distinguish between the accuracy of the estimates in cases 11 and 12, although slightly better estimates are obtained in case 11 where the data are concentrated in the region of most rapid change of x_1 and x_2.

Hougen and Watson [23] used numerical differentiation at each of the data points and then averaged the resulting values of k_1 and k_2, obtaining $k_1 = 348$ and $k_2 = 404$.

4. The effect of experimental error on the estimates. To examine the effect of larger experimental error on the estimates of k_1 and k_2, additional artificial error was introduced in the original experimental data. The values of k_1 and k_2 from case 2 were selected as the true values and were used to generate true values of x_1 and x_2.

TABLE 7.6
EFFECT OF ADDITIONAL ERRORS ON k_1 AND k_2

Case	M	α	ω	Final values k_1	Final values k_2
13	8	0.2	1	389.702	499.571
14	16	0.2	1	383.528	523.989
15	8	0.05	1	357.030	424.497
16	16	0.05	1	355.737	431.479
17	30	0.05	1	347.655	403.485
18	30	0.05	1000	347.421	403.932
19	8	0.02	1	351.126	411.403

Then experimental values of y_1 and y_2 were generated according to

$$y_1^*(t_i) = x_1(t_i)(1 + \alpha \sin \omega t_i)$$
$$y_2^*(t_i) = x_2(t_i)(1 + \alpha \cos \omega t_i)$$

where α and ω are constants governing the amplitude and frequency of the artificial error. With given values of α and ω, the cases in Table 7.6 were studied with k_1 and k_2 equal to 300 initially.

7.6.3 An extended Gauss-Newton method

In parameter estimation in algebraic models, Bard [3] found the Gauss-Newton method to be favored over simple steepest descent and the Newton-Raphson method. Therefore we might expect that an extension of the Gauss-Newton method to parameter estimation in O.D.E.s would also produce advantages (in convergence speed) over steepest descent and quasi-linearization. Such an extension can be derived very simply.

At any step i in the iteration (7.6.1) can be written as

$$S(\mathbf{k}) = \sum_{r=1}^{R} [\mathbf{y}_r - \mathbf{h}(t_r, \mathbf{x}^{(i)}(t_r; \mathbf{k}^{(i)})) - (\mathbf{h}(t_r, \mathbf{x}(t_r; \mathbf{k})) - \mathbf{h}(t_r, \mathbf{x}(t_r, \mathbf{k}^{(i)})))]^T \mathbf{Q}_r [\mathbf{y}_r - \mathbf{h}^{(i)} - (\mathbf{h} - \mathbf{h}^{(i)})] \tag{7.6.24}$$

where, to simplify notation, the arguments in the second bracketed term have been omitted. The term is identical to the first bracketed term.

We linearize $\mathbf{h}$ about $\mathbf{h}^{(i)}$ to get

$$\mathbf{h}(t, \mathbf{x}(t; \mathbf{k})) - \mathbf{h}(t, \mathbf{x}(t; \mathbf{k}^{(i)})) = \left[\frac{\partial \mathbf{h}}{\partial \mathbf{x}}\right]^{(i)} \mathbf{\Lambda}^{(i)} \delta \mathbf{k}^{(i)} + o(\delta k^{(i)}) \tag{7.6.25}$$

where $\mathbf{\Lambda}^{(i)}(t)$ is the matrix of sensitivity coefficients which are governed by (7.6.7). Substituting (7.6.25) into (7.6.24) and neglecting second-order terms, we get

$$S(\delta \mathbf{k}^{(i)}) = \sum_{r=1}^{R} \left[\mathbf{y}_r - \mathbf{h}^{(i)} - \left[\frac{\partial \mathbf{h}}{\partial \mathbf{x}}\right]^{(i)} \mathbf{\Lambda}^{(i)} \delta \mathbf{k}^{(i)}\right]^T \mathbf{Q}_r \cdot \left[\mathbf{y}_r - \mathbf{h}^{(i)} - \left[\frac{\partial \mathbf{h}}{\partial \mathbf{x}}\right]^{(i)} \mathbf{\Lambda}^{(i)} \delta \mathbf{k}^{(i)}\right] \tag{7.6.26}$$

where all time-varying terms are evaluated at t_r.

We now want to determine the correction $\delta \mathbf{k}^{(i)}$ such that

$$\frac{\partial S(\delta \mathbf{k}^{(i)})}{\partial (\delta k_j^{(i)})} = 0, \qquad j = 1, 2, \ldots, p \tag{7.6.27}$$

Solving the resulting linear algebraic equations yields the desired result:

$$\delta \mathbf{k}^{(i)} = \mathbf{K}^{(i)-1} \sum_{r=1}^{R} \mathbf{\Theta}^{(i)}(t_r)^T \mathbf{Q}_r [\mathbf{y}_r - \mathbf{h}(t_r, \mathbf{x}(t_r, \mathbf{k}^{(i)}))] \tag{7.6.28}$$

where

$$\mathbf{K}^{(i)} = \sum_{r=1}^{R} \mathbf{\Theta}^{(i)}(t_r)^T \mathbf{Q}_r \mathbf{\Theta}^{(i)}(t_r) \tag{7.6.29}$$

and

$$\mathbf{\Theta}^{(i)}(t) = \left[\frac{\partial \mathbf{h}}{\partial \mathbf{x}}\right]^{(i)} \mathbf{\Lambda}^{(i)}(t) \tag{7.6.30}$$

Equations (7.6.28)–(7.6.30) form the extended Gauss-Newton algorithm. As we noted previously, the basic idea behind the Gauss-Newton algorithm is that the nonlinear equations are linearized about each iteration and the resulting linear estimation problem is solved.

The algorithm is used as follows:

1. Guess the p initial values $k_1^{(0)}, \ldots, k_p^{(0)}$.
2. Integrate (7.1.22) and evaluate S by (7.6.1) or (7.6.2) depending on whether the data are discrete or continuous.
3. Integrate the sensitivity equations (7.6.7) from zero to t_R.
4. Compute $\mathbf{\Theta}(t)$, $\mathbf{K}$, and $\delta\mathbf{k}$ from (7.6.28)–(7.6.30).
5. Continue until $(|S^{(l+1)} - S^{(l)}|/|S^{(l)}|)$ is less than some preset criterion. (As an alternative one can use the convergence criterion for the quasi-linearization method based on the difference $|k_i^{(l+1)} - k_i^{(l)}|/|k_i^{(l)}|$.)

7.6.4 Convergence of parameter estimation methods

Two problems which often plague parameter estimation methods are (1) convergence to local minima of the objective function S and (2) lack of convergence due to ill-posedness, that is, when large changes in the parameter values cause only small changes in the objective function, resulting in a very flat $S(\mathbf{k})$ surface. In this section we shall present two possible ways of alleviating these problems.

To begin let us examine the properties of the extended Gauss-Newton method in the vicinity of a minimum of S, either local or global. For this purpose, we rewrite (7.6.26) as

$$\begin{aligned} S(\delta\mathbf{k}^{(i)}) = \sum_{r=1}^{R} [\mathbf{y}_r - \mathbf{h}^{(i)}]^T \mathbf{Q}_r [\mathbf{y}_r - \mathbf{h}^{(i)}] - 2\sum_{r=1}^{R} \mathbf{\Theta}^{(i)}(t_r)\mathbf{Q}_r[\mathbf{y}_r - \mathbf{h}^{(i)}]\,\delta\mathbf{k}^{(i)} \\ + \delta\mathbf{k}^{(i)T}\mathbf{K}^{(i)}\,\delta\mathbf{k}^{(i)} \end{aligned} \tag{7.6.31}$$

or equivalently

$$\begin{aligned} S(\delta\mathbf{k}^{(i)}) = \|\delta\mathbf{k}^{(i)} - \mathbf{K}^{(i)-1}\sum_{r=1}^{R} \mathbf{\Theta}^{(i)}(t_r)^T\mathbf{Q}_r[\mathbf{y}_r - \mathbf{h}^{(i)}]\|^2_{\mathbf{K}^{(i)}} \\ + \sum_{r=1}^{R} \|\mathbf{y}_r - \mathbf{h}^{(i)}\|^2_{\mathbf{Q}_r} - \|\sum_{r=1}^{R} \mathbf{\Theta}^{(i)}(t_r)^T\mathbf{Q}_r[\mathbf{y}_r - \mathbf{h}^{(i)}]\|^2_{\mathbf{K}^{(i)-1}} \end{aligned} \tag{7.6.32}$$

where we have used the norm notation $\|\mathbf{a}\|^2_A = \mathbf{a}^T\mathbf{A}\mathbf{a}$ for conciseness.

Equations (7.6.31) and (7.6.32) represent an approximation to the S-surface by a quadratic function of $\delta\mathbf{k}^{(i)}$ in the neighborhood of $\mathbf{k}^{(i)}$. The value of S corresponding to $\mathbf{k}^{(i)}$ is the second term on the right-hand side of (7.6.32) and can be denoted $S^{(i)}$. The minimum of $S(\delta\mathbf{k}^{(i)})$ occurs at $\delta\mathbf{k}^{(i)}$ given by (7.6.28). Since the third term on the right-hand side of (7.6.32) is positive, the value of S at the minimum of the quadratic approximation is decreased from $S^{(i)}$ by that amount. Normally, $\mathbf{k}^{(i)}$ will not be in the neighborhood of a true minimum of S, so that (7.6.28) provides a small step in the direction of decreasing S by approximating the actual S-surface by the quadratic form (7.6.26). If, however, the minimum of $S(\delta\mathbf{k}^{(i)})$ occurs at an actual minimum of S, then (7.6.32) is exact to $o(\delta k^{(i)})$ and the condition which holds at the minimum is

$$\sum_{r=1}^{R} \mathbf{\Theta}^{(i)}(t_r)^T \mathbf{Q}_r[\mathbf{y}_r - \mathbf{h}(t_r, \mathbf{x}(t_r; \mathbf{k}^{(i)}))] = 0 \tag{7.6.33}$$

The essential difficulty with convergence to a local minimum is embodied in (7.6.33). Since (7.6.33) holds at a local minimum, $\delta\mathbf{k}^{(i)}$ becomes zero when computed from (7.6.28). This is, of course, the problem encountered with all gradient-type methods. However, from (7.6.33) we can propose a trial-and-error method of probing for the global minimum of S. Local minima in S will depend for their existence and location on the weighting matrix $\mathbf{Q}_r$ as well as on the nature of the system. The usual procedure to certify that one has found the global minimum of S is to start the algorithm from a number of different initial guesses $\mathbf{k}^{(0)}$. If the dimension of $\mathbf{k}$ is large, this can be quite tedious. However, if the weighting matrix $\mathbf{Q}_r$ is changed, the location and number of local minima will, in general, change. Therefore we can start the iterations for our chosen $\mathbf{Q}_r$ and continue until $\|\delta\mathbf{k}^{(i)}\|$ is less than some criterion. Then we change $\mathbf{Q}_r$ such that the time weighting of the observations is different. If the algorithm had stopped at a local minimum, the change of $\mathbf{Q}_r$ should begin the algorithm moving again. Of course, whether the algorithm actually proceeds to another minimum after $\mathbf{Q}_r$ is changed depends on the depth and curvature of the local minimum. If the algorithm had stopped at the global minimum of S, $\|\delta\mathbf{k}^{(i)}\|$ should remain small even after $\mathbf{Q}_r$ has been changed.

The second problem we noted is poor convergence if the problem is ill-posed. When this situation occurs, any numerical scheme becomes extremely sensitive to various computational errors, such as truncation and rounding errors. To alleviate these difficulties, we can essentially extend the idea of Marquardt [35] to (7.6.28). Instead of using (7.6.28), in which $\mathbf{K}^{(i)}$ may be nearly singular, we use

$$\delta\mathbf{k}^{(i)} = (\mathbf{K}^{(i)T}\mathbf{K}^{(i)} + \nu\mathbf{I})^{-1}\mathbf{K}^{(i)T} \sum_{r=1}^{R} \mathbf{\Theta}^{(i)}(t_r)^T \mathbf{Q}_r[\mathbf{y}_r - \mathbf{h}(t_r, \mathbf{x}(t_r; \mathbf{k}^{(i)}))] \tag{7.6.34}$$

7.7 Algorithms for Parameter Estimation in Partial Differential Equations

Finally, we shall consider the case in which the model is given by a set of partial differential equations. The problem was defined in Section 7.1.2. We shall extend the methods of steepest descent and quasi-linearization to this case. This section has been adapted from Seinfeld [44] and Seinfeld and Chen [46].

7.7.1 The method of steepest descent

First we want to compute the components of the gradient of S in the space of $\mathbf{k}$. For simplicity, let us take $\mathbf{Q}$ as a diagonal matrix with elements q_i in all the definitions of the least squares criterion S. Consider the computation of $\partial S/\partial k_l$ in (7.1.31a), for example:

$$\begin{aligned}\frac{\partial S}{\partial k_l} &= \sum_{i=1}^{m_1} 2q_i e_i(T, 0)\frac{\partial e_i(T, 0)}{\partial k_l} \\ &= \sum_{i=1}^{m_1}\sum_{j=1}^{n} 2q_i e_i(T, 0)\left(\frac{\partial h_i}{\partial u_j}\right)\left(\frac{\partial u_j}{\partial k_l}\right)_{\substack{t=T\\x=0}}, \qquad l = 1, 2, \ldots, p\end{aligned} \tag{7.7.1}$$

For the least squares criteria (7.1.32a) and (7.1.33a), we obtain, likewise,

$$\begin{aligned}\frac{\partial S}{\partial k_l} &= \sum_{i=m_1+1}^{m_2} 2q_i e_i(T)\frac{\partial e_i(T)}{\partial k_l} \\ &= \sum_{i=m_1+1}^{m_2}\sum_{j=1}^{n} 2q_i e_i(T)\int_\Omega \left(\frac{\partial h_i}{\partial u_j}\right)\left(\frac{\partial u_j}{\partial k_l}\right)_{T=0} d\Omega\end{aligned} \tag{7.7.2}$$

and

$$\begin{aligned}\frac{\partial S}{\partial k_l} &= \sum_{i=m_2+1}^{m_3} 2q_i e_i(0)\frac{\partial e_i(0)}{\partial k_l} \\ &= \sum_{i=m_2+1}^{m_3}\sum_{j=1}^{n} 2q_i e_i(0)\int_0^T \left(\frac{\partial h_i}{\partial u_j}\right)\left(\frac{\partial u_j}{\partial k_l}\right)_{x=0} dt\end{aligned} \tag{7.7.3}$$

We can also compute the derivative of S with respect to the components of $\mathbf{k}$ for the integral average criteria (7.1.31a)–(7.1.33b). These derivatives are

$$\frac{\partial S}{\partial k_l} = \int_0^T \int_\Omega \sum_{i=1}^{m_1}\sum_{j=1}^{n} 2q_i e_i(t, x)\left(\frac{\partial h_i}{\partial u_j}\right)\left(\frac{\partial u_j}{\partial k_l}\right) d\Omega\, dt \tag{7.7.4}$$

$$\frac{\partial S}{\partial k_l} = \int_0^T \sum_{i=m_1+1}^{m_2}\sum_{j=1}^{n} 2q_i e_i(t)\int_\Omega \left(\frac{\partial h_i}{\partial k_j}\right)\left(\frac{\partial u_j}{\partial k_l}\right) d\Omega\, dt \tag{7.7.5}$$

$$\frac{\partial S}{\partial k_l} = \int_\Omega \sum_{i=m_2+1}^{m_3}\sum_{j=1}^{n} 2q_i e_i(x)\int_0^T \left(\frac{\partial h_i}{\partial u_j}\right)\left(\frac{\partial u_j}{\partial k_l}\right) dt\, d\Omega \tag{7.7.6}$$

Regardless of the type of least squares criterion, we need the term, $\partial u_j/\partial k_l$, $j = 1, 2, \ldots, n$, $l = 1, 2, \ldots, p$. These terms are again the *sensitivity coefficients* of the model output $u_j(t, x)$ to changes in the parameters k_l. These sensitivity coefficients can be denoted by

$$\lambda_{jl}(t, x) = \frac{\partial u_j}{\partial k_l} \tag{7.7.7}$$

and written as elements of a sensitivity matrix $\mathbf{\Lambda}(t, x)$. We now want to derive equations governing $\mathbf{\Lambda}(t, x)$. We begin by assuming that

$$\frac{\partial}{\partial t}\left(\frac{\partial u_j}{\partial k_l}\right) = \frac{\partial}{\partial k_l}\left(\frac{\partial u_j}{\partial t}\right)$$

Then differentiating (7.1.25) with respect to k_l we obtain

$$\begin{aligned}\frac{\partial \lambda_{jl}(t, x)}{\partial t} = \sum_{s=1}^{n}\left[\frac{\partial f_j}{\partial u_s}\lambda_{sl} + \frac{\partial f_j}{\partial(\partial u_s/\partial x)}\frac{\partial \lambda_{sl}}{\partial x} + \frac{\partial f_j}{\partial(\partial^2 u_s/\partial x^2)}\frac{\partial^2 \lambda_{sl}}{\partial x^2}\right] \\ + \frac{\partial f_j}{\partial k_l}\end{aligned} \tag{7.7.8}$$

Differentiating the general boundary condition (7.1.26) with respect to k_l,

$$\sum_{s=1}^{n}\left[\frac{\partial g_j}{\partial u_s}\lambda_{sl} + \frac{\partial g_j}{\partial(\partial u_s/\partial x)}\frac{\partial \lambda_{sl}}{\partial x}\right] + \frac{\partial g_j}{\partial k_l} = 0, \qquad x \in \partial\Omega \tag{7.7.9}$$

The initial condition for (7.7.8) is

$$\lambda_{jl}(0, x) = 0 \tag{7.7.10}$$

since the initial state $u(0, x)$ is assumed known. Solving (7.7.8)–(7.7.10), we can now compute explicitly any of the $\partial S/\partial k_l$.

The steepest descent algorithm consists of iteratively correcting an initial guess of k by means of

$$k_l^{(i+1)} = k_l^{(i)} - \gamma\left[\frac{\partial S}{\partial k_l}\right]_{\mathbf{k}^{(i)}}, \qquad l = 1, 2, \ldots, p \tag{7.7.11}$$

Summarizing, the algorithm is used as follows:

1. Guess the p initial values $k_1^{(0)}, \ldots, k_p^{(0)}$.
2. Integrate (7.1.25) subject to (7.1.26) and evaluate the appropriate e_j.
3. Integrate (7.7.8) subject to (7.7.9).
4. Evaluate $\partial S/\partial k_l$.
5. Form updated estimates of the k_l from (7.7.11).
6. Continue until $|k_l^{(i+1)} - k_l^{(i)}|/|k_l^{(i)}|$ is less than some preset criterion, ϵ.

7.7.2 Quasilinearization

By using finite difference approximations to the spatial derivatives in (7.1.25) and dividing the region Ω into N mesh points, (7.1.25) can be written

$$\frac{d\mathbf{U}}{dt} = \mathbf{F}(t, \mathbf{U}, \mathbf{k}) \tag{7.7.12}$$

where $\mathbf{U} = [\mathbf{u}(t, x_1)^T, \ldots, \mathbf{u}(t, x_N)^T]^T$, the nN-dimensional vector consisting of the state vector at each of the N mesh points. Since (7.7.12) is now an O.D.E., quasi-linearization, as in Section 7.6.2, can be used to estimate $\mathbf{k}$.

EXAMPLE 7.5. *Estimation of the Reaction Rate Constant for a First-Order Reaction in a Laminar Flow Tubular Reactor* [*44*]. Cleland and Wilhelm [10] studied the isothermal, liquid phase hydrolysis of acetic anhydride in a laminar flow tubular reactor. If axial diffusion is negligible compared to radial diffusion and there is no volume change on reaction, the mass balance for acetic anhydride is

$$v_0\left(1 - \left(\frac{r}{R}\right)^2\right)\frac{\partial c}{\partial z} = D\left(\frac{\partial^2 c}{\partial r^2} + \frac{1}{r}\frac{\partial c}{\partial r}\right) - kc \tag{A}$$

We let

$$u = \frac{c}{c_0}, \qquad x = \frac{r}{R}, \qquad t = \frac{z}{v_0}, \qquad \beta = \frac{D}{R^2}$$

and (A) becomes

$$(1 - x^2)\frac{\partial u}{\partial t} = \beta\left(\frac{\partial^2 u}{\partial x^2} + \frac{1}{x}\frac{\partial u}{\partial x}\right) - ku \tag{B}$$

The initial and boundary conditions for (B) are

$$\begin{aligned} u(0, x) &= 1 \\ \frac{\partial u}{\partial x} &= 0, \qquad x = 0, 1 \end{aligned} \tag{C}$$

The variables t and β are not the same as the authors used because we want to isolate k in (B). Cleland and Wilhelm studied the range of validity of (B) as a model for the reactor by comparing experimental conversions to those predicted from (B). In their work, values for the first-order reaction velocity constant k were determined beforehand from independent batch experiments. The problem we wish to consider is the opposite one, namely, the estimation of k from experimental measurements. We assume that (B) is a valid model for the reactor, and knowing β we desire to estimate k using the steepest descent algorithm developed above. We also assume that the measurements are exact; i.e., no experimental errors exist. The algorithm, of course, is unaffected by this assumption, which we make simply to test the convergence properties of the method.

To examine the convergence of the algorithm it is desirable to know the true value of k. Naturally in actual practice this will not be possible. The experimental values of $u(t, x)$ were generated by numerically integrating (B) from $t = 0$ to $t = T$ using the assumed true value of k. The experimentally observed output may assume any of the general forms (7.1.27)–(7.1.30). Two particular forms were used here. The first represents the integral average effluent conversion based on the volume flow rate,

$$y(t) = 4\int_0^1 x(1 - x^2)u(t, x)\, dx \tag{D}$$

the quantity measured experimentally by Cleland and Wilhelm, and the second represents the wall concentrations down the length of the reactor,

$$y(t) = u(t, 1) \tag{E}$$

The sensitivity coefficient $\lambda = \partial u/\partial k$ is governed by

$$(1 - x^2)\frac{\partial \lambda}{\partial t} = \beta\frac{\partial^2 \lambda}{\partial x^2} + \frac{1}{x}\frac{\partial \lambda}{\partial x} - k\lambda - u$$
$$\lambda(0, x) = 0 \tag{F}$$
$$\frac{\partial \lambda}{\partial x} = 0, \qquad x = 0, 1$$

It follows from (D) and (E) that

$$e(t) = y(t) - 4\int_0^1 x(1 - x^2)u(t, x)\, dx \tag{G}$$

and

$$e(t) = y(t) - u(t, 1) \tag{H}$$

respectively. Since the average concentration is only measured at $t = T$, the error criterion of (G) was chosen as

$$S = e(T)^2 \tag{I}$$

The measurements of $u(t, 1)$ can be made at all t, and so we selected

$$S = \int_0^T e(t)^2\, dt \tag{J}$$

corresponding to (H). The computation of $\partial S/\partial k$ in each case is straightforward.

We selected $\beta = 0.1$ and the true value of $k = 1$. Using these values, (B) was numerically integrated from $t = 0$ to $t = T$, $T = 1$. A discussion of the details of this computation can be found elsewhere [30].

Two basic questions are of interest in the application of the steepest descent algorithm. First, can we correctly identify the true value of k from either of the output observations (D) and (E)? Second, if the algorithm converges, what is the effect of the initial guess of k on the rate of convergence?

Consider first the output (D), for which $y(1) = 0.1378$, corresponding to $\beta = 0.1$ and $k = 1$. The steepest descent algorithm was applied with three different initial guesses of k and $\epsilon = 0.001$ with the following results:

Initial values			Final values		
k	γ	S	k	$S \times 10^7$	Iterations
0.5	10	0.03846	1.000061	0.2062	33
2.0	10	0.01175	1.000060	0.1962	11
0	0.2	0.5987	1.000060	0.1962	89

Convergence was obtained in each case studied, indicating that a unique value of k corresponds to each value of the integral average conversion, $y(1)$. Thus measurement of the integral average conversion from a laminar flow tubular reactor represents a valid method of obtaining first-order kinetic rate data.

The second type of experimental observation proposed, (E), is measurement of the wall concentrations down the length of the reactor. Admittedly this would be difficult to implement in practice; however, it was chosen to afford a comparison to effluent conversion output measurements. The steepest descent algorithm was used with the same three initial guesses of k as above and $\epsilon = 0.001$:

Initial values			Final values		
k	γ	S	k	$S \times 10^7$	Iterations
0.5	2	0.04524	1.00159	4.941	9
2.0	2	0.02984	1.00151	4.471	25
0	2	0.5137	1.00126	3.106	81

Again convergence was obtained in each case.

EXAMPLE 7.6. *Estimation of the Activation Energy for a Single Reaction* [46]. Consider an irreversible first-order reaction carried out in an adiabatic plug flow reactor. If physical properties are constant and the effect of radial gradients and axial diffusion are negligible, the transient concentration and temperature behavior of the reactor is described in dimensionless terms by

$$\frac{\partial u_1}{\partial t} + \frac{\partial u_1}{\partial x} = -\exp\left[\frac{ku_2}{\phi + u_2}\right]u_1 \tag{A}$$

$$\frac{\partial u_2}{\partial t} + \frac{\partial u_2}{\partial x} = \exp\left[\frac{ku_2}{\phi + u_2}\right]u_1 \tag{B}$$

where u_1 and u_2 are the dimensionless reactant concentration and temperature, respectively. At $t = 0$ the reactor is assumed to be in a steady state:

$$u_1(0, x) = u_{1_{ss}}(x) \tag{C}$$

$$u_2(0, x) = u_{2_{ss}}(x) \tag{D}$$

The inlet concentration and temperature are arbitrary time-varying functions:

$$u_1(t, 0) = u_{1_0}(t) \tag{E}$$

$$u_2(t, 0) = u_{2_0}(t) \tag{F}$$

We let $u_{1_{ss}}(x)$ and $u_{2_{ss}}(x)$ be the solutions corresponding to $u_{1_0} = 1$ and $u_{2_0} = 0$.

The problem is to estimate the activation energy k from measurements of only temperature, u_2, at R times and S' locations:

$$y_{1_{r,s}} = u_2(t_r, x_s) + \text{errors} \tag{G}$$

Artificial measurements are generated by

$$y_{1_{r,s}} = u_2^*(t_r, x_s)[1 + \sigma\,\text{GAUSS}(0, 0.3)] \tag{H}$$

where $u_2^*(t, x)$ is the solution of (A)–(F) with $\phi = 7.1203$ and the assumed true value of $k = 30.1588$ and GAUSS is a normally distributed random number routine with mean zero and standard deviation 0.3. We let

$$u_{1_0}(t) = 1 + 0.005t$$

$$u_{2_0}(t) = 0$$

to generate a transient response. The transient experiment is run from $t = 0$ to

$t = 0.6$ with measurements made only at the outlet of the reactor ($S' = 1$) with the dimensionless length $x = 0.12$.

Results of both steepest descent and quasilinearization are shown in Tables 7.7 and 7.8. In Table 7.7 we examine the effect of the level of measurement error

TABLE 7.7
EFFECT OF THE LEVEL OF MEASUREMENT ERROR ON ESTIMATED VALUES OF k, $\hat{k}$ (TRANSIENT CASE, TEMPERATURE MEASUREMENTS ONLY)

σ	Steepest descent	Quasilinearization
0.1		
$\hat{k}$	30.26595	30.27132
Iterations	5	4
% error	0.35	0.37
0.3		
$\hat{k}$	29.92751	30.48781
Iterations	5	4
% error	0.77	1.10
0.5		
$\hat{k}$	34.05261	29.77492
Iterations	8	4
% error	12.98	1.20

on the estimated value of k for $R = 30$, $N = 12$, $k^{(0)} = 15$, $\sigma = 2 \times 10^3$, and $\gamma = 0.1, 0.3, 0.5$. In Table 7.8 we examine the effect of R at $\sigma = 0.3$, $N = 12$,

TABLE 7.8
EFFECT OF NUMBER OF MEASUREMENTS R ON ESTIMATED VALUES OF k, $\hat{k}$ (TRANSIENT CASE, TEMPERATURE MEASUREMENTS ONLY)

R	Steepest descent	Quasilinearization
30		
$\hat{k}$	30.48145	30.48781
Iterations	5	4
% error	1.07	1.10
15		
$\hat{k}$	29.58272	29.58610
Iterations	7	4
% error	1.92	1.91
10		
$\hat{k}$	31.49580	31.53761
Iterations	9	4
% error	4.46	4.60

$k^0 = 15$, and $\gamma = 2 \times 10^3$. Steepest descent appears to be more sensitive to the level of errors as σ increases. As R is decreased, both methods yield almost identical results. In terms of computing time, quasilinearization takes roughly twice as long as steepest descent.

REFERENCES

1. Anderssen, A. S., and E. T. White, "Parameter Estimation by the Transfer Function Method," *Chem. Eng. Sci. 25*, 1015 (1970).
2. Angus, R. M., and L. Lapidus, "Characterization of Multiple Variable Linear Systems from Random Inputs," *AIChE J. 9*, 810 (1963).
3. Bard, Y., "Comparison of Gradient Methods for the Solution of Nonlinear Parameter Estimation Problems," *SIAM J. Numer. Anal. 7*, 157 (1970).
4. Bard, Y., and L. Lapidus, "Kinetics Analysis by Digital Parameter Estimation," *Catalysis Rev. 2*, 67 (1968).
5. Bellman, R., J. Jacquez, R. Kalaba, and S. Schwimmer, "Quasilinearization and the Estimation of Rate Constants from Raw Kinetic Data," *Math. Biosci. 1*, 71 (1967).
6. Bennett, A., and F. Goodridge, "Hydrodynamic and Mass Transfer Studies in Packed Absorption Columns. I. Axial Liquid Dispersion," *Trans. Inst. Chem. Eng. 48*, T232 (1970).
7. Buffham, B. A., "The Effects of Intrasolid Resistance and Axial Mixing on Transient Exchange in Packed Beds and the Unified Time Delay Model," *Chem. Eng. J. 2*, 71 (1971).
8. Cerro, R. L., and J. M. Smith, "Chromatography of Nonadsorbable Gases," *AIChE J. 16*, 1034 (1970).
9. Chao, R., and H. E. Hoelscher, "Simultaneous Axial Dispersion and Adsorption in a Packed Bed," *AIChE J. 12*, 271 (1966).
10. Cleland, F. A., and R. H. Wilhelm, "Diffusion and Reaction in Viscous-Flow Tubular Reactor," *AIChE J. 2*, 489 (1956).
11. Davidon, W. C., *Variable Metric Method for Minimization*, A.E.C. Res. Develop. Report ANL-5990, 1959.
12. Fletcher, R., and M. J. D. Powell, "A Rapidly Convergent Descent Method for Minimization," *Comp. J. 6*, 163 (1963).
13. Florentin, J. J., B. D. Hainsworth, J. B. Reswick, and J. H. Westcott, "Correlation Analysis of a Heat Exchanger" Joint Symposium on Instrumentation and Control, London, 1959.
14. Goldfeld, S. M., R. E. Quandt, and H. F. Trotter, "Maximization by Quadratic Hill-Climbing," *Econometrica 34*, 541 (1966).
15. Greenstadt, J., "On the Relative Efficiencies of Gradient Methods," *Math. Comp. 21*, 360 (1967).
16. Gunn, D. J., "The Transient and Frequency Response of Particles and Beds of Particles," *Chem. Eng. Sci. 25*, 53 (1970).
17. Grubner, D., "Statistical Moments Theory of Gas-Solid Chromatography," *Advan. Chromatog. 6*, 173 (1968).

18. Grubner, O., and D. W. Underhill, "Calculation of Adsorption Bed Capacity by the Theory of Statistical Moments," *Separation Sci. 5*, 555 (1970).

19. Himmelblau, D., *Process Analysis by Statistical Methods*, Wiley, New York, 1970.

20. Homan, C. J., and J. W. Tierney, "Determination of Dynamic Characteristics of Processes in the Presence of Random Disturbances," *Chem. Eng. Sci. 12*, 153 (1960).

21. Hooke, R., and T. A. Jeeves, "Direct Search Solution of Numerical and Statistical Problems," *J. Assoc. Comp. Mach. 8*, 212 (1961).

22. Hopkins, M. J., A. J. Sheppard, and P. Eisenklam, "The Use of Transfer Functions in Evaluating Residence Time Distribution Curves," *Chem. Eng. Sci. 24*, 1131 (1969).

23. Hougen, O., and K. M. Watson, *Chemical Process Principles*, vol. 3, Wiley, New York, 1948.

24. Hutchinson, A. W., and R. J. Shelton, "Measurement of Dynamic Characteristics of a Full-Scale Plant Using Random Perturbing Signals," *Trans. Inst. Chem. Eng. 45*, T334 (1967).

25. Jeffreson, C. P., "Dynamics of Packed Beds with Intraphase Heat or Mass Transfer," *Chem. Eng. Sci. 23*, 509 (1968); also *23*, 613 (1968).

26. Jeffreson, C. P., "A Unification of Dynamic Testing," *Chem. Eng. Sci. 25*, 1319 (1970).

27. Jenkins, G. M., and G. D. Watts, *Spectral Analysis and its Applications*, Holden-Day, Inc., San Francisco, 1967.

28. Johnson, J. L., L. T. Fan, and Y. Wu, "Comparison of Moments, *S*-Plane, and Frequency Response Methods for Analyzing Pulse Testing Data from Flow Systems," *Ind. Eng. Chem. Fund. 10*, 425 (1971).

29. Kucera, E., "Contributions to the Theory of Chromatography: Linear Non-Equilibrium Elution Chromatography," *J. Chromatog. 1*, 237 (1965).

30. Lapidus, L., *Digital Computation for Chemical Engineers*, McGraw-Hill, New York, 1962.

31. Lapidus, L., and J. H. Seinfeld, *Numerical Solution of Ordinary Differential Equations*, Academic Press, New York, 1971.

32. Lee, E. S., *Quasilinearization and Invariant Imbedding*, Academic Press, New York, 1968.

33. Levenberg, K., "A Method for the Solution of Certain Nonlinear Problems in Least Squares," *Quart. Appl. Math. 2*, 164 (1944).

34. Levenspiel, O., and K. B. Bischoff, "Patterns of Flow in Chemical Process Vessels," *Advan. Chem. Eng. 3*, 95 (1963).

35. Marquardt, D. W., "An Algorithm for Least Squares Estimation of Nonlinear Parameters," *J. Soc. Ind. Appl. Math. 11*, 431 (1963).

36. Neiman, R. E., D. G. Fischer, and D. E. Seborg, "A Review of Process Identification and Parameter Estimation Techniques," *Intern. J. Control 13*, 209 (1971).

37. Ostergaard, K., and M. L. Michelsen, "On the Use of the Imperfect Tracer Pulse Method for Determination of Hold-Up and Axial Mixing," *Can. J. Chem. Eng. 47*, 107 (1969).

38. Powell, M. J. D., "An Efficient Method for Finding the Minimum of a Function of Several Variables Without Calculating Derivatives," *Comp. J. 7*, 155 (1964).

39. Rosenbrock, H. H., "An Automatic Method for Finding the Greatest or Least Value of a Function," *Comp. J. 3*, 175 (1960).

40. Sagara, M., P. Schneider, and J. M. Smith, "The Determination of Heat-Transfer Parameters for Flow in Packed Beds Using Pulse Testing and Chromatography Theory," *Chem. Eng. J. 1*, 47 (1970).

41. Schiesser, W. E., and L. Lapidus, "Further Studies of Fluid Flow and Mass Transfer in Trickle Beds," *AIChE J. 7*, 163 (1961).

42. Schneider, P., and J. M. Smith, "Adsorption Rate Constants from Chromatography," *AIChE J. 14*, 762 (1968).

43. Schneider, P., and J. M. Smith, "Chromatographic Study of Surface Diffusion," *AIChE J. 14*, 886 (1968).

44. Seinfeld, J. H., "Identification of Parameters in Partial Differential Equations," *Chem. Eng. Sci. 24*, 65 (1969).

45. Seinfeld, J. H., "Nonlinear Estimation Theory," *Ind. Eng. Chem. 62*, 1, 32 (1970).

46. Seinfeld, J. H., and W. H. Chen, "Estimation of Parameters in Distributed Systems from Noisy Experimental Data," *Chem. Eng. Sci. 26*, 753 (1971).

47. Seinfeld, J. H., and G. R. Gavalas, "Analysis of Kinetic Parameters from Batch and Integral Reaction Experiments," *AIChE J. 16*, 644 (1970).

48. Suzuki, M., and J. M. Smith, "Kinetic Studies by Chromatography," *Chem. Eng. Sci. 26*, 221 (1971).

49. Van Swaaij, W. P. J., J. C. Charpentier, and J. Villermaux, "Residence Time Distribution in the Liquid Phase of Trickle Flow in Packed Columns," *Chem. Eng. Sci. 24*, 1083 (1969).

50. Villermaux, J., and W. P. J. Van Swaaij, "Modèles représentatifs de la distribution des temps de séjours dans un réacteur semi-infini à dispersion axiale avec zones stagnantes," *Chem. Eng. Sci. 24*, 1097 (1969).

51. Williams, J. A., R. J. Adler, and W. J. Zolner, III, "Parameter Estimation of Unsteady-State Distributed Models in the Laplace Domain," *Ind. Eng. Chem. Fund. 9*, 193 (1970).

52. Woodburn, E. T., R. P. King, and R. C. Everson, "Optimal Estimation of Process Parameters Using Pseudo-Random Binary Signals with Applications to Mixing in a Packed Tower," *Can. J. Chem. Eng. 47*, 301 (1969).

53. Zadeh, L., "From Circuit Theory to System Theory," *Proc. I.R.E. 50*, 856 (1962).

PROBLEMS

7.1 Consider the determination of the thermal diffusivity of a semiinfinite slab using frequency response. The slab shown in Figure P7.1 is subjected to temperature variation at the ($x = 0$)-face of $T(0, t) = A \cos \omega t$. If the temperature of the slab is governed by

$$\frac{\partial T}{\partial t} = k\frac{\partial^2 T}{\partial x^2}$$

$$T(x, 0) = 0$$

$$\frac{\partial T}{\partial x} = 0, \qquad x = l$$

$$T(0, t) = A \cos \omega t$$

determine the amplitude ratio and phase angle of a temperature measurement at $x = l$. *Note:* You must let t become large so that transients not associated with the frequency response can die out.

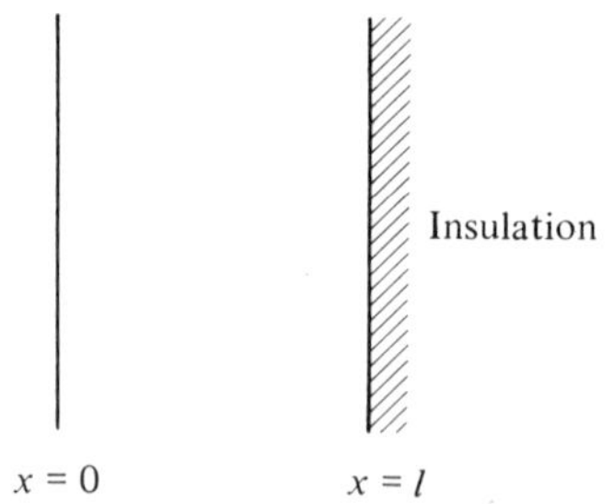

Figure P7.1. Slab of material occupying the region $x = 0$ to $x = l$ and insulated at the $x = l$ face.

7.2 The dispersion of solute from a steady point source on the axis of a packed bed can be described by

$$\bar{v}\frac{\partial c}{\partial z} = D_R\left[\frac{\partial^2 c}{\partial r^2} + \frac{1}{r}\frac{\partial c}{\partial r}\right]$$

if we neglect the effect of axial mixing. The boundary conditions are

$$\frac{\partial c}{\partial r} = 0, \qquad r = 0, R$$

a. Define an appropriate Peclet number Pe_R and place this equation in dimensionless form. Let c_0 be the concentration infinitely far downstream when the concentration is uniform across the tube.
b. Consider the problem of estimating Pe_R from an experimental test. Describe

an experimental procedure from which Pe_R can be estimated. Assume that concentration profiles can be measured at various distances down the reactor. (Recall Problem 6.9)

7.3 The dispersion of tracer NaCl in water in a 3-foot long, 4-inch-diameter column packed with spheres was studied by Schiesser and Lapidus [41]. A step or pulse of NaCl was introduced at the top of the column and the salt concentrations were monitored at the outlet of the column as a function of time. The data obtained for $\frac{1}{4}$-inch porous and nonporous alumina spheres are shown in Table P7.3. The water flow rate for these experiments was 9.28 gallons per minute-square foot. The internal holdup in the porous spheres, calculated from the difference in step responses, was 0.215 cubic foot per cubic foot of bed, or 79 percent of the pore volume.

TABLE P7.3[1]

Pulse response, Vc/Q (sec^{-1})			Step response, c/c_0		
Time (sec)	Porous	Nonporous	Time (sec)	Porous	Nonporous
12.5	0.0015	0.0015	1.0	1.000	1.000
13.5	0.0050	0.0050	3.0	1.000	1.000
14.5	0.0080	0.0100	5.0	1.000	1.000
15.5	0.0160	0.0190	7.0	1.000	1.000
16.5	0.0220	0.0380	9.0	1.000	1.000
17.5	0.0350	0.0620	11.0	1.000	1.000
18.5	0.0460	0.0850	13.0	1.000	1.000
19.5	0.0590	0.1000	15.0	0.990	—
20.5	0.0650	0.1030	17.0	0.950	0.950
21.5	0.0670	0.1010	19.0	0.820	0.880
22.5	0.0660	0.0910	21.0	0.630	0.750
23.5	0.0630	0.0780	23.0	0.620	0.450
24.5	0.0580	0.0650	25.0	0.500	0.290
25.5	0.0510	0.0530	27.0	0.410	0.190
26.5	0.0470	0.0420	29.0	0.340	0.120
27.5	0.0400	0.0330	31.0	0.290	0.070
29.0	0.0310	0.0240	33.0	0.260	0.050
31.0	0.0220	0.0140	35.0	0.220	0.040
33.0	0.0170	0.0090	37.0	0.210	0.030
35.0	0.0125	0.0070	39.0	0.200	0.020
37.0	0.0100	0.0050	41.0	0.190	0.020
39.0	0.0075	0.0040	43.0	—	0.010
41.0	0.0060	0.0030	45.0	0.180	0.010
43.0	0.0050	0.0020			
45.0	0.0030	0.0020			

Notation:

V = volumetric flow rate, ft^3 liquid/sec
c = concentration
Q = moles injected in a pulse input
t = time elapsed since injection of input

[1]Data of Schiesser and Lapidus [41].

Assume that the tracer concentration is well mixed radially so that only axial dispersion is important. Using moment matching,

a. Model this system using the axial-dispersed plug flow model; that is, determine the average residence time and the Peclet number.
b. Model the system with N identical ideally mixed vessels in series. Determine the size of the tanks (average residence time μ) and the number N.
c. Model the system with N identical ideally mixed vessels with backflow. Determine the size of the tanks, N, and b, the backflow parameter.
d. Compare and discuss.

7.4 We want to compare various models for axial dispersion in a flowing system in conjunction with parameter estimation in the transform domain [22]. Specifically, we want to consider the models in Figure P7.4.

a. Letting μ denote the average residence time in each total system, show that

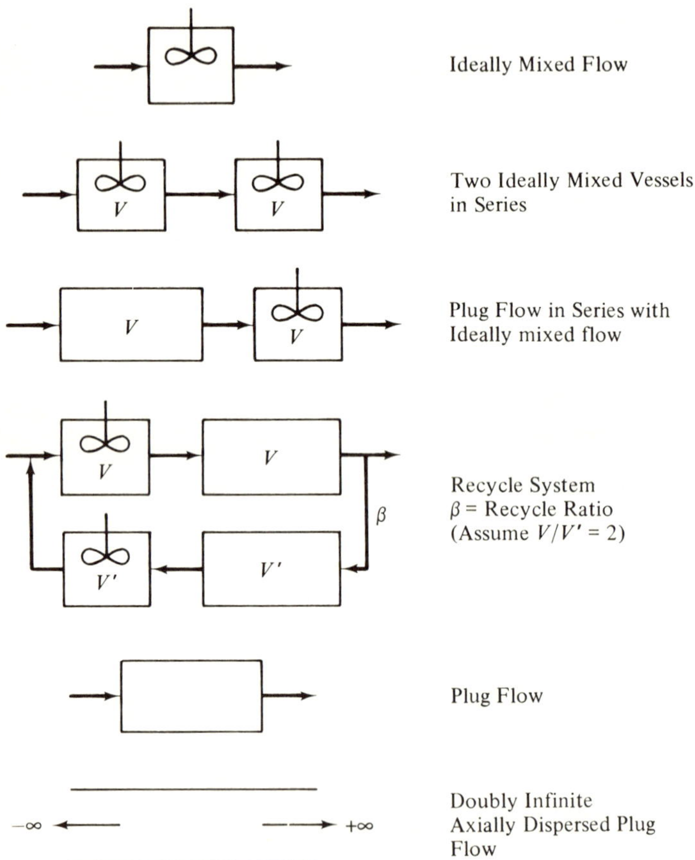

Figure P7.4. Models for axial dispersion in a flowing system.

the transfer functions of these six systems are

$$\frac{1}{1 + \mu s}$$

$$\frac{1}{(1 + \mu s)^2}$$

$$\frac{e^{-\mu s/2}}{(1 + \mu s/2)}$$

$$\frac{\exp(-\mu s/6)/[\mu s/6) + 1]}{2 - [\exp(-\mu s/3)/(1 + \mu s/6)^2]}, \qquad \beta = 1$$

$$\exp(\ \mu s)$$

$$\exp\left\{\frac{\text{Pe}}{2}\left[1 - \left(1 + \frac{4\mu s}{\text{Pe}}\right)^{1/2}\right]\right\}$$

b. Plot these six transfer functions vs. μs for values of μs from zero to 2.5. For the axial-dispersed plug flow model, use Pe = 2 and 20. From your plot suggest a lower limit on the value of μs below which values should not be employed to distinguish among these models.

c. Do you think it is necessary to prescribe an upper limit on the value of μs used for parameter estimation using s-plane matching of transfer function and data at real values of s?

7.5 A large number of simple exponential processes, such as the saturation of an ideally mixed vessel with tracer from a step function, are described by the relation

$$\eta = k_1(1 - e^{-k_2 t})$$

For the jth time measurement

$$y_j = k_1(1 - e^{-k_2 t_j}) + \epsilon_j$$

The only experimental variable is the elapsed time t_j, while there are two parameters and one output. Suppose the data shown in Table P7.5 were obtained.

a. Compute

$$S(k_1, k_2) = \sum_{j=1}^{6} [y_j - f(x_j, k_1, k_2)]^2$$

TABLE P7.5
EXPERIMENTAL DATA

t_j (min)	y_j
40	3.7
80	5.9
120	6.4
160	8.3
200	9.1
240	9.2

as a function of k_1 and k_2 over the ranges $8 \leq k_1 \leq 14$ and $0.003 \leq k_2 \leq 0.017$. Plot contours of constant S in the $(k_1 - k_2)$-plane. You should find the minimum at $k_1 = 10.5$ and $k_2 = 0.0104$, at which $S = 0.822$.

b. Determine $\hat{k}_1$ and $\hat{k}_2$ using steepest descent from $k_1^{(0)} = 14$, $k_2^{(0)} = 0.017$. Plot the course of the minimization in the $k_1 - k_2$ plane. Use $\lambda = 1$.
c. Repeat part b for the Newton-Raphson method. Use $\lambda = 1$.

7.6 Bellman et al. [5] have considered the estimation of the two rate constants k_1 and k_2 in the Bodenstein-Lindner model for the gas phase reaction of NO with O_2:

$$2NO + O_2 \underset{k_2}{\overset{k_1}{\rightleftharpoons}} 2NO_2$$

The model is

$$\frac{dx}{dt} = k_1(126.2 - x)(91.9 - x)^2 - k_2x^2, \qquad x(0) = 0$$

where x is the concentration of NO_2. Concentrations of NO_2 were measured experimentally at a number of times and were

t	0	1	2	3	4	5	6	7	9
y	0	1.4	6.3	10.5	14.2	17.6	21.4	23.0	27.0
t	11	14	19	24	29	39			
y	30.5	34.4	48.8	41.6	43.5	45.3			

Giving equal weighting to all observations,

a. Estimate k_1 and k_2 by the method of steepest descent.
b. Estimate k_1 and k_2 by quasilinearization.
c. Estimate k_1 and k_2 by the extended Gauss-Newton method.

In each case use as initial guesses $k_1^{(0)} = 10^{-6}$ and $k_2^{(0)} = 10^{-4}$. Compare your results to the values of Bodenstein and Lindner, $k_1 = 0.53 \times 10^{-5}$ and $k_2 = 0.41 \times 10^{-3}$.

d. To examine the convergence of the three algorithms, repeat the estimation with the methods with the following sets of initial guesses:
   (1) $k_1^{(0)} = 10^{-7}$, $k_2^{(0)} = 10^{-5}$.
   (2) $k_1^{(0)} = 10^{-4}$, $k_2^{(0)} = 10^{-2}$.
   (3) $k_1^{(0)} = 10^{-1}$, $k_2^{(0)} = 1$.
e. To examine the effect of the number of times at which data were taken, repeat the estimation with the methods using only the following sets of y-values:
   (1) $t = 0, 1, 2, 3, 4, 5, 6$.
   (2) $t = 0, 11, 14, 19, 24, 29, 39$.

   In this particular case does it matter if the data are collected early or late in the experiment?

7.7 Consider the following set of reactions:

$$A \underset{k_2}{\overset{k_1}{\rightleftharpoons}} B + C$$

$$B + B \xrightarrow{k_3} C$$

If x_1 = concentration of A and x_2 = concentration of B, the rate equations are

$$\frac{dx_1}{dt} = -k_1x_1 - \frac{3}{2}k_2x_1x_2 + \frac{3}{2}k_2x_2 - \frac{1}{2}k_2x_2^2, \qquad x_1(0) = 1$$

$$\frac{dx_2}{dt} = k_1x_1 + \frac{3}{2}k_2x_1x_2 - \frac{3}{2}k_2x_2 + \frac{1}{2}k_2x_2^2 - 2k_3x_2^2, \qquad x_2(0) = 0$$

Let us assume that the true values of k_1, k_2, and k_3 are

$$k_1 = 0.04, \qquad k_2 = 10^4, \qquad k_3 = 3 \times 10^7$$

Generate your own experimental data by integrating the rate equations numerically with the true k-values and then adding normal random error to the x_1- and x_2-values at integer values of t from zero to 30. The normal random error should have zero mean and normalized standard deviation of 0.1. Then using these data, estimate k_1, k_2, and k_3 using both steepest descent and quasilinearization. Select your own initial guesses. *Note:* The wide variation in the magnitude of the rate constants make this system highly "stiff"; that is, the eigenvalues of the O.D.E. differ greatly in magnitude. For a discussion of the numerical solution of this system, see Lapidus and Seinfeld [31].

7.8 A problem of considerable importance to the petroleum industry is *history matching*, the mathematical modeling of petroleum reservoirs. The term *history matching* is used because the objective is to estimate reservoir parameters so that the prior measured pressure histories at the wells are matched as closely as possible by the mathematical model. Obviously, this is precisely the parameter estimation problem we have been studying.

The ultimate objective of history matching is to determine reservoir properties so that future reservoir responses can be determined for various production policies. We want to be able to predict future reservoir pressures as a function of production rates to determine optimal production policies for the reservoir. To predict ahead in time we need to know as accurately as possible the properties of the reservoir that can be expected to remain generally constant in time. Knowing the properties of the reservoir, we can then use the model to explore various future production policies.

Let us consider the history-matching problem with reference to a simple reservoir model. We assume

(1) Two-dimensional reservoir, no property variations with thickness.
(2) Constant known compressibility, porosity, thickness, and viscosity. Unknown spatial variation of permeability.
(3) Single-phase system.
(4) The production (or injection) rates of each of M wells are known over a time period $0 \leq t \leq T$. The reservoir pressures at each of the M wells are measured over the time period $0 \leq t \leq T$.

From our assumptions the reservoir model is

$$c\phi h\frac{\partial p}{\partial t} = \frac{h}{\mu}\frac{\partial}{\partial x}\left[k(x, y)\frac{\partial p}{\partial x}\right] + \frac{h}{\mu}\frac{\partial}{\partial y}\left[k(x, y)\frac{\partial p}{\partial y}\right] + Q(x, y, t) \qquad \text{(A)}$$

where c, ϕ, h, μ, and k are constants and $Q(x, y, t)$ is the production or injection rate. If the M wells have coordinate locations $x_1, y_1; x_2, y_2; \ldots; x_M, y_M$, then

$$Q(x, y, t) = Q_i\,\delta(x - x_i)\,\delta(y - y_i), \qquad i = 1, 2, \ldots, M \tag{B}$$

The boundary conditions on (A) are

$$p(x, y, 0) = p_0, \qquad \text{all } x, y \tag{C}$$

$$p(x, y, t) = p_0', \qquad x, y \text{ on boundary} \tag{D}$$

The pressure observations at each of the M wells are denoted by $p_k^{\text{obs}}(t)$, $k = 1, 2, \ldots, M$. The measurements are related to the reservoir pressure p by

$$p_k^{\text{obs}}(t) = p(x_k, y_k, t) + \text{errors}, \qquad k = 1, 2, \ldots, M \tag{E}$$

The problem is to determine the reservoir permeability distribution $k(x, y)$ from the observed pressures. Let us consider the special case in which $k(x, y) = k$, a constant.

Assume that the reservoir occupies the region

$$0 \le x \le L_1$$
$$0 \le y \le L_2$$

Estimate k using the method of steepest descent for the following parameter values:

$$L_1 = 60{,}000 \text{ ft} \qquad M = 2\ (x_1 = y_1 = 10{,}000;$$
$$L_2 = 30{,}000 \text{ ft} \qquad x_2 = y_2 = 20{,}000)$$
$$c = 1.72 \times 10^{-5} \text{ psi}^{-1} \qquad p_0 = p_0' = 3000 \text{ psi}$$
$$\phi = 0.2 \qquad Q_i = -1000 \text{ ft}^3/\text{day}, \quad i = 1, 2$$
$$h = 1 \text{ ft}$$
$$\mu = 0.352 \text{ cp}$$

To do this assume that the true value of k is 0.05 millidarcies. Generate your own noisy data on the computer and then use these data to estimate k. Select your own levels of error. Be careful of your units! For aid on this question, see C. S. Mathews and D. G. Russell, *Pressure Buildup and Flow Tests in Wells*, Society of Petroleum Engineers, New York, 1967, Chap. 2.

7.9 Sagara et al. [40] have studied the determination of heat transfer parameters for flow in packed beds by relating the velocity and dispersion of a temperature pulse in a fluid flowing in a packed bed to specific heats and rate coefficients by the method of moments. The following assumptions were made:

(1) Adiabatic column.
(2) Negligible heat transfer from particle to particle by conduction.
(3) A single temperature T_i can be used to represent the energy transfer within the particle.
(4) Axial velocity and temperature profiles across the bed are uniform.
(5) Particles are spherical.
(6) The internal temperature T_i varies only with radial position in a particle.

The energy balance for the bed is

$$\frac{k_{ax}}{\alpha\rho_f c_f}\frac{\partial^2 T}{\partial z^2} - v\frac{\partial T}{\partial z} - \frac{\partial T}{\partial t} - \frac{3}{r_0}\left(\frac{1-\alpha}{\alpha}\right)\left(\frac{k_e}{\rho_f c_f}\right)\left(\frac{\partial T_i}{\partial r}\right)_{r=r_0} = 0 \qquad \text{(A)}$$

The energy balance for a particle is

$$\frac{\partial T_i}{\partial t} = \frac{k_e}{c_p \rho_p}\left[\frac{\partial T_i}{\partial r^2} + \frac{2}{r}\frac{\partial T_i}{\partial r}\right] \qquad \text{(B)}$$

Initial and boundary conditions for the experiments performed are

$$\begin{aligned}
T &= T_0 & z &= 0 & 0 \le t \le t_0 &\text{ (square wave)}\\
T &= 0 & z &= 0 & t > t_0 &\\
T &= 0 & t &= 0 & z > 0 &\\
T &= 0 & z &= \infty & t > 0 &\\
T_i &= 0 & t &= 0 & 0 \le r \le r_0 & \qquad \text{(C)}\\
\frac{\partial T_i}{\partial r} &= 0 & r &= 0 & t > 0 &\\
k_e\frac{\partial T}{\partial r} &= h_f(T - T_i) & r &= r_0 & t > 0 &
\end{aligned}$$

a. By Laplace-transforming (A) and (B) with respect to t and solving the resulting O.D.E. in z and then differentiating the L.T. according to

$$\mu_n' = \frac{(-1)^n}{m_0}\frac{d}{ds}\bar{T}(L, s)\bigg|_{s=0}$$

$$\mu_n' = \frac{m_n}{m_0} = \frac{1}{m_0}\int_0^\infty t^n T(L, t)\, dt \qquad \text{(D)}$$

show that the first two normalized noncentral moments are

$$\mu_1' = \frac{z}{v}(1 + \delta_0) + \frac{t_0}{2} \qquad \text{(E)}$$

$$\mu_2' = 2\frac{z}{v}\left[\delta_1 + K_{ax}(1 + \sigma_0)^2\left(\frac{1}{v}\right)^2\right] + \frac{1}{12}t_0^2 \qquad \text{(F)}$$

where

$$\delta_0 = \left(\frac{1-\alpha}{\alpha}\right)\sigma, \qquad \delta_1 = \left(\frac{1-\alpha}{\alpha}\right)\sigma\frac{r_0^2}{15}\left[\frac{1}{K_e} + \frac{5}{r_0 H_f}\right]$$

and

$$K_{ax} = \frac{k_{ax}}{\alpha\rho_f c_f}, \qquad K_e = \frac{k_e}{\rho_p c_p}, \qquad H_f = \frac{h_f}{\rho_p c_p}, \qquad \sigma = \frac{\rho_p c_p}{\rho_f c_f}$$

b. Moments can be calculated from the experimental temperature curves by (D) by numerically evaluating the integral. Then (E) and (F) can be used to evaluate the various rate parameters involved in the definition of K_{ax}, K_e, H_f, and σ. The authors show that the variance can be written as

$$\frac{\mu_2 - t_0^2/12}{2(z/v)} = \delta_s + \delta_f + \delta_e$$

where

$$\delta_s = \delta_0 \frac{r_0^2}{15} \frac{\rho_p c_p}{k_e} \qquad \text{(intraparticle conduction)}$$

$$\delta_f = \delta_0 \frac{r_0}{3} \frac{\rho_p c_p}{h_f} \qquad \text{(heat transfer between fluid and particle)}$$

$$\delta_e = \frac{d_p(1 + \delta_0)^2 \alpha}{v\text{Pe}} \qquad \text{(axial dispersion of heat)}$$

$$\text{Pe} = \frac{d_p \rho_f c_f v}{k_{ax}} \qquad \text{(axial Peclet number for heat transfer)}$$

A typical experimental input and output record is shown in Figure P7.9. Discuss how you would determine the various δs representing different contributions to heat transfer from experimental data like that in Figure P7.9. Consult the original article for a careful description of the experiments performed and an analysis of the data by moment matching.

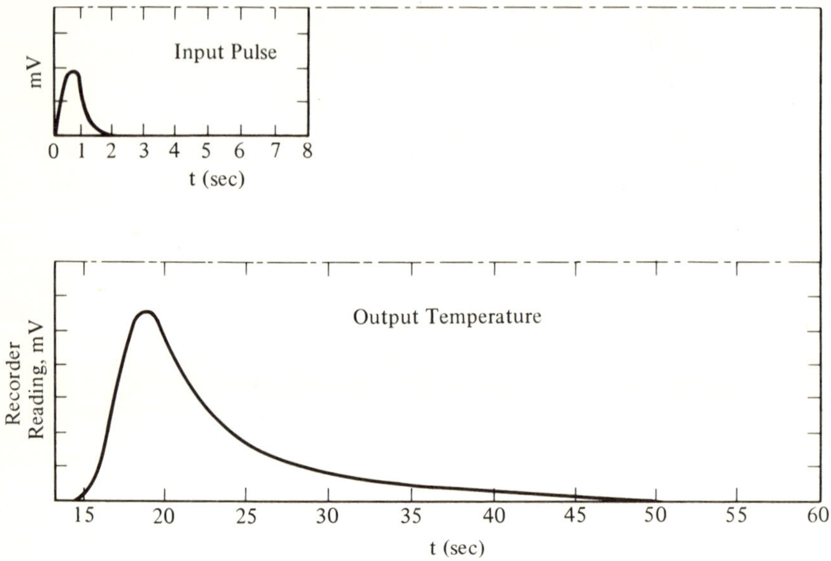

Figure P7.9. Experimental input and output data. [Reprinted from M. Sagara, P. Schneider and J. M. Smith, "The Determination of Heat-Transfer Parameters For Flow in Packed Beds Using Pulse Testing and Chromatographic Theory," *Chem. Eng. J.*, *1*, 47 (1970). Reprinted by permission of the copyright owner.]

Design of Experiments for Parameter Estimation

8

The parameter estimates we obtained in the last chapter are, in themselves, not the whole story. It is essential to know not only what the estimates are but also how reliable they are. The observations are random variables, and hence the estimates which are computed from them are also random variables; thus it is meaningful to try and estimate their probability distribution. Fortunately, it turns out that the distribution usually approaches the normal as the number of observations is increased. The means of the estimated distribution are the estimated values of the parameters; the covariance matrix of the distribution is a measure of the reliability of the estimates. This matrix expresses the manner in which variations in the observations would affect the parameter estimates.

The alternative approach to evaluating the reliability is by exploring the dependency of the objective function on the parameters. If the objective function is little affected by changes in a certain parameter, one would have doubts concerning its value. Since at the final estimates the objective function is at a minimum, its gradient vanishes and the effect of the parameters on the objective function is summarized in the second derivative matrix. Actually these two measures of reliability are equivalent; in fact, if **P** is the covariance matrix of the estimates, then it can be shown that

$$[\mathbf{P}^{-1}]_{ij} = \frac{\partial^2 \log L}{\partial k_i \partial k_j}$$

where L is the likelihood function.

The diagonal elements of **P** are the variances of the parameter estimates, and their square roots are the standard deviations. Off-diagonal elements indicate the interdependence of the estimates of the various parameters.

In this chapter we shall complete the treatment of parameter estimation by considering the computation of the reliability or accuracy of parameter estimates and the associated question of designing experiments so that optimally accurate estimates may be obtained, a particularly important aspect of which is the need to distinguish among rival models.

We shall consider only models that are, in general, nonlinear with respect to parameters, as was the case in Chapter 7. We have purposely avoided the issue of linear least squares because of its rather limited applicability and the fact that it is well covered elsewhere (see, for example, Himmelblau [9], Chaps. 4 and 5). In addition, this chapter is not intended as an exhaustive treatment of experimental design but rather as an introduction of one phase of experimental design dealing with parameter estimation. Again, for more complete coverage, we direct the reader to Himmelblau ([9], Chaps. 7 and 8).

8.1 Accuracy of Parameter Estimates

The parameter estimation problem, regardless of the form of the model, seeks the global minimum of $S(\mathbf{k})$. The accuracy or reliability of the estimates is encompassed in the covariance matrix **P** of the estimate $\hat{\mathbf{k}}$. Before we actually consider the computation of **P**, let us discuss qualitatively what effects might lead to large variances on some parameters (we say then that these parameters are "ill-determined"):

1. The model chosen is inappropriate. This case is usually marked by systematic deviations between model predictions and data (large residuals resulting from a consistently low or high deviation). The obvious remedy is to modify the model. Those terms whose parameters are most ill-determined should be prime targets for elimination or modification. The nature of the deviations of the residuals may also hold clues as to how the model should be changed. Hunter and Mezaki [11] have provided an excellent example of how analysis of residuals can be used for systematic model adaptation.

2. The measurement precision is low. This case is characterized by large random residuals, i.e., randomly high and low deviations. If no improvement of the measurement techniques is feasible, the only remedy is to make more

measurements. Unfortunately, a 10-fold increase in precision may require a 100-fold increase in the number of experiments. It should be noted, however, that the attainable precision in the estimates for a given number of experiments is maximized when the experimental conditions are properly chosen. This point is discussed below.

3. The experiments were not properly designed. This is the conclusion that must be reached if some parameters have large variances, even though the overall fit is good, i.e., even though the residuals are small. We shall cite several examples of how this may happen:

a. Suppose that we try to estimate both k_1 and k_2 in the reaction system

$$A \begin{array}{l} \xrightarrow{k_1} B \\ \xrightarrow[k_2]{} C \end{array} \tag{8.1.1}$$

from measurements of A only. It is only possible from these measurements to estimate $k_1 + k_2$ but not the individual values of k_1 and k_2. Thus we must devise an experiment such that if the measurements were totally without error it would be theoretically possible to recover each of the desired parameters. We could call this condition *parameter observability*. We shall develop precise conditions for parameter observability in this chapter.

b. Suppose that the model equation is

$$y_j = k_1 x_{1_j} + k_2 x_{2_j}$$

and that in all experiments it happened that $x_{1_j} \simeq x_{2_j}$. Then

$$y_j = (k_1 + k_2) x_{1_j}$$

would represent the data just as well as the original model. On this basis it is impossible to estimate k_1 and k_2 individually—only their sum.

c. Suppose that the objective function is a sum of several terms the orders of magnitudes of which vary greatly. Then the objective function will be far less sensitive to changes in those parameters associated with the smaller terms than those associated with the larger terms. The objective function should be redefined such that all terms are the same order of magnitude.

d. When two parameters appear in the model as a product of themselves or even as a product of functions of themselves, it is possible that an identical response can be obtained by widely varying values of the two parameters.

In each of these cases either proper experimental planning or proper model formulation would have eliminated the difficulties.

These examples illustrate the two distinct phases of *design* and *analysis*. The importance of proper design cannot be overstated. Very little useful information can be extracted from poorly planned data no matter how powerful the analysis might be. On the other hand, with a clever experimental design, much information may often be obtained with simple parameter estimation methods.

We shall now consider the actual computation of the covariance matrix $\mathbf{P}$ for a set of estimated parameters.

8.1.1 Algebraic models

Consider (7.1.6), our basic algebraic model. Let us assume that the m-dimensional error vectors, $\boldsymbol{\epsilon}_j, j = 1, 2, \ldots, M$, are independent and satisfy individually a normal distribution with mean zero and covariance $\mathbf{V}_j$, as in (7.1.13):

$$p(\boldsymbol{\epsilon}_j) = \frac{1}{(2\pi)^{m/2}\,|\mathbf{V}_j|^{1/2}} \exp\left(-\frac{1}{2}\boldsymbol{\epsilon}_j^T \mathbf{V}_j^{-1} \boldsymbol{\epsilon}_j\right) \tag{8.1.2}$$

We assume that the estimates $\hat{\mathbf{k}}$ have been determined to minimize the least squares criterion (7.1.7) with $\mathbf{Q}_j = \mathbf{V}_j^{-1}$, i.e., the maximum likelihood estimates:

$$S(\mathbf{k}) = \sum_{j=1}^{M} [\mathbf{y}_j - \mathbf{f}(\mathbf{x}_j, \mathbf{k})]^T \mathbf{V}_j^{-1} [\mathbf{y}_j - \mathbf{f}(\mathbf{x}_j, \mathbf{k})] \tag{8.1.3}$$

Our objective is to compute the covariance matrix of the $\mathbf{k}$, assuming that they also obey a normal distribution with mean $\mathbf{k}^*$, the true value of $\mathbf{k}$. Thus let

$$\hat{\mathbf{k}} = \mathbf{k}^* + \boldsymbol{\kappa}$$

where $\boldsymbol{\kappa}$ is the deviation from $\mathbf{k}^*$ of the $\hat{\mathbf{k}}$. We see that $\mathbf{f}$ is a nonlinear function of $\hat{\mathbf{k}}$ and that S is a nonlinear function of $\mathbf{f}$. Thus it will not be possible in general to recover the distribution of $\hat{\mathbf{k}}$ from that for the $\boldsymbol{\epsilon}_j$ because of all the nonlinear relationships. What we do to enable us to proceed is assume that $\hat{\mathbf{k}}$ is fairly close to $\mathbf{k}^*$ and then linearize (8.1.3) to first order about $\mathbf{k}^*$.

We found $\hat{\mathbf{k}}$ as that value that minimizes S. Since $\mathbf{k}^*$ is a fixed value, we can also state the minimization as having been carried out with respect to $\boldsymbol{\kappa}$. Therefore we want to find the value of $\boldsymbol{\kappa}$ such that

$$\frac{\partial}{\partial \kappa_i} \sum_{j=1}^{M} [\mathbf{y}_j - \mathbf{f}(\mathbf{x}_j, \mathbf{k}^* + \boldsymbol{\kappa})]^T \mathbf{V}_j^{-1} [\mathbf{y}_j - \mathbf{f}(\mathbf{x}_j, \mathbf{k}^* + \boldsymbol{\kappa})] = 0, \qquad i = 1, 2, \ldots, p \tag{8.1.4}$$

Linearizing $\mathbf{f}(\mathbf{x}_j, \mathbf{k}^* + \boldsymbol{\kappa})$ as

$$\mathbf{f}(\mathbf{x}_j, \mathbf{k}^* + \boldsymbol{\kappa}) = \mathbf{f}(\mathbf{x}_j, \mathbf{k}^*) + \left[\frac{\partial \mathbf{f}_j}{\partial \mathbf{k}}\right]_{\mathbf{k}^*} \boldsymbol{\kappa} \tag{8.1.5}$$

and replacing $\mathbf{y}_j$ by $\mathbf{f}(\mathbf{x}_j, \mathbf{k}^*) + \boldsymbol{\epsilon}_j$ in (8.1.4), we obtain

$$\frac{\partial}{\partial \kappa_i} \sum_{j=1}^{M} [\boldsymbol{\epsilon}_j - \mathbf{D}_j \boldsymbol{\kappa}]^T \mathbf{V}_j^{-1} [\boldsymbol{\epsilon}_j - \mathbf{D}_j \boldsymbol{\kappa}] = 0 \tag{8.1.6}$$

where the $m \times p$ Jacobian matrix

$$\mathbf{D}_j = \left\{\frac{\partial f_{l_j}}{\partial k_r}\right\}_{\mathbf{k}^*}, \qquad l = 1, 2, \ldots, m, \quad r = 1, 2, \ldots, p \tag{8.1.7}$$

From (8.1.6) we obtain

$$\sum_{j=1}^{M} \mathbf{D}_j \mathbf{V}_j^{-1} [\boldsymbol{\epsilon}_j^T - \mathbf{D}_j \boldsymbol{\kappa}] = 0 \tag{8.1.8}$$

Solving for $\boldsymbol{\kappa}$ in terms of the $\boldsymbol{\epsilon}_j$,

$$\left(\sum_{j=1}^{M} \mathbf{D}_j \mathbf{V}_j^{-1} \mathbf{D}_j^T \right) \boldsymbol{\kappa} = \sum_{j=1}^{M} \mathbf{D}_j \mathbf{V}_j^{-1} \boldsymbol{\epsilon}_j \tag{8.1.9}$$

Let us define the $m \times m$ matrix $\mathbf{H}$ by

$$\mathbf{H} = \sum_{j=1}^{M} \mathbf{D}_j \mathbf{V}_j^{-1} \mathbf{D}_j^T \tag{8.1.10}$$

If $\mathbf{H}$ is nonsingular, (8.1.9) relates the $\boldsymbol{\kappa}$ to the $\boldsymbol{\epsilon}_j$. Also, $\boldsymbol{\kappa}$ is a linear function of the $\boldsymbol{\epsilon}_j$, and since the density of the $\boldsymbol{\epsilon}_j$ is normal, that of $\boldsymbol{\kappa}$ will also be normal (see Chapter 4). The mean of $\boldsymbol{\kappa}$ will be zero and its covariance matrix is $\mathbf{P}$, the quantity we set out to determine. Thus

$$p(\boldsymbol{\kappa}) = \frac{1}{(2\pi)^{p/2} |\mathbf{P}|^{1/2}} \exp\left[-\frac{1}{2} \boldsymbol{\kappa}^T \mathbf{P}^{-1} \boldsymbol{\kappa} \right] \tag{8.1.11}$$

The final step is to determine $\mathbf{P}$, where

$$\mathbf{P} = E(\boldsymbol{\kappa}\boldsymbol{\kappa}^T) \tag{8.1.12}$$

From (8.1.9) we see that by multiplying (8.1.9) by its transpose and taking the expected value,

$$\mathbf{P} = \mathbf{H}^{-1} \tag{8.1.13}$$

In the case in which the errors of different experiments are identically distributed ($\mathbf{V}_j = \mathbf{V}$) and all measurements are independent ($\mathbf{V}$ is diagonal with elements v_i) and finally all variances are identical and equal to σ^2,

$$\mathbf{P} = \left[\frac{1}{\sigma^2} \sum_{j=1}^{M} \mathbf{D}_j \mathbf{D}_j^T \right]^{-1} \tag{8.1.14}$$

8.1.2 Ordinary differential equations

Now instead of (7.1.6) we assume that the model is (7.1.22) and (7.1.23). The objective is still to compute $\mathbf{P}$, the covariance matrix of $\boldsymbol{\kappa}$. This case is excellently covered by Rosenbrock and Storey ([17], Chap. 8) the details of which parallel those of Section 8.1.1.

We assume that the errors $\boldsymbol{\epsilon}_r$ in (7.1.23) are independent and normally distributed with mean zero and covariance matrix $\mathbf{V}_r$, as in (8.1.2). The parameters are assumed to have been determined to minimize

$$S(\mathbf{k}) = \sum_{r=1}^{R} [\mathbf{y}_r - \mathbf{h}(t_r, \mathbf{x}(t_r; \mathbf{k}))]^T \mathbf{V}_r^{-1} [\mathbf{y}_r - \mathbf{h}(t_r, \mathbf{x}(t_r; \mathbf{k}))] \tag{8.1.15}$$

where $\mathbf{x}(t; \mathbf{k})$ is the solution of (7.1.22).

We present only the formulas required to compute $\mathbf{P}$. As in the last section, if the estimate $\hat{\mathbf{k}}$ is close to $\mathbf{k}^*$ so that linearization is valid, $\boldsymbol{\kappa} = \hat{\mathbf{k}} - \mathbf{k}^*$ will be normally distributed with mean zero and covariance matrix

$$\mathbf{P} = \mathbf{H}^{-1} \tag{8.1.16}$$

where

$$\mathbf{H} = \sum_{r=1}^{R} \boldsymbol{\Lambda}_r^T \mathbf{G}_r^T \mathbf{V}_r^{-1} \mathbf{G}_r \boldsymbol{\Lambda}_r \tag{8.1.17}$$

$$\boldsymbol{\Lambda}_r = \left\{\frac{\partial x_{l_r}}{\partial k_\nu}\right\}_{\mathbf{k}^*}, \qquad l = 1, 2, \ldots, n, \quad \nu = 1, 2, \ldots, p \tag{8.1.18}$$

$$\mathbf{G}_r = \left\{\frac{\partial h_l}{\partial x_\nu}\right\}_{\mathbf{k}^*}, \qquad l = 1, 2, \ldots, m, \quad \nu = 1, 2, \ldots, p \tag{8.1.19}$$

The derivative matrix $\boldsymbol{\Lambda}$ must be computed as the solution of the sensitivity equations (7.6.7):

$$\dot{\boldsymbol{\Lambda}} = \mathbf{A}\boldsymbol{\Lambda} + \mathbf{B} \tag{8.1.20}$$

$$\boldsymbol{\Lambda}(0) = 0$$

where

$$\mathbf{A} = \left\{\frac{\partial f_l}{\partial x_\nu}\right\}_{\mathbf{k}^*}, \qquad l = 1, 2, \ldots, n, \quad \nu = 1, 2, \ldots, n \tag{8.1.21}$$

$$\mathbf{B} = \left\{\frac{\partial f_l}{\partial k_\nu}\right\}_{\mathbf{k}^*}, \qquad l = 1, 2. \ldots, n, \quad \nu = 1, 2, \ldots, p \tag{8.1.22}$$

The reader is strongly encouraged to work his way through Section 8.1.1, for the case considered in this section, to derive (8.1.16)–(8.1.22). The manipulations are straightforward but well worth the effort.

We might just note the differences between the algebraic case and this one. $\mathbf{H}$ becomes more complicated since the dimension m of the observation vector is not necessarily equal to the dimension n of the state vector. Also, the Jacobian matrix $\mathbf{D}$ in the last section can be computed analytically, whereas the Jacobian matrix $\boldsymbol{\Lambda}$ (the sensitivity matrix) must here be computed by the auxiliary differential equations (8.1.20).

As a practical matter, since we do not know $\mathbf{k}^*$ but only $\hat{\mathbf{k}}$, we must evaluate the derivatives in (8.1.18), (8.1.19), (8.1.21), and (8.1.22) at $\hat{\mathbf{k}}$ rather than $\mathbf{k}^*$.

Finally, Heineken et al. [8] have shown that as $R \longrightarrow \infty$ (continuous measurements) $\mathbf{H}$ defined by (8.1.17) approaches

$$R\mathbf{H}^* = \frac{R}{T - t_0} \int_{t_0}^{T} \mathbf{D}(t)^T \mathbf{G}(t)^T \mathbf{V}^{-1}(t) \mathbf{G}(t) \mathbf{D}(t)\, dt \tag{8.1.17a}$$

8.1.3 Partial differential equations

Finally, we wish to compute $\mathbf{P}$ in the case in which the model is defined by a set of partial differential equations, as in (7.1.25) and (7.1.26). We assume that the errors $\boldsymbol{\epsilon}_{r,s}$ are independent and normally distributed with mean zero and covariance matrix $\mathbf{V}_{r,s}$. The parameters are assumed to have been determined to minimize (7.1.36):

$$S(\mathbf{k}) = \sum_{r=1}^{R} \sum_{s=1}^{S'} [\mathbf{y}_{r,s} - \mathbf{h}(t_r, x_s, \mathbf{u}(t_r, x_s; \mathbf{k}))]^T \mathbf{V}_{r,s}^{-1} \cdot [\mathbf{y}_{r,s} - \mathbf{h}(t_r, x_s, \mathbf{u}(t_r, x_s; \mathbf{k}))] \tag{8.1.23}$$

where $\mathbf{u}(t, x; \mathbf{k})$ is the solution of (7.1.25) and (7.1.26).

We obtain for the covariance of $\boldsymbol{\kappa}$:

$$\mathbf{P} = \mathbf{H}^{-1} \tag{8.1.24}$$

where

$$\mathbf{H} = \sum_{r=1}^{R} \sum_{s=1}^{S'} \boldsymbol{\Lambda}_{r,s}^T \mathbf{G}_{r,s}^T \mathbf{V}_{r,s}^{-1} \mathbf{G}_{r,s} \boldsymbol{\Lambda}_{r,s} \tag{8.1.25}$$

with $\boldsymbol{\Lambda}(t, x)$ governed by (7.7.8)–(7.7.10) and

$$\mathbf{G}_{r,s} = \left\{\frac{\partial h_{l_{r,s}}}{\partial u_\nu}\right\}_{\mathbf{k}^*}, \qquad l = 1, 2, \ldots, m, \quad \nu = 1, 2, \ldots, n \tag{8.1.26}$$

The analogy to the O.D.E. case is obvious. The principal difference is that the sensitivity equations are P.D.E.s rather than O.D.E.s.

8.1.4 Confidence intervals

As we have stressed, the result of a parameter estimation method is precisely that, an *estimate* of the true value of the parameter, which is, of course, unknown. A measure of the accuracy of the estimate is usually given by its standard deviation.

The standard deviation of a parameter estimate can be interpreted as follows. If we perform a large number of replicate experiments, each time estimating the same parameter, we can consider the estimate as a random variable since the data from which it has been estimated contains random errors. The variance $\mathbf{P}$ of this random variable is the expected value of the square of the difference between the value and its mean, assumed to be the true value. The standard deviation is simply the square root of the variance. Then we ask, Given any one estimate of the parameter, what are the limits on either side of this estimate that insure that if this experiment is repeated many times the true value will lie within these limits a specified percentage of the time? These limits are called *confidence limits* and are simply a product of a fixed number (depending on the percentage) and the standard deviation. For example, if σ_k is the standard deviation for an estimate of k, the 95% confidence limits on $\hat{k}$ are $\hat{k} \pm 1.96\sigma_k$, where $\hat{k}$ is the estimate. Thus we can expect the true value of k, k^*, to lie within $\hat{k} \pm 1.95\sigma_k$ with 95% certainty.

The results of the analysis of parameter estimates are often presented pictorially. Consider as an example the one-dimensional parameter estimation problem shown in Figure 8.1, in which the error criterion S is plotted against the value of the parameter estimate for two cases: (1) curve A, a "good" experiment, and (2) curve B, a "poor" experiment.† The two curves

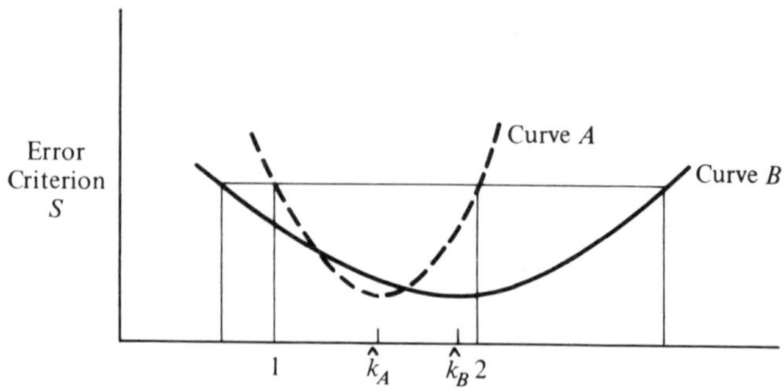

Figure 8.1. Comparison of two error criteria for estimation of a single parameter.

shown might arise from choices of different initial conditions for the experiment. The best estimate, $\hat{k}_A$ and $\hat{k}_B$, are those at the respective minima of the error criterion S. In a one-dimensional example the curvature of S in the vicinity of the minimum is related to the confidence (or degree of certainty) with which the parameter can be estimated. Because of the larger curvature of curve A, the true parameter value lies between points 1 and 2 with a higher probability than for the case of the "poor" experiment.

This same notion of steeper curvature in the S-curve implying better parameter estimates carries over to the multidimensional parameter estimation problem. For two parameters we can plot the sum of squares function $S(k_1, k_2)$ in the $k_1 - k_2$ plane. The plot is generally shown as lines of constant S values in the $k_1 - k_2$ plane, as illustrated in Figure 8.2. A flat minimum indicates that values of (k_1, k_2) almost as good as $(\hat{k}_1, \hat{k}_2)$ occur over wide ranges and that the parameters are estimated poorly. A sharp minimum indicates the reverse. If the S-surface is an obliquely oriented curved trough, $S(\mathbf{k})$ is clearly much less sensitive to changes in the parameters away from the minimum in some directions than in others. Surfaces such as that shown in Figure 8.2 are to be avoided, if possible. Their occurrence depends both on the model and on the experimental design. Such a surface implies that the joint posterior distribution of k_1 and k_2 exhibits dependence and specifically correlation.

After computing $\mathbf{P}$ and the standard deviations, confidence intervals can be calculated for each parameter. For two parameters it is customary to plot

†"Good" is used in the sense of rich in information content, "poor" in the sense of low in information content.

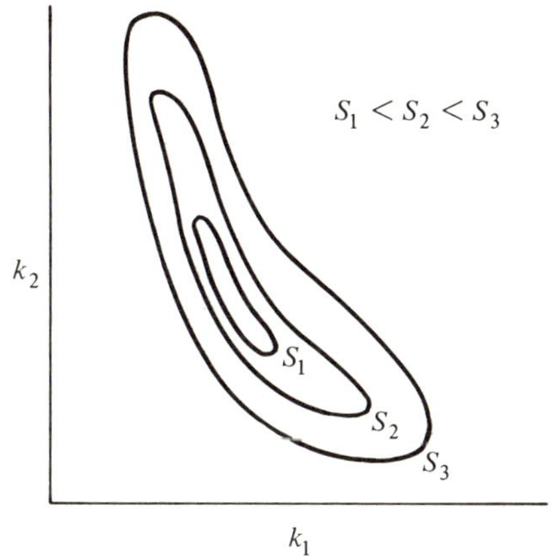

Figure 8.2. Sum of squares surfaces for two parameters.

the confidence regions, say at 95%, in the $k_1 - k_2$ plane. We will explain how to compute this region shortly. A sample two-dimensional confidence region is shown in Figure 8.3.

Very often the confidence regions assume the shape of ellipsoids, and, consequently, are called *error ellipsoids*. A convenient single-number measure for a set of confidence limits is the area of the error ellipsoid.† The smaller the area of this error ellipsoid, the higher the confidence in the parameter estimates, hence the desire to choose that set of experimental variables (for example, initial concentrations) that minimizes the volume.

In Figure 8.3 the individual distributions for the parameters k_1 and k_2 are shown along the k_1 and the k_2 axes. The figure serves to illustrate that when more than one parameter is estimated, the individual confidence limits can be misleading. For example, from the individual confidence distribution for k_1 we can see that a value $k_1 = 1.0$ would not of itself be regarded as unusual. Similarly, a value of $k_2 = 1.5$ individually is not exceptional. However, the pair of values (1.0, 1.5) correspond to a point well outside the 95% joint confidence region. The explanation (as is evident from the diagram) is that while *for some values* of k_2 a hypothesis that $k_1 = 1$ is not contradicted by this data and *for some values* of k_1 a hypothesis that $k_2 = 1.5$ is not contradicted; nevertheless, the hypothesis of the occurrence of these values together is contradicted at a fairly high significance level. Typically, if estimates are uncorrelated, the individual limits will provide a rough guide to the joint region; but when the estimates are highly correlated individual confidence

†These concepts apply equally well to an arbitrary number of parameters, in which case the error ellipsoid has a volume in the p-dimensional parameter space. With more than two parameters, the results are ordinarily not presented graphically, although projections of the volumes on certain planes is sometimes useful.

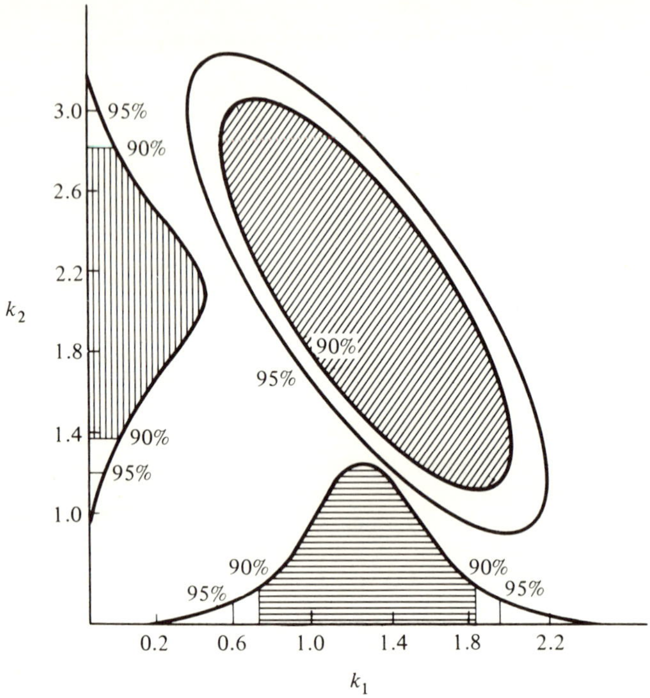

Figure 8.3. Error ellipsoids with individual distributions for estimation of two parameters.

distributions will have a very wide spread with the joint region contained in a rather narrow oblique band.

Let us now show how the individual confidence limits can be computed from the covariance matrix **P**. We consider the general case of p parameters.

If **b** is a unit p-vector, i.e., $b_1^2 + \cdots + b_p^2 = 1$, then the new random variable (scalar) $\mathbf{b}^T\boldsymbol{\kappa}$ will have a normal distribution with zero mean and variance

$$\sigma_b^2 = \mathbf{b}^T\mathbf{P}\mathbf{b} \tag{8.1.27}$$

Thus **b** is a unit vector in the p-dimensional space of the parameters, and $\mathbf{b}^T\boldsymbol{\kappa}$ is the projection of the vector $\boldsymbol{\kappa}$ on **b**. Now we say that the *confidence interval* $\gamma(\mathbf{b})$ will be such that the true value will lie in

$$\mathbf{b}^T\hat{\mathbf{k}} - \gamma(\mathbf{b}) \leq \mathbf{b}^T\mathbf{k}^* \leq \mathbf{b}^T\hat{\mathbf{k}} + \gamma(\mathbf{b})$$

a certain fraction of the time in a large series of similar experiments.

The confidence interval $\gamma(\mathbf{b})$ associated with the projection $\mathbf{b}^T\boldsymbol{\kappa}$ is

$$\gamma(\mathbf{b}) = \beta\sigma_b$$

where β depends on the confidence level. At the 95 percent level $\beta = 1.96$, and so

$$\mathbf{b}^T\hat{\mathbf{k}} - 1.96\sigma_b \leq \mathbf{b}^T\mathbf{k}^* \leq \mathbf{b}^T\hat{\mathbf{k}} + 1.96\sigma_b$$

will be true 95 percent of the time. The confidence interval for a parameter k_i is found by letting $b_i = 1$, $b_j = 0$, $j \neq i$. Then σ_b is just the standard deviation of k_i, i.e., the square root of the ith diagonal element of **P**.

The region in p-dimensional space surrounding $\hat{\mathbf{k}}$ containing the confidence intervals for all values of **b** on the unit hypersphere centered at $\hat{\mathbf{k}}$ is called the *confidence region hypervolume* (or hyperellipse).

The hypervolume is related to **P** as follows [17]. **P** will be positive-definite with positive eigenvalues and real and orthogonal eigenvectors. The hyperellipse defined by $\mathbf{z}^T\mathbf{P}\mathbf{z} = 1$ has the eigenvectors of **P** as its principal axes and each principal semidiameter equal to the square root of the corresponding eigenvalue of **P**. It follows that

$$\lambda_{\min} \leq \sigma_b \leq \lambda_{\max}$$

The ratio $\lambda_{\max}/\lambda_{\min}$ determines the shape of the hyperellipse. If this ratio is large, there may be certain parameters which are ill-determined, corresponding to the eigenvalues that are excessively large. Thus the the confidence intervals for individual elements of **k** may be less informative than those for the eigenvectors of **P**. These eigenvectors are linear combinations of the $\hat{k}_i$ corresponding to the principal diameters of the hyperellipse.

A question raised earlier was that of parameter observability. We are now in a position to prescribe precise conditions for this condition, which, of course, is the first aspect of an experimental program that should be checked. If any of the elements of **P** are infinite, then certain parameters cannot be determined. Since $\mathbf{P} = \mathbf{H}^{-1}$, if **H** is singular, such a condition will result. Therefore a necessary and sufficient condition for parameter observability is that **H** be nonsingular. For example, if one proposes to measure only A in the system

$$A \xrightarrow{k_1} B \xrightarrow{k_2} C$$

only k_1 can be estimated and not k_2. We leave it as an exercise for the reader to demonstrate that if we tried to estimate both k_1 and k_2 from measurements of A only that **H** becomes singular in this case.

EXAMPLE 8.1. Assuming that the measurements in Example 7.4 have independent identically distributed errors with variance σ^2, let us compute the confidence intervals of these estimates. **P** can be computed from (8.1.17):

$$\mathbf{P} = \left[\frac{1}{\sigma^2}\sum_{r=1}^{R}\mathbf{\Lambda}_r^T\mathbf{\Lambda}_r\right]^{-1}$$

The results of the analysis for several of the previous cases are presented in Table 8.1. The results of case 2 were taken to be optimal estimates k_1^*, k_2^*.

By comparing cases 2 and 6 and 2 and 7 it is seen that better estimates are achieved by observing x_2 alone than x_1 alone. It is difficult to attribute this result

TABLE 8.1
95 PERCENT CONFIDENCE INTERVALS FOR k_1 AND k_2 [47, CHAPTER 7]

Case	P_{11}	P_{12}	P_{22}	$1.96\sigma_1$	$1.96\sigma_2$
2	0.428	−0.488	13.3	1.28	7.13
6	5.55	−49.2	507.0	4.62	44.2
7	6.85	14.4	39.3	5.14	12.3
9	0.614	−0.781	23.6	1.53	9.52
10	0.659	−0.650	21.6	1.59	9.10
15	533.0	−391.0	9626.0	45.3	192.0
16	90.4	−99.9	2645.0	18.6	101.0

to the nature of x_2 (diphenyl) as an intermediate product or simply to the manner in which x_2 enters r_1 and r_2. When fewer data points are taken, cases 9 and 10, the confidence intervals are very close to those of case 2. This indicates that it is better to observe more state variables at fewer values of t than fewer states at more values of t.

Finally, the effect of experimental error is considered. The confidence intervals are much larger, as can be predicted directly wherein σ_b is proportional to σ. It has been shown by Heineken et al. [8] that at constant level of error the confidence limits for parameter estimates asymptotically decrease as R^{-1}. This is verified in this example by comparing the confidence limits of cases 15 and 16, where R was increased from 8 to 16.

8.2 Design of Experiments for Parameter Estimation

In designing an experiment the most fundamental consideration is the objective of the experiment. Very often the objective will dictate the type of measurements to be made, the variables to change, etc. We shall consider the design of experiments where parameter estimation is the objective, either as an end in itself or for model discrimination.

Hunter et al. [12] have given some examples of the effect of the objective on experimental planning. For the chemical reaction $A \xrightarrow{k_1} B \xrightarrow{k_2} C$ it is sometimes desirable to obtain estimates as precise as possible for both k_1 and k_2. However, we can also think of situations where it would be desirable to obtain a relatively more precise estimate of k_1 than k_2, or vice versa. Suppose that the development group in a company is supplying A, which contains a very small concentration of an impurity which acts as a catalyst for the reaction $B \longrightarrow C$, to another group in the same company. It is felt that the impurity can be removed, and in the meantime k_1 must be precisely estimated for process design purposes. On the other hand, say that we are studying a two-step organic synthesis carried out in a batch system. Past experience indicates that

$k_1 \gg k_2$. The constant k_2 must be precisely estimated, perhaps, so that the time necessary for 95 percent conversion to C can be calculated.

A number of questions must be faced, including

1. Which variables should be changed?
2. By how much should the variables be changed?
3. In what form should the variables be considered? (For example, should we think in terms of time and temperature or of log time and reciprocal temperature?)

The basic motivation for using optimal statistical designs in planning experiments is to save time and money in meeting the objectives of the experimental program.

In attempting to systematize the planning of experiments we are led into the area of information theory, since the purpose of an experimental program is to gain information. In the case of parameter estimation the relevant information is contained in the posterior distribution of the parameters. When one tries to quantify the notion of the amount of information contained in a given distribution, the following relation can be obtained [1]:

$$I = E[\log p] = \int p \log p \, d\mathbf{k}$$

where I is the measure of the information content of p. If p is a multivariate normal distribution with covariance matrix $\mathbf{P}$, then the information is given by

$$I = \text{constant} - \tfrac{1}{2} \log |\mathbf{P}|$$

To maximize I, we would want to plan our experiments to minimize the determinant of $\mathbf{P}$, where $\mathbf{P}$ is the expected variance of the posterior distribution of the parameter estimates. Thus we see that a distribution with a small variance contains more information than one with a large variance.

Suppose that we wish to plan a series of M experiments. Our current information on the values of the parameters is summarized in a prior distribution $p_0(\mathbf{k})$, for example, a normal distribution with covariance matrix $\mathbf{P}_0$. In many cases, $p_0(\mathbf{k})$ will actually be the posterior distribution obtained by estimating the parameters on the basis of experiments conducted to date. The expected covariance of the posterior distribution is given approximately by (8.1.10) and (8.1.13):

$$[\mathbf{P}^{-1}]_{ij} = [\mathbf{P}_0^{-1}]_{ij} + \sum_{\mu=1}^{M} \sum_{l=1}^{m} \sum_{r=1}^{m} \frac{\partial f_{l\mu}}{\partial k_i} \frac{\partial f_{r\mu}}{\partial k_j} [\mathbf{V}^{-1}]_{lr} \tag{8.2.1}$$

where $\mathbf{V}$ is the covariance matrix of the observations for each experiment. The derivatives $\partial f_{l\mu}/\partial k_i$ are to be evaluated at the current estimates of the $\mathbf{k}$s. The simplest case occurs when there is only one observed variable per experiment. Then l and r may be eliminated, $\mathbf{V}$ becomes the scalar σ^2, and

(8.2.1) becomes [see (8.1.14)]

$$[\mathbf{P}^{-1}]_{ij} = [\mathbf{P}_0^{-1}]_{ij} + \frac{1}{\sigma^2}\sum_{\mu=1}^{M}\frac{\partial f_\mu}{\partial k_i}\frac{\partial f_\mu}{\partial k_j} \tag{8.2.2}$$

We see from (8.2.2) that the determinant of **P** is, in general, a rather complicated function of the x_{i_μ}, the experimental conditions which are to be determined for the M experiments. The experimental design criterion is to choose x_{i_μ}, $i = 1, 2, \ldots, n$, $\mu = 1, 2, \ldots, M$, to minimize the determinant of **P**, $|\mathbf{P}|$, or, equivalently, maximize $|\mathbf{P}^{-1}|$. This criterion was first used by Box and Lucas [5] and Draper and Hunter [6, 7]. The criterion for maximizing $|\mathbf{P}^{-1}|$ is equivalent to attempting to minimize the p-dimensional hypervolume of the confidence region for the parameters. The remainder of this section will be devoted to illustrations of the use of this criterion in planning experiments for parameter estimation.

Hunter et al. [12] give a simple example as an illustration of the design technique. In the reaction

$$A \xrightarrow{k_1} B \xrightarrow{k_2} C$$

suppose that it is desirable to obtain precise estimates of both k_1 and k_2 from measurements of B. If $[A]_0 = 1$, $[B]_0 = [C]_0 = 0$, then

$$[B] = \frac{k_1}{k_1 - k_2}(e^{-k_2 t} - e^{-k_1 t}) \tag{8.2.3}$$

Considering the R.H.S. of (8.2.3) as f, $\mathbf{D}_j$ is a two-dimensional row vector with elements

$$\frac{\partial f_j}{\partial k_1} = \frac{\{k_1(k_1 - k_2)t_j + k_2\}e^{-k_1 t_j} - k_2 e^{-k_2 t_j}}{(k_1 - k_2)^2}$$

$$\frac{\partial f_j}{\partial k_2} = \frac{\{-k_1(k_1 - k_2)t_j + k_1\}e^{-k_2 t_j} - k_1 e^{-k_1 t_j}}{(k_1 - k_2)^2}$$

If the best two experiments ($M = 2$) are sought, i.e., the best two values of t_1 and t_2, we need to maximize

$$\left|\sum_{j=1}^{2}\mathbf{D}_j\mathbf{D}_j^T\right|$$

To evaluate this determinant we need preliminary estimates of k_1 and k_2. Say that these are 0.2 and 0.7, respectively. The maximization can then be shown to yield $t_1 = 1.59$ and $t_2 = 5.89$ (the reader is encouraged to verify this). Thus we would try to measure the concentration of B as close as possible to these two times.

EXAMPLE 8.2. *The Accuracy of Kinetic Parameters Estimated from Batch and Integral Reactor Data* [*16*]. The estimation of rate constants from experimental concentration time measurements in batch and tubular flow reactors is an important problem in chemical engineering. We want to consider the analysis of the accuracy of kinetic parameter estimates for the four reaction schemes

$$A \xrightarrow{k_1} B$$

$$A \underset{k_2}{\overset{k_1}{\rightleftharpoons}} B$$

$$A \begin{array}{l} \xrightarrow{k_1} B \\ \xrightarrow[k_2]{} C \end{array}$$

$$A \xrightarrow{k_1} B \xrightarrow{k_2} C$$

The particular objective is the optimal design of experiments to estimate these kinetic parameters. To do this we shall compute the dimensionless variances of the parameter estimates as a function of times of observation, number of data points R, level of error, etc.

In the analysis to follow we shall make the following assumptions:

1. The initial concentrations are known without error.
2. The initial charge in each case consists of pure A.
3. The concentrations are measured directly.
4. Unless otherwise indicated, the covariance matrix of experimental errors, $\mathbf{V}(t)$, has the simple form $\sigma^2\mathbf{I}$.

We shall examine the effect of the following on the confidence intervals of the rate constant estimates:

1. How many data points R should be taken and over what range in order to attain a given accuracy of parameter estimates?
2. What effect does the level of experimental error have on the problem?
3. Above what value of R can (8.1.17a) be taken as an adequate representation of (8.1.17)?
4. How many of the n independent concentrations should be measured, and which ones should be chosen in the event of a choice being available?
5. What advantage can be gained by measuring more than the independent number of concentrations?
6. Instead of spacing the observation points equally in the time interval (t_0, T), might it not be advantageous to space the same number of points in a different manner? In particular, what is the effect of spacing them equally in the conversion interval $(1, \xi)$ rather than the corresponding time interval (t_0, T)?
7. If concentration time data from runs at different temperatures are used to determine the rate constants and activation energies, how accurate would these estimates be? At how many temperatures should observations be taken in order to achieve a given accuracy?

Questions 1 and 2 are answered for all the four reaction schemes. Question 3 is considered for the schemes $A \rightleftharpoons B$ and $A \longrightarrow B \longrightarrow C$. Questions 4 and 5 are thoroughly discussed for the scheme $A \begin{array}{l} \rightarrow B \\ \rightarrow C \end{array}$, and question 5 is taken up for the scheme $A \longrightarrow B$, where the matrix of experimental errors $\mathbf{V}(t)$ has a form different from the $\sigma^2\mathbf{I}$ usually assumed. The simple scheme $A \longrightarrow B$ is also used to illustrate how the same techniques can be used to answer the more complicated questions 6 and 7.

a. *Results.* Following Heineken et al., dimensionless parameters of the type $\psi_k = \sigma_k^2 R/k^2\sigma^2$ are used to correlate the results. Here, k is the mean value, σ_k^2 the variance of an estimated parameter, R the number of data points, and σ^2 the variance of the dimensionless forms of the experimentally measured concentrations. The ψ_k are obtained as diagonal elements of the matrix $\mathbf{P}R/\sigma^2k^2$, where $\mathbf{P} \simeq \mathbf{H}^{*-1}/R$ for large R and $\mathbf{P} = \mathbf{H}^{-1}$ in general. The ψ_k parameter contains information regarding the accuracy of an estimate (the larger the value of k^2/σ_k^2, the more accurate is the estimate) for any given value of the experimental error. It also incorporates the number of observations R and can be used to decide what value of R is necessary in order to get an acceptable accuracy of the estimate. Plots of ψ_k are therefore a concise and convenient representation of the results of the study. They can be used to plan kinetic experiments which seek to obtain accurate estimates of parameters when approximate values are known from a less sophisticated analysis.

Figure 8.4 presents $\psi_1 = \sigma_{k_1}^2 R/k_1^2\sigma^2$ and $\psi_2 = \sigma_{k_2}^2 R/k_2^2\sigma^2$ vs. k_1T for the isothermal scheme $A \underset{k_2}{\overset{k_1}{\rightleftharpoons}} B$. The curve of $\psi_1 = \sigma_{k_1}^2 R/k_1^2\sigma^2$ for the scheme $A \overset{k_1}{\longrightarrow} B$ is also shown for comparison. Figures 8.5 and 8.6 show the same type of plots for

$$A \overset{k_1}{\longrightarrow} B \overset{k_2}{\longrightarrow} C \quad \text{and} \quad A \begin{smallmatrix} \overset{k_1}{\nearrow} B \\ \underset{k_2}{\searrow} C \end{smallmatrix}$$

when the number of components measured is the same as the number of independent components (= 2), and these are chosen to be A and B.

Figure 8.7 illustrates some typical results when the exact formula $\mathbf{P} = \mathbf{H}^{-1}$ is used for computation instead of the approximation $\mathbf{P} = \mathbf{H}^{*-1}/R$ in the isothermal scheme $A \underset{k_2}{\overset{k_1}{\rightleftharpoons}} B$. Similar curves for $A \overset{k_1}{\longrightarrow} B \overset{k_2}{\longrightarrow} C$ are shown in Figure 8.8. Figure 8.9 is a cross-plot of Figure 8.8 in which the asymptotic approach to the approximate formula as R increases is clearly seen.

Figure 8.10 is an attempt to answer the question of how many components and which ones should be measured in the isothermal scheme

$$A \begin{smallmatrix} \overset{k_1}{\nearrow} B \\ \underset{k_2}{\searrow} C \end{smallmatrix}$$

It indicates specifically the influence of changes in the $\mathbf{G}$ matrix on the accuracy of estimation, the $\mathbf{D}$ matrix remaining the same and the $\mathbf{V}$ matrix continuing to be $\sigma^2\mathbf{I}$. This important question is raised again for the isothermal scheme $A \overset{k_1}{\longrightarrow} B$, but this time the $\mathbf{G}$ matrix is kept constant at column $[1 \quad -1]$. The source of variation now is the matrix of experimental errors $\mathbf{V}$. $\mathbf{V}$ is taken to be

$$\begin{bmatrix} \sigma_A^2 & 0 \\ 0 & \sigma_B^2 \end{bmatrix}$$

which represents independent measurements on A and B with different variances. This, of course, is a more realistic assumption than $\mathbf{V} = \sigma^2\mathbf{I}$. $s = \sigma_A/\sigma_B$ is allowed to vary, and the results are shown in Figure 8.11.

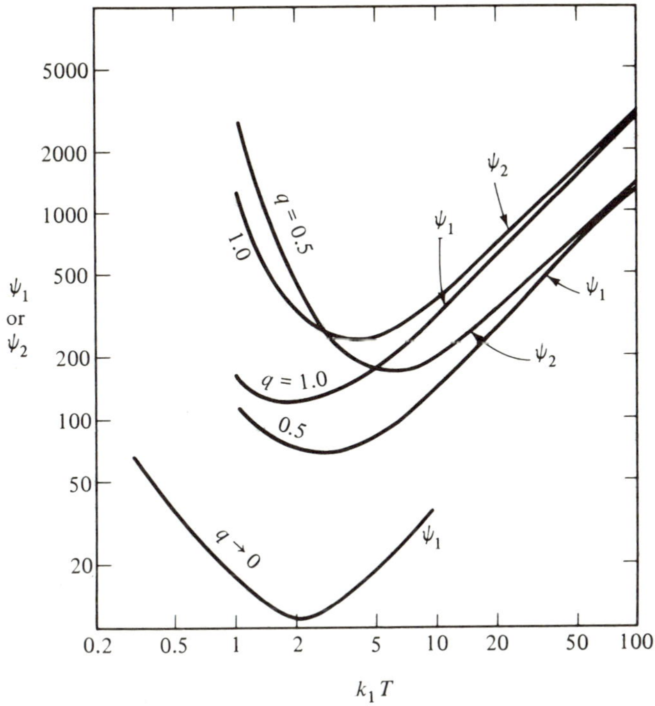

Figure 8.4. Dimensionless variance of parameter estimates vs. dimensionless time in the reaction scheme $A \underset{2}{\overset{1}{\rightleftarrows}} B$; only component A is measured and the number of observations, R, is large. [Reprinted from A. L. Ravimohan, W. H. Chen, and J. H. Seinfeld, "The Accuracy of Kinetic Parameters Estimated from Batch and Integral Reactor Data," *Can. J. Chem. Eng. 48*, 420 (1970). Reprinted by permission of the copyright owner.]

In Figure 8.12, the effect of unequal division of the time interval is explored for the isothermal scheme $A \xrightarrow{k_1} B$. When the time interval (t_0, T) is equally divided by R points, it can be shown that

$$\psi_1 = \frac{\sigma_{k_1}^2}{k_1^2} \frac{R}{\sigma_A^2} = \frac{R}{\sum_1^R (k_1 t_r) \exp(-2k_1 t_r)} \tag{A}$$

where

$$t_r = t_0 + \frac{r(T - t_0)}{R}, \qquad r = 1, \ldots, R$$

When the conversion interval $(1, \xi)$ is equally divided by R points, the corresponding result is

$$\psi_1 = \frac{\sigma_{k_1}^2}{k_1^2} \frac{R}{\sigma_A^2} = \frac{R}{\sum_1^R (x_r \ln x_r)^2} \tag{B}$$

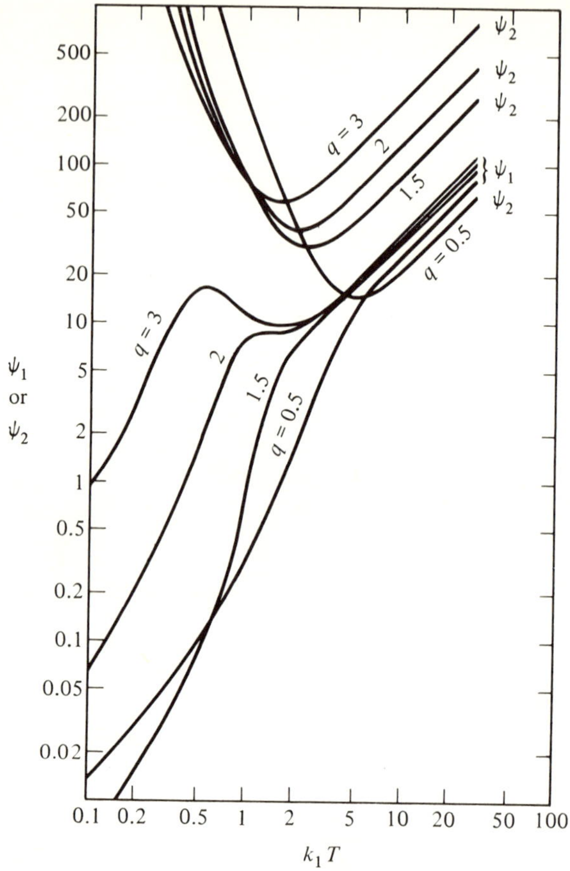

Figure 8.5. Dimensionless variance of parameter estimates vs. dimensionless time in the reaction scheme $A \xrightarrow{1} B \xrightarrow{2} C$; components A and B are measured and the number of observations, R, is large. [Reprinted from A. L. Ravimohan, W. H. Chen and J. H. Seinfeld, "The Accuracy of Kinetic Parameters Estimated from Batch and Integral Reactor Data," *Can. J. Chem. Eng. 48*, 420 (1970). Reprinted by permission of the copyright owner.]

where

$$x_r = 1 - \frac{r(1-\xi)}{R}, \qquad r = 1, \ldots, R$$

and

$$\xi = \exp\{-k_1(T - t_0)\}$$

As R becomes large, (8.1.17) and (8.1.17a) should converge to the same value.

In Figure 8.13, the dimensionless variances of the frequency factor and activation energy estimates are plotted against the number of temperatures at which

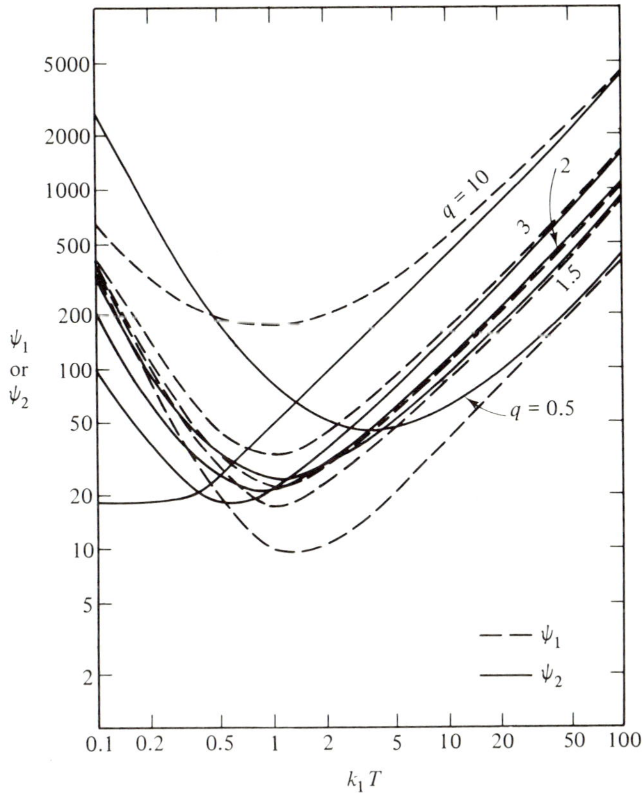

Figure 8.6. Dimensionless variance of parameter estimates vs. dimensionless time in the reaction scheme

$$A \xrightarrow{1} B, \quad A \xrightarrow{2} C$$

Components A and B are measured and the number of observations, R, is large. [Reprinted from A. L. Ravimohan, W. H. Chen and J. H. Seinfeld, "The Accuracy of Kinetic Parameters Estimated from Batch and Integral Reactor Data," *Can. J. Chem. Eng. 48*, 420 (1970). Reprinted by permission of the copyright owner.]

observations are taken:

$$\psi_G = \frac{\sigma_G^2 R}{G^2 \sigma^2} = \frac{h_{22}}{G^2 \sigma^2 |\mathbf{H}^*|}$$

$$\psi_E = \frac{\sigma_E^2 R}{E^2 \sigma^2} = \frac{h_{11}}{E^2 \sigma^2 |\mathbf{H}^*|}$$

where

$$\mathbf{H}^* = \sum_{l=0}^{L} \left\{ \frac{1}{\sigma^2 (T - t_0)} \int_{t_0}^{T} \begin{bmatrix} D_1^2 & D_1 D_2 \\ D_1 D_2 & D_2^2 \end{bmatrix}_l dt \right\} \tag{C}$$

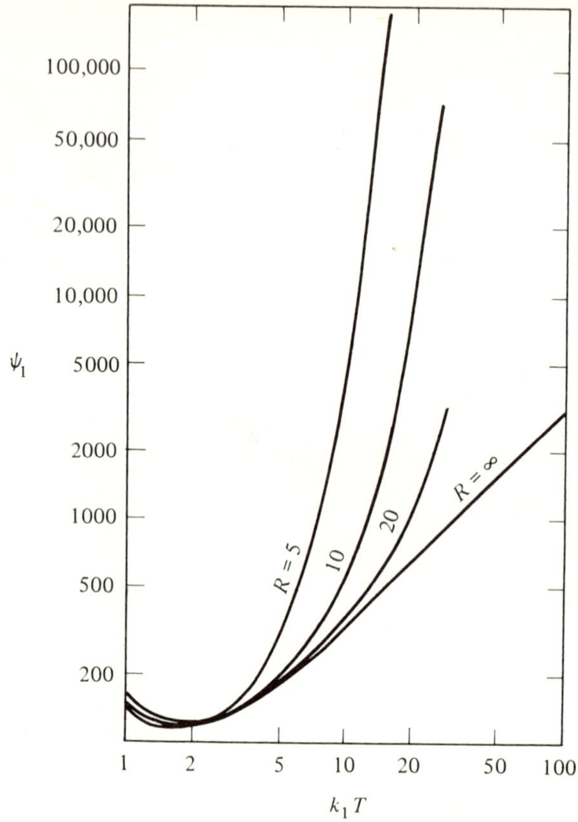

Figure 8.7. Effect of number of observations, R, on the dimensionless variance of k_1 estimate vs. dimensionless time for the reaction scheme $A \underset{2}{\overset{1}{\rightleftharpoons}} B$; only component A is measured and $q = k_2/k_1 = 1$. [Reprinted from A. L. Ravimohan, W. H. Chen, and J. H. Seinfeld, "The Accuracy of Kinetic Parameters Estimated from Batch and Integral Reactor Data," *Can. J. Chem. Eng. 48*, 420 (1970). Reprinted by permission of the copyright owner.]

It must be noted that (C) represents a hybrid expression which is $\mathbf{H}^*$ from the point of view of each run but is actually $\mathbf{H}$ from the point of view of $\sum_{l=0}^{L}$. Hence, $\mathbf{P} = \mathbf{H}^{*-1}/R$, and no division by L is necessary. Very large values of L are experimentally unrealistic, as the number of temperatures at which isothermal runs are conducted seldom exceeds 10.

b. *Discussion of results.* Figure 8.4: For each value of $q = k_2/k_1$ there exists a minimum value of ψ_1 and ψ_2 corresponding to a particular k_1T. As q decreases, the value of k_1T at which the minimum occurs at first increases, because as k_2/k_1

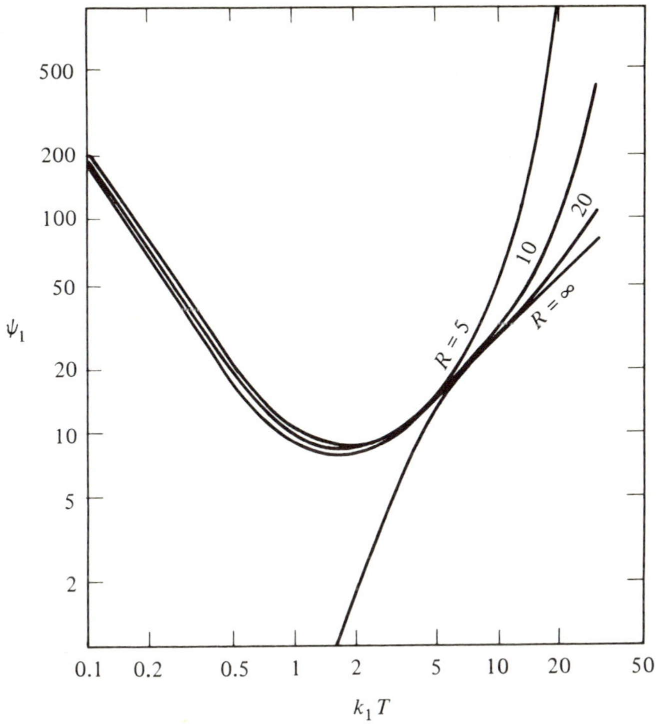

Figure 8.8. Effect of number of observations, R, on the dimensionless variance of k_1 estimate vs. dimensionless time for the reaction scheme $A \underset{2}{\overset{1}{\rightleftharpoons}} B$; components A and B are measured and $q = k_2/k_1 = 0.5$. [Reprinted from A. L. Ravimohan, W. H. Chen and J. H. Seinfeld, "The Accuracy of Kinetic Parameters Estimated from Batch and Integral Reactor Data," *Can. J. Chem. Eng.* *48*, 420 (1970). Reprinted by permission of the copyright owner.]

decreases, more B is present at equilibrium and it takes longer for the reaction to achieve equilibrium. To obtain maximum accuracy it is necessary in this case to continue taking measurements for a longer time. However, when k_2/k_1 becomes very small, the amount of A present at equilibrium tends to zero. Because of the error in measuring A, there is not much point in taking a large number of observations when its concentration is very low. Hence the optimum value of $k_1 T$ starts decreasing as k_2/k_1 approaches zero.

For small values of $k_1 T$, k_1 can be estimated more accurately than k_2. In this range, the smaller the value of q, the greater the accuracy that can be obtained in ψ_1. Conversely, in this range, the smaller the value of q, the lesser the accuracy that can be obtained in ψ_2, because the smaller the value of q, the more significant is the forward reaction. If, on the other hand, measurements are continued for a long

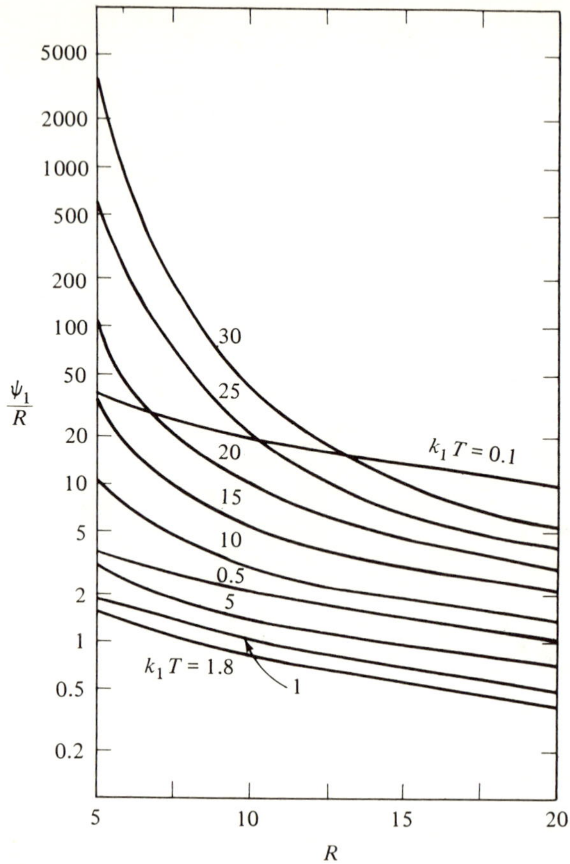

Figure 8.9. Effect of number of observations, R, on the dimensionless variance of k_1 estimate vs. dimensionless time for the reaction scheme $A \xrightarrow{1} B \xrightarrow{2} C$; components A and B are measured and $q = k_2/k_1 = 0.5$ (cross-plot of Figure 8.8). [Reprinted from A. L. Ravimohan, W. H. Chen and J. H. Seinfeld, "The Accuracy of Kinetic Parameters Estimated from Batch and Integral Reactor Data," *Can. J. Chem. Eng. 48*, 420 (1970). Reprinted by permission of the copyright owner.]

time (high k_1T), this difference is narrowed down, because more of the observations are now being taken close to equilibrium and are effectively being wasted.

Figure 8.5: For small values of k_1T (small extent of reaction) the dimensionless variance ψ_1 tends toward zero, because only the first step in the reaction is important and ψ_1 can be estimated accurately by a large number of measurements carried out over the interval $(0, T)$. As q increases, the rate of the second reaction relative to the first increases and ψ_1 increases for small values of k_1T, because it becomes harder to estimate k_1 from observations over the fixed interval $(0, T)$ due to the increasing effect of the second reaction.

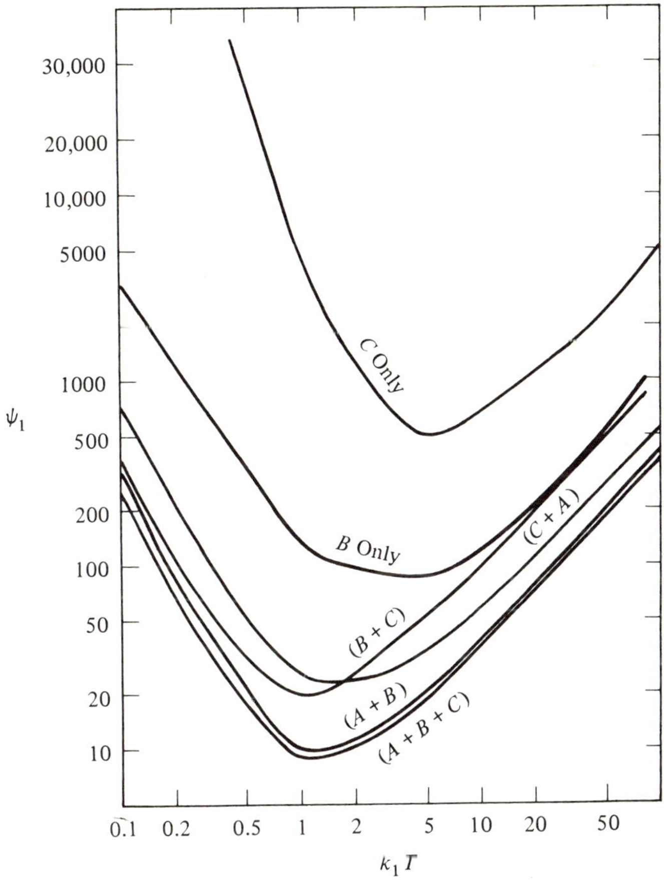

Figure 8.10. Effect of number of components measured on the dimensionless variance of k_1 estimate vs. dimensionless time in the reaction scheme

$$A \overset{1}{\longrightarrow} B, \quad A \overset{2}{\longrightarrow} C$$

The number of observations is large and $q = k_2/k_1 = 0.5$. [Reprinted from A. L. Ravimohan, W. H. Chen and J. H. Seinfeld, "The Accuracy of Kinetic Parameters Estimated from Batch and Integral Reactor Data," *Can. J. Chem. Eng. 48*, 420 (1970). Reprinted by permission of the copyright owner.]

For small values of k_1T, ψ_2 increases rapidly because of the inability to observe the effect of the second reaction, which is necessary to estimate k_2. Conversely for small k_1T, as q increases, ψ_2 decreases, because the second reaction becomes important.

For large values of k_1T, both ψ_1 and ψ_2 increase with k_1T, the more accurate estimates corresponding to smaller values of q. Smaller values of q correspond

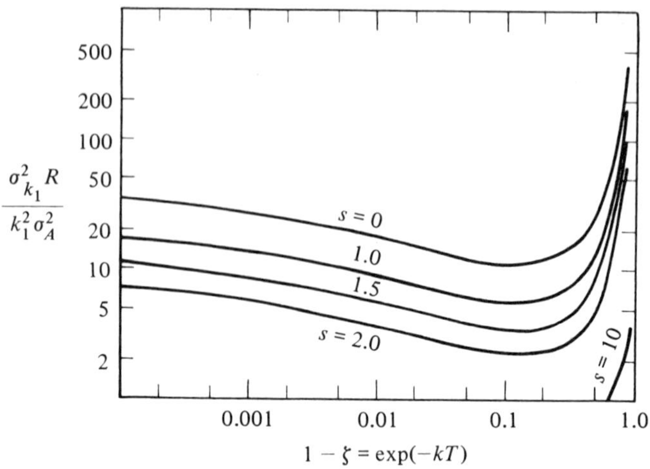

Figure 8.11. Effect of measuring more than the number of independent components on the dimensionless variance of parameter estimate vs. conversion in the reaction scheme $A \xrightarrow{1} B$; the covariance of errors is

$$\mathbf{V}_r = \begin{bmatrix} \sigma_A^2 & 0 \\ 0 & \sigma_B^2 \end{bmatrix}$$

[Reprinted from A. L. Ravimohan, W. H. Chen and J. H. Seinfeld, "The Accuracy of Kinetic Parameters Estimated from Batch and Integral Reactor Data," *Can. J. Chem. Eng. 48*, 420 (1970). Reprinted by permission of the copyright owner.]

to a longer time necessary for complete conversion to C and the ability to obtain more information on a fixed interval $(0, T)$. As a result, the ψ_2-curves exhibit a minimum coupled with a reversal of the dependence on q. For $q = 3$ the ψ_1-curve also exhibits a minimum (a relative minimum) reflecting the trade-off between the increase in ψ_1 at high values of k_1T, the decrease in ψ_1 at low values of k_1T, and the effect of a large k_2/k_1, which causes an increase in ψ_1 for moderate values of k_1T.

Figure 8.6: For very small and very large values of k_1T, both ψ_1 and ψ_2 become large. In the former case it is because neither reaction has proceeded to any appreciable extent. In the latter case, the low accuracy is due to the fact that most of the measurements are being wasted in measuring values of A and B, which have, for all practical purposes, long since reached their final values. The trade-off between these two considerations results in a minimum in both the ψ_1- and ψ_2-curves for all values of $q = k_2/k_1$.

As q increases, the value of k_1T corresponding to the minimum value of the ψ_1- and ψ_2-curves decreases. This is a direct consequence of measurements being taken on A and B only. The higher the value of q for a given k_1, the more important the second reaction becomes. It is then advantageous to take measurements on A and B before all the A is converted to C by the second reaction and the value of B becomes static. The optimum value of k_1T therefore becomes smaller as q increases.

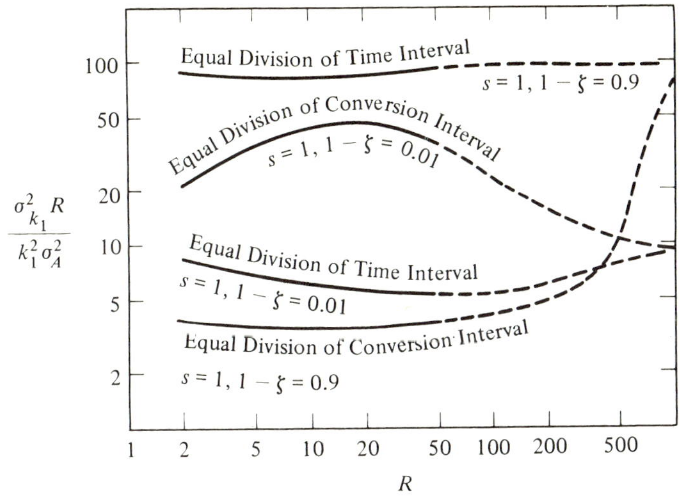

Figure 8.12. Effect of distribution of experimental observations on the dimensionless variance of parameter estimate vs. the number of observations in the reaction scheme $A \xrightarrow{1} B$. [Reprinted from A. L. Ravimohan, W. H. Chen and J. H. Seinfeld, "The Accuracy of Kinetic Parameters Estimated from Batch and Integral Reactor Data," *Can. J. Chem. Eng. 48*, 420 (1970). Reprinted by permission of the copyright owner.]

For values of q less than about 2, ψ_1 is less than ψ_2; i.e., k_1 is estimated more accurately than k_2. However, for higher values of q, the situation is exactly the opposite. This effect is important for low values of k_1T and is again a consequence of the fact that we are observing only A and B. We may think of the measurements on A as determining $k_1 + k_2$ and the measurements on B as determining k_1. As q increases, it becomes less and less profitable to measure B, as the predominant reaction is the second one. Hence the k_1 estimates start getting less and less accurate. However, under these circumstances, $k_1 + k_2$ becomes very nearly the same as k_2 itself, and therefore the accuracy of the k_2-estimate does not suffer as much. For low values of q, on the other hand, the situation is exactly the opposite; k_2-estimates are poor because we are effectively taking no observations on the second reaction.

Figure 8.7: In general, as R is increased, the accuracy of the estimates of k_1 and k_2 is increased.

For small values of k_1T, smaller values of ψ_1 and ψ_2 are obtained with fewer measurements, because ψ_1 and ψ_2 depend on the product of σ_i^2 and R, and as R decreases, the decrease in R more than compensates for the increase in σ_i^2. For large values of k_1T, as R decreases, σ_i^2 increases faster than R decreases and ψ_1 and ψ_2 increase. At large values of kT, ψ_1 and ψ_2 depend very strongly on R, and it is in this range that R should be made as large as possible.

Figures 8.8 and 8.9: For small values of k_1T and $R = 5$, 10, and 20, ψ_1 and ψ_2 are not strongly dependent on R. If $R = \infty$, a significant decrease in ψ_1 and ψ_2 occurs. For large values of k_1T, ψ_1 and ψ_2 are strong functions of R. Thus, if the

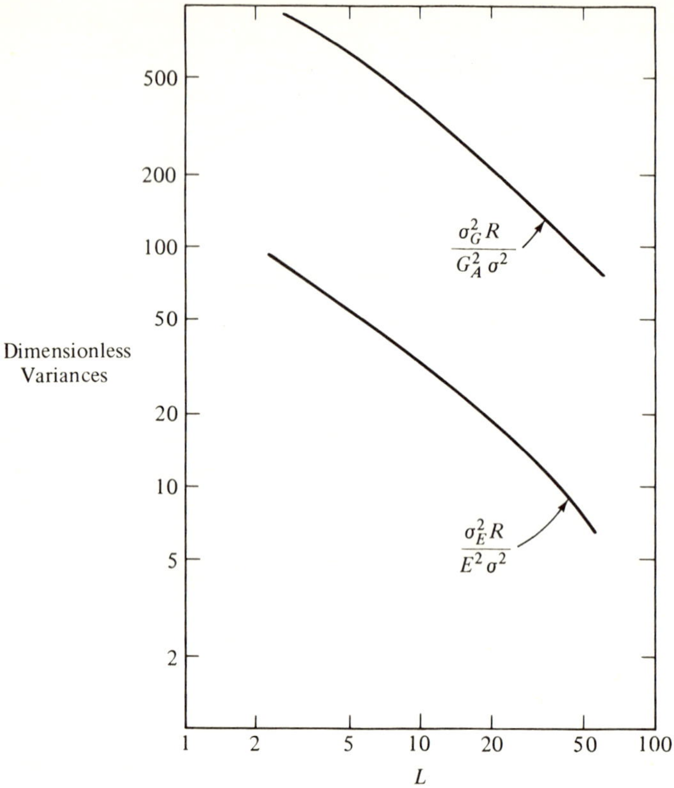

Figure 8.13. Effect of number of isothermal runs on the dimensionless variances of the frequency factor and activation energy estimates in the scheme $A \xrightarrow{G_A \exp(-E/\theta)} B$ with $G_A(T) = 10$, $\theta(0)/E = 1.5$. [Reprinted from A. L. Rovimohan, W. H. Chen and J. H. Seinfeld, "The Accuracy of Kinetic Parameters Estimated from Batch and Integral Reactor Data," *Can. J. Chem. Eng. 48*, 420 (1970). Reprinted by permission of the copyright owner.]

reaction is allowed to proceed substantially toward the complete conversion to C, it is important for a fixed value of T that one make as many measurements as possible. If the reaction is stopped early, then the number of observations is not as important. These conclusions are clearly visible in Figure 8.9 in which ψ_1/R is plotted against R for constant values of $k_1 T$.

Figure 8.10: The number of independent components in a reaction scheme must be equal to the number of independent reactions. Also, they must be chosen in such a way that the stoichiometric matrix **A** relating them to the extents of the independent reactions is nonsingular:

$$\begin{pmatrix} C_1 \\ \cdot \\ \cdot \\ \cdot \\ C_n \end{pmatrix} = \mathbf{A} \begin{pmatrix} \xi_1 \\ \cdot \\ \cdot \\ \cdot \\ \xi_n \end{pmatrix}$$

In many problems of kinetic parameter estimation, there is a choice regarding which components are to be measured.

In the present study, this question was thoroughly examined for the isothermal scheme

$$A \begin{array}{l} \xrightarrow{k_1} B \\ \xrightarrow[k_2]{} C \end{array}$$

Figure 8.10 is an example of the large number of plots that were obtained for various values of k_2/k_1. The only conclusion that is valid for all these plots is that the most accurate estimates of k_1 and k_2 are obtained when all three components are measured. However, a certain combination of two components always comes very close to the $(A + B + C)$-curve. Hence it would appear that there is little advantage in measuring more than the independent number of concentrations when the error matrix $\mathbf{V}$ has the form $\sigma^2\mathbf{I}$.

For specific guidance in an actual problem, it is necessary to look at the plot for the relevant value of k_2/k_1. Consider, for example, $k_2/k_1 = 0.5$, which is illustrated in Figure 8.10. Further, let us confine our attention to the ψ_1-curves, which give information regarding the accuracy of the k_1-estimate. Here, the best two components to measure are obviously A and B. Measuring C and A is better for high values of k_1T, but B and C is better for lower values of k_1T. Near the optimum values of k_1T, there is little choice between these two combinations.

The best single component to measure is B. Measurement of C alone is likely to lead to extremely inaccurate estimates. Of course, measurement of A alone will not give us k_1 and k_2 separately, a fact that is strikingly evident in the singularity of the $\mathbf{D}^T\mathbf{G}^T\mathbf{G}\mathbf{D}$ matrix (and hence the $\mathbf{H}$ matrix) for that case.

Figure 8.11: For a given σ_A and a given number of observations R, the accuracy of estimation improves as s increases. This is correct as increasing s means increasing the accuracy of measuring B. The curve for $s = 0$ corresponds to extremely imprecise measurements on B, which is equivalent to measurement of A alone.

Measuring more than the number of independent components is advantageous under all circumstances except $s = 0$. Further, for every value of n, there is a minimum which represents the point at which the reaction must be skipped to get maximum accuracy of estimate.

As $T \longrightarrow t_0$ ($\xi \longrightarrow 1$), all the curves go up to infinity. This is obvious, as there can be no estimates when the reaction is not carried out at all.

Figure 8.12: The limiting value plotted in Figure 8.12 is strictly valid for a very large number of observations. This limit is approached rapidly from the lower side when the time interval (t_0, T) is equally divided. However, if the same number of observations are taken at intervals of equal conversion, the results are profoundly different. For high values of ξ (e.g., 0.9), it is possible to achieve much greater accuracies by this device. For low values of ξ (e.g., 10^{-2}), on the other hand, this must not be done, as errors increase greatly. In fact, for low ξ, the limit as $R \longrightarrow \infty$ is approached from the upper side. We therefore conclude that taking a finite number of observations at equal conversion intervals will be useful when the reaction has to be stopped far from completion. The physical meaning of this is as follows: When concentrations are changing rapidly, observations must be taken more frequently, but when they are changing slowly, the time period between observations can be lengthened. The experimental kineticist very often makes use of this trick while collecting his data.

Figure 8.13: The accuracy of estimation G_A and E improves almost in direct proportion to the number of temperatures at which observations are taken.

In the limited range considered, the activation energy is estimated more accurately than the frequency factor. This would indicate that nonlinear regression of the pooled data yields a better estimate of E than that obtained by regressing the estimated rate constants at different temperatures. In such cases, the technique presented here could be invaluable in assigning confidence limits to the parameters obtained by the nonlinear analysis.

8.3 Design of Experiments for Model Discrimination

In the last section we have considered the question of designing an experiment to obtain the best estimates of parameters in a given model. Often, however, one has several models in contention, each of which has unknown parameters. Therefore an important question is, Can we design a set of experiments so that we can discriminate among rival models in some optimal fashion?

We shall consider only two rival models. As was the case with parameter estimation, it is the experimenter's task at each stage to seek out that experiment which is likely to yield the greatest amount of relevant information. Suppose that we have a scalar variable Z, which we write as a capital letter to signify that it is a random variable. If hypothesis 1 is true, then Z is distributed according to the density $p_1(z)$; if hypothesis 2 is true, z is distributed according to $p_2(z)$. Kullback [15] has suggested that the quantity

$$\ln \frac{p_1(z)}{p_2(z)}$$

is a measure of the favorability of hypothesis 1 over hypothesis 2. The expected information in favor of 1 is

$$\int_{-\infty}^{\infty} p_1(z) \ln \frac{p_1(z)}{p_2(z)}\, dz$$

Kullback suggested that the measure of total information

$$I = \int_{-\infty}^{\infty} \left[p_1(z) \ln \frac{p_1(z)}{p_2(z)} + p_2(z) \ln \frac{p_2(z)}{p_1(z)} \right] dz \tag{8.3.1}$$

be maximized.

Suppose that we have performed $N - 1$ experiments and have fitted the data to both models 1 and 2, yielding parameter estimates $\hat{\mathbf{k}}^{(1)}$ and $\hat{\mathbf{k}}^{(2)}$. Let $\mathbf{x}_N$ represent the values of the independent variables for a proposed Nth experiment. We can compute the predicted values $\mathbf{y}_N^{(1)}$ and $\mathbf{y}_N^{(2)}$ of the observed variables for the proposed experiments based on the estimated parameters, $\hat{\mathbf{k}}^{(1)}$ and $\hat{\mathbf{k}}^{(2)}$, respectively, from

$$\mathbf{y}_N^{(1)} = \mathbf{f}(\mathbf{x}_N, \hat{\mathbf{k}}^{(1)}), \qquad \mathbf{y}_N^{(2)} = \mathbf{f}(\mathbf{x}_N, \hat{\mathbf{k}}^{(2)})$$

Let the covariance matrices of these two predictions be $\mathbf{V}^{(1)}$ and $\mathbf{V}^{(2)}$, respectively; that is,

$$E\{[\mathbf{y}_N - \mathbf{y}_N^{(i)}][\mathbf{y}_N - \mathbf{y}_N^{(i)}]^T\} = \mathbf{V}^{(i)}, \qquad i = 1, 2 \tag{8.3.2}$$

Furthermore, let $\mathbf{U}^{(i)} = [\mathbf{V}^{(i)}]^{-1}$. For the random variable Z in (8.3.1) we substitute the two residuals in (8.3.2) and assume that p_1 and p_2 are each normal distributions. Then (8.3.1) becomes

$$\begin{aligned} I = -m &+ \tfrac{1}{2} tr[\mathbf{U}^{(2)}\mathbf{V}^{(1)}] + \tfrac{1}{2} tr[\mathbf{U}^{(1)}\mathbf{V}^{(2)}] \\ &+ \tfrac{1}{2}[\mathbf{y}_N^{(1)} - \mathbf{y}_N^{(2)}]^T[\mathbf{U}^{(1)} + \mathbf{U}^{(2)}][\mathbf{y}_N^{(1)} - \mathbf{y}_N^{(2)}] \end{aligned} \tag{8.3.3}$$

where $tr[\cdot]$ means the trace of a matrix.

Application of (8.3.3) requires estimates of the two covariance matrices $\mathbf{V}^{(1)}$ and $\mathbf{V}^{(2)}$, which are measures of the uncertainties in the predicted values $\mathbf{y}_N^{(1)}$ and $\mathbf{y}_N^{(2)}$. There are two sources for these uncertainties:

1. The uncertainty in the values of the estimated parameters, $\mathbf{k}^{(1)}$ and $\mathbf{k}^{(2)}$, measured by the covariance matrices of the estimates, $\mathbf{P}^{(1)}$ and $\mathbf{P}^{(2)}$.
2. The difference between the measured values of the variables $\mathbf{y}_N^{(1)}$ and $\mathbf{y}_N^{(2)}$ and their true values, due to experimental errors. A measure of these errors is the covariance matrix of the residuals $\mathbf{\Gamma}^{(i)}$. (We have normally used $\mathbf{V}$ for this, but we now use $\mathbf{\Gamma}$ since $\mathbf{V}$ is being used for the total residual of each model.)

The matrices $\mathbf{P}^{(i)}$ and $\mathbf{\Gamma}^{(i)}$ are obtained as by-products of estimating the parameters in each of the two models, based on experiments already conducted. Now, assuming that these two sources cause independent errors, we have approximately

$$\mathbf{V}^{(i)} = \mathbf{\Gamma}^{(i)} + \left[\frac{\partial \mathbf{y}_N^{(i)}}{\partial \mathbf{k}^{(i)}}\right] \mathbf{P}^{(i)} \left[\frac{\partial \mathbf{y}_N^{(i)}}{\partial \mathbf{k}^{(i)}}\right]^T, \qquad i = 1, 2 \tag{8.3.4}$$

In component form,

$$V_{\alpha\beta}^{(i)} = \Gamma_{\alpha\beta}^{(i)} + \sum_{l=1}^{p} \sum_{r=1}^{p} \frac{\partial y_{\alpha N}^{(i)}}{\partial k_l^{(i)}} P_{lr}^{(i)} \frac{\partial y_{\beta N}^{(i)}}{\partial k_r^{(i)}}, \qquad \alpha, \beta = 1, 2, \ldots, m$$

In the case of $m = 1$ (scalar observation), $V^{(i)} = \sigma_i^2$, $\Gamma^{(i)} = s_i^2$ and

$$\sigma_i^2 = s_i^2 + \sum_{l=1}^{p} \sum_{r=1}^{p} P_{lr}^{(i)} \frac{\partial y_N^{(i)}}{\partial k_l^{(i)}} \frac{\partial y_N^{(i)}}{\partial k_r^{(i)}}$$

Then (8.3.3) becomes

$$I = \frac{1}{2}\left[\frac{(\sigma_1^2 - \sigma_2^2)^2}{\sigma_1^2 \sigma_2^2} + (y_N^{(1)} - y_N^{(2)})^2\left(\frac{1}{\sigma_1^2} + \frac{1}{\sigma_2^2}\right)\right] \tag{8.3.5}$$

If σ_1 and σ_2 do not vary much from one set of experimental conditions to another, the information is essentially proportional to $(y_N^{(1)} - y_N^{(2)})^2$, i.e., the square of the difference between the predicted values under the two hypotheses. The experiment to be performed is the one for which this difference is the greatest—a commonsense result. This result was first obtained by Box and Hill [4], who have been the pioneers in the application of sequential design for model discrimination.

EXAMPLE 8.3. Bard and Lapidus [2] have employed the above technique on a computer-simulated experiment in which a computer subroutine replaced the laboratory equipment. The selected experimental conditions (values of the independent variables) are fed as input to the subroutine, which computes the hypothetical results of the experiment, using one of the alternative models. The subroutine then adds to these results a pseudorandom variable having prescribed statistical properties (e.g., variance). These results are accepted by the parameter estimation and experimental design programs as though they had been produced by an actual experiment.

A system selected for simulation was that of the catalytic hydrogenation of mixed isooctenes, described by Hougen and Watson [10] and further analyzed by Blakemore and Hoerl [3]. The latter authors reduced the number of acceptable rate equations to two:

$$\text{Model 1:} \quad y^{(1)} = \frac{k_1^{(1)} x_1 x_2}{(1 + k_2^{(1)}\sqrt{x_1} + k_3^{(1)} x_2 + k_4^{(1)} x_3)^3}$$

$$\text{Model 2:} \quad y^{(2)} = \frac{k_1^{(2)} x_1 x_2}{(1 + k_2^{(2)} x_1 + k_3^{(2)} x_2 + k_4^{(2)} x_3)^2}$$

where y is the rate of the reaction, and x_1, x_2, and x_3 are, respectively, the partial pressures of hydrogen, isooctene, and isooctane. The available experimental data were insufficient for preferring one of these models. For the computer-simulated experiments, the partial pressures of the reactants were confined to the same region as in the data used by Blakemore and Hoerl, namely,

$$0.1 \leq x_1 \leq 2.5$$

$$0.1 \leq x_2 \leq 3.0$$

$$0.05 \leq x_3 \leq 2.7$$

For the first six experiments, the fractional design given in Table 8.2 was used and the results (i.e., the rate of the reaction y) were computed assuming that model 1 was correct. Subsequent experiments were chosen so as to maximize the

TABLE 8.2

Experiment	x_1	x_2	x_3
1	0.1	1.55	1.375
2	2.5	1.55	1.375
3	1.3	0.1	1.375
4	1.3	3.0	1.375
5	1.3	1.55	0.05
6	1.3	1.55	2.7

information measure given by (8.3.5) [in this case, at least, precisely the same results are obtained by maximizing $(y_N^{(1)} - y_N^{(2)})^2$]. The results of those experiments were computed assuming that model 2 was correct; in this way one desired to find out how soon the sequence of experiments would overcome the "misleading" results of the first six experiments and choose model 2 as the correct one. A total of 27 experiments were "performed." Also, 27 experiments constituting a $3 \times 3 \times 3$ factorial design were carried out. Table 8.3 compares the levels of confidence in

TABLE 8.3

	Confidence in model 2 after 27 experiments (%)	
Experimental errors (%)	Factorial design	Sequential design
1	60	99.8
3	50	98
6	50	80
10	50	50

model 2 achieved after 27 experiments at various levels of simulated experimental error. As expected, the discrimination power of each procedure diminishes with increasing experimental error; the factorial designs fail completely to discriminate at 3 percent error, whereas the sequential design is spectacularly better than that of the factorial.

It is interesting to list the number of experiments that were needed to reach a 90 percent confidence level by the sequential design. These were 15, 17, and 30 for experimental errors of 1, 3, and 6 percent, respectively. In contrast, the factorial design never exceeded a 60 percent level.

The above results show that sequential experimental design holds great promise. Whether this promise will be fulfilled in practice is not yet known. The main theoretical question that arises when these methods are applied in practice arises from the fact that rarely is any of the proposed mechanisms exactly correct, and therefore one should not exaggerate the confidence placed on any model selected as a result of analyzing the data obtained in a series of experiments. Clearly, if the correct model is not among those proposed, it cannot be the one selected. Frequently, however, the data themselves may lead to systematic modifications of the model, until a suitable model is found. Such an analysis is described in detail by Hunter and coworkers [11, 13, 14].

REFERENCES

1. Ash, R., *Information Theory*, Wiley-Interscience, New York, 1965.
2. Bard, Y., and L. Lapidus, "Kinetics Analysis by Digital Parameter Estimation," *Catalysis Rev. 2*, 67 (1968).
3. Blakemore, J. W., and A. E. Hoerl, "Fitting Nonlinear Reaction Rate Equations to Data," *Chem. Eng. Progr. Symp. Ser. 59*, 14 (1963).
4. Box, G. E. P., and W. J. Hill, "Discrimination Among Mechanistic Models," *Technometrics 9*, 57 (1967).
5. Box, G. E. P., and H. L. Lucas, "Design of Experiments in Nonlinear Situations," *Biometrika 46*, 77 (1959).
6. Draper, N. R., and W. G. Hunter, "Design of Experiments for Parameter Estimation in Multiresponse Situations," *Biometrika 53*, 525 (1966).
7. Draper, N. R., and W. G. Hunter, "The Use of Prior Distributions in the Design of Experiments for Parameter Estimation in Nonlinear Situations," *Biometrika 54*, 147 (1967).
8. Heineken, F. G., H. M. Tsuchiya, and R. Aris, "On the Accuracy of Determining Rate Constants in Enzymatic Reactions," *Math. Biosci. 1*, 115 (1967).
9. Himmelblau, D. M., *Process Analysis by Statistical Methods*, Wiley, New York, 1970.
10. Hougen, O. A., and K. M. Watson, *Chemical Process Principles*, part 3, Wiley, New York, 1947.
11. Hunter, W. G., and R. Mezaki, "A Model Building Technique for Chemical Engineering Kinetics," *AIChE J. 10*, 315 (1964).
12. Hunter, W. G., W. J. Hill, and T. L. Henson, "Designing Experiments for Precise Estimation of All or Some of the Constants in a Mechanistic Model," Report 172, Department of Statistics, University of Wisconsin, Madison, Aug. 1968.
13. Kittrell, J. R., W. G. Hunter, and C. C. Watson, "Nonlinear Least Squares Analysis of Catalytic Rate Models," *AIChE J. 11*, 6, 1051 (1965).
14. Kittrell, J. R., W. G. Hunter, and C. C. Watson, "Obtaining Precise Estimates for Nonlinear Catalytic Rate Models," *AIChE J. 12*, 1, 5 (1966).
15. Kullback, S., *Information Theory and Statistics*, Wiley, New York, 1959.
16. Ravimohan, A. L., W. H. Chen, and J. H. Seinfeld, "The Accuracy of Kinetic Parameters Estimated from Batch and Integral Reactor Data," *Can. J. Chem. Eng. 48*, 420 (1970).

17. Rosenbrock, H. H., and C. Storey, *Computational Techniques for Chemical Engineers*, Pergamon, Elmsford, N.Y., 1966.

PROBLEMS

8.1 The following reaction system has received much interest:

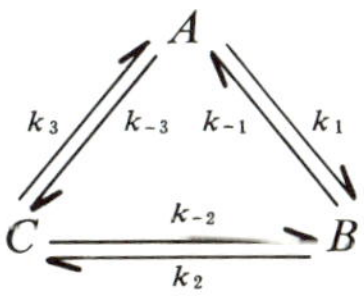

Suppose that we know the three equilibrium constants K_1, K_2, and K_3 and desire to estimate k_1, k_2, and k_3 (the forward rate constants) in this system. We measure the concentrations of A and B at three times starting from $[A]_0 = 1$, $[B]_0 = [C]_0 = 0$.

a. At what values of time should the three observations be made if both species are measured with error of variance σ^2?

b. How would these change if the variance of errors in A-measurements is 10 times as great as the variance of errors in B-measurements?

8.2 Consider a slab the temperature of which is governed by the heat equation

$$\frac{\partial u(x,t)}{\partial t} = k\frac{\partial^2 u(x,t)}{\partial x^2}$$

$$u(x,0) = \phi(x)$$

$$u(0,t) = u(1,t) = 0$$

Our objective is to measure $u(x, t)$ at two locations continuously in time, that is,

$$y(x_r, t) = u(x_r, t) + \epsilon_r(t), \qquad r = 1, 2$$

in order to estimate k. To minimize the variance of the estimate of k, where should the two thermocouples be placed if the initial temperature distribution is

a. $\phi(x) = \sin 2\pi x$

b. $\phi(x) = \sin \pi x$

What peculiarities do you note about case b? Explain.

If the two measurements must be located at 0.25 and 0.75, what would be the optimal $\phi(x)$ for estimating k with minimum variance?

8.3 Extend Example 8.2 to the reaction

$$S + E \underset{k_{-1}}{\overset{k_1}{\rightleftharpoons}} C \xrightarrow{k_2} P + E$$

This is a classic enzyme reaction mechanism. Compare your analysis with Heineken et al. [8].

8.4 The problem of estimating reservoir properties from well pressure data is an important one in the petroleum industry (see Problem 7.8). In general this determination is carried out by a parameter estimation routine, such as in Problem 7.8. In this problem we consider two aspects, namely the determination of the accuracy of estimated reservoir parameters and the design of a test procedure to determine parameters with minimum error. The common test procedure is to produce the well for a time observing the pressure decrease and then close up the well and observe the pressure build-up.

We consider the transient pressure behavior in a radial reservoir containing a single phase fluid of small and constant compressibility. We assume that the reservoir is finite and that the centrally located well produces from time 0 to t_f and then is shut in (no oil is produced) from t_f to T. In dimensionless form the pressure behavior in the reservoir satisfies

$$\frac{\partial p(x, t)}{\partial t} = \frac{1}{k_2} \exp\left\{-2 \ln\left(\frac{r_e}{r_w}\right)x\right\} \frac{\partial}{\partial x}\left[k_1 \frac{\partial p}{\partial x}\right]$$

where

x = dimensionless distance from well $= \ln (r/r_w)/\ln (r_e/r_w)$
(r_e = radius of reservoir, r_w = radius of well bore)
$r_e/r_w = 100$.
t = dimensionless time $= t'/T$
p = dimensionless pressure $= p'/p_{\text{initial}}$
k_1 = dimensionless permeability (constant)
k_2 = dimensionless porosity (constant)

The initial and boundary conditions are

$$p(x, 0) = 1 \qquad 0 \leq x \leq 1$$

$$\left.\frac{\partial p}{\partial x}\right|_{x=0} = \frac{1}{2\pi k_1} \qquad 0 \leq t \leq \alpha \qquad \alpha = \frac{t_f}{T}$$

$$\left.\frac{\partial p}{\partial x}\right|_{x=0} = 0 \qquad \alpha < t \leq 1$$

$$\left.\frac{\partial p}{\partial x}\right|_{x=1} = 0 \qquad 0 \leq t \leq 1$$

where α is the fraction of the total well test time that the well is produced.

a. If the observations consist of measurements of the well pressure $p(0, t)$ at R_1 values of time, the data being corrupted with errors of mean zero and standard deviation σ_p, show that the variance of the error in the estimates of k_1 and k_2 are given by

$$\sigma_i^2 = \frac{\sigma_p^2}{\sum_{r=1}^{R_1} \lambda_i^2(0, t_r)} \qquad i = 1, 2$$

where $\lambda_1(x, t) = \partial p/\partial k_1$ and $\lambda_2(x, t) = \partial p/\partial k_2$.

b. Consider k_2 as known. Plot the dimensionless standard deviation $\psi_1 = \sigma_1/k_1\sigma_p$ vs. k_1 as a function of α for a selected value of k_2. Is there an optimum value of α for estimating k_1?

c. Consider k_1 as known. Plot the dimensionless standard deviation $\psi_2 = \sigma_2/k_2\sigma_p$ vs. k_2 as a function of α for a selected value of k_1. Is there an optimum value of α for estimating k_2?

d. Consider both k_1 and k_2 as unknown. Plot ψ_1 and ψ_2 vs. α for fixed k_1 and k_2. Is there an optimum α for estimating both parameters?

NOTE: The range of variation of k_1 and k_2 are to be taken as:

k_1	0.01 – 1.0
k_2	10^{-6} – 0.01

Process Identification: Linear Systems† 9

In this and the final chapter we shall consider the problem of *process identification*, namely the construction of a model for a system directly from its input/output data. This chapter is devoted to the construction of general linear models, whereas Chapter 10 will treat nonlinear models. Both chapters focus on the so-called black box approach, wherein we have no prior knowledge about the system—only its input/output data from some experiment.

We shall outline a technique operating in the time domain which generates a linear model based only on a given set of input/output data. The method will not only furnish a linear model but will also develop a so-called *minimal model*, that is, the linear model consisting of the minimum number of system parameters necessary to generate the output data given the input data.

In developing the material of this chapter we shall carefully introduce the dual concepts of controllability and observability which serve as basic foundations for the model development. A specific algorithm due to Ho and Kalman [7] will be extensively detailed along with extensions to suboptimal algorithms and methods which apply to real systems. The connections to moment analysis and transfer functions will also be developed.

†This chapter was written in collaboration with Dr. Robert Rossen during the development of his Ph.D. thesis [18].

9.1 Further Linear Theory

Here we shall begin to construct a theory which will form the basis of methods for the construction of a mathematical description of the internal operation of a linear system. These methods require only the output response to a particular set of input functions, an impulse in each of the input (or control) variables, to yield appropriate input/output data. As pointed out in Chapter 2, the linear, constant, finite dimensional, continuous, or discrete dynamical systems can be represented as either

$$\begin{aligned} \dot{\mathbf{x}}(t) &= \mathbf{F}\mathbf{x}(t) + \mathbf{G}\mathbf{u}(t) \\ \mathbf{y}(t) &= \mathbf{H}\mathbf{x}(t) \end{aligned} \tag{9.1.1}$$

or as

$$\begin{aligned} \mathbf{x}(k+1) &= \mathbf{\Phi}\mathbf{x}(k) + \mathbf{G}\mathbf{u}(k) \\ \mathbf{y}(k) &= \mathbf{H}\mathbf{x}(k), \qquad k = 0, 1, 2, \ldots \end{aligned} \tag{9.1.2}$$

Alternatively, we may simply write (9.1.2) as

$$\begin{aligned} \mathbf{x}(t+1) &= \mathbf{\Phi}\mathbf{x}(t) + \mathbf{G}\mathbf{u}(t) \\ \mathbf{y}(t) &= \mathbf{H}\mathbf{x}(t) \end{aligned} \tag{9.1.2a}$$

and retain the symbol k for other matters. In these equations $\mathbf{x}(t)$, $\mathbf{u}(t)$, and $\mathbf{y}(t)$ are the state $(n \times 1)$, control, or input $(r \times 1)$ and output $(m \times 1)$ vectors, respectively.

The triple of constant matrices $\{\mathbf{F}, \mathbf{G}, \mathbf{H}\}$, or $\{\mathbf{\Phi}, \mathbf{G}, \mathbf{H}\}$, which represents the internal operation of the linear system, can be used to determine, a priori, the system's response to any input. Such a representation is called a *realization* of the system. Any system has an infinite number of realizations which will predict the identical response for any particular input.

We can illustrate the last point in a simple fashion. Let us define a new vector $\boldsymbol{\eta}$ such that

$$\boldsymbol{\eta} = \mathbf{T}\mathbf{x} \quad \text{or} \quad \mathbf{x} = \mathbf{T}^{-1}\boldsymbol{\eta}$$

where $\mathbf{T}$ is *any* $(n \times n)$ nonsingular square matrix. Substitution into (9.1.1) yields

$$\begin{aligned} \mathbf{T}^{-1}\dot{\boldsymbol{\eta}} &= \mathbf{F}\mathbf{T}^{-1}\boldsymbol{\eta} + \mathbf{G}\mathbf{u} \\ \mathbf{y} &= \mathbf{H}\mathbf{T}^{-1}\boldsymbol{\eta} \end{aligned}$$

or, upon rearrangement,

$$\begin{aligned} \dot{\boldsymbol{\eta}} &= \mathbf{T}\mathbf{F}\mathbf{T}^{-1}\boldsymbol{\eta} + \mathbf{T}\mathbf{G}\mathbf{u} \\ \mathbf{y} &= \mathbf{H}\mathbf{T}^{-1}\boldsymbol{\eta} \end{aligned} \tag{9.1.3}$$

The effect of input $\mathbf{u}(t)$ on $\mathbf{y}(t)$ will be the same for this new system, (9.1.3), as for the original system, (9.1.1). Thus the triple $\{\mathbf{T}\mathbf{F}\mathbf{T}^{-1}, \mathbf{T}\mathbf{G}, \mathbf{H}\mathbf{T}^{-1}\}$ will also be a realization of (9.1.1) and the predicted responses using this realization will be identical to those predicted by $\{\mathbf{F}, \mathbf{G}, \mathbf{H}\}$. Since there are an infinite number of nonsingular matrices **T**, there are an infinite number of such realizations.

This can be further seen by noting that the transfer function matrix for (9.1.1) is (see Section 2.3)

$$\bar{\mathbf{Y}}(s) = \mathbf{H}(s\mathbf{I} - \mathbf{F})^{-1}\mathbf{G}$$

while that for (9.1.3) is

$$\bar{\mathbf{Y}}_1(s) = \mathbf{HT}^{-1}(s\mathbf{I} - \mathbf{TFT}^{-1})^{-1}\mathbf{TG}$$

However, some algebraic manipulation shows that $\bar{\mathbf{Y}}(s) = \bar{\mathbf{Y}}_1(s)$. Thus the two systems are related by the linear transformation given by **T**, are algebraically equivalent, and have the same transfer function matrix.

The dimension of the realization will be defined as the dimension, n, of the **F** (or $\mathbf{\Phi}$) matrix. Different realizations of the same system can be of different dimension, but for convenience in computation as well as for simplicity of the solution, we would like to obtain a realization of the smallest dimension, a *minimum dimension* or *minimum realization*.

The solution to (9.1.1) has previously been presented as (with $t_0 = 0$)

$$\mathbf{x}(t) = \exp[\mathbf{F}t]\mathbf{x}(0) + \int_0^t \exp[\mathbf{F}(t - \tau)]\mathbf{Gu}(\tau)\, d\tau \tag{9.1.4}$$

If the initial condition is

$$\mathbf{x}(0) = \mathbf{0} \tag{9.1.5}$$

and, since $\mathbf{y}(t) = \mathbf{Hx}(t)$, we find that

$$\mathbf{y}(t) = \mathbf{Hx}(t) = \mathbf{H}\int_0^t \exp[\mathbf{F}(t - \tau)]\mathbf{Gu}(\tau)\, d\tau \tag{9.1.6}$$

This is the system's response to any input $\mathbf{u}(\tau)$. To obtain the response to an impulse in one of the input variables, we can substitute $\delta\ (t = 0)$ for the appropriate element in the vector $\mathbf{u}(\tau)$. When this is done for each input element, the results are combined to obtain the impulse or pulse response function matrix:

$$\mathbf{Y}(t) = \mathbf{H}\exp[\mathbf{F}t]\mathbf{G} \tag{9.1.7}$$

In the L.T. domain,

$$\bar{\mathbf{Y}}(s) = \mathbf{H}\int_0^\infty \int_0^t \exp[\mathbf{F}(t - \tau)]\mathbf{Gu}(\tau)\, d\tau \exp(-st)\, dt \tag{9.1.8}$$

and the equivalent result is

$$\bar{\mathbf{Y}}(s) = \mathbf{H}(s\mathbf{I} - \mathbf{F})^{-1}\mathbf{G} \tag{9.1.9}$$

where $\bar{\mathbf{Y}}(s)$ is the transform of $\mathbf{Y}(t)$. A similar approach can be taken for the discrete case and it can be shown that the resulting functions are

$$\mathbf{Y}_t = \mathbf{H\Phi}^T\mathbf{G} \tag{9.1.10}$$

and, in the z-domain,

$$\bar{\mathbf{Y}}(z) = \mathbf{H}(z\mathbf{I} - \mathbf{F})^{-1}\mathbf{G} \tag{9.1.11}$$

As a basis for mathematical model building, the impulse is attractive because it is simple enough to be implemented in most physical systems and because, for stable systems, the pulse response damps out, returning the system to the initial state. Other inputs tend to alter the steady state of the system, thereby disturbing the system they were meant to model.

9.1.1 Markov parameters

Each of these relations can be characterized by a sequence of constant matrices $\mathbf{Y}_k$, where $\mathbf{Y}_k = \mathbf{H}\mathbf{F}^k\mathbf{G}$, $k = 0, 1, 2, \ldots$. For the continuous case, these matrices, known as *Markov parameters*, can be obtained from the experimental input/output data by a power series time fit of the expansion of (9.1.7) in terms of t^k:

$$\mathbf{Y}(t) = \sum_{k=0}^{\infty} \mathbf{Y}_k \frac{t^k}{k!} \tag{9.1.12}$$

If the L.T. is given, the parameters can be obtained by an expansion of (9.1.9) in powers of $1/s$:

$$\bar{\mathbf{Y}}(s) = \sum_{k=1}^{\infty} \mathbf{Y}_{k-1} s^{-k} \tag{9.1.13}$$

Likewise, for the discrete case, the expansion

$$\bar{\mathbf{Y}}(z) = \sum_{k=1}^{\infty} \mathbf{Y}_{k-1} z^{-k} \tag{9.1.14}$$

can be used if the z-transform is available. The discrete time solution case is easily converted into Markov parameters by the simple relation $\mathbf{\Phi} = \mathbf{F}$.

The $\mathbf{Y}_k$ are useful because they can convert each of these separate cases into a problem of the same form. Other input responses can be represented in terms of the Markov parameters. That is, if two solutions are characterized by the same set of $\mathbf{Y}_k$, $k = 0, 1, 2, \ldots$, they are equivalent and will predict exactly the same response to inputs other than the impulse function. For example, the step response and the ramp response can be treated in the same manner as the impulse response. In each case, the output can be represented as a power series in time (for the continuous case) where the coefficients are related to the Markov parameters. The $\mathbf{Y}_k$ can be used to calculate, a priori, the system's response to these inputs. Therefore any triple which produces the same sequence of $\mathbf{Y}_k$ as the system in question will be a realization of that system.

For the remainder of this chapter, only the continuous time case will be discussed. It should be remembered that the discrete case is analogous and that the treatments will be applicable to the theory of discrete time minimum realizations $\{\mathbf{\Phi}, \mathbf{G}, \mathbf{H}\}$ as well as continuous time solutions $\{\mathbf{F}, \mathbf{G}, \mathbf{H}\}$.

9.2 Controllability, Observability, and Minimum Realizations

Underlying many, if not all, of the important concepts of linear realizations are the theories of controllability and observability. The important work of Kalman [10] and Gilbert [6] should be consulted for a definitive analysis; here we shall merely present a short summary of certain important results. To do this we shall use the theorem approach so that the main features can be highlighted; we shall, however, say a few further words which are more qualitative and descriptive of the results.

9.2.1 Controllability

A system is said to be *completely controllable* or *completely state-controllable* if any state of the system can be reached from any initial state of the system in a finite interval of time by the use of some control action. From (9.1.4) we can see that the system is completely state-controllable if (and only if)

$$\mathbf{x}(t_0) = \exp[\mathbf{F}(t_0 - t_f)]\mathbf{x}(t_f) - \int_{t_0}^{t_f} \exp[\mathbf{F}(t_0 - \tau)]\mathbf{G}\mathbf{u}(\tau)\, d\tau \qquad (9.2.1)$$

for some $\mathbf{u}(\boldsymbol{\tau})$. The necessary and sufficient conditions for complete state controllability for (9.1.1) can be stated in the following theorem:

THEOREM 9.1. Any linear, finite dimensional, constant time, constant dynamical system is completely state-controllable iff the block matrix

$$[\mathbf{G} \quad \mathbf{FG} \quad \mathbf{F}^2\mathbf{G} \quad \cdots \quad \mathbf{F}^{n-1}\mathbf{G}]$$

has rank n.

The proof of this theorem is dependent on the Cayley-Hamilton theorem which states that a square matrix $\mathbf{F}$ satisfies its characteristic equation. In other words, the characteristic equation for $\mathbf{F}$ is

$$|\mathbf{F} - \lambda\mathbf{I}| = (-1)^n(\lambda^n + c_{n-1}\lambda^{n-1} + \cdots + c_2\lambda^2 + c_1\lambda + c_0) = 0$$

Thus the matrix $\mathbf{F}$ satisfies

$$\mathbf{F}^n + c_{n-1}\mathbf{F}^{n-1} + c_{n-2}\mathbf{F}^{n-2} + \cdots + c_1\mathbf{F} + c_0\mathbf{I} = \mathbf{0}$$

or

$$\sum_{k=0}^{n} c_k\mathbf{F}^k = \mathbf{0}, \qquad c_n = 1 \qquad (9.2.2)$$

Consider now $\mathbf{F}^T$, postmultiplication by $\exp[\mathbf{F}^T t]\boldsymbol{\zeta}$, where $\boldsymbol{\zeta}$ is a nonzero vector, and premultiplication by $\mathbf{G}^T$ yields

$$\sum_{k=0}^{n} \mathbf{G}^T c_k \mathbf{F}^{Tk} \exp(\mathbf{F}^T t)\boldsymbol{\zeta} = \mathbf{0} \tag{9.2.3}$$

However,

$$\frac{d^k}{dt^k} \exp(\mathbf{F}^T t) = \mathbf{F}^{Tk} \exp(\mathbf{F}^T t)$$

and thus

$$\sum_{k=0}^{n} c_k \left[\frac{d^k}{dt^k} \mathbf{G}^T \exp(\mathbf{F}^T t)\boldsymbol{\zeta} \right] = \mathbf{0} \tag{9.2.4}$$

If we define $\tilde{\mathbf{x}}(t_f) = \mathbf{x}(t_f) - \exp[\mathbf{F}(t - t_0)]\mathbf{x}(t_0)$ so that $\tilde{\mathbf{x}}(t_0) = \mathbf{0}$, then

$$\tilde{\mathbf{x}}(t_f) = \int_{t_0}^{t_f} \exp[\mathbf{F}(t_f - \tau)]\mathbf{G}\mathbf{u}(\tau)\, d\tau \tag{9.2.5}$$

For the system to be completely controllable it is necessary and sufficient that every point $\tilde{\mathbf{x}}(t_f)$ be within reach of the origin. This will be true iff the scalar product of $\tilde{\mathbf{x}}(t_f)$ with every fixed-length nonzero vector $\boldsymbol{\zeta}$ can be made arbitrarily large by some choice of $\mathbf{u}(\tau)$. If we choose $\mathbf{u}(\tau)$ so that

$$\mathbf{u}(\tau) = k\mathbf{G}^T \exp[\mathbf{F}(t_f - \lambda)]\boldsymbol{\zeta}$$

where k is positive, then the scalar product will be

$$\boldsymbol{\zeta}^T \tilde{\mathbf{x}}(t_f) = k \int_{t_0}^{t_f} [\mathbf{G}^T \exp(\mathbf{F}(t_f - \tau))\boldsymbol{\zeta}]^T [\mathbf{G}^T \exp(\mathbf{F}(t_f - \tau))\boldsymbol{\zeta}]\, d\tau \tag{9.2.6}$$

Thus the sufficient condition for complete controllability in $[t_0, t_f]$ is that

$$\mathbf{G}^T \exp[\mathbf{F}(t_f - \tau)]\boldsymbol{\zeta} \neq \mathbf{0} \tag{9.2.7}$$

For this condition to hold on any interval, it is necessary and sufficient that at least one of the initial conditions for (9.2.4) be nonzero; that is,

$$\left[\frac{d^k}{dt^k} \mathbf{G}^T \exp[\mathbf{F}^T t]\boldsymbol{\zeta} \right]_{t=0} = \mathbf{G}^T \mathbf{F}^{Tk} \exp(\mathbf{F}^T t)\boldsymbol{\zeta} \neq \mathbf{0}$$

for at least one $k = 0, 1, 2, \ldots, n - 1$. This is equivalent to the requirement that for each $\boldsymbol{\zeta} \neq \mathbf{0}$ at least one of the vectors

$$\mathbf{G}^T\boldsymbol{\zeta},\ \mathbf{G}^T\mathbf{F}^T\boldsymbol{\zeta},\ \ldots,\ \mathbf{G}^T(\mathbf{F}^T)^{n-1}\boldsymbol{\zeta}$$

be nonzero. This means that for each $\boldsymbol{\zeta} \neq \mathbf{0}$,

$$\begin{bmatrix} \mathbf{G}^T \\ \mathbf{G}^T\mathbf{F}^T \\ \cdot \\ \cdot \\ \cdot \\ \mathbf{G}^T(\mathbf{F}^T)^{n-1} \end{bmatrix} \boldsymbol{\zeta} \neq \mathbf{0} \tag{9.2.8}$$

This can be true only if all the columns of (9.2.8) are linearly independent; i.e., the rank is n. This proves Theorem 9.1.

We can also describe complete controllability in a simpler manner. A system is said to be completely controllable if it is not possible to find a coordinate system in which the state variables are separated into two groups such that the second group is not affected by the first group or by the inputs to the system. One example of an uncontrollable two-dimensional system is

$$\begin{bmatrix} \dot{x}_1 \\ \dot{x}_2 \end{bmatrix} = \begin{bmatrix} f_{11} & f_{12} \\ 0 & f_{22} \end{bmatrix} \begin{bmatrix} x_1 \\ x_2 \end{bmatrix} + \begin{bmatrix} g_1 & 0 \\ 0 & 0 \end{bmatrix} \begin{bmatrix} u_1 \\ u_2 \end{bmatrix}$$

$$y = [h_1 \quad h_2] \begin{bmatrix} x_1 \\ x_2 \end{bmatrix}$$

where neither u_1 nor u_2 can affect x_2 either directly or indirectly through coupling with x_1. On this basis controllability is a property of the system which depends only on **F** and **G** and is independent of the system outputs. Figure 9.1(a) illustrates this uncontrollable system.

Since $\mathbf{x}(t)$ cannot in general be measured directly, it will be useful to introduce the concept of *output controllability*. A linear system is said to be completely output-controllable if any output $\mathbf{y}(t_0)$ can be transferred to any other output $\mathbf{y}(t_f)$ in a finite interval of time $[t_0, t_f]$ by some control $\mathbf{u}(\tau)$.

THEOREM 9.2. Any linear, finite dimensional, continuous time, constant dynamical system is completely output-controllable if the block matrix

$$[\mathbf{HG} \quad \mathbf{HFG} \quad \mathbf{HF}^2\mathbf{G} \quad \cdots \quad \mathbf{HF}^{n-1}\mathbf{G}]$$

is of rank n.

The proof is analogous to that of Theorem 9.1.

EXAMPLE 9.1. The difference between state controllability and output controllability can be illustrated by the example shown in Figure 9.1(b) whose state equations are

$$\dot{x}_1 = u(t)$$
$$\dot{x}_2 = u(t)$$

and whose output is

$$y(t) = x_1 + x_2$$

Thus, for example, if the initial state is $x_1 = x_1(0)$ and $x_2 = x_2(0)$, the state $x_1 = x_1(0)$, $x_2 = -x_2(0)$ cannot be reached using any $u(t)$. The system is therefore not state-controllable. However, any output $y(t_f)$ can be reached from $y(0) = x_1(0) + x_2(0)$. Then

$$y(t_f) = x_1(t_f) + x_2(t_f) = x_1(0) + \int_0^{t_f} u(t)\,dt + x_2(0) + \int_0^{t_f} u(t)\,dt$$

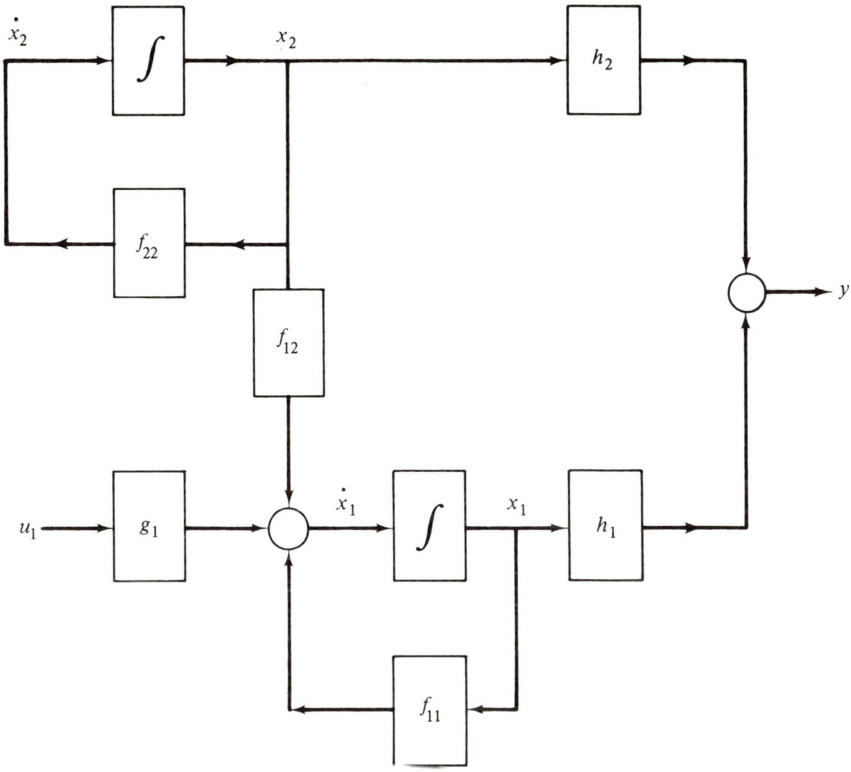

(a) Uncontrollable System

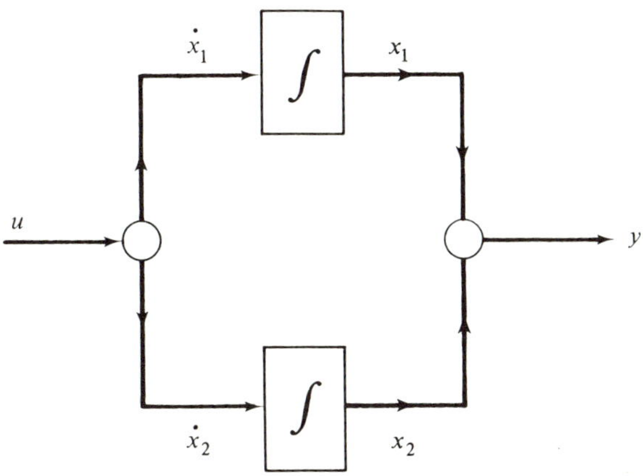

(b) Output Controllable System

Figure 9.1. Various input/output systems illustrating the features of controllability and observability.

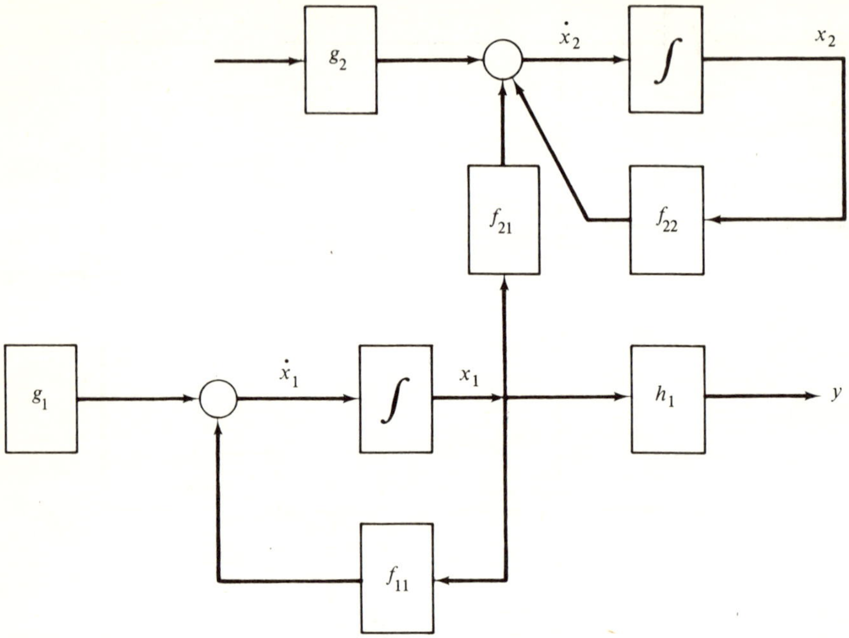

(c) Unobservable System

Figure 9.1. Continued

Therefore $y(t_f) = y(0) + 2\int_0^{t_f} u(t)\,dt$. The output $y(t_f)$ can be reached in finite time by the action of the control

$$u(t) = \frac{y(t_f) - y(0)}{2}$$

for the time interval $0 \leq t \leq 1$. The reader can readily see that this system would be output-controllable but not state-controllable.

Output controllability is important because we can always measure $\mathbf{y}(t)$; if $\mathbf{x}(t)$ cannot be measured directly, then it must be obtained by a knowledge of $\mathbf{y}(t)$. This leads us to the concept of observability.

9.2.2 Observability

A system is said to be *completely observable* if every change in the state of the system, $\mathbf{x}(t)$, results in some change in the output of the system, $\mathbf{y}(t)$. An unforced system (no control) can be described by the differential equations

$$\begin{aligned}\dot{\mathbf{x}}(t) &= \mathbf{F}\mathbf{x}(t)\\ \mathbf{y}(t) &= \mathbf{H}\mathbf{x}(t)\end{aligned} \tag{9.2.9}$$

Such a system is completely observable if for every t_0 and $t_f > t_0$, $\mathbf{x}(t_0)$ can be determined from the knowledge of $\mathbf{y}(t)$ for $t_0 \leq t \leq t_f$. The conditions for observability are stated in Theorem 9.3.

THEOREM 9.3. An unforced linear, finite dimensional, continuous time, constant dynamical system is completely observable iff the composite block matrix

$$[\mathbf{H}^T \quad \mathbf{F}^T\mathbf{H}^T \quad \cdots \quad (\mathbf{F}^T)^{n-1}\mathbf{H}^T]$$

is of rank n.

The proof of Theorem 9.3 is also analogous to the proof of Theorem 9.1.

From Theorems 9.1 and 9.3, it can be seen that a linear system is completely controllable and completely observable iff both matrices

$$[\mathbf{G} \quad \mathbf{FG} \quad \mathbf{F}^2\mathbf{G} \quad \cdots \quad (\mathbf{F}^T)^{n-1}\mathbf{G}]$$

and

$$[\mathbf{H}^T \quad \mathbf{F}^T\mathbf{H}^T \quad \cdots \quad (\mathbf{F}^T)^{n-1}\mathbf{H}^T]$$

are of rank n.

EXAMPLE 9.2. To illustrate this point we shall consider the circuit shown in Figure 9.2 [10]. Here $x_1(t)$ is the magnetic flux in the inductor and $x_2(t)$ the charge

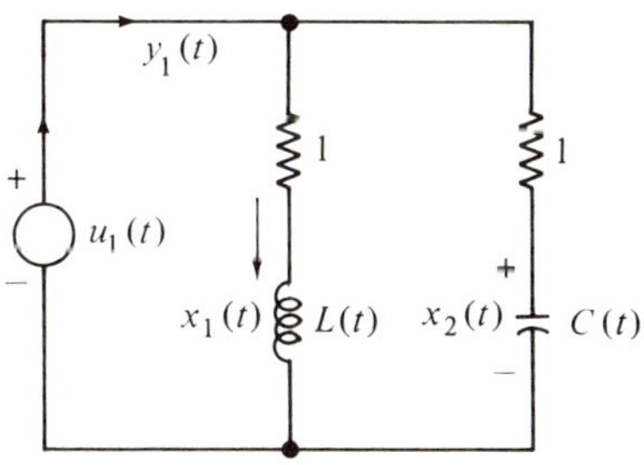

Figure 9.2. Circuit diagram for the system in Example 9.2.

on the capacitor, while $y_1(t)$ is the current in the system and $u_1(t)$ is the voltage source. We also require that $L(t)/c(t) = 1$.

The system is described by the differential equations

$$\frac{dx_1}{dt} = -\frac{1}{L(t)}x_1 + u_1(t)$$

$$\frac{dx_2}{dt} = -\frac{1}{c(t)}x_2 + u_1(t)$$

$$y_1(t) = \frac{1}{L(t)}x_1 - \frac{1}{c(t)}x_2 + u_1(t)$$

It appears at just a glance that the elements x_1 and x_2 can be controlled and observed. However, if we form the matrices

$$[\mathbf{G} \quad \mathbf{FG}] = \begin{bmatrix} 1 & \frac{1}{L} \\ 1 & \frac{1}{c} \end{bmatrix}$$

and

$$[\mathbf{H}^T \quad \mathbf{F}^T\mathbf{H}^T] = \begin{bmatrix} \frac{1}{L} & -\frac{1}{L^2} \\ -\frac{1}{c} & \frac{1}{c^2} \end{bmatrix}$$

we see that they are both of rank 1 and thus the system is neither controllable nor observable. This can be seen by defining a new set of state variables

$$\bar{x}_1 = \frac{x_1 + x_2}{2}$$

$$\bar{x}_2 = \frac{x_1 - x_2}{2}$$

so that since $L(t) = c(t)$, the state equations become

$$\frac{d\bar{x}_1}{dt} = -\frac{1}{L(t)}\bar{x}_1 + u_1(t)$$

$$\frac{d\bar{x}_2}{dt} = -\frac{1}{L(t)}\bar{x}_2$$

$$y_1(t) = \frac{\bar{x}_2}{L(t)} + u_1(t)$$

From these equations it is clear that $\bar{x}_1$ is not observable and that $\bar{x}_2$ is not controllable.

Just as we gave a simple definition for complete controllability we can also do the same for complete observability. A system is said to be completely observable if it is not possible to find a coordinate system in which the state variables are separated into two groups such that the second group does not affect either the first group or the outputs of the system. A general system equation of second order which is unobservable is given by

$$\begin{bmatrix} \dot{x}_1 \\ \dot{x}_2 \end{bmatrix} = \begin{bmatrix} f_{11} & 0 \\ f_{21} & f_{22} \end{bmatrix} \begin{bmatrix} x_1 \\ x_2 \end{bmatrix} + \begin{bmatrix} g_1 & 0 \\ 0 & g_2 \end{bmatrix} \begin{bmatrix} u_1 \\ u_2 \end{bmatrix}$$

$$y = [h_1 \quad 0] \begin{bmatrix} x_1 \\ x_2 \end{bmatrix}$$

Figure 9.1(c) shows a simple unobservable system since x_2 cannot influence x_1 or y. As such, one can see that observability depends only on system output and is independent of inputs.

Figure 9.3 shows a composite system for which the state and output equa-

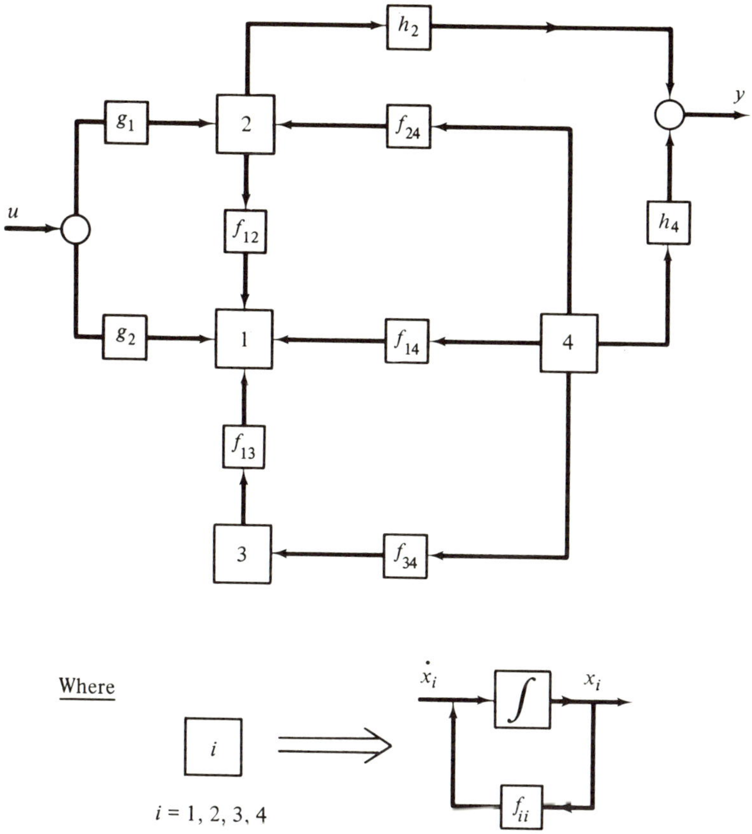

Figure 9.3. Grouping of systems to illustrate controllability and observability.

tions are

$$\begin{bmatrix} \dot{x}_1 \\ \dot{x}_2 \\ \dot{x}_3 \\ \dot{x}_4 \end{bmatrix} = \begin{bmatrix} f_{11} & f_{12} & f_{13} & f_{14} \\ 0 & f_{22} & 0 & f_{24} \\ 0 & 0 & f_{33} & f_{34} \\ 0 & 0 & 0 & f_{44} \end{bmatrix} \begin{bmatrix} x_1 \\ x_2 \\ x_3 \\ x_4 \end{bmatrix} + \begin{bmatrix} g_1 \\ g_2 \\ 0 \\ 0 \end{bmatrix} u$$

$$y = [0 \quad h_2 \quad 0 \quad h_4] \begin{bmatrix} x_1 \\ x_2 \\ x_3 \\ x_4 \end{bmatrix}$$

In this case the sections marked as 1–4 in Figure 9.3 are

1. Completely controllable but unobservable.

2. Completely controllable and observable.
3. Uncontrollable and unobservable.
4. Uncontrollable but observable.

9.2.3 The construction of a minimum realization from a transfer function

If a realization is completely observable and completely controllable, then it is of minimum dimension and is called a *minimum realization*. This necessary and sufficient condition was presented and proved by Kalman [10] and has been an important axiom in several algorithms devised to obtain realizations of minimum dimensions. The first two such algorithms were presented simultaneously by Kalman and by Gilbert [6]. Gilbert proposed a sequence of linear algebra computations which resulted in the following map: transfer function matrix $\longrightarrow$ state variable differential equations. In his algorithm, heavy emphasis was placed on the residue matrices of the poles of the transfer function matrix, $\bar{\mathbf{Y}}(s)$. If s_i is a pole of $\bar{\mathbf{Y}}(s)$, then the residue matrix corresponding to that pole is defined to be

$$\mathbf{R}(i) = \lim_{s \to s_i} (s - s_i)\bar{\mathbf{Y}}(s) \tag{9.2.10}$$

Gilbert also defined r_i to be the rank of $\mathbf{R}(i)$. The dimension of the minimum realization of $\bar{\mathbf{Y}}(s)$ was proved to be

$$\sum_{i=1}^{\text{no. of poles}} r_i = n \tag{9.2.11}$$

Since $\mathbf{R}(i)$ is of rank r_i, there are r_i independent columns in $\mathbf{R}(i)$, which can be combined to form the $r \times r_i$ matrix $\mathbf{H}(i)$. Then if we compute the $r_i \times m$ matrix $\mathbf{G}(i)$ by the relation

$$\mathbf{G}(i) = [\mathbf{H}(i)^T\mathbf{H}(i)]^{-1}\mathbf{H}(i)^T\mathbf{R}(i) \tag{9.2.12}$$

$\mathbf{R}(i)$ will equal $\mathbf{H}(i)\mathbf{G}(i)$. Gilbert showed that if q is the number of poles, $\bar{\mathbf{Y}}(s)$ would have the irreducible realization

$$\mathbf{F} = \begin{bmatrix} s_1\mathbf{I}_{r_1} & \mathbf{0} \\ \mathbf{0} & s_q\mathbf{I}_{r_q} \end{bmatrix}$$

$$\mathbf{G} = \begin{bmatrix} \mathbf{G}(1) \\ \cdot \\ \cdot \\ \cdot \\ \mathbf{G}(q) \end{bmatrix}$$

and

$$\mathbf{H} = [\mathbf{H}(1) \quad \cdots \quad \mathbf{H}(q)]$$

where $\mathbf{I}_{r_i}$ is an $r_i \times r_i$ unit matrix.

Kalman's algorithm adopted a different approach to obtain a minimum realization of the transfer function $\bar{\mathbf{Y}}(s)$. First, a realization $\{\mathbf{F}, \mathbf{G}, \mathbf{H}\}$ which is not necessarily minimal, is constructed. To construct this realization the least common denominator is found for each column (or row) in $\bar{\mathbf{Y}}(s)$. Each column is then treated as a separate realization problem of the form

$$\begin{bmatrix} \dfrac{a_{n1}s^{n-1} + \cdots + a_{11}}{s^n + b_n s^{n-1} + \cdots + b_1} \\ \cdot \\ \cdot \\ \cdot \\ \dfrac{a_{nm}s^{n-1} + \cdots + a_{1m}}{s^n + b_n s^{n-1} + \cdots + b_1} \end{bmatrix} \tag{9.2.13}$$

and the solution to the smaller problem is

$$\mathbf{F}_k = \begin{bmatrix} 0 & 1 & 0 & \cdots & 0 \\ 0 & 0 & 1 & & 0 \\ \cdot & \cdot & \cdot & & \cdot \\ \cdot & \cdot & \cdot & & \cdot \\ \cdot & \cdot & \cdot & & \cdot \\ 0 & 0 & 0 & & 1 \\ -b_1 & -b_2 & -b_3 & \cdots & -b_n \end{bmatrix} \tag{9.2.14}$$

$$\mathbf{G}_k = \begin{bmatrix} 0 \\ 0 \\ \cdot \\ \cdot \\ \cdot \\ 1 \end{bmatrix}, \qquad \mathbf{H}_k = \begin{bmatrix} a_{11} & \cdots & a_{1n} \\ \cdot & & \cdot \\ \cdot & & \cdot \\ \cdot & & \cdot \\ a_{r1} & \cdots & a_{rn} \end{bmatrix}$$

The solution to the larger problem is then the combination of the smaller problems:

$$\mathbf{F} = \begin{bmatrix} \mathbf{F}_1 & 0 & \\ 0 & \mathbf{F}_2 \, . & 0 \\ 0 & & \dot{} \, \mathbf{F}_n \end{bmatrix}, \quad \mathbf{G} = \begin{bmatrix} \mathbf{G}_1 & 0 & \\ 0 & \mathbf{G}_2 \, . & 0 \\ 0 & & \dot{} \, \mathbf{G}_n \end{bmatrix}, \quad \mathbf{H} = [\mathbf{H}_1 \quad \cdots \quad \mathbf{H}_n] \tag{9.2.15}$$

This realization is then reduced by finding its controllable and observable parts and using linear algebra techniques to determine an irreducible realization.

In 1966, Ho and Kalman proposed another algorithm for minimum realizations using modular theory. This method will be discussed in detail in the next section of this chapter, since it is a most important procedure.

In 1968, Rosenbrock [17] and Mayne [15] independently introduced methods which arrived at realizations of minimum dimension by a reduction of higher-order realizations. Both methods rely on the determination of the controllable and observable parts of the higher-order realization. Mayne's algorithm is essentially a modification of Kalman's treatment in [10]. Mayne uses the Kalman procedure to construct an observable *or* controllable realization. At this time we shall detail the method used by Mayne to find the observable part of a controllable realization.

Given the realization $\{\mathbf{F}_c, \mathbf{G}_c, \mathbf{H}_c\}$ of dimension n, construct the matrix $\mathbf{V}$ in the following manner. Let $\mathbf{h}_1, \ldots, \mathbf{h}_m$ denote the rows of $\mathbf{H}_c$. Then $\mathbf{h}_1 = \mathbf{v}_1$ and

$$\mathbf{V}^T = [\mathbf{v}_1 \quad \mathbf{v}_2 \quad \cdots \quad \mathbf{v}_{n_k}]$$

where n_k is the rank of $[\mathbf{H}_c^T \;\; \mathbf{F}_c^T\mathbf{H}_c^T \;\; \mathbf{F}_c^{T^2}\mathbf{H}_c^T \;\; \cdots \;\; \mathbf{F}_c^{T^{n-1}}\mathbf{H}_c^T]$. If $\mathbf{F}_c^T\mathbf{v}_j$ is linearly independent of $\mathbf{v}_1 \cdots \mathbf{v}_j$, then $\mathbf{v}_{j+1} = \mathbf{F}_c^T\mathbf{v}_j$. If not and if $\mathbf{v}_j = (\mathbf{F}_c^T)^\xi \mathbf{h}_q$, $\mathbf{v}_{j+1} = \mathbf{h}_i$, where $\mathbf{h}_i$ is the first row of $\mathbf{h}_{q+1} \cdots \mathbf{h}_n$, which is linearly independent on $\mathbf{v}_1 \cdots \mathbf{v}_j$. The process is continued until $j = n_k$ yields the matrix $\bar{\mathbf{Y}}$. Find an $\mathbf{S}$ matrix such that

$$\mathbf{VS} = \mathbf{I}_{n_k}$$

i.e., $\mathbf{S}$ is the pseudoinverse of $\mathbf{V}$. The realization

$$\{\mathbf{F} = \mathbf{VF}_c\mathbf{S},\ \mathbf{G} = \mathbf{VG}_c,\ \mathbf{H} = \mathbf{H}_c\mathbf{S}\}$$

will then be a minimum realization and the dimension will be n_k. The procedure for obtaining a minimum realization from any observable realization is analogous.

In 1969, Wolovich and Falb [23] derived yet another method of obtaining a minimum realization from the transfer function. Their procedure was analogous to those of Mayne and Kalman in that they proposed a method of obtaining a controllable realization and then of finding the observable part of that realization. Suppose that the $r \times m$ transfer function matrix, $\bar{\mathbf{Y}}(s)$, has elements $\bar{Y}_{ij}(s) = n_{ij}(s)/d_{ij}(s)$, where $n_{ij}(s)$ and $d_{ij}(s)$ are relatively prime and assume that $d_{ij}(s)$ is of higher degree than $n_{ij}(s)$. The polynomials $n_{ij}(s)$ and $d_{ij}(s)$ are relatively prime if there is no root common to both. This restriction guarantees that there is no number that is both a zero and a pole of the quotient $n_{ij}(s)/d_{ij}(s)$.

As in the algorithm developed by Kalman, the Wolovich and Falb algorithm will operate on the transfer function one column at a time. The resulting realization of the system will be controllable but not necessarily observable. The Wolovich and Falb algorithm first calculates the least common multiple of the denominator polynomials in each column of $\bar{\mathbf{Y}}(s)$ and calls it $g_j(s)$ for the jth column. Then h_j is defined as the degree of $g_j(s)$ and $n_1 = \sum_{j=1}^m h_j$ with $r_k = \sum_{j=1}^k h_j$. A new array of numerators is then defined as $n_{ij}^*(s) = n_{ij}(s)g_j(s)/d_{ij}(s)$. The transfer functions will then be $\bar{Y}_{ij}(s) = n_{ij}^*(s)/g_j(s)$. The polynomials $g_j(s)$ and $n_{ij}^*(s)$ will be of the form

$$\begin{aligned} g_j(s) &= s^{h_j} + \gamma_{j1}s^{h_j-1} + \cdots + \gamma_{jh_j} \\ n_{ij}^*(s) &= \alpha_{ij1}s^{h_j-1} + \alpha_{ij2}s^{h_j-2} + \cdots + \alpha_{ijh_j} \end{aligned} \tag{9.2.16}$$

where $g_i(s)$ is of higher degree than n_{ij}^*.

Every element in each column of $\bar{\mathbf{Y}}(s)$ will then have the same denominator. Wolovich and Falb used the same techniques developed by Kalman, and discussed above, to compute the controllable realization. They then formed the $n_1 \times m$ matrix, $\mathbf{G}_c$, with zero elements except for the elements $(\mathbf{G}_c)_{r_k,k} = 1$, $k = 1, \ldots, m$, and the $r \times n_1$ matrix $\mathbf{H}_c$, where

$$\mathbf{H}_c = \begin{bmatrix} \alpha_{11h_1} & \alpha_{11(h_1-1)} & \cdots & \alpha_{111} & \alpha_{12h_2} & \cdots & \alpha_{121} & \cdots & \alpha_{1m1} \\ \alpha_{21h_1} & \alpha_{21(h_1-1)} & \cdots & \alpha_{211} & \alpha_{22h_2} & \cdots & \alpha_{221} & \cdots & \alpha_{2m1} \\ \cdot & \cdot & & \cdot & \cdot & & \cdot & & \cdot \\ \cdot & \cdot & & \cdot & \cdot & & \cdot & & \cdot \\ \cdot & \cdot & & \cdot & \cdot & & \cdot & & \cdot \\ \alpha_{r1h_1} & \alpha_{r1(h_1-1)} & \cdots & \alpha_{r11} & \alpha_{r2h_2} & \cdots & \alpha_{r21} & \cdots & \alpha_{rm1} \end{bmatrix}$$

Then if

$$\mathbf{F}_c = \begin{bmatrix} \mathbf{F}_{c,1} & & \\ & \mathbf{F}_{c,2} & 0 \\ 0 & & \mathbf{F}_{c,m} \end{bmatrix}$$

where $\mathbf{F}_{c,j}$ is the companion matrix corresponding to $g_j(s)$,

$$\mathbf{F}_{c,j} = \begin{bmatrix} 0 & 1 & 0 & \cdots & 0 \\ 0 & 0 & 1 & \cdots & 0 \\ \cdot & \cdot & \cdot & & \cdot \\ \cdot & \cdot & \cdot & & \cdot \\ \cdot & \cdot & \cdot & & \cdot \\ 0 & 0 & 0 & \cdots & 1 \\ -\gamma_{jh_1} & -\gamma_{jh_1-1} & & \cdots & -\gamma_{j1} \end{bmatrix}$$

he proved that $\{\mathbf{F}_c, \mathbf{G}_c, \mathbf{H}_c\}$ will be a controllable realization of $\bar{\mathbf{Y}}(s)$. This realization can then be reduced to a minimum realization by finding its observable part as in the Mayne solution. Wolovich and Falb define a similarity transformation which separates the realization into its observable part and its unobservable part.

EXAMPLE 9.3. Let us construct a minimum realization for a system whose transfer function is

$$\bar{\mathbf{Y}}(s) = \begin{bmatrix} \dfrac{1}{s+0.1} & 0 & \dfrac{2s+0.4}{(s+0.1)(s+0.3)} \\ 0 & \dfrac{2s+0.6}{(s+0.4)(s+0.2)} & 0 \\ \dfrac{1}{s+0.5} & \dfrac{1}{s+0.2} & 0 \end{bmatrix}$$

Then, from the Wolovich and Falb algorithm we see that

$$\begin{aligned}
s_1 &= s^2 + 0.6s + 0.05 \\
s_2 &= s^2 + 0.6s + 0.08 \\
s_3 &= s^2 + 0.4s + 0.03 \\
n_{11}^* &= s + 0.5 \\
n_{21}^* &= 0 \\
n_{31}^* &= s + 0.1 \\
n_{12}^* &= 0 \\
n_{22}^* &= 2s + 0.6 \\
n_{32}^* &= s + 0.4 \\
n_{13}^* &= 2s + 0.4 \\
n_{23}^* &= n_{33}^* = 0
\end{aligned}$$

The realization is then

$$\mathbf{H}_c = \begin{bmatrix} 0.5 & 1 & 0 & 0 & 0.4 & 2 \\ 0 & 0 & 0.6 & 2 & 0 & 0 \\ 0.1 & 1 & 0.4 & 1 & 0 & 0 \end{bmatrix}$$

$$\mathbf{G}_c = \begin{bmatrix} 0 & 0 & 0 \\ 1 & 0 & 0 \\ 0 & 0 & 0 \\ 0 & 1 & 0 \\ 0 & 0 & 0 \\ 0 & 0 & 1 \end{bmatrix}$$

$$\mathbf{F}_c = \begin{bmatrix} 0 & 1 & 0 & 0 & 0 & 0 \\ -0.05 & -0.6 & 0 & 0 & 0 & 0 \\ 0 & 0 & 0 & 1 & 0 & 0 \\ 0 & 0 & -0.08 & -0.6 & 0 & 0 \\ 0 & 0 & 0 & 0 & 0 & 1 \\ 0 & 0 & 0 & 0 & -0.03 & -0.4 \end{bmatrix}$$

From this controllable realization, a minimum realization is found as shown earlier:

$$\mathbf{H} = \begin{bmatrix} 0 & 1 & 0 & 0 & 0 \\ 0 & 0 & 0 & 1 & 0 \\ 0 & 0 & 0 & 1.5 & 1 \end{bmatrix}$$

$$\mathbf{G} = \begin{bmatrix} 0.3 & 0 & 0.4 \\ 1 & 0 & 2 \\ 0 & 0.6 & 0 \\ 0 & 2 & 0 \\ 1 & -2 & 0 \end{bmatrix}$$

$$\mathbf{F} = \begin{bmatrix} 0 & -0.03 & 0 & 0 & 0 \\ 1 & -0.4 & 0 & 0 & 0 \\ 0 & 0 & 0 & -0.08 & 0 \\ 0 & 0 & 1 & -0.6 & 0 \\ 0 & 0 & 0 & -0.15 & -0.5 \end{bmatrix}$$

9.2.4 Computer algorithms

Recently, computer programs, utilizing these algorithms in the construction of minimum realizations, have appeared in the literature. In his manual for the ASP matrix manipulation program [13], Kalman (with Engler) included a chapter on the use of the computer program to obtain minimum realization by Ho's algorithm. Kalman outlined the steps to be taken to obtain a solution and then solved several sample problems. The listing of the program is included in the discussion. Then, in 1968, the theory derived by Wolovich and Falb was incorporated into a computer program for the construction of a minimum realization $\{\mathbf{F}, \mathbf{G}, \mathbf{H}\}$ from the system's transfer function. This program was given by Wolovich and Bonna [22] and provides a very thorough documentation of the algorithm. Block diagrams are presented for each of the subroutines used in the program and a complete discussion is given for the operation of the algorithm. In 1969, Roberts [16] published a discussion of the computer program which he derived based on the theory presented by Mayne. The program was developed for the IBM 7094 computer. The overall structure of the program is given and the subroutines which are required are mentioned but the flow diagrams or listings are not available in the paper.

9.3 The Algorithm of Ho and Kalman

As an aid in understanding the theory of minimum realizations, we shall now examine in some detail the algorithm presented by Ho and Kalman [7]. This is one of the simplest and most useful algorithms that have been proposed for constructing realizations of minimum dimension. The method assumes that

1. The system is linear.
2. The system has constant coefficients.
3. (9.1.1) or (9.1.2) are valid model forms.
4. The system is noise-free.
5. The initial state of the system is $\mathbf{x}(0) = \mathbf{0}$.

The impulse response data can be given in the time, the s-, or the z-domain since the use of Markov parameters allows a direct correlation among the three domains.

The problem can be stated as

> Given a sequence of $r \times m$ constant matrices, $\mathbf{Y}_k$, $k = 0, 1, 2, \ldots$, find a triple $\{\mathbf{F}, \mathbf{G}, \mathbf{H}\}$ of constant matrices such that
>
> $$\mathbf{Y}_k = \mathbf{H}\mathbf{F}^k\mathbf{G}, \qquad k = 0, 1, 2, \ldots$$

The solution to this problem follows from

> This problem will have a solution of finite dimension iff there exists a d and constants α_i such that
>
> $$\mathbf{Y}_{d+j} = \sum_{i=1}^{d} \alpha_i \mathbf{Y}_{d+j-i} \qquad \text{for all } j \geq 0 \tag{9.3.1}$$
>
> where d is the degree of the annihilating polynomial of $\mathbf{F}_{\min}$ (see below).

The necessity of (9.3.1) is a consequence of the existence of such a polynomial for a finite dimensional matrix. Given a finite dimensional $\mathbf{F}$, there exists a $\mathbf{\Psi}(\mathbf{F})$ such that

$$\mathbf{\Psi}(\mathbf{F}) = \sum_{i=0}^{q} \beta_i \mathbf{F}^{q-i} = \mathbf{0} \tag{9.3.2}$$

Since $\mathbf{Y}_k = \mathbf{H}\mathbf{F}^k\mathbf{G}$, it follows that

$$\mathbf{0} = \mathbf{H}\mathbf{F}^j\mathbf{\Psi}(\mathbf{F})\mathbf{G} = \sum_{i=0}^{q} \beta_i \mathbf{H}\mathbf{F}^{q-i+j}\mathbf{G} \tag{9.3.3}$$

and (9.3.1) holds for $d = q$ and $\alpha_i = -\beta_i/\beta_0$. The sufficiency of (9.3.1) follows in the solution of the problem. The proof will show that a realization exists by using (9.3.1) to find such a realization.

We assume that the value of d is known. If the impulse response function is given in the s- or z-domain, the value of d is determined easily from the poles of the elements of the transform matrix [11]. For a rational matrix $\bar{\mathbf{Y}}(s)$, the value of d is simply the degree of the least common polynomial denominator of $\bar{\mathbf{Y}}(s)$. $\bar{\mathbf{Y}}(s)$ is said to be a rational matrix iff each element $Y_{ij}(s)$ can be expressed as a ratio of polynomials in s. If the input/output description is obtained in the time domain, as experimental data, numerical methods or curve-fitting techniques must be used to determine d [7]. Extensions of this algorithm deal with this problem in greater detail [21] and they will be mentioned later (see Section 9.6).

We begin by forming the generalized Hankel matrix, a $d \times d$ block matrix composed of the Markov parameters,

$$\mathbf{S}_d = \begin{bmatrix} \mathbf{Y}_0 & \mathbf{Y}_1 & \cdots & \mathbf{Y}_{d-1} \\ \mathbf{Y}_1 & \mathbf{Y}_2 & \cdots & \mathbf{Y}_d \\ \cdot & \cdot & & \cdot \\ \cdot & \cdot & & \cdot \\ \cdot & \cdot & & \cdot \\ \mathbf{Y}_{d-1} & \mathbf{Y}_d & \cdots & \mathbf{Y}_{2d-2} \end{bmatrix} = [\mathbf{Y}_{i+j-2}]$$

and by defining a set of matrices $\tau^l \mathbf{S}_d$ formed by shifting the elements in the Hankel matrix to obtain

$$\tau^l \mathbf{S}_d = [\mathbf{Y}_{i+j+l-2}]$$

The last equation defines τ.

The rank of $\mathbf{S}_d$, s, is equal to the dimension, n_0, of the minimum realization of the system. To confirm this point, consider a minimum realization, $\{\mathbf{F}, \mathbf{G}, \mathbf{H}\}$. Then form $\mathbf{S}_r = \mathbf{V}^T\mathbf{W}$, where

$$\begin{aligned} \mathbf{V} &= [\mathbf{H}^T \quad \mathbf{F}\mathbf{H}^T \quad \cdots \quad (\mathbf{F}^T)^{d-1}\mathbf{H}^T] \\ \mathbf{W} &= [\mathbf{G} \quad \mathbf{F}\mathbf{G} \quad \cdots \quad \mathbf{F}^{d-1}\mathbf{G}] \end{aligned} \tag{9.3.4}$$

Since $\mathbf{F}$ is $n_0 \times n_0$, we can see that rank $\mathbf{V} \leq n_0$ and rank $\mathbf{W} \leq n_0$. Therefore rank $\mathbf{S}_d \leq n_0$. However, if rank $\mathbf{S}_d < n_0$. The algorithm would give rise to a realization of dimension less than n_0. Since $\{\mathbf{F}, \mathbf{G}, \mathbf{H}\}$ is assumed to be of minimum dimension, a contradiction develops and rank $\mathbf{S}_d = n_0$. Thus, if the algorithm does indeed construct a realization of dimension s, that dimension is minimal.

It remains to show that a realization of dimension s can be constructed and to give the steps necessary for that construction. After the formation of the block matrix $\mathbf{S}_d$, matrices $\mathbf{P}$ and $\mathbf{Q}$ are found such that

$$\mathbf{P}\mathbf{S}_d\mathbf{Q} = \begin{bmatrix} \mathbf{I}_s & \mathbf{0} \\ \mathbf{0} & \mathbf{0} \end{bmatrix} = \mathbf{J} = \mathbf{U}_s^T\mathbf{U}_s \tag{9.3.5}$$

where $\mathbf{I}_s$ is an $s \times s$ unit matrix and $\mathbf{U}_s = [\mathbf{I}_s \ \mathbf{0}]$. This can be accomplished by standard linear algebra techniques. The matrix $\mathbf{E}_r$ is then defined to be the $1 \times d$ block matrix $[\mathbf{I}_r \ \mathbf{0}_r \ \mathbf{0}_r \ \cdots \ \mathbf{0}_r]$. A similar definition is made for $\mathbf{E}_s$. The realization can then be constructed by use of the following theorem.

THEOREM 9.4. Whenever a finite dimensional realization exists, a realization of dimension s can be constructed by forming

$$\begin{aligned} \mathbf{F}_s &= \mathbf{U}_s[\mathbf{J}\mathbf{P}(\tau\mathbf{S}_d)\mathbf{Q}\mathbf{J}]\mathbf{U}_s^T \\ \mathbf{G} &= \mathbf{U}_s[\mathbf{J}\mathbf{P}\mathbf{S}_d\mathbf{E}_m^T] \\ \mathbf{H} &= [\mathbf{E}_r\mathbf{S}_d\mathbf{Q}\mathbf{J}]\mathbf{U}_s^T \end{aligned} \tag{9.3.6}$$

where the block matrices are computed as shown above.

The preceding proof indicates that the realization will be minimal.

The verification that (9.3.6) is in fact a realization of the system involves a theoretical trick, the definition of a $d \times d$ *block companion matrix*, $\mathbf{M}$, where

$$\mathbf{M} = \begin{bmatrix} \mathbf{0}_r & \mathbf{I}_r & \mathbf{0}_r & \cdots & \mathbf{0}_r \\ \mathbf{0}_r & \mathbf{0}_r & \mathbf{I}_r & \cdots & \mathbf{0}_r \\ \cdot & \cdot & \cdot & & \cdot \\ \cdot & \cdot & \cdot & & \cdot \\ \cdot & \cdot & \cdot & & \cdot \\ \mathbf{0}_r & \mathbf{0}_r & \mathbf{0}_r & \cdots & \mathbf{I}_r \\ \alpha_d \mathbf{I}_r & \alpha_{d-1}\mathbf{I}_r & \alpha_{d-2}\mathbf{I}_r & \cdots & \alpha_1 \mathbf{I}_r \end{bmatrix}$$

Using (9.3.1), it is apparent that

$$\tau^k \mathbf{S}_d = \mathbf{M}^k \mathbf{S}_d, \qquad k = 0, 1, 2, \ldots \tag{9.3.7}$$

We can also define another $d \times d$ block companion matrix, $\mathbf{N}$, such that $\mathbf{N}$ is the transpose of the block pattern in $\mathbf{M}$ but with $m \times m$ blocks. Then

$$\tau^k \mathbf{S}_d = \mathbf{S}_d \mathbf{N}^k, \qquad k = 0, 1, 2, \ldots \tag{9.3.8}$$

Finally, $\mathbf{S}^\#$ is defined to be the product $\mathbf{QJP}$. $\mathbf{S}^\#$ is the pseudoinverse of $\mathbf{S}_d$, an inverse defined for nonsquare matrices such that

$$\mathbf{S}_d = \mathbf{S}_d \mathbf{S}^\# \mathbf{S}_d \tag{9.3.9}$$

For simplicity, the subscript d will be omitted for the remainder of the proof. With these definitions, observe that

$$\mathbf{Y}_k = \mathbf{E}_r(\tau^k \mathbf{S})\mathbf{E}_m^T$$

and from (9.3.8)

$$\mathbf{Y}_k = \mathbf{E}_r \mathbf{S} \mathbf{N}^k \mathbf{E}_m^T$$

Substituting (9.3.9), we get

$$\mathbf{Y}_k = \mathbf{E}_r \mathbf{S}\mathbf{S}^\# \mathbf{S} \mathbf{N}^k \mathbf{E}_m^T$$

Then, using (9.3.7) and (9.3.9), we see that

$$\mathbf{Y}_k = \mathbf{E}_r \mathbf{S}\mathbf{S}^\# \mathbf{M}^k \mathbf{S} \mathbf{E}_m^T$$

and that

$$\mathbf{Y}_k = \mathbf{E}_r \mathbf{S}\mathbf{S}^\# \mathbf{M}^k \mathbf{S}\mathbf{S}^\# \mathbf{S} \mathbf{E}_m^T$$

The definition of $\mathbf{S}^\#$ yields

$$\mathbf{Y}_k = \mathbf{E}_r \mathbf{SQJPM}^k \mathbf{SQJPSE}_m^T \tag{9.3.10}$$

However, observe that

$$\mathbf{JPM}^k \mathbf{SQJ} = (\mathbf{JPMSQJ})^k \tag{9.3.11}$$

and since $\mathbf{JJ} = \mathbf{J}$,

$$\mathbf{Y}_k = (\mathbf{E}_r \mathbf{SQJ})(\mathbf{JPMSQJ})^k(\mathbf{JPSE}_m^T) \tag{9.3.12}$$

Thus (9.3.6) is indeed a realization and it is therefore of minimum dimension.

It should be noted that although the knowledge of d is important in the algorithm, the values of the α_i for (9.3.1) are never used in the construction of (9.2.6). The **M** matrix is essential to the verification of (9.3.6) but it is also never used in the construction.

This algorithm is one method of obtaining an internal description of a linear system from its impulse response function. Several other methods involve different approaches to the minimum realization problem, but the overall influence of controllability and observability [i.e., (9.3.6)] is always present. The Ho algorithm is one of the simplest methods of computing a realization and also has the advantage of being easily adapted to the proof of some interesting theoretical properties of minimum realizations.

EXAMPLE 9.4. To illustrate the preceding method, we shall attempt to find a triple $\{\mathbf{F}, \mathbf{G}, \mathbf{H}\}$ which is a minimum realization of a 2×2 transfer function matrix:

$$\bar{\mathbf{Y}}(s) = \frac{1}{\psi(s)}\begin{bmatrix} s^2 + 6 & s^2 + s + 4 \\ 2s^2 - 7s - 2 & s^2 - 5s - 2 \end{bmatrix}$$

$$\psi(s) = s^3 + 2s^2 - s - 2$$

There are three distinct poles of the matrix indicating that $d = 3$ and therefore six Markov parameters will be required for the Ho method. If $\bar{\mathbf{Y}}(s)$ is expanded in powers of $1/s$, it is found that

$$\mathbf{Y}_0 = \begin{bmatrix} 1 & 1 \\ 2 & 1 \end{bmatrix}, \qquad \mathbf{Y}_2 = \begin{bmatrix} 11 & 7 \\ 22 & 13 \end{bmatrix}, \qquad \mathbf{Y}_4 = \begin{bmatrix} 51 & 31 \\ 102 & 61 \end{bmatrix}$$

$$\mathbf{Y}_1 = \begin{bmatrix} -2 & -1 \\ -11 & -7 \end{bmatrix}, \qquad \mathbf{Y}_3 = \begin{bmatrix} -22 & -13 \\ -51 & -31 \end{bmatrix}, \qquad \mathbf{Y}_5 = \begin{bmatrix} -102 & -61 \\ -211 & -117 \end{bmatrix}$$

The Hankel matrix, $\mathbf{S}_d$, is formed,

$$\mathbf{S}_d = \begin{bmatrix} 1 & 1 & -2 & -1 & 11 & 7 \\ 2 & 1 & -11 & -7 & 22 & 13 \\ -2 & -1 & 11 & 7 & -22 & -13 \\ -11 & -7 & 22 & 13 & -51 & -31 \\ 11 & 7 & -22 & -13 & 51 & 31 \\ 22 & 13 & -51 & -31 & 102 & 61 \end{bmatrix}$$

and when **P** and **Q** are found as prescribed in (9.3.5), a resulting realization is found to be

$$\mathbf{F} = \begin{bmatrix} -2 & 1 & 0 \\ 7 & -2 & 28 \\ 1 & -\frac{5}{14} & 2 \end{bmatrix}$$

$$\mathbf{H} = \begin{bmatrix} 1 & 0 & 0 \\ 2 & -1 & 0 \end{bmatrix}$$

$$\mathbf{G} = \begin{bmatrix} 1 & 1 \\ 0 & 1 \\ 0 & 0 \end{bmatrix}$$

This triple is a realization of $\bar{\mathbf{Y}}(s)$. That is, $\bar{\mathbf{Y}}(s) = \mathbf{H}(s\mathbf{I} - \mathbf{F})^{-1}\mathbf{G}$, which can be readily shown. Using Theorems 9.1 and 9.2, it can also be proved that $\{\mathbf{F}, \mathbf{G}, \mathbf{H}\}$ are observable and controllable and that the triple is therefore a minimum realization.

9.4 Exact Linear System Realizations

We have already mentioned in Section 9.1 that a given linear system can be represented by an infinite number of triples $\{\mathbf{F}, \mathbf{G}, \mathbf{H}\}$. It is also true that there are an infinite number of realizations of the system which are of minimum dimension. These realizations are then equivalent and any one of them will suffice to define the system.

9.4.1 Equivalent minimum realizations

Stated in another fashion, we shall consider two minimal realizations $\{\mathbf{F}_1, \mathbf{G}_1, \mathbf{H}_1\}$ and $\{\mathbf{F}_2, \mathbf{G}_2, \mathbf{H}_2\}$ of the same $\mathbf{Y}_k$. These two realizations must be isomorphic. That is, there exists a nonsingular matrix $\mathbf{T}$ such that

$$\begin{aligned} \mathbf{F}_2 &= \mathbf{T}\mathbf{F}_1\mathbf{T}^{-1} \\ \mathbf{G}_2 &= \mathbf{T}\mathbf{G}_1 \\ \mathbf{H}_2 &= \mathbf{H}_1\mathbf{T}^{-1} \end{aligned} \tag{9.4.1}$$

and this $\mathbf{T}$ is given by

$$\mathbf{T} = (\mathbf{V}_2\mathbf{V}_2^T)^{-1}\mathbf{V}_2\mathbf{V}_1^T \tag{9.4.2}$$

where $\mathbf{V}_i$ is defined by (9.3.4).

This may be proved using the ideas developed in Ho's algorithm (Section 9.3). Each of the block matrices $\{\mathbf{V}_1, \mathbf{W}_1, \mathbf{V}_2, \mathbf{W}_2\}$ associated with the two minimal realizations have n_0 rows and rank n_0. Thus $\mathbf{V}_i\mathbf{V}_i^T$ and $\mathbf{W}_i\mathbf{W}_i^T$, $i = 1$, 2, are all nonsingular. By the definition of the Hankel matrix, $\mathbf{V}_1^T\mathbf{W}_1 = \mathbf{S}_d = \mathbf{V}_2^T\mathbf{W}_2$. Premultiplication by $\mathbf{V}_2$ and then by $(\mathbf{V}_2\mathbf{V}_2^T)^{-1}$ yields

$$(\mathbf{V}_2\mathbf{V}_2^T)^{-1}\mathbf{V}_2\mathbf{V}_1^T\mathbf{W}_1 = \mathbf{W}_2 \tag{9.4.3}$$

We define

$$\begin{aligned} \mathbf{T} &= (\mathbf{V}_2\mathbf{V}_2^T)^{-1}\mathbf{V}_2\mathbf{V}_1^T \\ \mathbf{U} &= \mathbf{W}_1\mathbf{W}_2^T(\mathbf{W}_2\mathbf{W}_2^T)^{-1} \end{aligned} \tag{9.4.4}$$

and see that $\mathbf{T}$ is nonsingular because

$$\mathbf{T}\mathbf{U} = (\mathbf{V}_2\mathbf{V}_2^T)^{-1}\mathbf{V}_2\mathbf{V}_1^T\mathbf{W}_1\mathbf{W}_2^T(\mathbf{W}_2\mathbf{W}_2^T)^{-1} = \mathbf{I} \tag{9.4.5}$$

We observe that $\mathbf{U} = \mathbf{T}^{-1}$ and therefore that

$$\mathbf{V}_1^{-1}\mathbf{U} = \mathbf{V}_1^T\mathbf{T}^{-1} = \mathbf{V}_2^T \tag{9.4.6}$$

and that

$$\mathbf{W}_1 = \mathbf{T}^{-1}\mathbf{W}_2 \tag{9.4.7}$$

Since

$$\tau\mathbf{S}_d = \mathbf{V}_1^T\mathbf{F}_1\mathbf{W}_1 = \mathbf{V}_2^T\mathbf{T}\mathbf{F}_1\mathbf{T}^{-1}\mathbf{W}_2 \tag{9.4.8}$$

and

$$\tau\mathbf{S}_d = \mathbf{V}_2^T\mathbf{F}_2\mathbf{W}_2 \tag{9.4.9}$$

we can readily see that $\mathbf{F}_2 - \mathbf{T}\mathbf{F}_1\mathbf{T}^{-1}$ for any two minimum realizations. Similar verifications exist for the relations involving **G** and **H** [7].

It can also be shown that if $\{\mathbf{F}_1, \mathbf{G}_1, \mathbf{H}_1\}$ is a minimum realization of $\{\mathbf{Y}_k\}$ and **T** is any nonsingular matrix, then (9.4.1) is also a minimum realization of $\{\mathbf{Y}_k\}$. The proof is short. Since **T** is nonsingular and of dimension n_0, $\mathbf{I} = \mathbf{T}^{-1}\mathbf{T}$. Thus

$$\begin{aligned} \mathbf{H}_2\mathbf{F}_2^k\mathbf{G}_2 &= \mathbf{H}_1\mathbf{T}^{-1}(\mathbf{T}\mathbf{F}_1^k\mathbf{T}^{-1})^k\mathbf{T}\mathbf{G}_1 \\ &= \mathbf{H}_1\mathbf{F}_1^k\mathbf{G}_1 \end{aligned} \tag{9.4.10}$$

9.4.2 Diagonal minimum realizations

Any minimum realization can be obtained from any other minimum realization through the choice of an appropriate **T**. It would be of special interest if a minimum realization could be found such that $\mathbf{F}_2$ would be a diagonal matrix. Premultiplication by $\mathbf{T}^{-1}$ in (9.4.1) would then yield

$$\mathbf{F}_1\mathbf{T}^{-1} = \mathbf{T}^{-1}\mathbf{F}_2 = \mathbf{T}^{-1}\mathbf{D} \tag{9.4.11}$$

where **D** is a diagonal minimal realization. Such a **D** exists whenever $\mathbf{F}_1$ is nonsingular and has distinct eigenvalues and **D** is formed by the eigenvalues of $\mathbf{F}_1$. $\mathbf{T}^{-1}$ is merely the matrix of eigenvectors of $\mathbf{F}_1$. Thus the triple of $\{\mathbf{D}, \mathbf{G}_2, \mathbf{H}_2\}$ can be quickly obtained from any minimum realization $\{\mathbf{F}, \mathbf{G}, \mathbf{H}\}$ using linear algebra techniques. The eigenvalues can be found analytically by solving the characteristic equation of $\mathbf{F}_1$ and the corresponding eigenvectors can be found by solving (9.4.11) for $\mathbf{T}^{-1}$. The eigenvectors will not be unique but can be found if some additional assumption is made about the vector [i.e., $\sum_i (T_{ij}^{-1})^2 = 1$ or $T_{1j}^{-1} = 1$]. It should also be noted that all minimum realizations have the same set of eigenvalues and that these eigenvalues are parameters of the system itself. A system can have only one such diagonal **F**(or **D**) matrix.

The solution of the linear O.D.E.s when **F** is the diagonal matrix **D** is [see (9.1.6) and (9.1.7)]

$$\mathbf{y}(t) = \mathbf{H}\int_0^t \exp[\mathbf{D}(t - \tau)]\mathbf{G}\mathbf{u}(\tau)\, d\tau \tag{9.4.12}$$

and the impulse response function matrix becomes

$$\mathbf{Y}(t) = \mathbf{H} \exp[\mathbf{D}t]\mathbf{G} \tag{9.4.13}$$

The necessary and sufficient condition for an asymptotically stable impulse response is that the real parts of the eigenvalues of the **F** matrix must all be negative. This condition ensures that the impulse response will damp out to zero as $t \longrightarrow \infty$. This also ensures that the transient behavior of other input function responses will damp out with time.

Assume that the **D** matrix has diagonal elements, λ_i, $i = 1, 2, \ldots, n$, such that each λ_i is real and negative. Assume also that at least one eigenvalue, λ_j, is much larger in absolute magnitude than the smallest eigenvalue, $|\lambda_i|_{\min}$. If there is a wide range in the magnitudes of the eigenvalues of a matrix, the matrix is said to be stiff [31, Chapter 7]. Such matrices can be difficult to handle numerically. Calculation of determinants, minors, and inverses can result in large truncation errors. If the larger eigenvalues could be discarded, it would lead to a system of smaller dimension which could be treated more easily, would require less computation time and storage space, and would not have the truncation drawbacks of stiff systems. The essential question, then, is whether the simplification of the system will significantly decrease the accuracy of response predictions made from the realization, thereby negating the model's usefulness as an internal description of the system.

Davison [3] has made some studies in this area and concluded that the dimension of the system could be reduced significantly by this technique. The simplified realization, of course, no longer models the original system exactly. In fact, as $k \longrightarrow \infty$, it is apparent that the $\mathbf{Y}_k$ become dominated by the largest eigenvalues so that the elimination of these eigenvalues will drastically change the $\mathbf{Y}_k$. The resulting sequence of $\mathbf{Y}_k$ will, however, model the system's impulse and step responses very well at large times because the $e^{\lambda_i t}$ terms, when λ_i is very large and negative, become negligible at such large times. In the step response of the realization, there will be a small error in the steady-state values; i.e., the simplified system will not have the same steady-state value as the original system. Davison deals with this problem by defining a new set of state variables $\mathbf{x}_1 = \mathbf{D}\mathbf{x}$, where **D** is a diagonal matrix picked so that the steady-state response is identical to that of the original system [4]. He then demonstrates the technique by reducing a ninth-order system to a fourth-order system with a negligible loss of accuracy.

The technique retains the dominant modes (i.e., eigenvalues closest to zero) of the original system and they respond in the same proportion for given forcing function inputs. The simplified system model thus reacts in the same general way that the original system reacts. The main discrepancy occurs at small times when the large eigenvalues have their greatest effect. If, for example, one is interested in the initial slope of the step response, the simplification technique would be undesirable because the value of $\mathbf{Y}_0$ can be altered significantly by the simplifications (remember that $\mathbf{Y}_0$ is the initial value of the impulse response functions). It is important to point out that, although the terms involving the largest negative eigenvalues will die out fastest at large times, they may be the most important terms at very small times. The procedure by Davison ignores possible deviations around $t = 0$ in an attempt to get a good overall fit of the system response.

EXAMPLE 9.5. The system examined by Davison was a boiler system ($n = 9$) described by (9.1.1) with $\mathbf{H} = \mathbf{I}$ and†

$$\mathbf{F} = \begin{bmatrix} -3.925042 & -0.00315463 & 0 & 0 \\ 0 & 0.0000402537 & 0 & 0 \\ 0 & & & \\ 367.8026 & -3.051455 & 3.030725 & 0 \\ 0 & -0.00377204 & 0 & 0 \\ 0 & & & \\ 27.43095 & 0.0786945 & -0.0596402 & 0 \\ 0 & -0.000281321 & 0 & 0 \\ 0 & & & \\ -0.0646776 & -0.0000519827 & 0 & -0.2558487 \\ -0.00000335212 & 0.000000360429 & 0.000063356996 & 0.00019394473 \\ 0 & & & \\ 3851.207 & 17.25898 & -12.81674 & -12{,}617.88 \\ -2.911237 & -0.1051962 & 12.72677 & 43.08814 \\ 0 & & & \\ 22434.67 & 18.03119 & 0 & -35.601 \\ -0.000104 & -0.4136099 & 90.07017 & 56.89565 \\ 0 & & & \\ 0 & 0 & 0.002339145 & 0 \\ 0 & 0.000222186 & -0.202623 & 0 \\ 0 & & & \\ 0 & 0 & 0 & -1.272016 \\ 0.00100585 & 0.00007862347 & 0 & -0.071713 \\ 0 & & & \\ -2.20356 & -0.001771046 & 0 & 0 \\ -0.000110586 & 0.00001383153 & 0.001489088 & 0.006018683 \\ 0 & & & \end{bmatrix}$$

The eigenvalues of $\mathbf{F}$ are

$$\begin{array}{l} 0 \\ -7.8514 \times 10^{-3} \\ -8.9048 \times 10^{-3} \\ -9.8154 \times 10^{-2} \\ -2.3753 \times 10^{-1} \\ -3.2726 \times 10^{-1} \\ -2.9418 \\ -(3.6348 \pm 9.290i) \times 10^{-1} \end{array}$$

After the simplification of the system, $\mathbf{F}^*$ was obtained as

$$\mathbf{F}^* = \begin{bmatrix} -8.3069 \times 10^{-3} & -1.6203 & 7.2786 \times 10^{-7} & 0 \\ -2.2844 \times 10^{-7} & -9.7077 & 3.1713 \times 10^{-9} & 0 \\ 4.8761 \times 10^{-1} & 2.9968 & -9.5266 \times 10^{-3} & 0 \\ -1.1029 \times 10^{-5} & -3.4090 & 1.7019 \times 10^{-7} & 0 \end{bmatrix}$$

†*Note:* Elements of the matrix are in order row by row.

and the new system, which retained only x_2, x_4, x_6, and x_9, was

$$\dot{\mathbf{x}}^* = \mathbf{F}^*\mathbf{x} + \mathbf{G}^*\mathbf{u}$$

The $\mathbf{G}^*$ were the elements of $\mathbf{G}$ corresponding to the retained elements of $\mathbf{x}$. The new system was found to predict essentially the same step response for a $\mathbf{D}$ of

$$\mathbf{D} = \begin{bmatrix} 1.03 & 0 & 0 & 0 \\ 0 & 2.44 & 0 & 0 \\ 0 & 0 & 1.03 & 0 \\ 0 & 0 & 0 & 1.00 \end{bmatrix}$$

The response for both systems, the ninth- and fourth-order ones, to a step change is shown in Figure 9.4. The resulting curve is essentially identical for the two systems.

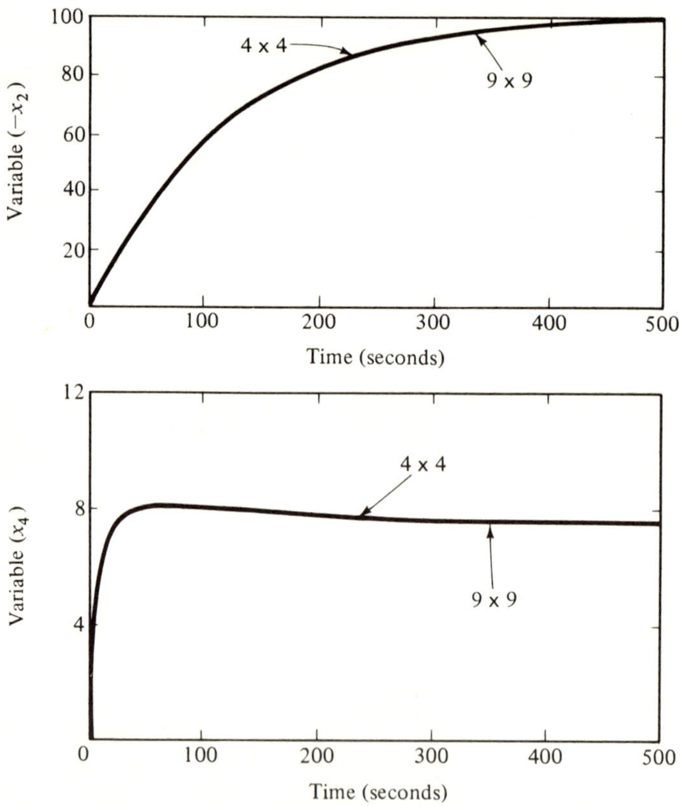

Figure 9.4. Davison's reduced system. [Reprinted from E. J. Davison, "A New Method for Simplifying Large Linear Dynamic Systems," *IEEE Trans. Auto. Cont. 13*, 214 (1968). Reprinted by permission of the copyright owner.]

9.4.3 Connection to moment analysis

The minimum realization algorithm presented by Ho and Kalman provides a method of obtaining a realization of a system from its pulse response function in terms of Markov parameters. These parameters can be difficult to measure when the data are in the form of an experimental time plot of the impulse response. The Markov parameters, coefficients in a negative power series expansion of $\bar{\mathbf{Y}}(s)$ in terms of s^{-k}, can be interpreted as time derivatives of the impulse response matrix. Bruni et al. [1] suggested that an expansion in positive powers of s be used to construct a realization. The coefficients in this expansion would be the matrices $\mathbf{HF}^{-i}\mathbf{G}$, and these parameters could be treated in the Ho algorithm to get a triple of matrices $\{\mathbf{F}^{-1}, \mathbf{G}, \mathbf{H}\}$, from which the actual realization could be easily obtained. The real advantage of this change would be in the case for which the external description of the system is in the form of an empirical time response curve because the $\mathbf{HF}^{-i}\mathbf{G}$ correspond to the moments of the impulse response. To show this we rewrite (9.1.6) as

$$\mathbf{Y}(t) = \mathbf{H}\exp[\mathbf{F}t]\mathbf{G} \tag{9.4.14}$$

for the impulse response matrix $\mathbf{Y}(t)$ and define the moments of this response by

$$\boldsymbol{\mu}'_k = \int_0^\infty t^k \mathbf{Y}(t)\,dt, \qquad k = 0, 1, 2, \ldots \tag{9.4.15}$$

Substituting (9.4.14) into (9.4.15), carrying out the integration by parts k times, and assuming that the system is asymptotically stable, i.e.,

$$\lim_{t \to \infty} \exp[\mathbf{F}t] = \mathbf{0} \tag{9.4.16}$$

the following relations are found:

$$\boldsymbol{\mu}'_k = (-1)^{k+1}k!\,\mathbf{HF}^{-k-1}\mathbf{G}, \qquad k = 0, 1, 2, \ldots \tag{9.4.17}$$

Thus

$$\begin{aligned} \boldsymbol{\mu}'_1 &= \mathbf{HF}^{-2}\mathbf{G} \\ \boldsymbol{\mu}'_2 &= 2\mathbf{HF}^{-3}\mathbf{G} \\ &\;\;\vdots \\ \boldsymbol{\mu}'_n &= n\mathbf{HF}^{-(n+1)}\mathbf{G} \end{aligned} \tag{9.4.18}$$

connect the non-central impulse moments with the triple $\{\mathbf{F}, \mathbf{G}, \mathbf{H}\}$. If we further define

$$\boldsymbol{\mu}'^*_k = \frac{(-1)^k}{(k-1)!}\boldsymbol{\mu}'_{k-1}, \qquad k = 1, 2, \ldots$$

and

$$\boldsymbol{\mu}_0'^* = \mathbf{Y}(t)\Big|_{t=0} = \lim_{s\to\infty} s\bar{\mathbf{Y}}(s)$$

then (9.4.15) can be written as

$$\boldsymbol{\mu}_k'^* = \mathbf{HF}^{-k}\mathbf{G}, \qquad k = 0, 1, 2, \ldots \tag{9.4.19}$$

In the presence of noise, the computation of moments would be preferable to the determination of local time derivatives or the determination of a power series fit to the time curve. Bruni and his coworkers proved that the solution by this method is indeed a minimum realization.

The calculation of the true realization from the triple $\{\mathbf{F}^{-1}, \mathbf{G}, \mathbf{H}\}$ involves an inversion of the $\mathbf{F}^{-1}$ matrix and, of course, an assumption that $\mathbf{F}$ is nonsingular so that $\mathbf{F}^{-1}$ does exist. This is always a valid assumption when the impulse response, $\mathbf{Y}(t)$, is asymptotically stable and if $\mathbf{F}$ is minimal. The Bruni modification cannot be used if $\mathbf{F}$ is singular because the moments would then not be finite and the treatment of the matrices $\mathbf{HF}^{-i}\mathbf{G}$ could not be possible. The applicability of the approach can be determined by an examination of the asymptotic stability of the pulse response. In the case where Bruni's adaptation cannot be used, the Ho algorithm can still be applied to the Markov parameters $\mathbf{HF}^{-i}\mathbf{G}$ to obtain a realization.

It is important to note that when the pulse response is asymptotically stable, a combination of the two approaches might be employed and indeed may prove to be of significant value. For example, the parameters

$$\ldots, \mathbf{HF}^{-1}\mathbf{G}, \mathbf{HG}, \mathbf{HFG}, \mathbf{HF}^2\mathbf{G}, \ldots$$

might be used to construct a realization. Ho's method is applicable for any sequence of $2d$ parameters of the form $\mathbf{HF}^i\mathbf{G}$, where $i = \ldots, -2, -1, 0, 1, 2, \ldots$. The sequence need not consist entirely of nonnegative values of i. In the example sequence shown above, the realization determined from Ho's algorithm would be $\{\mathbf{F}, \mathbf{G}, \mathbf{HF}^{-1}\}$, which can easily be converted to the true realization $\{\mathbf{F}, \mathbf{G}, \mathbf{H}\}$.

EXAMPLE 9.6. At this point we shall present a detailed example of the procedure one might use to construct a minimum realization of a linear system operating at a stable set of steady-state conditions.

Consider a CSTR in which the first-order reactions

$$A \xrightarrow{k_1} B \xrightarrow{k_2} C$$

are occurring. Assume that the volume of fluid in the reactor, V, is constant and that the inlet stream contains reactants A and B and product C, with flow rates F_A, F_B, and F_C. The total inlet flow rate is F, a constant. The effluent concentrations of A, B, and C are X_A, X_B, and X_C, respectively.

The first step in modeling any such system is the definition of the variables to be included in the model. For this particular example, we would like to determine the relationship between the flow rates of the reactants, F_A, F_B, and F_C (inputs), in the inlet stream and the concentrations, X_A, X_B, and X_C (outputs), in the effluent.

To conform to the Kalman-Ho requirement of $\mathbf{x}(0) = \mathbf{0}$, we shall define our input and output variables to be deviation variables:

$$\begin{aligned} u_1 &= F_A - F_A^* & y_1 &= X_A - X_A^* \\ u_2 &= F_B - F_B^* & y_2 &= X_B - X_B^* \\ u_3 &= F_C - F_C^* & y_3 &= X_C - X_C^* \end{aligned}$$

where the asterisk refers to the steady-state values. We then introduce an impulse in each of the input variables and measure the moments or Markov parameters in each of the output variables.

For this example, assume steady-state values and parameters of

$$\begin{aligned} F_A^* &= 1 & k_1 &= 1 \\ F_B^* &= 1 & k_2 &= 2 \\ F_C^* &= 0 & X_A^* &= 0.20 \\ F &= 4 & X_B^* &= 0.20 \\ V &= 1 & X_C^* &= 0.10 \end{aligned}$$

Each of the impulses yields a set of responses in the y_i which go to zero as $t \longrightarrow \infty$. The moments can be obtained from these curves as described earlier. The Markov parameters can be obtained by fitting each curve with a polynomial in time (see Section 9.1.1).

In this problem we shall use the areas under the impulse curves, $\mathbf{HF}^{-1}\mathbf{G}$, and the first three Markov parameters. These responses would normally be obtained experimentally, but for the present numerical example they may be calculated from the system equations

$$V\frac{dX_A}{dt} = -k_1 X_A + F_A - FX_A$$

$$V\frac{dX_B}{dt} = -k_2 X_B + k_1 X_A + F_B - FX_B$$

$$V\frac{dX_C}{dt} = k_2 X_B + F_C - FX_C$$

The impulse response is thus the response in the X_i to an impulse in F_i. If sufficient care were taken in the experimental measurements, we would find that

$$\mathbf{HF}^{-1}\mathbf{G} = \begin{bmatrix} -0.2 & 0 & 0 \\ -0.03333 & -0.166667 & 0 \\ -0.0166667 & -0.08333 & -0.25 \end{bmatrix}$$

$$\mathbf{HG} = \begin{bmatrix} 1 & 0 & 0 \\ 0 & 1 & 0 \\ 0 & 0 & 1 \end{bmatrix}$$

$$\mathbf{HFG} = \begin{bmatrix} -5 & 0 & 0 \\ 1 & -6 & 0 \\ 0 & 2 & -4 \end{bmatrix}$$

$$\mathbf{HF}^2\mathbf{G} = \begin{bmatrix} 25 & 0 & 0 \\ -11 & 36 & 0 \\ 2 & -20 & 16 \end{bmatrix}$$

The realization may already be obvious for this example, but let us continue to follow the Ho-Kalman algorithm.

The $\mathbf{S}_d$ matrix is formed as

$$\mathbf{S}_d = \begin{bmatrix} -\frac{1}{5} & 0 & 0 & 1 & 0 & 0 \\ -\frac{1}{30} & -\frac{1}{6} & 0 & 0 & 1 & 0 \\ -\frac{1}{60} & -\frac{1}{12} & -\frac{1}{4} & 0 & 0 & 1 \\ 1 & 0 & 0 & -5 & 0 & 0 \\ 0 & 1 & 0 & 1 & -6 & 0 \\ 0 & 0 & 1 & 0 & 2 & -4 \end{bmatrix}$$

and it follows that

$$\mathbf{J} = \mathbf{PS}_d\mathbf{Q} = \begin{bmatrix} 1 & 0 & 0 & 0 & 0 & 0 \\ 0 & 1 & 0 & 0 & 0 & 0 \\ 0 & 0 & 1 & 0 & 0 & 0 \\ 0 & 0 & 0 & 0 & 0 & 0 \\ 0 & 0 & 0 & 0 & 0 & 0 \\ 0 & 0 & 0 & 0 & 0 & 0 \end{bmatrix}$$

and

$$\mathbf{E}_d = \begin{bmatrix} 1 & 0 & 0 & 0 & 0 & 0 \\ 0 & 1 & 0 & 0 & 0 & 0 \\ 0 & 0 & 1 & 0 & 0 & 0 \end{bmatrix}$$

According to the algorithm, matrices **P** and **Q** are found to be

$$\mathbf{P} = \begin{bmatrix} -5 & 0 & 0 & 0 & 0 & 0 \\ 1 & -6 & 0 & 0 & 0 & 0 \\ 0 & 2 & -4 & 0 & 0 & 0 \\ 5 & 0 & 0 & 1 & 0 & 0 \\ -1 & 6 & 0 & 0 & 1 & 0 \\ 0 & -2 & 4 & 0 & 0 & 1 \end{bmatrix}$$

$$\mathbf{Q} = \begin{bmatrix} 1 & 0 & 0 & 5 & 0 & 0 \\ 0 & 1 & 0 & -1 & 6 & 0 \\ 0 & 0 & 1 & 0 & -2 & 4 \\ 0 & 0 & 0 & 1 & 0 & 0 \\ 0 & 0 & 0 & 0 & 1 & 0 \\ 0 & 0 & 0 & 0 & 0 & 1 \end{bmatrix}$$

Define

$$\tau \mathbf{S}_d = \begin{bmatrix} 1 & 0 & 0 & -5 & 0 & 0 \\ 0 & 1 & 0 & 1 & -6 & 0 \\ 0 & 0 & 1 & 0 & 2 & -4 \\ -5 & 0 & 0 & 25 & 0 & 0 \\ 1 & -6 & 0 & -11 & 36 & 0 \\ 0 & 2 & -4 & 2 & -20 & 16 \end{bmatrix}$$

$$\mathbf{u}_S = \begin{bmatrix} 1 & 0 & 0 & 0 & 0 & 0 \\ 0 & 1 & 0 & 0 & 0 & 0 \\ 0 & 0 & 1 & 0 & 0 & 0 \end{bmatrix}$$

From (9.3.6) we find that the system can be described by the realization

$$\mathbf{F} = \begin{bmatrix} -5 & 0 & 0 \\ 1 & -6 & 0 \\ 0 & 2 & -4 \end{bmatrix}$$

$$\mathbf{G} = \begin{bmatrix} 1 & 0 & 0 \\ 0 & 1 & 0 \\ 0 & 0 & 1 \end{bmatrix}$$

$$\mathbf{HF}^{-1} = \begin{bmatrix} -\frac{1}{5} & 0 & 0 \\ -\frac{1}{30} & -\frac{1}{6} & 0 \\ -\frac{1}{60} & -\frac{1}{12} & -\frac{1}{4} \end{bmatrix}$$

The **H** matrix is then obtained from

$$\mathbf{H} = \mathbf{HF}^{-1}\mathbf{F} = \begin{bmatrix} 1 & 0 & 0 \\ 0 & 1 & 0 \\ 0 & 0 & 1 \end{bmatrix}$$

9.5 Suboptimal Realizations

The algorithms that have been proposed for finding minimum dimension realizations have different degrees of complexity but tend to require a considerable amount of cumbersome mathematical manipulation. It is therefore of some value to investigate simple methods of constructing suboptimal solutions to the realization problem. Such a solution might require significantly less work while still yielding a realization to the system which could, if necessary, be reduced to minimum size by the methods suggested by

Mayne [15] and Rosenbrock [17]. Kalman [10], in his treatment discussed in the previous section, presented a method of obtaining a realization of a transfer function $\bar{\mathbf{Y}}(s)$, where the dimension of the realization is equal to

$$N_0 = \min\left\{\sum_{i=1}^{r} \alpha_i, \sum_{j=1}^{m} \beta_j\right\} \tag{9.5.1}$$

where α_i and β_i are the number of distinct poles in the ith row and jth column, respectively. N_0 is an upper bound for n_0, the dimension of the minimum realization. Roveda and Schmid [20] recently proposed another very simple method for calculating a suboptimal solution of dimension

$$N_0^* = \sum_{k=1}^{q} r_k^* \tag{9.5.2}$$

where q is the number of distinct poles of $\bar{\mathbf{Y}}(s)$ and r_k^* the minimum number of rows and columns of the residue matrix of the kth pole, s_k, of $\bar{\mathbf{Y}}(s)$ that contain all the nonzero elements of that residue matrix. The residue matrix of the kth pole of $\bar{\mathbf{Y}}(s)$ is obtained by taking the limit as s approaches s_k of $(s - s_k)\bar{\mathbf{Y}}(s)$. It can be shown that $n_0 \leq N_0^* \leq N_0$, so that N_0^* is more restrictive than N_0 as an upper bound on N_0. N_0^* has been shown to be the minimum dimension for realizations having an invariant structure with respect to the coefficients of the elements, $\bar{Y}_{ij}(s)$. That is, if the numerical coefficients were altered randomly, the N_0^* would not change. It is the least upper bound on the dimension of minimum realization of all transfer functions with the distribution of poles found in $\bar{\mathbf{Y}}(s)$. The algorithms for constructing a suboptimal solution to the realization problem are less cumbersome than those used in the construction of a minimum realization. Roveda and Schmid presented an algorithm by which the matrices could be obtained by inspection from the residue matrices of $\bar{\mathbf{Y}}(s)$ with respect to the poles of $\bar{\mathbf{Y}}(s)$. Suboptimal methods of this type are useful in obtaining a quick and easy solution to the realization problem.

9.6 Minimum Partial Realizations

The minimum realization algorithms discussed previously in this chapter have treated essentially the same problem. They deal with linear, finite dimensional, constant dynamical systems for which either the transfer function or the sequence of Markov parameters is known exactly. Suppose, however, that only partial information about the system is available. Can the theory of minimum realizations still be employed to obtain a useful model for the system? One method of approach to this problem involves the choice of the form of the transfer function a priori, and then the constants in each element of the transfer function matrix are determined by parameter estimation (Chapter 7). Any of the minimum realization algorithms can then be used to obtain a realization for this transfer function. This method can be useful but suffers from the need to fix a priori the form of the matrix $\bar{\mathbf{Y}}(s)$ and thus restricts the flexibility of the model. It is, then, desirable to have a method for finding an approximate realization to the system which could be obtained directly from experimental data in the same way that the Ho and Kalman algorithm is used to find an exact realization for completely specified systems.

The Ho-Kalman algorithm requires that $2d$ Markov parameters or $2d$ moment matrices be specified exactly. Consider, however, a system which is not finite dimensional but rather is a distributed parameter system. In this case, any finite sequence of parameters $\{\mathbf{Y}_0, \mathbf{Y}_1, \ldots, \mathbf{Y}_{N_0}\}$ would provide insufficient data for the solution to the problem. Similarly, it may be that the system is finite dimensional but that too few parameters are available for an exact solution. In addition, it is likely that there will be some measurement error in the $\mathbf{Y}_k$ which have been obtained empirically. Such errors lead to difficulties in the determination of d, the degree of the annihilating polynomial of $\mathbf{F}_{\min}$. Also, the operation of the algorithm would be hindered because the rows and columns in the Hankel matrix would not cancel exactly, as they would do in the noise-free, completely accurate case. In these frequently encountered instances, the Ho-Kalman algorithm by itself is not sufficient to obtain a realization for the system.

With these thoughts in hand we shall now consider a sequence of $r \times m$ matrices $\mathbf{Y}_k$, $k = 0, 1, 2, \ldots, N_0 - 1$. We shall then ask if it is possible to construct a realization which will have the identical initial N_0 Markov parameters $\mathbf{Y}_k$. Further, if such a realization can be constructed, we shall ask how this construction can be accomplished. The answer to these questions resides in a method called the minimum partial realization algorithm, which is most useful in modeling systems for which exact linear models cannot be obtained from experimental data. As such, the algorithm is extremely important.

9.6.1 Tether's minimum partial realization algorithm

Solutions to the questions raised above were proposed by Tether [21] and Kalman [12]. We shall examine the algorithm presented by Tether in some detail but we shall need to clarify the basic concepts of partial realizations and minimum partial realizations. The triple $\{\mathbf{F}, \mathbf{G}, \mathbf{H}\}$ is said to be a partial realization of order N_0 of $\{\mathbf{Y}_k\}$ iff $\mathbf{Y}_k = \mathbf{H}\mathbf{F}^k\mathbf{G}$ for $k = 0, 1, \ldots, N_0 - 1$. In addition, $\{\mathbf{F}, \mathbf{G}, \mathbf{H}\}$ is said to be a minimum partial realization of order N_0 iff the dimension of $\mathbf{F}$ is minimal among all other partial realizations as defined above. In the modeling of continuous time systems, the partial realization concept is equivalent to matching the first N_0 terms of the power series expansion of $\bar{\mathbf{Y}}(s)$ or $\mathbf{Y}(t)$ as shown in (9.1.12) and (9.1.13). Likewise, for the discrete time system, if the input/output data are given in the z-

domain, the method is equivalent to matching the first N_0 terms of the power series expansion of $\mathbf{T}(z)$ in (9.1.14). Also, in the discrete case, if the initial conditions are zero, the output $\mathbf{y}(k)$ is related to the input $\mathbf{u}(k)$ via the Markov parameters in the following manner:

$$\begin{aligned} \mathbf{y}(1) &= \mathbf{Y}_0\mathbf{u}(0) \\ \mathbf{y}(2) &= \mathbf{Y}_1\mathbf{u}(0) + \mathbf{Y}_0\mathbf{u}(1) \\ &\vdots \\ \mathbf{y}(N_0) &= \mathbf{Y}_{N_0-1}\mathbf{u}(0) + \cdots + \mathbf{Y}_0\mathbf{u}(N_0 - 1) \end{aligned} \tag{9.6.1}$$

Thus the partial realization for the discrete system models the system exactly for the first N_0 time steps.

There always exists a partial realization which fits the given sequence of Markov parameters. This is sufficiently important that we state this feature in concise terms, namely,

> Every finite sequence of N_0 $r \times m$ constant matrices $\{\mathbf{Y}_0, \mathbf{Y}_1, \ldots, \mathbf{Y}_{N_0-1}\}$ admits an extension sequence $\{\mathbf{Y}_{N_0}, \mathbf{Y}_{N_0+1}, \cdots\}$ for which a completely controllable and completely observable partial realization $\{\mathbf{F}, \mathbf{G}, \mathbf{H}\}$ of order N_0 exists via the Ho-Kalman algorithm.

The proof of this statement can be found in [21] and asserts the existence of a partial realization. Let the dimension of the realization be n'. Then, since there are only a finite number of integers which are less than or equal to n', there must be one realization whose dimension is less than or equal to all the others. Therefore a minimum partial realization also exists.

As in the Ho-Kalman algorithm, we can reduce both the discrete time case and the continuous time case to the same problem by the use of Markov parameters. Then the statement of the minimum partial realization problem becomes

> Given a finite sequence $\{\mathbf{Y}_0, \mathbf{Y}_1, \mathbf{Y}_2, \ldots, \mathbf{Y}_{N_0-1}\}$ of constant $r \times m$ matrices with real elements, find a triple of constant matrices $\{\mathbf{F}, \mathbf{G}, \mathbf{H}\}$ such that
>
> $$\mathbf{Y}_k = \mathbf{H}\mathbf{F}^k\mathbf{G}, \qquad k = 0, 1, 2, \ldots, N_0 - 1$$
>
> and $\mathbf{F}$ is of minimum dimension.

We should say something about the uniqueness of the minimum partial realization because if this realization is unique, then it is a true minimum realization in the Ho-Kalman sense. A minimum realization is said to be unique if all other minimum realizations can be obtained from the given realization by a transformation involving some nonsingular matrix $\mathbf{T}$ in the manner described in Section 9.4. A minimal partial realization will be unique (in the same sense) iff the extension sequence defined by $\mathbf{Y}_k = \mathbf{HF}^k\mathbf{G}$, $k = N_0, \ldots$, is unique.

The central theme of the Tether algorithm for solving the above-stated problem will shortly be given as Theorem 9.5. Before giving this theorem, however, we need to introduce some notation and specify certain concepts. Thus we define a notational change for the Hankel matrix

$$\mathbf{H}_{N',N} = \begin{bmatrix} \mathbf{Y}_0 & \cdots & \mathbf{Y}_{N-1} \\ \cdot & & \cdot \\ \cdot & & \cdot \\ \cdot & & \cdot \\ \mathbf{Y}_{N'-1} & \cdots & \mathbf{Y}_{N'+N-2} \end{bmatrix} \tag{9.6.2}$$

and let $\rho\mathbf{H}_{N',N} = \text{rank}\,(\mathbf{H}_{N',N})$. Now we can state the following:

> Given a finite sequence of $r \times m$ constant matrices $\{\mathbf{Y}_k\}$, $k = 0, 1, 2, \ldots, N_0 - 1$, satisfying
>
> $$\rho\mathbf{H}_{N',N} = \rho\mathbf{H}_{N'+1,N} = \rho\mathbf{H}_{N',N+1}$$
>
> for some N', N such that $N' + N = N_0$, the extension of the sequence $\{\mathbf{Y}_0, \ldots, \mathbf{Y}_{N_0-1}\}$ to $\{\mathbf{Y}_0, \ldots, \mathbf{Y}_{N_0-1}, \mathbf{Y}_{N_0}, \ldots\}$ for which $\rho\mathbf{H}_{m',m} = \rho\mathbf{H}_{N',N}$ where $m' + m = N_0 + k$, $k = 0, 1, 2, \ldots$, is unique.

This statement is proved in [21] and leads directly to the criterion for the existence of a unique minimum partial realization, namely,

> Let $\{\mathbf{Y}_0, \ldots, \mathbf{Y}_{N_0-1}\}$ be an arbitrary finite sequence of $r \times m$ real matrices and let $\mathbf{H}_{i,j}$, $i + j \leq N_0$, be a corresponding Hankel matrix. Then a minimum partial realization given by Ho's algorithm is unique and realizes the sequence up to and including the N_0th term iff there exist positive integers N' and N such that
>
> 1. $N' + N = N_0$.
> 2. $\rho\mathbf{H}_{N',N} = \rho\mathbf{H}_{N'+1,N} = \rho\mathbf{H}_{N',N+1}$.

This criterion is nothing more than a direct consequence of the Ho-Kalman algorithm which we can employ with

$$n_0 = \rho\mathbf{H}_{N',N}$$

using $\mathbf{H}_{N',N}$ as defined in (9.6.2) and defining $\tau\mathbf{H}_{N',N}$ as before. The resulting minimal partial realization is unique because the extension sequence $\mathbf{Y}_k = \mathbf{HF}^k\mathbf{G}$, $k = N_0, N_0 + 1, \ldots$, generated by the realization satisfies

$$\rho\mathbf{H}_{N'+i,N+j} = \rho\mathbf{H}_{N',N}$$

for all $i, j \geq 0$. Thus by the criterion above the extension sequence is unique and therefore the partial realization is unique.

If conditions 1 and 2 are not satisfied, Ho's algorithm requires that new matrices $\{\mathbf{Y}_{N_0}, \ldots, \mathbf{Y}_{P_0-1}\}$ must be specified until $\rho\mathbf{H}_{N',N} = \rho\mathbf{H}_{N'+1,N} = \rho\mathbf{H}_{N',N+1}$, where $N' + N = P_0$. These matrices may be completely or partially arbitrary, and therefore since **F**, **G**, and **H** are functions of $\{\mathbf{Y}_{N_0}, \ldots, \mathbf{Y}_{P_0-1}\}$, they may not be unique.

The minimum dimension for a partial realization of the system in the above criterion is obviously

$$n_0 = \rho\mathbf{H}_{N',N} \tag{9.6.3}$$

If the above criterion is not satisfied, the lower bound on the minimum dimension is the rank of the matrix

$$\begin{bmatrix} \mathbf{Y}_0 & \mathbf{Y}_1 & \mathbf{Y}_2 & \cdots & \mathbf{Y}_{N_0-2} & \mathbf{Y}_{N_0-1} \\ \mathbf{Y}_1 & \mathbf{Y}_2 & \mathbf{Y}_3 & \cdots & \mathbf{Y}_{N_0-1} & * \\ \vdots & \vdots & & & & \\ \mathbf{Y}_{N_0-2} & \mathbf{Y}_{N_0-1} & * & \cdots & * & * \\ \mathbf{Y}_{N_0-1} & * & * & \cdots & * & * \end{bmatrix}$$

where the asterisk ($*$) signifies a blank to be filled in at a later time, because the rank cannot decrease as the $*$ are filled. The minimum dimension can be found by counting the number of linearly independent rows or by counting the number of linearly independent columns. Tether showed that the minimum dimension could then be computed with the aid of the following statement, which he proved:

> Let $\{\mathbf{Y}_0, \mathbf{Y}_1, \ldots, \mathbf{Y}_{N_0-1}\}$ be a finite sequence of $r \times m$ constant matrices with real elements and let $\{\mathbf{F}, \mathbf{G}, \mathbf{H}\}$ be a partial realization whose first N_0 Markov parameters are equal to the given sequence. Then the dimension of the minimum partial realization, $n_{\min}$, satisfies the following inequality:

$$n_{\min} \geq n(N_0) \equiv \sum_{j=1}^{N_0} \rho\mathbf{H}_{j,N_0+1-j} - \sum_{j=1}^{N_0} \rho\mathbf{H}_{j,N_0-j} \tag{9.6.4}$$

Thus $n(N_0)$ is the lower bound for the dimension of the partial realization for any sequence of Markov parameters. But does a realization of the sequence exist such that its dimension is $n(N_0)$?

Let $N'(N_0)$ be defined to be the smallest integer such that every row of the block row $[\mathbf{Y}_{N'(N_0)}, \cdots, \mathbf{Y}_{N_0-1}]$ is linearly dependent on the rows of the Hankel matrix, $\mathbf{H}_{N'(N_0),N_0-N'(N_0)}$. Then let $N(N_0)$ be the first integer so that the columns of $[\mathbf{Y}_{N(N_0)}, \ldots, \mathbf{Y}_{N_0-1}]$ are linearly dependent on the columns of $\mathbf{H}_{N_0-N(N_0),N(N_0)}$. It has been shown that any extension of the sequence whose realization is of dimension $n(N_0)$ satisfies

$$\rho \mathbf{H}_{N'(N_0),N(N_0)} = \rho \mathbf{H}_{N'(N_0)+1,N(N_0)} = \rho \mathbf{H}_{N'(N_0),N(N_0)+1} \tag{9.6.5}$$

for the extension.

It should be noted that $N'(N_0)$ and $N(N_0)$ are invariants of the sequence of $\mathbf{Y}_k$ and are related to the controllability and observability of the system realization. The realization, therefore, has the property that

$$\rho[\mathbf{G} \quad \mathbf{FG} \quad \mathbf{F}^2\mathbf{G} \quad \cdots \quad \mathbf{F}^{N-1}\mathbf{G}] = n(N_0)$$

and

$$\rho[\mathbf{H} \quad \mathbf{HF} \quad \mathbf{HF}^2 \quad \cdots \quad \mathbf{HF}^{N'-1}]^T = n(N_0) \tag{9.6.6}$$

The groundwork is now set for the central theorem in Tether's algorithm.

THEOREM 9.5. Let $\{\mathbf{Y}_0, \ldots, \mathbf{Y}_{N_0-1}\}$ be a fixed partial sequence of $r \times m$ constant matrices with real coefficients and let $n(N_0)$, $N(N_0)$, $N'(N_0)$ be the integers defined above. Then

1. $n(N_0)$ is the dimension of the minimum partial realization.
2. $N(N_0)$ and $N'(N_0)$ are the smallest integers such that (9.6.4) holds simultaneously for all minimal extensions.
3. There is a minimal extension of order $P(N_0) = N(N_0) + N'(N_0)$ for which $n(N_0)$ is the dimension of the realization computed by Ho's algorithm but which is in general not unique.
4. Every extension to $P(N_0)$ is uniquely determined thereafter.

The proof of properties 1, 2, and 4 follows readily from the material we have already stated if property 3 is true. Tether indicates that property 3 is indeed true by outlining a method for the construction of such a realization. We shall not give the details of the method here, but logically the extension sequence is found by constructing the needed $\mathbf{Y}_k$ matrices so that the rank of the matrix

$$\begin{bmatrix} \mathbf{Y}_0 & \mathbf{Y}_1 & \cdots & \mathbf{Y}_N \\ \mathbf{Y}_1 & \mathbf{Y}_2 & \cdots & \mathbf{Y}_{N+1} \\ \vdots & \vdots & & \vdots \\ \mathbf{Y}_{N_0-N} & \mathbf{Y}_{N_0-N+1} & \cdots & \mathbf{Y}_{N_0} \\ \vdots & \vdots & & \vdots \\ \mathbf{Y}_{N'} & \mathbf{Y}_{N'+1} & \cdots & \end{bmatrix}$$

remains the same as succeeding $\mathbf{Y}_k$ are added. This is accomplished by finding all the rows which are linearly dependent and filling them so as to retain their linear dependency. Once the sequence is completed, the minimum partial realization can be constructed using the Ho-Kalman algorithm.

This method can thus be employed to find approximate realizations for systems when only partial input/output information is available. Such a technique is very useful since the analyses of most systems have inherent measurement errors and the presence of noise in the output variables.

EXAMPLE 9.7. As an example of the use of Tether's method we shall examine the following problem.

Let

$$\mathbf{Y}_0 = \begin{bmatrix} 1 & 1 \\ 0 & 0 \end{bmatrix} \qquad \mathbf{Y}_1 = \begin{bmatrix} 4 & 3 \\ 0 & 0 \end{bmatrix}$$

$$\mathbf{Y}_2 = \begin{bmatrix} 10 & 7 \\ 1 & 1 \end{bmatrix} \qquad \mathbf{Y}_3 = \begin{bmatrix} 22 & 15 \\ 3 & 3 \end{bmatrix}$$

For the dimension of the minimal partial realization we compute

$$\rho \begin{bmatrix} 1 & 1 & 4 & 3 & 10 & 7 & 22 & 15 \\ 0 & 0 & 0 & 0 & 1 & 1 & 3 & 3 \end{bmatrix} = \rho \mathbf{H}_{1,4} = 2$$

$$\rho \begin{bmatrix} 1 & 1 & 4 & 3 & 10 & 7 \\ 0 & 0 & 0 & 0 & 1 & 1 \end{bmatrix} = \rho \mathbf{H}_{1,3} = 2$$

$$\rho \begin{bmatrix} 1 & 1 & 4 & 3 & 10 & 7 \\ 0 & 0 & 0 & 0 & 1 & 1 \\ 4 & 3 & 10 & 7 & 22 & 15 \\ 0 & 0 & 1 & 1 & 3 & 3 \end{bmatrix} = \rho \mathbf{H}_{2,3} = 4$$

$$\rho \begin{bmatrix} 1 & 1 & 4 & 3 \\ 0 & 0 & 0 & 0 \\ 4 & 3 & 10 & 7 \\ 0 & 0 & 1 & 1 \end{bmatrix} = \rho \mathbf{H}_{2,2} = 3$$

$$\rho \begin{bmatrix} 1 & 1 & 4 & 3 \\ 0 & 0 & 0 & 0 \\ 4 & 3 & 10 & 7 \\ 0 & 0 & 1 & 1 \\ 10 & 7 & 22 & 15 \\ 1 & 1 & 3 & 3 \end{bmatrix} = \rho \mathbf{H}_{3,2} = 4$$

$$\rho \begin{bmatrix} 1 & 1 \\ 0 & 0 \\ 4 & 3 \\ 0 & 0 \\ 10 & 7 \\ 1 & 1 \end{bmatrix} = \rho \mathbf{H}_{3,1} = \rho \mathbf{H}_{4,1} = 2$$

Then $n(4) = 2 + 4 - 2 + 3 - 2 + 2 - 2 = 5$. Since $\rho\mathbf{H}_{4,1} = \rho\mathbf{H}_{3,1}$ and $\rho\mathbf{H}_{1,4} = \rho\mathbf{H}_{1,3}$, $N' = 3$ and $N = 3$ and there exists a $\mathbf{Y}_4$ and $\mathbf{Y}_5$ such that the rank of $\mathbf{H}_{N',N}$ remains the same. To assure that $\mathbf{H}_{4,3}$ and $\mathbf{H}_{3,4}$ are of rank 3, the matrices must be

$$\mathbf{Y}_4 = \begin{bmatrix} 46 & 31 \\ x_1 & x_1 \end{bmatrix}$$

and

$$\mathbf{Y}_5 = \begin{bmatrix} 94 & 63 \\ x_2 & x_2 \end{bmatrix}$$

where x_1 and x_2 are arbitrary elements. Then, using the Ho algorithm, we obtain

$$\mathbf{F} = \begin{bmatrix} 4 & -1 & 0 & 0 & 0 \\ 6 & -1 & 0 & 0 & 0 \\ 1 & 0 & -1 & 1 & 0 \\ 0 & 1 & x_1 - 13 & 4 & -6x_1 + x_2 + 27 \\ 0 & 0 & 1 & 0 & 0 \end{bmatrix}$$

$$\mathbf{G} = \begin{bmatrix} 1 & 1 \\ 0 & 1 \\ 0 & 0 \\ 0 & 0 \\ 0 & 0 \end{bmatrix} \qquad \mathbf{H} = \begin{bmatrix} 1 & 0 & 0 & 0 & 0 \\ 0 & 0 & 0 & 0 & 1 \end{bmatrix}$$

which corresponds to the first four Markov parameters. We can readily see that the realization depends on the choice of x_1 and x_2 and that it is therefore not unique. The arbitrary aspect of the solution can lead to difficulties in the stability of resulting realizations. A study of this problem can be found in [18, 19].

9.7 Comparison with Other Identification Methods

We have shown that the theory of minimum realizations can be useful in obtaining a description of the internal workings of a linear system. Other methods for accomplishing this same result have also been proposed, and it will be helpful to place the minimum realization in proper perspective in relation to these other methods. In particular, the two most frequently used techniques are moment and frequency response analysis, both of which we have discussed previously (see also [2, 8, 14]).

If all systems were linear, constant coefficient, and finite dimensional and if all measurements could be made to infinite accuracy, the minimum realization algorithms would be superior to the frequency response methods and moment methods for the identification of unknown systems. The minimum realization solution is an exact solution, while the others are approximations to the system. In real systems, however, measurements are not exact and, in some cases, only a finite number of moments or Markov parameters are available. If this number is insufficient for a Ho type of solution, then an exact solution would be impossible. How would the different types of solutions compare in this case?

Let us assume that we have a linear, finite dimensional, constant dynamical system and that only a few moments are known and that these moments may contain measurement errors. Given a fixed number of exact moments, the minimum realization method can extract more information, by simultaneous treatment of all the input/output pairs, than could be extracted by moment methods. It has been shown [18] that, in this case, the minimum partial realization is generally a more accurate representation of the system than can be obtained using moment methods. Thus, if the measured moments contain errors, the minimum realization approach would be expected to generate a more accurate model of the approximate system (i.e., the measured moments) by fitting the given measured moments exactly. The model constructed with moment methods is essentially an approximate model of approximate data, while the partial realization is a more accurate model of approximate data. The partial realization would then be expected to generate more accurate predictions of future system behavior in response to other forcing functions. This result is reasonable because the minimum realization algorithms were devised for linear systems, while the other methods were devised for all varieties of systems.

The Bruni adaptation of the Ho-Kalman algorithm is probably the best way to obtain a model for a linear system from the moments [1]. It should also be noted that the usual result of moment analysis and frequency response analysis of experimental data is the transfer function of the system. To find a set of state variable equations to get this transfer function exactly, one can make use of the minimum realization techniques. In this way, the state variable equations can be found to exactly fit the transfer function. We have shown in Chapter 7 how the moments of a system can be obtained from the frequency response data. We showed that conventional methods of least squares polynomial curve fitting in the frequency domain allow estimates to be made of the moments of the impulse response of a linear system where the magnitude and phase angle of the transfer function were expanded as even and odd power series, respectively, in frequency. The coefficients of the powers of the frequency are easily converted to the central moments of an impulse response. Thus the moments required for Ho's algorithm could be obtained from frequency response data by conventional techniques. Frequency response analysis and moment methods, then, can be converted back and forth and the information necessary for the more accurate minimum realization algorithms can be obtained readily from either. Figure 9.5 shows a schematic for a

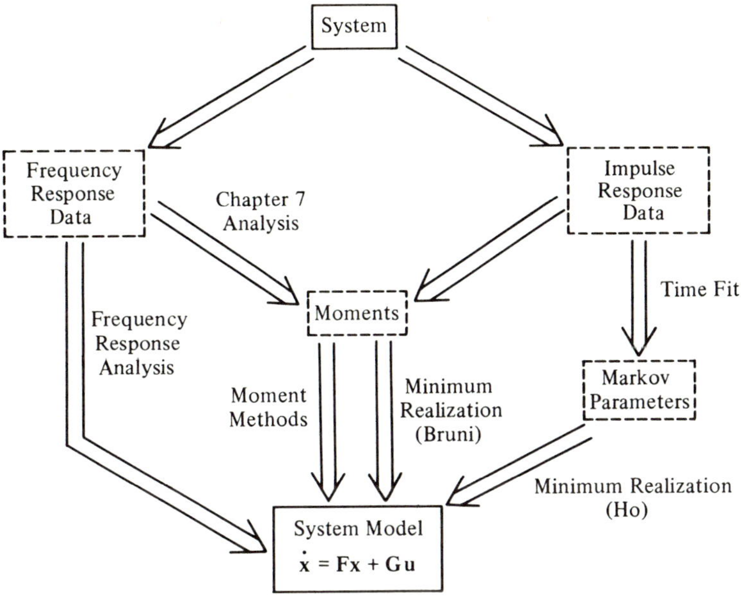

Figure 9.5. Flow of information in linear model analysis. [Reprinted from R. H. Rossen and L. Lapidus, "Minimum Realizations and System Modeling," *AIChE J. 18*, 673, 881 (1972). Reprinted by permission of the copyright owner.]

possible flow of information in model analysis. The most advantageous sequence of steps to be used in the analysis of any linear system may involve, as the final step, a solution using minimum realization algorithms.

EXAMPLE 9.8. As a comparison of Tether's partial realization algorithm with moment methods, we shall present some results obtained by Rossen [18] in the study of a system with five state variables and three input and three output variables. The original system was described by (9.1.1) with the matrices

$$\mathbf{F} = \begin{bmatrix} -0.5 & & & & \\ & -0.4 & & \mathbf{0} & \\ & & -0.3 & & \\ & \mathbf{0} & & -0.2 & \\ & & & & -0.1 \end{bmatrix}$$

$$\mathbf{G} = \begin{bmatrix} 1 & 0 & 0 \\ 1 & 0 & 1 \\ 1 & 0 & 0 \\ 1 & 1 & 0 \\ 1 & 0 & 1 \end{bmatrix} \qquad \mathbf{H} = \begin{bmatrix} 1 & 0 & 1 & 1 & 0 \\ 0 & 0 & 0 & 1 & 1 \\ 1 & 1 & 1 & 1 & 1 \end{bmatrix}$$

For simplicity, $\mathbf{F}$ is chosen to be a diagonal matrix. Recall that any triple can be transformed into a triple with a diagonal $\mathbf{F}$ matrix if the original $\mathbf{F}$ matrix has distinct eigenvalues.

The matrices $\mathbf{G}$ and $\mathbf{H}$ are also carefully selected. Recall that the transfer function for a linear system is $\bar{\mathbf{Y}}(s) = \mathbf{H}(s\mathbf{I} - \mathbf{F})^{-1}\mathbf{G}$. In this instance, therefore,

$$\bar{\mathbf{Y}}(s) = \begin{bmatrix} \dfrac{1}{(s+0.5)(s+0.3)(s+0.2)} & \dfrac{1}{s+0.2} & 0 \\ \dfrac{1}{(s+0.2)(s+0.1)} & \dfrac{1}{s+0.2} & \dfrac{1}{s+0.1} \\ \dfrac{1}{(s+0.5)(s+0.4)(s+0.3)(s+0.2)(s+0.1)} & \dfrac{1}{s+0.2} & \dfrac{1}{(s+0.2)(s+0.4)} \end{bmatrix}$$

This transfer function allows the simultaneous study of problems of several different dimensions. The element $\bar{Y}_{11}(s)$ can be viewed as the transfer function of a 1×1 system with three eigenvalues:

$$\mathbf{F} = \begin{bmatrix} -0.5 & 0 & 0 \\ 0 & -0.3 & 0 \\ 0 & 0 & -0.2 \end{bmatrix}$$

$$\mathbf{G} = [1.0 \quad 1.0 \quad 1.0]^T \qquad \mathbf{H} = [1.0 \quad 1.0 \quad 1.0]$$

That is, for this smaller system, $\mathbf{H}(s\mathbf{I} - \mathbf{F})^{-1}\mathbf{G} = \bar{Y}_{11}(s)$. The effect of $u_1(t)$ on the measurement variable $y_1(t)$ in the system described by $\bar{\mathbf{Y}}(s)$ will be identical to the corresponding effect in the three-eigenvalue system. Likewise, the element $\bar{Y}_{12}(s)$ can be viewed as a smaller system dependent on only one eigenvalue, while $\bar{Y}_{31}(s)$ can be considered to represent a system of five eigenvalues. In this way, the study of this one system can yield valuable information about systems of various dimensions.

Three mathematical representations will be constructed via Tether's method. Note that the partial minimum realizations can be constructed either from the moments or from the Markov parameters of the impulse response. For this example problem, one model will be constructed from the first three moments and two additional models will be constructed from the first three Markov parameters. In this way, the advantages of each procedure can be studied when the models are compared with each other and with the models obtained from the moment methods. Each of the models formulated by the minimum partial realization approach and each of the models formulated by the moment approach will be used to predict the response of the system to a step input forcing function. The predicted responses will be compared with the actual system response for each element of the transfer function matrix.

The moment methods used were those given by Gibilaro and Lees [5], Kropholler [14], and Buffham and Gibilaro [2]. The form of the transfer function was picked a priori and a least squares type of fit was obtained for the first three moments for each of the elements of the transfer function. The explicit transfer functions chosen were

1. Given above.
2. $\dfrac{\exp(-\tau_1 s)}{(1 + \tau_2 s)}\tau_3.$
3. $\dfrac{\exp(-\tau_1 s)}{(1 + \tau_2 s)(1 + \tau_3 s)}\tau_4.$
4. $\dfrac{1}{(1 + \tau_1 s)(1 + \tau_2 s)}\tau_3.$
5. $\dfrac{1 + [(\tau_1\tau_2)/(\tau_1 + \tau_2)]^2}{(1 + \tau_1 s)(1 + \tau_2 s)}\tau_3.$

In addition to calculating moments from these transfer functions, three mathematical representations were constructed via Tether's method. One model was constructed from the first three moments and two additional models were constructed from the first three Markov parameters.

One typical set of results is shown in Figure 9.6 with the models obtained from

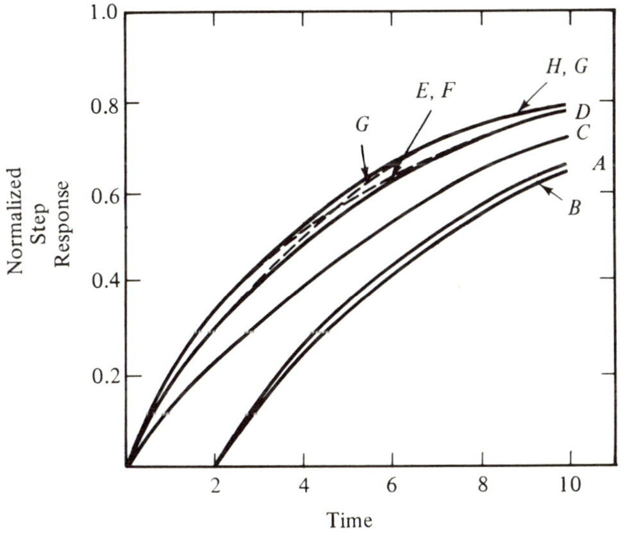

Figure 9.6. Step response for linear Example 9.7. [Reprinted from R. H. Rossen and L. Lapidus, "Minimum Realizations and System Modeling," *AIChE J. 18*, 673, 881 (1972). Reprinted by permission of the copyright owner.]

the moments of 2 through 5 above, labeled *A*, *B*, *C*, and *D* respectively; the two partial realizations constructed from the Markov parameters are designated as *E* and *F* and the partial realization constructed from the moments designated as *G*. The actual step response is assigned the letter *H*. The best overall model is the Tether method *G*, which utilizes the moments as suggested in Section 9.4.3.

The study indicated that for elements depending on only one eigenvalue, the system can be accurately represented by any of the methods. For more complex system elements, the best moment method for the linear system was found to be case 5. This method, however, does not utilize information which can be obtained by treating more than one element simultaneously, as in the Tether algorithm. Predictions of the element depending on all five eigenvalues was off for all moment methods tested. The Tether method which used Markov parameters provided a very good fit near $t = 0$ but tended to be in error at larger times. The Tether method using moments, on the other hand, approximated the actual response closely (within 2 percent) throughout the entire response curve.

The advantage of the minimum realization algorithms over moment methods is the information gained by the interaction (of state variables) between different input and output variables. The Kalman-Tether approach will have the greatest advantage for the systems with the larger number of eigenvalues. In his study of linear systems, Rossen [18] made some interesting discoveries concerning the shape and size of the systems. If we define

$$\begin{aligned} r &\equiv \text{number of input variables} \\ m &\equiv \text{number of output variables} \\ n &\equiv \text{minimum number of state variables} \\ \xi &\equiv \min(r, m) \end{aligned}$$

we find that $n/\xi + 2$ moments will be required for an exact solution, while an exact solution by moment methods would require n moments and then will be exact only if the correct form was originally chosen for the transfer function. For linear systems, as would be expected, the most advantageous choice for a transfer function to fit the data by moment analysis is of the form

$$\bar{Y}(s) = \sum_{i=1}^{k} \frac{\alpha_i}{1 + \beta_i s}$$

This is the general form of the linear transfer function. The constants must be determined empirically. It is, however, exact only for systems of dimension $n \leq k$ and such an exact solution would require the accurate measurement of $2n$ moments. The partial minimum realization solution, however, can be used to construct an exact model using as few as $2n/\xi$ moments. The number $2n$ is an upper bound for the number of moments which are required for a solution. It is clear that $2n/\xi$ is a lower bound, because if $2n/\xi$ moments are used, the smaller dimension (number of rows or columns) of the Hankel matrix is $2n/\xi \times \xi/2 = n$. Specifically, if N_0 is the number of moments required for an exact realization, then

$$\frac{2n}{\xi} \leq N_0 \leq 2n$$

The Tether solution, then, requires less information than the moment method, because the partial minimum realization technique operates on all the input/output pairs simultaneously. In this way, valuable information concerning the interaction of these pairs is utilized—information which is not available to the moment methods. For linear systems, this additional information means that fewer moments are required for an exact model or that a more comprehensive and a more accurate model can be constructed from a given number of moments.

It was observed that if $\epsilon > 0$ and $r = m + \epsilon$ or $m = r + \epsilon$, the solution methods will be mirror images of each other because the same form of information will be available to the method. Also, given any $r \times m$ system where $r = m$ if an even number of moments is known, the Kalman solution will fit all the moments exactly without the Tether estimation of additional moments. If the number of moments is odd, any one additional moment will lead to a solution which is an exact fit of the known moments. It was also observed that the Tether algorithm will often result in a model which contains one or more eigenvalues with positive real parts. Such a model is, of course, unstable and the solution is not useful. Rossen, however, has devised a technique for obtaining a stable model from Tether's unstable one [18, 19].

9.8 Nonlinear System Modeling

One of the primary contributions of minimal partial realization theory is that it opens the door to analysis of a wide variety of systems. Not only can the technique be applied to linear systems when only partial information is known, but it can also be used to construct models of nonlinear systems [9, 19] This can be a powerful tool in the analysis of chemical engineering problems.

From the input/output data for a nonlinear system, a linear model can be determined by minimal partial realization techniques. This model would approximate the initial or transient response of the original system to various inputs. The linear model could be used to predict the response of the process to changes in set point operating conditions, for example, based on the system's response to impulse functions. The analysis would be strictly of a black box nature; nothing need be known a priori about the internal operation of the system. One need not make a guess at the differential equations governing the system.

The minimal partial realization technique can also be used to determine a possible nonlinear model of the system by observing the effects of various inputs on the elements of **F**, **G**, and **H**. There appears to be great opportunity for significant advancement in this type of approach to nonlinear problems.

Another adaptation of the theory to the study of nonlinear problems is in the area of adaptive methods of analysis. Here the estimates of the matrices **F**, **G**, and **H** would be changed continuously with time. When the linear model no longer holds for the system, a new one could be calculated by the algorithm discussed previously.

9.9 The Adsorption Column with Axial Diffusion

As a final illustration of the application of the partial realization approach to physical systems, we shall consider the one-dimensional adsorption column with axial diffusion defined in Section 3.6.4. The equations for this system model are given as (3.6.51)–(3.6.53) with the final analytical solution corresponding to a step function input given in (3.6.63). To obtain the physical "response" of this system, relating the concentration of the adsorbate in the effluent to the concentration of the adsorbate in the inlet, it is necessary to impose an impulse forcing function in the inlet concentration and observe the response in the effluent. From this response curve, the moments or Markov parameters can be computed. The Markov parameters are obtained by fitting the response curve with a polynomial in the independent variable, time. The moments would be obtained as in Section 9.4.3.

These parameters can then be used as input to the Ho algorithm and a

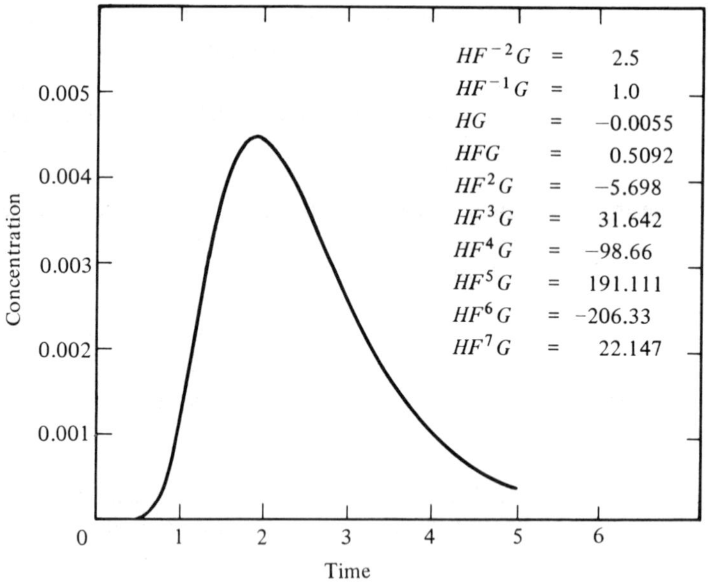

Figure 9.7. Impulse response for a one-dimensional packed bed. [Reprinted from R. H. Rossen and L. Lapidus, "Minimum Realizations and System Modeling," *AIChE J. 18*, 673, 881 (1972). Reprinted by permission of the copyright owner.]

partial realization, {**F**, **G**, **H**}, can be constructed. This realization will relate the concentration in the effluent to that in the inlet. It should be noted here that the partial realization analysis recognizes only one independent variable, time. Therefore the model discussed above could not be used to predict concentration variations at other points in the column unless the impulse response was measured at those points.

Assume that the adsorption column is characterized by the following parameters (see Section 2.6.4):

$$c_0 = 0.1 \qquad \frac{v}{t} = 1$$

$$c_i = 0 \qquad v = 20$$

$$\gamma = \alpha = 5.0 \qquad t_1 = 0.1$$

The impulse response is obtained by changing the inlet concentration from c_i to c_0 for $0 \leq t \leq t_1$. This response is shown in Figure 9.7. The Markov parameters are determined from this curve. In accordance with the ideas expressed in Section 9.4.3, we can use the parameters listed in Figure 9.7 to construct a model for the system. Note that the partial sequence is $\mathbf{Y}_i$, $i = -1, 0.1, \ldots, 6$. The use of Ho's algorithm on this sequence generates the realization {**F**, **G**, $\mathbf{HF}^{-1}$}, which can be converted to {**F**, **G**, **H**} as discussed in Section 9.4.3.

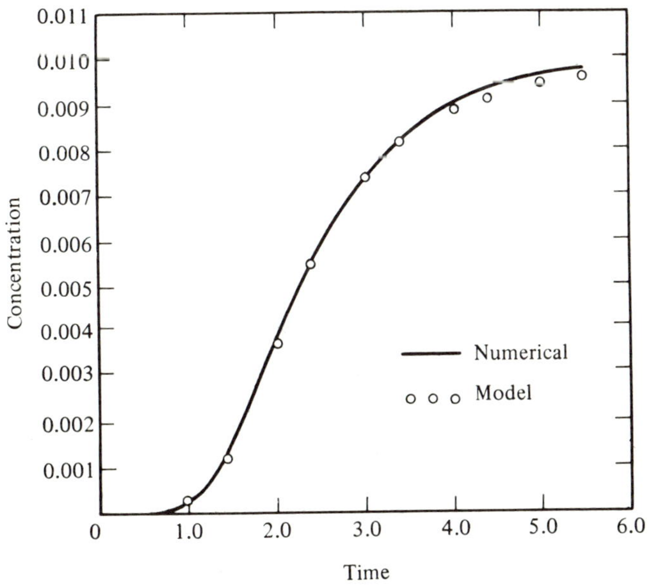

Figure 9.8. Packed bed step response. [Reprinted from R. H. Rossen and L. Lapidus, "Minimum Realizations and System Modeling," *AIChE J. 18*, 673, 881 (1972). Reprinted by permission of the copyright owner.]

For these parameters, the Tether algorithm produces the realization

$$\mathbf{F} = \begin{bmatrix} 0.0055 & -0.5092 & 0 & 0 \\ 1.0 & -11.20 & -62.7 & 0 \\ 0 & -11.20 & 3.27 & -8.12 \\ 0 & 0 & 1.0 & 2.65 \end{bmatrix}$$

$$\mathbf{G} = [1 \quad 0 \quad 0 \quad 0]^T$$

$$\mathbf{H} = [-0.0055 \quad 0.5092 \quad 0 \quad 0]$$

To test the accuracy of this realization model, we shall use it to predict the behavior of the effluent concentration in response to a step of 0.1 in the inlet concentration. This prediction will then be compared with the "actual" system step response as determined analytically (see Section 2.6.4). The plots of both step responses are shown in Figure 9.8, indicating success in the model or realization construction.

REFERENCES

1. Bruni, C., A. Isidori, and A. Raberti, "A Method of Realization Based on the Moments of the Impulse Response Matrix," *IEEE Trans. Auto. Cont. 14*, 203 (1968).
2. Buffham, B. A., and L. G. Gibilaro, "A Unified Time Delay Model for Dispersion in Flowing Media," *Chem. Eng. J. 1*, 31 (1970).
3. Davison, E. J., "A Method for Simplifying Linear Dynamic Systems," *IEEE Trans. Auto. Cont. 11*, 93 (1966).
4. Davison, E. J., "A New Method for Simplifying Large Linear Dynamic Systems," *IEEE Trans. Auto. Cont. 13*, 214 (1968).
5. Gibilaro, L. G., and F. P. Lees, "The Reduction of Complex Transfer Function Models to Simple Models Using the Method of Moments," *Chem. Eng. Sci. 24*, 55 (1969).
6. Gilbert, E. G., "Controllability and Observability in Multivariable Control Systems," *SIAM J. Control 1*, 1227 (1963).
7. Ho, B. L., and R. E. Kalman, "Effective Construction of Linear State-Variable Models from Input/Output Functions," *Regelungstechnik 14*, 545 (1966).
8. Jeffreson, C. P., "Dynamic Testing—A Unification," *Chem. Eng. Sci. 25*, 1319 (1970).
9. Kallina, C., "An Approach to the Regulation Problem for Nonlinear Chemical Processes," *JACC Preprints 11*, 329 (1970).
10. Kalman, R. E., "Mathematical Description of Linear Dynamical Systems," *SIAM J. Control 1*, 52 (1963).

11. Kalman, R. E., "On Canonical Realizations," *Proc. Allerton Conf. Circuit System Theory 2*, 32 (1964).

12. Kalman, R. E., "On Partial Realizations of a Linear Input/Output Map," in *Guillemin Anniversary Volume*, Holt, Rinehart and Winston, New York, 1970, pp. 385–407.

13. Kalman, R. E., and T. S. Engler, "Computation of A Minimal Realization," *A User Manual for ASP C*, 1966, Chap. XV.

14. Kropholler, H. W., "Determination of Relative Variance and Other Moments from Generalized Flow Networks or System Transfer Functions," *IEC Fund. 9*, 329 (1970).

15. Maync, D. Q., "Computational Procedure for the Minimal Realization of Transfer-Function Matrices," *Proc. IEE 115*, 1363 (1968).

16. Roberts, D. M., "A Computer Program For the Minimal Realization of Transfer Function Matrices," *IEEE Trans. Auto. Cont. 14*, 412 (1969).

17. Rosenbrock, H. H., "Computation of Minimal Representations of a Rational Transfer-Function Matrix," *Proc. IEE 115*, 325 (1958).

18. Rossen, R. H., "Minimum Partial Realizations of Chemical Engineering Systems," Ph.D. thesis, Princeton University, Princeton, N.J., 1972.

19. Rossen, R. H., and L. Lapidus, "Minimum Realizations and System Modeling," *AIChE J. 18*, 673, 881 (1972).

20. Roveda, C., and R. Schmid, "An Upper Bound on the Dimension of Minimal Realizations of Linear Time-Invariant Dynamical Systems," *IEEE Trans. Auto. Cont. 15*, 639 (1970).

21. Tether, A., "Construction of Minimum Linear State Variable Models From Finite Input/Output Data," *IEEE Trans. Auto. Cont. 15*, 427 (1970).

22. Wolovich, W. A., and J. Bonna, *Minimal Realization*, computer program available from W. A. Wolovich of Brown University, Providence, R. I. (1970).

23. Wolovich, W. A., and P. L. Falb, "On the Structure of Multivariable Systems," *SIAM J. Control 7*, 437 (1969).

PROBLEMS

9.1 What are the first four moments of the system

$$\bar{\mathbf{Y}}(s) = \begin{bmatrix} \dfrac{1}{(s+1)(s+3)} & \dfrac{1}{(s+4)} & \dfrac{1}{(s+1)(s+2)} \\ \dfrac{1}{(s+3)(s+5)} & \dfrac{1}{(s+3)(s+4)} & \dfrac{1}{(s+2)} \\ \dfrac{1}{(s+2)(s+5)} & \dfrac{1}{(s+1)(s+3)(s+5)} & \dfrac{1}{(s+3)(s+4)} \end{bmatrix}$$

What are the system's first four Markov parameters?

9.2 Is the following system controllable? Observable?

$$\mathbf{F} = \begin{bmatrix} 1 & -4 & 1 \\ 2 & -5 & 1 \\ 2 & -4 & 0 \end{bmatrix}, \qquad \mathbf{G} = \begin{bmatrix} 4 & 1 \\ 3 & 1 \\ 3 & 2 \end{bmatrix}, \qquad \mathbf{H} = \begin{bmatrix} 2 & -4 & 1 \\ 0 & -1 & 1 \end{bmatrix} \tag{A}$$

9.3 Is (A) in Problem 9.2 stable? Why?

9.4 Given the system

$$\begin{bmatrix} \dot{x}_1 \\ \dot{x}_2 \end{bmatrix} = \begin{bmatrix} -3 & 1 \\ -2 & 1.5 \end{bmatrix} \begin{bmatrix} x_1 \\ x_2 \end{bmatrix} + \begin{bmatrix} 1 \\ 4 \end{bmatrix} u$$

show that it is not completely state-controllable by comparing [**FG G**]. Then show that the transfer function is

$$\bar{\mathbf{Y}}(s) = \frac{s + 2.5}{(s + 2.5)(s - 1)}$$

What do you conclude from $\bar{\mathbf{Y}}(s)$?

9.5 Given the system

$$\begin{bmatrix} \dot{x}_1 \\ \dot{x}_2 \\ \dot{x}_3 \end{bmatrix} = \begin{bmatrix} 2 & 0 & 0 \\ 0 & 2 & 0 \\ 0 & 3 & 1 \end{bmatrix} \begin{bmatrix} x_1 \\ x_2 \\ x_3 \end{bmatrix} + \begin{bmatrix} 1 \\ 1 \\ 1 \end{bmatrix} u$$

$$y = [1 \quad 1 \quad 1] \begin{bmatrix} x_1 \\ x_2 \\ x_3 \end{bmatrix}$$

show that the system is not completely observable.

9.6 Using the Ho-Kalman algorithm, find a minimum realization of

$$\bar{\mathbf{Y}}(s) = \begin{bmatrix} \dfrac{s + 4}{(s + 1)(s + 2)} & \dfrac{s + 1}{(s + 2)(s + 3)} \\ \dfrac{1}{(s + 2)} & \dfrac{1}{(s + 1)(s + 3)} \end{bmatrix} \tag{A}$$

9.7 Using the algorithm by Wolovich and Falb, calculate a minimum realization of (A) in Problem 9.6.

9.8 Show that the solutions to Problems 9.6 and 9.7 are equivalent system models.

9.9 How many additional moments must be filled in to get a Tether solution matching the first three moments of

$$\bar{\mathbf{Y}}(s) = \begin{bmatrix} \dfrac{s+4}{(s+1)(s+3)} & \dfrac{s+8}{(s+2)(s+3)(s+4)(s+5)} & \dfrac{1}{(s+3)} \\ \dfrac{s+3}{(s+1)(s+5)(s+7)} & \dfrac{s+4}{(s+1)(s+2)(s+3)(s+5)} & \dfrac{1}{(s+1)(s+4)} \\ \dfrac{s+1}{(s+3)(s+4)(s+7)} & \dfrac{(s+2)(s+3)}{(s+1)(s+4)(s+5)} & \dfrac{s+7}{(s+2)(s+7)(s+8)} \end{bmatrix}$$

Process Identification: Nonlinear Systems

10

We shall now consider the identification of nonlinear systems, the most difficult problem in the hierarchy of parameter estimation and identification problems. By a nonlinear system we simply mean that the input/output properties of the system do not obey the conditions for linearity, namely, the superposition principle. We shall assume that the system is to be viewed as a black box about which we have no prior knowledge except for the fact that it is a nonlinear transformation of the input to the output. For the system S (see Figure 10.1) with a system functional F ($y = F(u)$), the identification problem can be stated as follows: Given a system $y = F(u)$, choose a set of inputs $\{u\}$ such that the input/output pairs $\{u_i, y_i\}$ will allow the determination of F. The specification of the system as nonlinear could be the result of unsuccessful tries at linear models or the result of general knowledge that the types of processes occurring in the system (such as chemical reactions) are inherently nonlinear.

Historically, techniques of linear system theory have been used extensively in the study of nonlinear systems. There are basically two reasons for this: first is the simplicity and completeness of linear theory, and second is the near absence of powerful and general nonlinear system techniques. Linearizing techniques based on small perturbations have been used, often yielding useful results. Phase plane and describing function methods are two special-

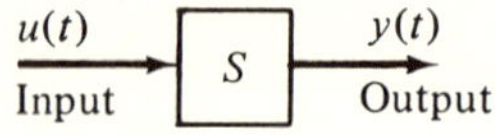

Figure 10.1. Nonlinear transformation of input to output.

ized techniques that have had considerable use. However, all these methods have serious limitations and are applicable to rather narrow classes of nonlinear systems.

A general theory for the identification of nonlinear systems was initiated by Wiener in 1942 [24] and has since been developed and tested in a number of fields. Two special approaches based on Wiener's theory will be the subject of this chapter.

The identification of nonlinear systems is a popular topic of current research. We shall not attempt to provide a survey of this field here. For the reader who wishes to pursue this topic beyond the material in this chapter, we recommend the excellent reviews of Bekey [3], Nieman et al. [17], Cuenod and Sage [8], and the book by Sage and Melsa [20]. These references contain a number of techniques that we shall not touch upon here but that are of importance for both linear and nonlinear identification.

10.1 The Wiener Identification Theory

The basic problem with which we are faced is the determination of a mathematical representation of a nonlinear system that reveals certain basic functional characteristics of the system. Consider the system S shown in Figure 10.1. At any time t, the system can be considered to be a *functional* F whose value is $y(t)$, the value of the output at this time, and whose argument is $[u(\tau), \tau \leq t]$, a function that describes the whole past input to the system. (A *functional* is a mapping from a space of functions into the real numbers.) Thus the system can be expressed mathematically as

$$y(t) = F[t; u(\tau), \tau \leq t]$$

The convolution integral of a linear system

$$y(t) = \int_0^t K(t - \tau)u(\tau)\,d\tau$$

is an example of a specific functional.

Volterra suggested that the input/output relationship of a nonlinear system be represented as a power series with functionals as terms:

$$y(t) = K_0 + \int_{-\infty}^{\infty} K_1(\tau)u(t - \tau)\,d\tau$$

$$+ \int_{-\infty}^{\infty} \int_{-\infty}^{\infty} K_2(\tau_1, \tau_2)u(t - \tau_1)u(t - \tau_2)\, d\tau_1\, d\tau_2$$
$$+ \int_{-\infty}^{\infty} \int_{-\infty}^{\infty} \int_{-\infty}^{\infty} K_3(\tau_1, \tau_2, \tau_3)u(t - \tau_1)u(t - \tau_2)u(t - \tau_3)\, d\tau_1\, d\tau_2\, d\tau_3$$
$$+ \cdots \qquad (10.1.1)$$

where the integral kernels $K_1(\tau)$, $K_2(\tau_1, \tau_2)$, ... are zero for any of their arguments less than zero to satisfy the requirement of physical realizability. We see that the second term describes the linear behavior and that the higher-order integrals are generalizations of the linear convolution integral.

Wiener suggested that a hierarchy of functionals of increasing order which are orthogonal to each other with respect to a Gaussian white noise input be constructed, the sum of which characterizes the system. Wiener showed that the relationship between the input and output can be written as

$$y(t) = \sum_{n=0}^{\infty} G_n[K_n, u(t)]$$

where $\{G_n\}$ is a set of orthogonal functions and $\{K_n\}$ is the set of kernels of (10.1.1). Each K_n is a symmetric function of its arguments. The first four Wiener functionals are

$$G_0[K_0, u(t)] = K_0$$
$$G_1[K_1, u(t)] = \int_0^{\infty} K_1(\tau)u(t - \tau)\, d\tau$$
$$G_2[K_2, u(t)] = \int_0^{\infty} \int_0^{\infty} K_2(\tau_1, \tau_2)u(t - \tau_1)u(t - \tau_2)\, d\tau_1\, d\tau_2 - \alpha \int_0^{\infty} K_2(\tau_1, \tau_1)\, d\tau_1$$
$$G_3[K_3, u(t)] = \int_0^{\infty} \int_0^{\infty} \int_0^{\infty} K_3(\tau_1, \tau_2, \tau_3)u(t - \tau_1)u(t - \tau_2)u(t - \tau_3) \cdot d\tau_1\, d\tau_2\, d\tau_3 - 3\alpha \int_0^{\infty} \int_0^{\infty} K_3(\tau_1, \tau_2, \tau_2)u(t - \tau_1)\, d\tau_1\, d\tau_2$$

where α is the power density spectrum of the white noise $u(t)$.

Therefore the system S is characterized by a set of integral kernels $\{K_i\}$ which can be considered as generalized impulse responses of the system. To see this and also gain some insight into the meaning of the kernels, consider a system S which is completely described by the linear and quadratic terms of the series

$$y(t) = \int_0^{\infty} K_1(\tau)u(t - \tau)\, d\tau + \int_0^{\infty} \int_0^{\infty} K_2(\tau_1, \tau_2)u(t - \tau_1)u(t - \tau_2)\, d\tau_1\, d\tau_2$$

The response of S to an impulse at $t = 0$, i.e., $u(t) = \delta(t)$, is

$$y(t) = K_1(t) + K_2(t, t)$$

The response of S to an impulse at $t = t_0$, i.e., $u(t) = \delta(t - t_0)$, is

$$y(t) = K_1(t - t_0) + K_2(t - t_0, t - t_0)$$

The response of S to impulses at $t = 0$ and $t = t_0$, i.e., $u(t) = \delta(t) + \delta(t - t_0)$, is

$$y(t) = K_1(t) + K_1(t - t_0) + K_2(t, t) + K_2(t, t - t_0) + K_2(t - t_0, t) + K_2(t - t_0, t - t_0)$$

If we subtract from the response to two impulses the separate responses to the two impulses when each alone excited the system, we are left with

$$K_2(t, t - t_0) + K_2(t - t_0, t)$$

which, since K_2 is symmetric, reduces to $2K_2(t, t - t_0)$. Thus the second-order kernel $K_2(\tau_1, \tau_2)$ is a quantitative measure of the nonlinear cross-correlation between the two impulses and $2K_2(t, t - t_0)$ represents the deviation from pure linear superposition of inputs, that is, the deviation of the response to the sum of two impulses from the sum of the responses to each impulse separately. The value of $K_2(\tau_1, \tau_2)$ gives the nonlinear contribution due to interaction of portions of the input signal τ_1 and τ_2 time units in the past.

Wiener's theory is limited to stable systems with finite memories. These systems possess the necessary property that the present output is influenced to an arbitrarily small extent by that portion of the past of the input beyond some arbitrarily large but finite time. This means, generally, that in such systems the influence of the past input on the present output attentuates with time. Excluded are systems whose characteristics change with time and systems with infinite memory such as oscillators. In addition, attention is restricted to nonlinear systems operating on continuous, real-valued functions of time, thereby generating continuous, real-time functions as outputs. All physical systems with finite physical inputs, both in amplitude and duration, are included in this classification. Therefore, compared with other methods of nonlinear system analysis such as the phase plane and describing function techniques, the Weiner method has a much greater range of applicability and it can, in principle, describe nearly all physical systems.

The key aspect of Wiener's theory is the use of white noise as the input used to characterize the system. With white Gaussian noise as the test signal, there is a finite, nonzero probability that any given time function will be approximated arbitrarily closely over any finite time interval by some sample of this noise. Because white noise possesses a flat frequency spectrum over all frequencies, the system is effectively being exposed to every physical input. Two systems which respond identically to white Gaussian noise are said to be equivalent and will have the same response to any given input. Fortunately, white Gaussian noise, while not physically realizable, can be approximated to almost any degree of accuracy by the output of an electronic or computational device termed a shot-noise generator [9]. The shot effect, output of a shot-noise generator, is, then, the actual physical input postulated as the universal probe for nonlinear systems.

Let us now develop the fundamentals of the Wiener theory of identification. Given a time-invariant system S, we can write the general relation

$$y(t) = F[u(\tau), \tau \leq t]$$

where t does not enter explicitly in the functional F because of the assumption of time invariance. Thus the present output $y(t)$ is a functional of the past and present values of the input. Let us relabel $t = 0$ as the present time with the more remote past becoming increasingly negative. This nonlinear transformation or mapping can be performed in two successive operations.

First, the past and present of the input are characterized by a set of coefficients $\{c_n\}$:

$$[u(t), t \leq 0] \xrightarrow{M} \{c_n\}$$

where the $\{c_n\}$ are an ordered set of real numbers obtained from the mapping M. Second, the $\{c_n\}$ are transformed into the present value of the output by mapping N:

$$\{c_n\} \xrightarrow{N} y(0)$$

Consider the first step. The function $[u(\tau), \tau \leq t]$ can be expanded in a series of orthonormal functions $\phi_n(t)$. Wiener selected these as Laguerre functions, and so

$$u(-t) = \sum_{n=0}^{\infty} c_n \phi_n(t), \qquad t \geq 0 \tag{10.1.2}$$

The $\{c_n\}$ are the desired set of coefficients characterizing the past of the input. The Laguerre functions form a complete orthonormal system in the interval $(0, \infty)$ and are defined by

$$\phi_n(t) = \frac{e^{t/2}}{n!} \frac{d^n}{dt^n}(t^n e^{-t}), \qquad t \geq 0, \ n = 0, 1, 2, \ldots \tag{10.1.3}$$

Because the Laguerre coefficients do form an orthonormal system,

$$\int_0^{\infty} \phi_m(t)\phi_n(t)\, dt = \begin{cases} 0, & m \neq n \\ 1, & m = n \end{cases} \tag{10.1.4}$$

we obtain the $\{c_n\}$ by

$$c_n = \int_0^{\infty} u(-t)\phi_n(t)\, dt \tag{10.1.5}$$

We shall now show how the c_n can be simply generated. We note that the Laguerre functions $\phi_n(t)$ are related to the Laguerre polynomials,

$$L_n(t) = \frac{1}{(n-1)!} e^t \frac{d^{n-1}}{dt^{n-1}}(t^{n-1}e^{-t}) \tag{10.1.6}$$

by

$$\phi_n(t) = \begin{cases} e^{-(t/2)}L_n(t), & t \geq 0 \\ 0, & t < 0 \end{cases} \tag{10.1.7}$$

To illustrate how these c_n can be generated from a series of Laguerre functions we need the L.T. of a Laguerre function. The general form is

$$\phi_k(t) \xrightarrow{L} \frac{(s - \frac{1}{2})^k}{(s + \frac{1}{2})^{k+1}} \tag{10.1.8}$$

To obtain (10.1.8), consider the case of (10.1.7) for $n = 1$ and 2:

$$\phi_1(t) = e^{-t/2}e^t = e^{t/2}$$

$$\phi_2(t) = e^{-t/2}e^t \frac{d}{dt}[te^{-t}] = e^{-t/2}(-t + 1)$$

However, from Table 3.1

$$e^{at} \xrightarrow{L} \frac{1}{s - a}; \qquad \frac{t^{n-1}e^{at}}{\Gamma(n)} \xrightarrow{L} \frac{1}{(s - a)^n}$$

and thus

$$\phi_1(t) \xrightarrow{L} \frac{1}{s + \frac{1}{2}}; \qquad \phi_2(t) \xrightarrow{L} \frac{(s - \frac{1}{2})}{(s + \frac{1}{2})(s + \frac{1}{2})}$$

which are the cases $k = 0, 1$ in (10.1.8). Obviously this could be continued to yield (10.1.8) inductively. Now if we make up a series cascade with $u(t)$ as input, it follows that the response $r_k(t)$ to the input $u(t)$ is given in the transform domain as

$$L[r_k(t)] = L[u(t)]\frac{(s - \frac{1}{2})^k}{(s + \frac{1}{2})^{k+1}} \qquad k = 0, 1, 2, \ldots \tag{10.1.9}$$

In other words, $L[\phi_k(t)]$ is the transfer function for $k + 1$ sections and $L[r_k(t)]$ is the response to the input of that section.

However, we may now invert (10.1.9) using the convolution theorem to yield

$$r_k(t) = \int_{-\infty}^{t} u(\tau)\phi_k(t - \tau)\, d\tau$$

Substituting $t' = t - \tau$ and letting the measurement of $r_k(t)$ occur at $t = 0$ yields

$$r_k(0) = \int_0^\infty u(-t')\phi_k(t')\, dt' \tag{10.1.10}$$

which is identical to (10.1.5). Thus

$$r_k(0) = c_k \tag{10.1.11}$$

In general, the output of the $(k+1)$st section of the Laguerre cascade at any time t is equal to the kth Laguerre coefficient of the past of the input $u(t)$ up to time t. To be more specific, the outputs $r_k(t)$ satisfy the following set of *linear* differential equations:

$$\frac{dr_0}{dt} + \frac{1}{2}r_0 = u(t)$$
$$\frac{dr_k}{dt} + \frac{1}{2}r_k = u(t) - \sum_{\nu=0}^{k-1} r_\nu(t), \qquad k = 1, 2, \ldots, n-1 \tag{10.1.12}$$

with initial conditions

$$r_0(0) = r_1(0) = \cdots r_{n-1}(0) = 0 \tag{10.1.13}$$

As such, the generation of the $r_k(t)$ or c_k can be carried out easily on a digital or an analog computer. Further, this shows how the Laguerre coefficients can be obtained by using $u(t)$ as input to a chain of linear Laguerre functions. By sampling the output of the chains at the proper time, normally at $t = 0$, the coefficients $c_k(0)$ may be evaluated directly.

Although an infinite number of Laguerre coefficients are required to completely characterize the past of the system input, in practice only a finite number can be evaluated. Because the Laguerre functions decay in almost exponential fashion with increasing argument, any finite sum of these functions evaluated at a suitably large value of the argument will be negligibly small. This erroneously implies that in the finite Laguerre function expansion of the past of the input there exists some finite time in the past beyond which the input all but vanishes. Thus Wiener's method is restricted to systems with arbitrarily large but finite memories. Unstable systems and limit-cycle oscillators are not in this classification, because once these systems are excited, the output will exist for all future time independent of the input.

The Laguerre coefficients of the past of any integrable square input are identically equal to the outputs of the Laguerre cascade. Each coefficient, $c_n(t)$, for a white Gaussian input is a sample function of a continuous state, stationary random process and has the following three statistical properties:

1. Each c_n has a Gaussian distribution.
2. The coefficients $\{c_n\}$, evaluated at the same time, are statistically independent random variables.
3. The coefficients $\{c_n\}$ all have the same variance.

The validity of these assertions has been demonstrated [11]. It has also been shown that if for convenience the coefficients are assigned a unit variance, then the $\{c_n\}$ are standardized random variables; i.e., they each have a mean value of zero and a variance of unity.

The present value of the system output $y(t)$ is now to be evaluated in terms of the past and present input as given by the Laguerre coefficients. This is the second step of the analysis. To do this we shall use Hermite functions because they are orthogonal with respect to a white Gaussian random process and form an orthonormal set over the interval $(-\infty, +\infty)$ [23].

The nth normalized Hermite polynomial is defined as

$$H_n(z) = (-1)^n(2^n n! \pi^{1/2})^{-1/2} e^{z^2} \frac{d^n}{dz^n} e^{-z^2}, \qquad n = 0, 1, 2, \ldots \tag{10.1.14}$$

The corresponding Hermite functions are

$$\psi_n(z) = e^{-z^2/2} H_n(z) \tag{10.1.15}$$

The Hermite functions constitute an orthonormal system over the infinite interval $(-\infty, +\infty)$; hence

$$\int_{-\infty}^{\infty} \psi_m(z)\psi_n(z)\, dz = \begin{cases} 0 & \text{for } m \neq n \\ 1 & \text{for } m = n \end{cases}$$

The multivariable Hermite function expansion of the Laguerre coefficients of the past of the time function $u(t)$ is

$$\begin{aligned} y(t) &= \lim_{n \to \infty} \left\{ \sum_{i=0}^{\infty} \sum_{j=0}^{\infty} \cdots \sum_{k=0}^{\infty} a_{i,j,\ldots,k} \psi_i(c_0)\psi_j(c_1) \cdots \psi_k(c_n) \right\} \\ &= \lim_{n \to \infty} \left\{ \sum_{i=0}^{\infty} \sum_{j=0}^{\infty} \cdots \sum_{k=0}^{\infty} a_{i,j,\ldots,k} H_i(c_0)H_j(c_1) \cdots H_k(c_n) \right. \\ &\quad \left. \cdot \exp\left[-\tfrac{1}{2}(c_0^2 + c_1^2 + \cdots + c_n^2)\right] \right\} \end{aligned} \tag{10.1.17}$$

There are $n + 1$ summations indicated in (10.1.17), one for each Laguerre coefficient. This is the desired general input/output relation for a time-invariant, nonlinear system.

This equation can be simplified in appearance by the substitutions,

$$V(\alpha) = H_i(c_0)H_j(c_1) \cdots H_k(c_n)$$

and

$$A_\alpha = a_{i,j,\ldots,k}$$

to yield

$$y(t) = \lim_{n \to \infty} \left\{ \prod_{\nu=0}^{n} \exp\left(-\frac{c_\nu^2}{2}\right) \sum_{\alpha} A_\alpha V(\alpha) \right\} \tag{10.1.18}$$

The index of summation α extends over all permutations of the indices $i, j, \ldots, k$, i.e., over all permutations of the integers 0, 1, 2, . . . to infinity. The infinite set of Hermite expansion coefficients $\{A_\alpha\}$ characterizes, or serves to identify, the unknown nonlinear system because the system response to any physical input can be evaluated through (10.1.18) once the $\{A_\alpha\}$ are known.

In any practical application of Wiener's theory, only a finite number of Laguerre coefficients can be generated, and for each coefficient only a finite number of Hermite functions can be constructed. Suppose that in (10.1.17) the number of Laguerre coefficients is limited to $n + 1$ and that each Hermite function index can take on only $p + 1$ values, ranging from zero to p. Equation (10.1.18) now becomes a finite sum:

$$\hat{y}(t) = \prod_{\nu=0}^{n} \exp\left(-\frac{c_\nu^2}{2}\right) \sum_{\alpha \in \gamma} A_\alpha V(\alpha) \tag{10.1.19}$$

where $\hat{y}(t)$ is the finite approximation of $y(t)$. For any bounded, continuous input to the system, i.e., any physical input, (10.1.19) yields a bounded continuous output $\hat{y}(t)$; hence, the aforementioned restriction to real physical systems becomes necessary. Because the Hermite functions form a complete orthonormal set on the interval $(-\infty, +\infty)$, $\hat{y}(t)$ can be made to approximate $y(t)$ within any desired degree of accuracy by increasing the number of Laguerre coefficients and the number of Hermite functions in the expansion. The index α ranges over the members of the set γ, where γ includes all permutations of the integers 0, 1, 2, . . . , p taken $n + 1$ at a time. There are, therefore, $(p + 1)^{n+1}$ Hermite coefficients to be evaluated in the finite Hermite function expansion.

A procedure whereby the parameters $\{A_\alpha\}$ characterizing a given nonlinear system can be evaluated experimentally will now be developed. In the experimental method the unknown system is subjected to a stationary white Gaussian signal to expose it effectively to all inputs.

It is possible to evaluate the $\{A_\alpha\}$ in a number of ways. Harris and Lapidus [12, 13] and Roy and DeRusso [18] use a procedure which we shall merely explain; the details are excessive and involve considerable manipulation. In effect a time domain error $\bar{e}^2$, a weighted mean square error between $y(t)$ and $\hat{y}(t)$, is minimized with respect to $A(\beta)$, where $A(\beta)$ is any of the coefficients $\{A_\alpha\}$. The weighting factor is $\prod_{r=0}^{n} \exp(-c_\nu^2/2)$ and the minimization is achieved by differentiating $\bar{e}^2$ with respect to A_β and setting the result to zero. Then by invoking the ergodic theorem, i.e., that a time average can be replaced by an ensemble average, and using the orthogonality properties of the Hermite polynomials, there results

$$A_\beta = (2\pi)^{n/2}\overline{y(t)V(\beta)} \tag{10.1.20}$$

Here $\overline{y(t)V(\beta)}$ denotes the time average value of $y(t)V(\beta)$. This, in turn, yields an explicit value for $\bar{e}^2_{\min}$, which can be reduced to any preassigned level by increasing the number of Laguerre coefficients and Hermite functions, for it tends in the limit to zero. Owing to the orthogonality of the model, the $\{A_\alpha\}$-values are determined independently and have the important property of finality; i.e., improving the characterization by using $n + 1$ Hermite expansion coefficients will not necessitate reevaluating the previously determined n coefficients.

We also wish to point out that while we have used the Laguerre and Hermite functions, this is not the only combination which can be developed.

In fact, Brick [5] has indicated that many other orthogonal functions can be substituted, but this combination seems to have the most desirable properties and leads to the simplest results.

10.1.1 Implementation of Wiener's method

Figures 10.2 and 10.3 show schematic diagrams for arranging the flow of information and the sequential calculations required for implementing the above concepts.

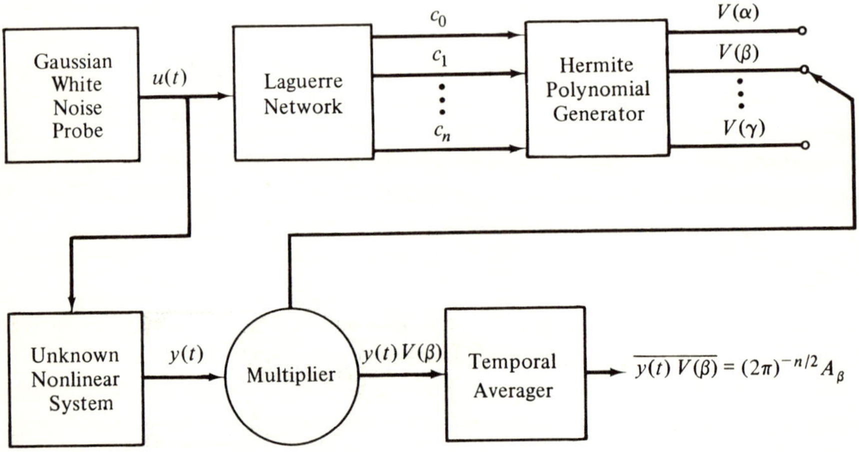

Figure 10.2. Characterization of a nonlinear system according to Wiener's method. [Reprinted from G. Harris and L. Lapidus, "The Identification of Nonlinear Systems," *Ind. Eng. Chem. 59*, No. 6, 66 (1967). Reprinted by permission of the copyright owner.]

The Hermite polynomial generator consists of np piecewise linear function generators to yield the np polynomials $\{H_j(c_i)\}$ and p^n multipliers to form the $\{V(\alpha)\}$ from the Hermite polynomials. p^n integrators are required for the time-weighting process to provide the p^n characterizing coefficients $\{A_\alpha\}$. This much analog equipment is required to perform the characterization of a given nonlinear system so that the $\{c_n\}$ may be computed simultaneously in the least amount of time. If the time of identification is of no concern, a considerable reduction of equipment is possible. If each Laguerre coefficient is evaluated serially, then only n function generators, one multiplier and one integrator, are used. Synthesis, however, will necessitate, in addition to np function generators and p^n multipliers to form the $\{V(\alpha)\}$, an exponential function generator consisting of many operational units (there are several different ways to construct this generator) and p^n amplifiers for introducing the $\{A_\alpha\}$, one summer and one multiplier.

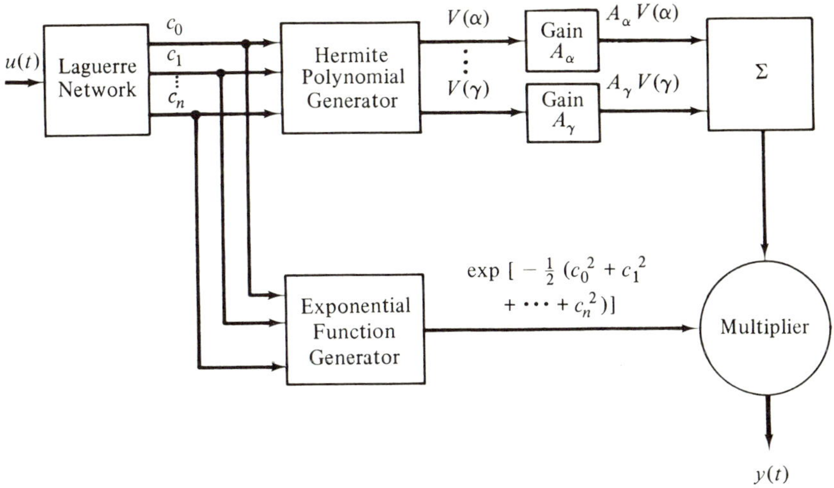

Figure 10.3. Synthesis of a nonlinear system according to Wiener's method. [Reprinted from G. Harris and L. Lapidus, "The Identification of Nonlinear Systems," *Ind. Eng. Chem.* 59, No. 6, 66 (1967). Reprinted by permission of the copyright owner.]

Because of the excessive demands on hardware, the characterization and synthesis of nonlinear systems entirely by analog computer means is beyond the scope of present-day analog computing facilities. Unfortunately, purely digital methods of implementing the Wiener theory also have serious drawbacks. In particular, the number of coefficients needed to characterize almost any system, linear or nonlinear, is extremely large. If n coefficients are used in the Laguerre expansion to describe the past of the input at any time and p coefficients are used to expand the system functional in terms of Hermite functions, then the number of coefficients needed to characterize the system is p^n. Even for a simple nonlinear system n and p must be approximately 10–20, giving the number of coefficients of 10^{10} to 20^{20} to be evaluated. The computation required for the evaluation of the characterizing coefficients is extremely long, particularly since the computation must be performed serially. When the response to a particular input is desired, the computation is again quite long because of the large number of coefficients and the repeated Laguerre and Hermite expansions.

In addition to these practical computational drawbacks, we can make some general observations on the Wiener method. It would be desirable to assign some physical significance to the coefficients $\{a_{i,j,\ldots,k}\}$ that would reveal some features of the structure of the system. This is generally not possible. We must regard the coefficients $\{a_{i,j,\ldots,k}\}$ as purely formal mathematical quantities, from which it is not possible to extract physical properties of the system they characterize. A linear system which is characterized very simply by the classic linear theory is characterized very cumbersomely by the Wiener method. A large number of coefficients $\{a_{i,j,\ldots,k}\}$ is needed to identify a linear system. This is due to the fact that a large number of Hermite polynomials is needed so that their sum will cancel out the effect of the highly nonlinear exponential transformation. The method is so general that it is not capable of recognizing a simple situation, such as a linear system, and treating it accordingly. It is very difficult to incorporate into Wiener's method any prior information about the system so as to reduce the number of the characterizing coefficients.

All these difficulties encountered in the practical application of the theory point out the need for other formulations of the white-noise theory if it is to be made a working tool for nonlinear system identification.

10.2 Use of Two-Level Inputs

Because of the difficulties in implementing the Wiener theory outlined in Section 10.1, we shall introduce here what may be termed a first-order approximation to the full theory. This approach uses a special two-level input which is randomly (or nonrandomly) switched between its two levels; also an alternative orthogonal expansion is used instead of the Hermite functions. This procedure is termed first order since the full Gaussian white-noise probe is approached as the number of levels of input is increased from two toward infinity.

10.2.1 The two-level input function

We shall allow the input function $u(t)$ to switch between two levels denoted by -1 and $+1$ (equivalent to a binary yes–no sequence). These numerical values are chosen for convenience since a simple change of variable can convert any alternative finite values to ± 1. A typical sequence for this input is shown in Figure 10.4, where $t = 0$ is the present time and the plot shows the input as a function of a sampling period T into the past. If T_m is an effective estimate of the system's memory, then $T_m = (n - 1)T$, where n is the sampling index into the past. Frequently T_m is taken as equal to T_s, the system settling time, the time required for the system response to come within a specified proximity of the system's final value under the action of an impulse or step input. Obviously as $T \longrightarrow 0$ or $n \longrightarrow 0$ the discrete set of $u(nT) = u_n$ approaches $u(t)$.

The n sampled values of $u(t)$ characterizing the past input can be very conveniently stored in a digital storage device which is tapped (or available for reading out) at n points. In Figure 10.4 the integer values of $-t/T$ are the

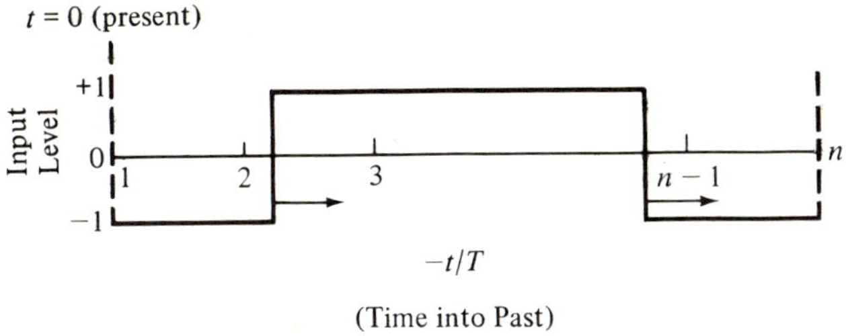

Figure 10.4. Schematic of two-level input.

tap numbers or locations in the storage device. It is apparent that each tap can exist in either of the two levels ± 1 and thus the entire memory device will be able to accommodate 2^n distinct binary sequences or configurations.

10.2.2 Orthogonal expansion of the output

At any given time each tap can exist at only one level so that at each instant one and only one input set can exist in the memory device. If an expansion of the system output in terms of the tap levels is used, the model will be orthogonal in time. This means that each term in the expansion represents the system output for one particular input configuration; there is one term for each input state. An orthogonal expansion is required so that each term can be evaluated independently of all other terms. Thus we write

$$\hat{y}(t) = \sum_{i=1}^{2} \sum_{j=1}^{2} \cdots \sum_{k=1}^{2} a_{i,j,\ldots,k} \psi_i(B_1) \psi_j(B_2) \cdots \psi_k(B_n) \qquad (10.2.1)$$

There are n summations in (10.2.1), one for each tap.

The functions $\psi_1(B)$ and $\psi_2(B)$ are given by

$$\psi_1(B) = \begin{cases} 1, & \text{if } B = -1 \\ 0, & \text{if } B = +1 \end{cases}$$

$$\psi_2(B) = \begin{cases} 0, & \text{if } B = -1 \\ 1, & \text{if } B = +1 \end{cases}$$

or

$$\psi_1(B) = \frac{1 - B}{2}$$

$$\psi_2(B) = \frac{1 + B}{2}$$

At any time t' the input state exists in one of the 2^n distinct states as given by the sequence of tap levels $\{B_i\}$, $i = 1, 2, \ldots, n$, in (10.2.1). Consider a specific term (of which there are 2^n) denoted by the indices $i', j', \ldots, k'$ in the above expansion:

$$a_{i',j',\ldots,k'}\psi_{i'}(B_1)\psi_{j'}(B_2)\cdots\psi_{k'}(B_n)$$

This term is nonzero whenever

$$\psi_{i'}(B_1) = \psi_{j'}(B_2) = \cdots = \psi_{k'}(B_n) = 1$$

and when this occurs, all other terms must necessarily vanish. The input configuration consisting of the tap levels $\{B_i\}$ defines a unique input space, which is denoted by $\{i', j', \ldots, k'\}$. The model output $\hat{y}(t')$ is given by the coefficient $a_{i',j',\ldots,k'}$ assigned to that particular section. Since the model output is given by a sequence of numbers, $\hat{y}(t)$ is a staircase approximation to the actual system output. As the number of taps in the input memory is increased (the input is sampled more frequently), the input characterization becomes more accurate and $\hat{y}(t)$ approximates $y(t)$ more closely.

By letting

$$\Phi(\alpha) = \psi_i(B_1)\psi_j(B_2)\cdots\psi_k(B_n)$$

and

$$A_\alpha = a_{i,j,\ldots,k}$$

(10.2.1) can be written as

$$\hat{y}(t) = \sum_\alpha A_\alpha \Phi(\alpha) \tag{10.2.2}$$

The index of summation α ranges over the 2^n cells of the input space until the one which is occupied is found. Suppose that this is the βth cell. Then $\Phi(\beta) = 1$, $\hat{y}(t) = A_\beta$, and the remaining $(2^n - 1)\Phi$s are zero.

10.2.3 Evaluation of A_α

The coefficients $\{A_\alpha\}$ can be evaluated in the same manner as outlined previously, namely by minimizing the mean square error $\bar{e}^2$ between $y(t)$ and $\hat{y}(t)$. For the error defined as

$$\bar{e}^2 = \lim_{T\to\infty} \frac{1}{2T} \int_{-T}^{T} [y(t) - \hat{y}(t)]^2\, dt \tag{10.2.3}$$

differentiation with respect to the βth coefficient, A_β, gives

$$\frac{\partial \bar{e}^2}{\partial A_\beta} = \lim_{T\to\infty} \frac{1}{2T} \int_{-T}^{T} - 2\Phi(\beta)[y(t) - \sum_\alpha A_\alpha \Phi(\alpha)]\, dt$$

where (10.2.2) has been used. The error is a minimum provided

$$\frac{\partial \bar{e}^2}{\partial A_\beta} = 0$$

for all β. Hence

$$\overline{\Phi(\beta) \sum_\alpha A_\alpha \Phi(\alpha)} = \overline{y(t)\Phi(\beta)}$$

Since the Φ functions are orthogonal, they satisfy the relationship

$$\sum_{\alpha} \Phi(\alpha)\Phi(\beta) = \Phi(\beta)$$

thus yielding

$$A_\beta = \frac{\overline{y(t)\Phi(\beta)}}{\overline{\Phi(\beta)}} \tag{10.2.4}$$

If a weighting function $w(t)$ had been used in (10.2.3), then

$$A_\beta = \frac{\overline{y(t)w(t)\Phi(\beta)}}{\overline{w(t)\Phi(\beta)}} \tag{10.2.5}$$

These are the desired expressions relating the orthogonal coefficients to certain experimentally measurable time averages of the system's inputs and outputs.

10.2.4 Further observations

A number of further observations can be made regarding the development just completed. These involve certain simplifications, alternative interpretations, and general comments.

When $w(t) = 1$ the quantity $\overline{\Phi(\beta)}$ may be interpreted as the probability that the βth cell in the input space is occupied at any instant since

$$\overline{\Phi(\beta)} < 1$$

and

$$\sum_{\alpha} \overline{\Phi(\alpha)} = 1$$

Hence the coefficient A_α in (10.2.4) is simply equal to the average value of the system output $y(t)$ over the time that the input configuration β exists, this average taken over an infinitely long observation period. That is, A_β is the conditional mean of $y(t)$ assuming that the βth cell in the input space is occupied, or

$$A_\beta = [\overline{y(t)} \,|\, \Phi(\beta) = 1]$$

By use of the Student t-distribution test it is possible to place percentage confidence limits on whether sufficient averaging has occurred to ensure a relatively constant value of A_α. In this way a measure of the error due to the necessary finite sampling of the input rather than the theoretical infinite sampling can be established.

The specific weighting factor $w(t)$ can be suggested by whatever a priori information one has about the actual behavior of the system. For a time-varying system, as an example, a nonnegative monotonic nondecreasing weighting function may be suggested, while a time-invariant system would suggest a $w(t) = 1$. Harris and Lapidus [12] show, as an illustration, a possible weighting function for the case where the system always moves between two steady-state conditions corresponding to the inputs $u = +1$ and $u = -1$.

Although only input memories with uniformly spaced taps (the input is sampled at equal time intervals T) have been considered here, a tapered-tap arrangement (a nonuniformly time-sampled input) can be expected to produce a smaller mean square error term than will uniform spacing. This statement is valid for systems in which the influence of the past input on the present output attenuates with time. An input memory whose distribution of taps generally diminishes with distance down the memory (in the direction of the more distant past) will be more efficient than one with uniform tap spacing. This is not to say that it is always desirable to have the closest tap spacing at the very beginning of the memory. Systems with appreciable time delays could best be identified by using memories with the closest tap spacing in the vicinity of the delay time. Such a possible tap spacing is shown in Figure 10.5.

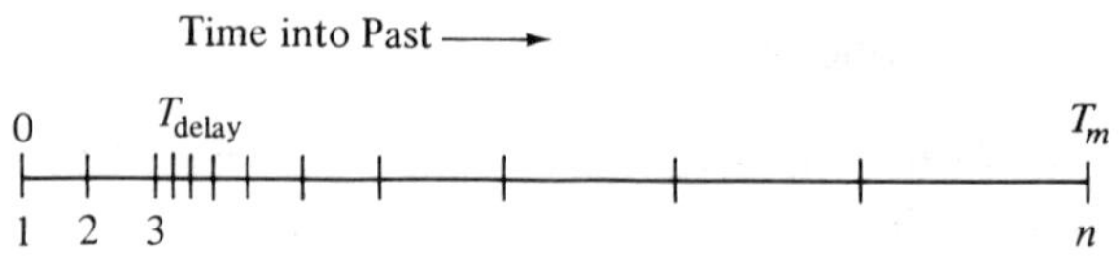

Figure 10.5. Possible tap spacing along memory for a system with a time delay.

The extension of the present development beyond the scalar input/scalar output is of interest. When there are multiple outputs, $y(t)$ is a vector, and there will be 2^n coefficients $\{A_\beta\}$ corresponding to the 2^n cells of the input space for each output variable. For q outputs there are $q2^n$ coefficients to be evaluated. This is not a serious increase in the number of coefficients. When there are p inputs, however, the problem becomes significant; now there will be $q2^{np}$ coefficients. Such a number, even for modest n and p, can tax the memory of any computer. Thus it may be necessary to restrict the analysis to $p = 1$ to maintain a degree of practicality.

The evaluation of $q2^n$ coefficients, where n is of the order of 20 or more (more than a million coefficients), is still not, however, a simple matter, and it may be necessary to restrict the number of model coefficients further. Such a restriction may be that the time between consecutive input switches, or the switching time T_{sw}, is not less than half the effective duration of the system's memory T_m, that is,

$$T_{sw} \geq \tfrac{1}{2}T_m$$

Therefore no more than two switches can occur during a time interval T_m, thereby drastically reducing the number of allowable input configurations. For this reason, the results to be discussed shortly are limited to systems whose inputs satisfy the above constraint.

The initial assumption that 2^n distinct input states exist implies that the input can switch at least as fast as T_m/n, since 2^n wave forms can be distinguished in the input memory. This assumption is unreasonable in many control system applications (when n is of the order of 20 or more). In most chemical process control systems, for example, such rapid switching is impractical due to the general sluggishness of these processes. It is expected, however, that a minimum switching time does exist, although admittedly it may not be as large as $\frac{1}{2}T_m$.

Under the assumption that the minimum switching time is $\frac{1}{2}T_m$, the number of allowable input tap configurations can be calculated directly since the input can switch at most twice during a time interval T_m. In fact, for as many as 71 taps only 1402 configurations are allowable.

Note also that while we have inferred that the binary input be randomly switched, an a priori deterministic type of switching could be employed. The influence of such switching is difficult, if not impossible, to ascertain theoretically.

Finally, we need to say a few words about the minimum identification time; i.e., how long shall we need to probe, even a time-invariant system, with the coefficient averaging to completely identify a specific system? For an input memory with n uniformly spaced taps, the input will change at most twice during a time interval $T_m/(n-1)$, the time between consecutive taps. This is shown in Figure 10.6. Wave form A in passing continuously to wave form

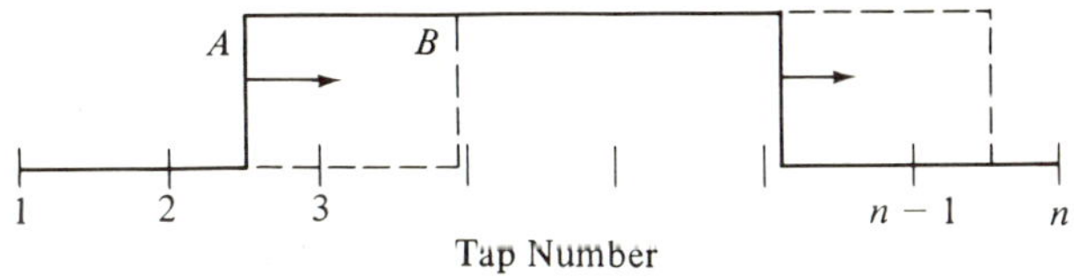

Solid line (A) – Initial Input Waveform.
Dashed line (B)– Input waveform time $T_m/(n-1)$ later.

Figure 10.6. Typical input waveform.

B will produce two different configurations. If, however, the initial wave form has but one switch, only one configuration will change, and for a wave form with no switch, no new configuration will result. Thus $n-1$ is a conservative estimate of the number of input configurations occurring in time T_m. On this basis the theoretical minimum identification time is

$$T_{\min} = \frac{r_m}{n-1} T_m$$

where r_m is the median of a random sample of r configurations. r_m can be calculated and it can then be shown that $T_{\min}/T_m$ increases almost linearly with the number of taps [11].

10.2.5 Identification of a CSTR

To illustrate the feasibility and defects of the probing identification algorithm just presented we shall consider the CSTR with nonlinear reaction detailed by, among others, Lapidus and Luus [14]. In this system model the outlet CSTR concentration and temperature, C and T, are taken as the output (or state) variables. The specific equations are (note the correspondence to those in Section 5.4.8)

$$\frac{dC}{dt} = \frac{F}{V}(C_0 - C) - kC$$

$$\frac{dT}{dt} = \frac{F}{V}(T_0 - T) + \left(\frac{-\Delta H}{\rho C_p}\right)kC - \frac{UF_c}{V\rho C_p(K + F_c)}(T - T_c)$$

where C_0, T_0 = influent concentration and temperature
F = flow rate
V = CSTR volume
ρ, C_p = fluid density and heat capacity
$-\Delta H$ = heat of reaction
U = heat transfer coefficient
K = heat transfer parameter
T_c = inlet coolant temperature
F_c = coolant water flow rate

k, the reaction rate constant, is given by the usual nonlinear Arrhenius expression

$$k = k_0 \exp \frac{-E}{RT}$$

The input used to probe the CSTR is either the inlet feed temperature, T_0, or the coolant water flow rate, F_c. When the feed temperature is considered to be the input, the resulting system equations are governed by two coupled time-invariant O.D.E.s; when the coolant water flow rate is used as input, an equivalent set of time-varying O.D.E.s results. In either case two stable states and one unstable equilibrium state exist for the system.

Further, when the inlet feed temperature was used as input, the two levels $u(t) = \pm 1$ were obtained by specifying an upper and lower temperature range for the input (based on a preliminary examination of the system response) and transforming these values to ± 1; when the coolant flow rate was used for the input the natural range of no cooling (the cooling valve closed) and maximum cooling (the cooling valve completely open) was chosen to correspond to $u = -1$ and $u = +1$, respectively.

The overall identification procedure involved a simulation of the CSTR system on a digital computer and the subsequent probing with the appropriate two-level input to evaluate the coefficients $\{A_\alpha\}$. With these coefficients, a black box model of the system could be constructed and the output $\hat{\mathbf{y}}(t)$ of the identified model compared with the computed process output $\mathbf{y}(t)$. In other words, the system simulation is used to replace the real physical system in the identification scheme.

The explicit determination of the $\{A_\alpha\}$ was carried out as follows:

1. Based on a preliminary examination of the system response, a choice for T_m was made.
2. At time $t = 0$, the system was taken as at one of the steady states, corresponding, for example, to the input $u(t) = -1$.
3. At time $t = 0$, the input was changed to $u(t) = +1$.
4. At time $T_m/2 < t < T_m$, the input was changed again to $u(t) = -1$.
5. This switching of the input was continued for a total time of as much as $100T_s$.
6. As the switching occurs, the coefficients $\{A_\alpha\}$ were continuously evaluated using (10.2.4) or (10.2.5). At the same time $\mathbf{y}(t)$ was calculated and then compared with the predicted staircase function $\hat{\mathbf{y}}(t)$.

In particular we shall discuss the time-invariant results here, e.g., probing with the inlet feed temperature; the time-varying results are qualitatively equivalent and may be found in Harris and Lapidus [13].

Typical results for the temperature and the concentration outputs for a period of over one switch are shown in Figures 10.7 and 10.8. In this particular case 31 tapered taps were used and the input switching was random (generated by a digital random number generator). By a close consideration of the steps of the outputs it is possible to ascertain the placement of the taps. In the first 50 seconds, with the first switch corresponding to $T_{sw} = 295.6$ seconds, the model output steps 14 times.

As might be expected, there are differences in the temperature and concentration results. The temperature model staircase output is an excellent fit to the simulated system results (the smooth curve); by contrast, the concentration response is not as good. In fact, the mean square error is 0.79 percent of the maximum range over which the temperature can vary for Figure 10.7, but it is 1.7 percent for Figure 10.8 using an equivalent maximum concentration range. Other results not shown here indicate that the value of T_m of 300 seconds, the type of tapered tap arrangement, and the particular form of the equations may be responsible for the difference in the two responses.

The effect of varying the number of taps in the input on the temperature output is shown in Figure 10.9. It is apparent that significant decreases in the mean square error can be achieved by increasing the number of taps. A maximum number of 51 was used in this investigation, but there is no reason to expect that a further decrease in the error would not result from a further increase in the taps. These results hold for either a random or a nonrandom type of input with uniformly spaced taps.

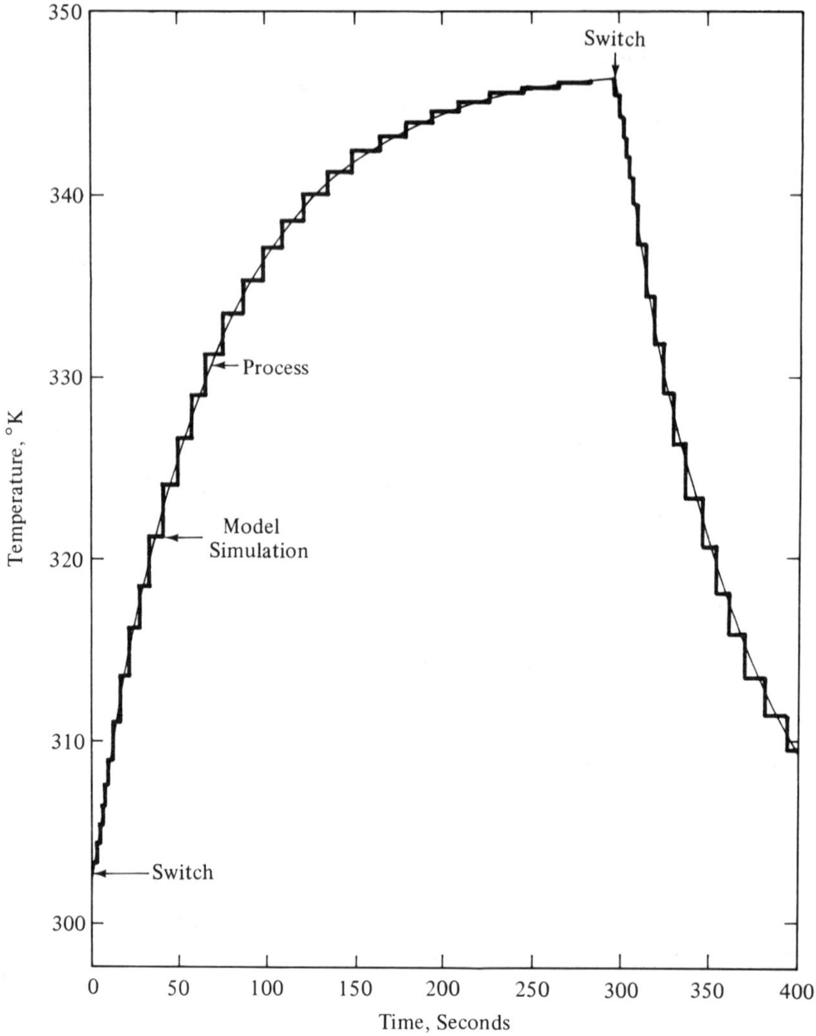

Figure 10.7. Temperature simulation of CSTR system (31 tapered taps, $T_m = 300$ seconds, random input). [Reprinted from G. Harris and L. Lapidus, "The Identification of Nonlinear Reaction Systems with Two-Level Inputs," *AIChE J. 13*, 291 (1967). Reprinted by permission of the copyright owner.]

Other results which the numerical experiments on this CSTR system suggest are that a tapered tap distribution can always be found which is better (lower mean square error) than the equivalent number of uniform taps. In many cases the error can be decreased by a factor of approximately 2–3. For example, a 31-tap model with uniform spacing had mean square errors in its temperature responses of 0.52. With a tapered-tap model the equivalent

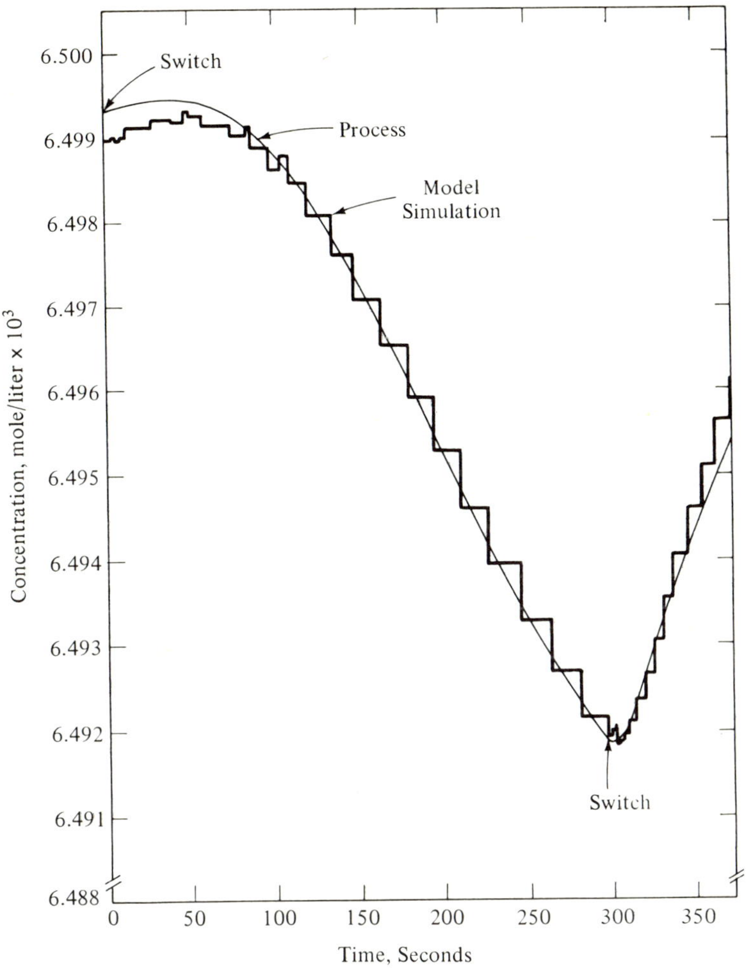

Figure 10.8. Concentration simulation of CSTR system (31 tapered taps, $T_m = 300$ seconds, random input). [Reprinted from G. Harris and L. Lapidus, "The Identification of Nonlinear Reaction Systems with Two-Level Inputs," *AIChE J. 13*, 291 (1967). Reprinted by permission of the copyright owner.]

results were 0.26. Further, the accuracy of the simulations was not significantly improved by the use of small confidence limits on the model coefficients. The minimum identification time of models with prescribed confidence limits of ± 1.0 varied from $61T_s$ for a 10-tap model to $72T_s$ for a 16-tap model. In no instance could identification of the system be achieved with a randomly switched input; a maximum identification time of $100T_s$ was permitted. The occurrence of all allowable input configurations seems to be the criterion of

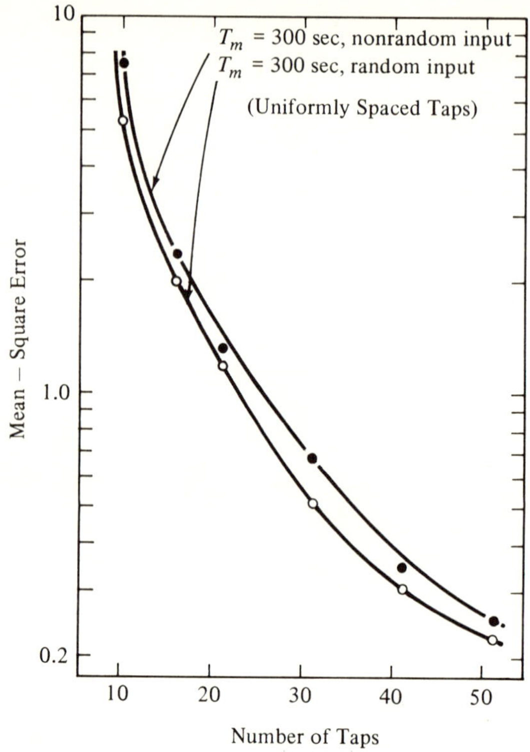

Figure 10.9. Accuracy of CSTR simulation. [Reprinted from G. Harris and L. Lapidus, "The Identification of Nonlinear Reaction Systems with Two-Level Inputs," *AIChE J. 13*, 291 (1967). Reprinted by permission of the copyright owner.]

identification; the criterion is not the satisfaction of imposed confidence limits.

Thus by a choice of the number of input taps, by the explicit distribution of taps, and by varying the switching time, a most acceptable identification can be achieved. This is especially true for the temperature output but less so for the concentration output. Three aspects of the identification technique can be suggested as most attractive. First, model parameters can be adjusted to give a small error in the simulation of a specified system output at the expense of correspondingly larger errors in the other model outputs. Second, errors in the model performance at discrete times are not propagated; the model output at any time depends solely on the state of the input and not at all on its previous value. Third, the model is expected to be relatively insensitive to noise in the measured system outputs since the characterizing coefficients are the result of an averaging process.

The extension of these results to more than two-level inputs, to the use of different error criteria, to a more rapid switching, and to multivariable systems seems apparent as the next step in the use of the method. Certain of these extensions are not trivial since the number of coefficients may increase in a startling manner. Nevertheless, the theory itself seems most viable and computer implementation is feasible.

Finally, we should point out that in qualitative terms this identification algorithm matches an input pattern of on–off or high–low states with an output pattern. The computer, then, is essentially a device for identifying patterns, and as such the algorithm is related to the problem of pattern recognition [19].

10.3 Identification Via a Finite Volterra Series

We shall now consider a method of nonlinear identification related directly to the infinite Volterra series (10.1.1). However, a feasible algorithm can be achieved only by truncating the series after a finite number of terms. We restate that the Volterra series expansion for a nonlinear system takes the form

$$y(t) = \sum_{i=0}^{\infty} y_i(t) \tag{10.3.1}$$

where

$$\begin{aligned} y_i(t) = \int_0^{\infty} \cdots \int_0^{\infty} K_i(\tau_1, \tau_2, \ldots, \tau_i) u(t - \tau_1) u(t - \tau_2) \cdots \\ \cdot u(t - \tau_i)\, d\tau_1\, d\tau_2 \cdots d\tau_i \end{aligned} \tag{10.3.2}$$

where we have changed the lower limit to zero since, by definition, $K_i = 0$ if any of its arguments is less than zero. Let us also assume that $K_0 = 0$.

We can consider a discretized version of (10.3.2). Assume that the output is observed at a series of discrete times, $t_k, t_{k-1}, \ldots$. Then the Volterra series can be written for scalar input and output as

$$\begin{aligned} y(t_k) = \sum_{\gamma=0}^{\infty} K_{1\gamma} u(t_{k-\gamma}) + \sum_{\gamma_1=0}^{\infty} \sum_{\gamma_2=0}^{\infty} K_{2\gamma_1\gamma_2} u(t_{k-\gamma_1}) u(t_{k-\gamma_2}) \\ + \sum_{\gamma_1=0}^{\infty} \sum_{\gamma_2=0}^{\infty} \sum_{\gamma_3=0}^{\infty} K_{3\gamma_1\gamma_2\gamma_3} u(t_{k-\gamma_1}) u(t_{k-\gamma_2}) u(t_{k-\gamma_3}) + \cdots \end{aligned} \tag{10.3.3}$$

This equation can be extended to an m-dimensional input vector $(u_1, u_2, \ldots, u_m)$ and written as

$$\begin{aligned} y(t_k) = \sum_{\gamma=0}^{\infty} \sum_{i=1}^{m} K_{1\gamma} u_i(t_{k-\gamma}) + \sum_{\gamma_1=0}^{\infty} \sum_{\gamma_2=0}^{\infty} \sum_{i_1=1}^{m} \sum_{i_2=1}^{m} K_{2\gamma_1\gamma_2} \\ \cdot u_{i_1}(t_{k-\gamma_1}) u_{i_2}(t_{k-\gamma_2}) \\ + \sum_{\gamma_1=0}^{\infty} \sum_{\gamma_2=0}^{\infty} \sum_{\gamma_3=0}^{\infty} \sum_{i_1=1}^{m} \sum_{i_2=1}^{m} \sum_{i_3=1}^{m} K_{3\gamma_1\gamma_2\gamma_3} \\ \cdot u_{i_1}(t_{k-\gamma_1}) u_{i_2}(t_{k-\gamma_2}) u_{i_3}(t_{k-\gamma_3}) + \cdots \end{aligned} \tag{10.3.4}$$

This result is still too complicated for practical use when written as an infinite series, and so we shall truncate it to obtain workable formulas. Convergence of the truncated forms has been discussed by Christensen [7].

There are two rather obvious and straightforward ways to truncate the series of (10.3.3):

1. Replace $\sum_{\gamma_i=0}^{\infty}$ by $\sum_{\gamma_i=0}^{M}$. This in effect replaces the time periods over which $y(t_k)$ is to be evaluated to a finite integer M. In other words, the input past is now restricted to a finite number of time periods back from t_k.
2. Restrict the maximum number of indices. This is equivalent to restricting the order of any term and thus the number of kernels.

With these changes the equations are in a manageable form and an identification algorithm can be specified. In particular, we wish to show that with the use of the measured past input and the output itself we can construct an equivalent truncated Volterra series which is a least squares output fit to the data. Of special importance will be the retention, based on statistical analysis, of only those terms in the series expansion which are meaningful or important. Thus we shall now develop an algorithm which includes a means for retaining only the important terms in the series expansion.

10.3.1 Numerical algorithm

At time $t_k = kT$ we measure the input vector $\mathbf{u}_k = [u_{1_k}, \ldots, u_{m_k}]^T$ and the output scalar. The input vectors from the previous time periods are joined together to form an augmented $m(M+1)$-dimensional input vector:

$$\mathbf{z}_k = \begin{bmatrix} \mathbf{u}_k \\ \mathbf{u}_{k-1} \\ \cdot \\ \cdot \\ \cdot \\ \mathbf{u}_{k-M} \end{bmatrix}$$

We want to express the output at time t_k as a polynomial in the elements of $\mathbf{z}_k$; that is,

$$\hat{y}_k = \boldsymbol{\theta}^T \mathbf{w}_k = \sum_{j=0}^{l} \theta_j w_{k_j} \tag{10.3.5}$$

where $\hat{y}_k$ is the predicted value of y_k, $\boldsymbol{\theta}$ is an $(l+1)$-vector of coefficients to be determined, and $\mathbf{w}_k$ is an $(l+1)$-vector consisting of all products of the powers of the elements of $\mathbf{z}_k$. Let us see how all the $\mathbf{w}_k$ can be generated. The vector $\mathbf{z}_k$ has $K = m(M+1)$ elements. A term w_{k_j} is specified as an expression of the form

$$w_{k_j} = \prod_{i=1}^{K} z_{k_i}^{a_{ij}} \tag{10.3.6}$$

where z_{k_i} is the ith element of $\mathbf{z}_k$ and $a_{ij} = 0, 1, 2, \ldots$. The relations (10.3.5) and (10.3.6) really say that $\hat{y}_k$ is expressed as a polynomial in the past and present values of the inputs with $\boldsymbol{\theta}$ as coefficients.

To define a term in (10.3.6) we must specify the exponents a_{ij}. The order of a term w_{k_j} will be the sum of its exponents. The terms are generated as follows. For $j = 0$, we have the zeroth order term:

$$w_{k_0} = 1$$

For $j = 1$, the first-order terms are (using $K = 4$ as an illustration)

$$w_{k_1} = z_{k_1} z_{k_2} z_{k_3} z_{k_4}$$

where

$$[a_{42}] = \begin{bmatrix} 1 & 0 & 0 & 0 \\ 0 & 1 & 0 & 0 \\ 0 & 0 & 1 & 0 \\ 0 & 0 & 0 & 1 \end{bmatrix}$$

For $j = 2$, the second-order terms are (still using $K = 4$)

$$w_{k_2} = z_{k_1}^2 (z_{k_1} z_{k_2})(z_{k_2}^2)(z_{k_1} z_{k_3})(z_{k_2} z_{k_3})(z_{k_3}^2)(z_{k_1} z_{k_4})(z_{k_3} z_{k_4})(z_{k_4}^2)$$

where

$$[a_{42}] = \begin{bmatrix} 2 & 0 & 0 & 0 \\ 1 & 1 & 0 & 0 \\ 0 & 2 & 0 & 0 \\ 1 & 0 & 1 & 0 \\ 0 & 1 & 1 & 0 \\ 0 & 0 & 2 & 0 \\ 1 & 0 & 0 & 1 \\ 0 & 1 & 0 & 1 \\ 0 & 0 & 0 & 2 \end{bmatrix}$$

Thus for $j = 2$ and $K = 4$ there are 10 combinations of the elements of the $\mathbf{z}_k$, and if we include up to $j = 2$, $l = 3$ (0, 1, 2); that is,

$$\begin{aligned} \hat{y}_k &= \theta_0 w_{k_0} + \theta_1 w_{k_1} + \theta_2 w_{k_2} \\ &= \theta_0 + \theta_1 w_{k_1} + \theta_2 w_{k_2} \end{aligned}$$

Therefore $\hat{y}_k$ is a weighted (by the θs) linear combination of all possible products of the past and present input data. This is a standard form for polynomial least squares fitting of data.

As with the other nonlinear identification methods, the problem with this approach is that as the number of back time periods increases, the number of polynomial terms and thus the dimension of $\boldsymbol{\theta}$ increases drastically. For example, the number of possible terms up to the lth order is

$\binom{K+l}{l}$. For practical reasons, then, it is necessary to consider at one time only a limited number, say j, of the total terms that can represent $\hat{y}_k$. Therefore we divide the total number of terms $\mathbf{w}_k$ into two categories:

$\mathbf{w}_k$ a set of p terms which are included in $\hat{y}_k$ (*in* terms)

$\mathbf{v}_k$ a set of $j-p$ terms which are not included in $\hat{y}_k$ (*out* terms) but may be candidates for being included in

We then proceed in cycles. In each cycle, the θ coefficients for the *in* terms are evaluated, as well as the correlation between the *out* terms and the regression residuals. *In* terms which seem superfluous are dropped, and promising *out* terms are turned into *in* terms. The remaining *out* terms are dropped, and new *out* terms are introduced from the list of as yet untested terms. The current equation is tested for its prediction ability. If there is no improvement over the previous cycle, terms dropped in that cycle are restored.

To start the calculation we include the first p terms as *in* and the succeeding $j-p$ as *out*. Then we carry out a linear regression between the $y(t_k)$ and the inputs to determine $\boldsymbol{\theta}$. We shall simply outline the algorithm without justifying each step. The reader may consult one of the references previously given on estimation and regression for this.

We note that

$$\hat{y}(t_k) = \theta_0 + \theta_1 w_{k_1} + \theta_2 w_{k_2} + \cdots$$

and let

$$\mathbf{M}(t_k) = \begin{bmatrix} 1 \\ w_{k_1} \\ w_{k_2} \\ \cdot \\ \cdot \\ \cdot \\ w_{kl} \end{bmatrix}, \qquad \boldsymbol{\theta} = \begin{bmatrix} \theta_0 \\ \theta_1 \\ \theta_2 \\ \cdot \\ \cdot \\ \cdot \\ \theta_l \end{bmatrix}$$

In this terminology the least squares fit for $\boldsymbol{\theta}$ becomes

$$\boldsymbol{\theta} = \left(\sum_{k=1}^{n} \mathbf{M}(t_k)\mathbf{M}(t_k)^T\right)^{-1} \sum_{k=1}^{n} \mathbf{M}(t_k)y(t_k) \tag{10.3.7}$$

where n is the number of input past (or data) values.

For $k = 1, 2, \ldots, n$ we

1. Obtain data for regression:
   a. Obtain $\mathbf{u}_k$, y_k, and form $\mathbf{z}_k$.
   b. Compute $\mathbf{w}_k$ and $\mathbf{v}_k$.
   c. Accumulate the values

$$\mathbf{B} = \sum_k \mathbf{w}_k \mathbf{w}_k^T \qquad \mathbf{b} = \sum_k y_k \mathbf{w}_k$$

$$\mathbf{C} = \sum_k \mathbf{w}_k \mathbf{v}_k^T \qquad \mathbf{c} = \sum_k y_k \mathbf{v}_k$$

$$\alpha = \sum_k y_k^2 \qquad \beta_i = \sum_k v_{ki}^2$$

2. Start regression calculations. Let $\mathbf{D} = \text{diag}\,[B_{ii}]^{1/2}$ and compute $\mathbf{E} = \mathbf{D}^{-1}\mathbf{B}\mathbf{D}^{-1}$. $\mathbf{E}$ has unit elements on the main diagonal. Find the eigenvalues, λ_i, and the eigenvectors, $\mathbf{p}_i$, of $\mathbf{E}$. Let

$$\mu_i = \begin{cases} \lambda_i^{-1}, & \text{if } \lambda_i > \epsilon_1 \\ 0, & \text{if } \lambda_i \leq \epsilon_1 \end{cases}$$

where ϵ_1 is a preset tolerance factor. Form

$$\mathbf{B}^{\#} = \mathbf{D}^{-1}\left(\sum_i \mu_i \mathbf{p}_i \mathbf{p}_i^T\right)\mathbf{D}^{-1}$$

Matrix $\mathbf{B}^{\#}$ is an approximation to the pseudoinverse of $\mathbf{B}$. The scaling accomplished by the multiplication by $\mathbf{D}^{-1}$ improves the accuracy of the computations. This is equivalent to the inverse matrix in (10.3.7).

3. Complete regression calculations. Compute [see (10.3.7)]

$$\boldsymbol{\theta} = \mathbf{B}^{\#}\mathbf{b} \qquad \text{(regression coefficients)}$$

$$\sum = \alpha - \boldsymbol{\theta}^T\mathbf{b} \qquad \text{(sum of squares of residuals)}$$

$$\sigma = \left(\frac{1}{n-m}\sum\right)^{1/2} \qquad \text{(standard deviation of residuals)}$$

$$t_i = \frac{|\theta_i|}{\sigma[B_{ii}^{\#}]^{1/2}} \qquad (t\text{-statistics for regression coefficients } \theta_i)$$

$$\rho_i = \frac{[\mathbf{c} - \mathbf{C}^T\boldsymbol{\theta}]_i}{\sum^{1/2}[\beta_i - C_{ii}]^{1/2}} \qquad \text{(correlations between residuals and } v_i)$$

Note that we compute a statistical assessment for the coefficients that are *in* and a correlation to those that are *out*.

With this information in hand we now test the predictive ability of the current model. This is done by

1. Select $k = n + 1, n + 2, \ldots, n + N$.
2. Obtain $\mathbf{u}_k$, y_k, and form $\mathbf{z}_k$.
3. Compute $\mathbf{w}_k$ and

$$\hat{y}_k = \boldsymbol{\theta}^T\mathbf{w}_k$$

$$r_k = y_k - \hat{y}_k$$

4. Accumulate $S = \sum_k r_k^2$ (sum of squares of residuals).

To evaluate the goodness of the predictions we

1. Compute the root mean square prediction error:

$$s = \left(\frac{S}{N}\right)^{1/2}$$

2. This current s is compared to the value of s obtained in the previous cycle of the whole identification.
3. If it is larger, we restore to **w** all terms dropped at the end of the previous cycle and go to step 5. Otherwise we go to step 4.
4. Remove from **w** all terms for which $t_i < \epsilon_2$ (a predetermined constant).
5. Transfer from **v** to **w** all terms for which $\rho_i > \epsilon_3$ (a predetermined constant).
6. Let m^* be the new dimension of **w**. Bring in $j - m^*$ new terms to constitute a new vector, **v**.
7. Repeat the entire cycle.

To summarize, starting with specified **w**- and **v**-vectors the parameters in (10.3.5) are calculated using a fixed amount of input past data. Since this is carried out using linear parameter estimation or regression, the statistical accuracy of the parameters can be estimated. The future prediction using the calculated model is then evaluated. Depending on the result of this prediction a new **w**-vector may be constructed by adding terms from **v** and adding new previously unused terms to **v**. In the limit as the cycles proceed the identification will converge to a **w** which yields the minimum prediction error. This is the best model in the least squares sense which has the form of (10.3.5). Extensions to multiple outputs are possible as well as revisions on the manner in which terms are discarded from **w**.

10.4 Slurry Flow in an Ore-Crushing Rod Mill

Bard and Lapidus [1] considered a system representing the flow of slurry through an ore-crushing rod mill to test this identification algorithm. Here we wish to present selected results of this calculation.

The system model represented the flow of slurry through the rod mill as equivalent to a set of four CSTRs arranged in series. A balance on the volume of slurry in the kth-tank gives

$$A_k \frac{dL_k}{dt} = V_{k-1} + Q_{k-1} - V_k - Q_k \tag{10.4.1}$$

and that on the solids

$$\frac{dS_k}{dt} = Q_{k-1} - Q_k \tag{10.4.2}$$

where L_k = slurry level above bottom of CSTR k
V_k = flow rate of liquid out of CSTR k

Q_k = flow rate of solid out of CSTR k
S_k = volume of solids in CSTR k
A_k = area of CSTR k

In addition, the effluent from CSTR k equals the slurry in the CSTR:

$$\frac{Q_k}{V_k + Q_k} = \frac{S_k}{A_k L_k} \tag{10.4.3}$$

Given initial values of L_k, S_k, V_k, and Q_k at $t = 0$ and inputs V_0 and Q_0, (10.4.1)–(10.4.3) can be integrated forward in time to yield all concentrations in any of the CSTRs.

The output was defined as

$$y(t) = \int_{t-2}^{t} Q_4(\lambda)\, d\lambda - 5 \tag{10.4.4}$$

using a fluctuation around a mean input slurry composition of 5. The input values were

$$\begin{aligned} V_0 &= 10 + \sum_{k=1}^{4} \sin(c_k t + d_k) \\ Q_0 &= 5 + \sum_{k=1}^{4} \sin(e_k t + f_k) \end{aligned} \tag{10.4.5}$$

with c_k, d_k, e_k, and f_k random numbers such that V_0 and Q_0 were random oscillating inputs.

Note that because of (10.4.4) the equations are nonlinear and that this may be thought of as a two-input (10.4.5)/one-output (10.4.4) model.

The input past was constrained to 12 sampling periods and thus $\mathbf{z}_k$ was a vector with 24 elements. The total number of terms in $\mathbf{w}$ and $\mathbf{v}$ was constrained to 56; at the start $\mathbf{w}$ contained 25 terms and $\mathbf{v}$ contained the remaining 31. These initial terms were the lower powers in the input products, but as the calculation proceeded, higher powers were included. Prediction for one time period ahead was used.

In cycle 1 of the calculation (using 25 terms in $\mathbf{w}$ and 31 in $\mathbf{v}$) the prediction standard deviation was 0.045; after 12 full cycles as many as 250 terms had been tried in $\mathbf{w}$, and with 42 terms left undiscarded, the prediction standard deviation was down to 0.0160. In this final phase of the calculation the inputs which were left *in* had standard deviations ranging from 4.796 ± 0.007 to 0.000545 ± 0.004094. As expected, the major terms remaining at the end of 12 cycles were the input past values closer to the time of prediction.

Figure 10.10 shows the final true (the system model was integrated) and predicted or identified values for the output. The fit is excellent, and at least for this example the identification algorithm seems to be a most viable one. However, as with the Wiener-type approach the model which ultimately results from this type of prediction does not shed any light on the real nature of the process being identified.

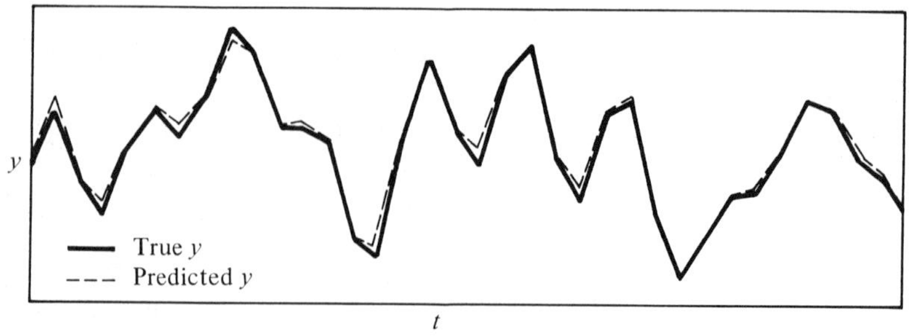

Figure 10.10. True and predicted response of slurry reactors. [Reprinted from Y. Bard and L. Lapidus, "Nonlinear System Identification," *Ind. Eng. Chem. Fund.* 9, 628 (1970). Reprinted by permission of the copyright owner.]

10.5 Summary

The underlying concept behind the use of white noise to identify a nonlinear system is that it tests a system exhaustively with a very large number of inputs and allows the formulation of a system representation which includes the response of the system to all these inputs. In a sense, the "average" system functional is determined.

In this chapter we have presented the basic Wiener identification theory, using Laguerre and Hermite function expansions, and two identification techniques, based on the general Wiener theory but more advantageous from an applications point of view, namely, the two-level input and finite Volterra series methods. There are several other formulations of the white-noise theory [2, 4, 6, 10, 15, 16, 21, 22, 25–27] which we have not presented here. From a computational standpoint, the two methods presented were chosen because of their efficiency.

Nevertheless, the nonlinear identification methods available are generally difficult and involved. The design of experiments and use of a nonlinear identification method should be preceded by a careful analysis of the problem. Once a particular system has been chosen for study, several preliminary decisions have to be made. The input and output of the system have to be defined. Then the time invariance of the system must be demonstrated. A system is time-invariant if the system response is independent of the particular time at which the input is applied. This can be ascertained easily by performing several simple experiments repeatedly at different times and comparing the results. It must also be demonstrated that the system will receive no other time-varying inputs during the course of the testing. The amplitude range of the white noise should be chosen such that it covers the most important region of response of the system. The temporal length of the experiment must be decided, together with the amount of computation required to analyze the data.

In the estimate of the temporal length of the experiment we shall need to consider the memory of the system. The memory T_m of a system has been loosely defined as the length of time that is required for the effect of the input at $t - T_m$ on the output at time t to become zero. A simple two-pulse experiment can be performed to measure T_m. The system is subjected to an impulse at time t and an impulse at time $t + \tau$. The separation τ is increased until the response of the system to the second impulse is identical to the response to the first impulse. Then τ is taken equal to T_m.

REFERENCES

1. Bard, Y., and L. Lapidus, "Nonlinear System Identification," *Ind. Eng. Chem. Fund. 9*, 628 (1970).
2. Barrett, J. F., "The Use of Functionals in the Analysis of Nonlinear Physical Systems," *J. Electron. Control 15*, 567 (1963).
3. Bekey, G. A., "System Identification—An Introduction and a Survey," *Simulation 13*, 151 (1970).
4. Bose, A. G., *A Theory of Nonlinear Systems*, Technical Report 309, Research Laboratory of Electronics, M.I.T., Cambridge, Mass., 1956.
5. Brick, D. B., "On the Applicability of Wiener's Canonical Expansions," *IEEE Trans. System Sci. Cybernetics 4*, 29 (1968).
6. Brilliant, M. B., *Theory of the Analysis of Nonlinear Systems*, Technical Report 345, Research Laboratory of Electronics, M.I.T., Cambridge, Mass., 1958.
7. Christensen, G. S., "On the Convergence of the Volterra Series," *IEEE Trans. Auto. Cont. 13*, 736 (1968).
8. Cuenod, M., and A. P. Sage, "Comparison of Some Methods Used for Process Identification," *Automatica 4*, 235 (1968).
9. Daly, K. C., and J. B. Thomas, "The Simulation of Shot Processes," *Simulation 13*, 249 (1970).
10. George, D. A., *Continuous Nonlinear Systems*, Technical Report 355, Research Laboratory of Electronics, M.I.T., Cambridge, Mass., 1958.
11. Harris, G. H., "The Identification of Nonlinear Systems with Two-Level Inputs," Ph.D. dissertation, Department of Chemical Engineering, Princeton University, Princeton, N. J., 1964.
12. Harris, G. H., and L. Lapidus, "The Identification of Nonlinear Systems," *Ind. Eng. Chem. 59*, No. 6, 66 (1967).
13. Harris, G. H., and L. Lapidus, "The Identification of Nonlinear Reaction Systems with Two-Level Inputs," *AIChE J. 13*, 291 (1967).

14. Lapidus, L., and R. Luus, *Optimal Control of Engineering Processes*, Ginn/Blaisdell, Waltham, Mass., 1967.

15. Lee, Y. W., and M. Schetzen, *Measurement of the Kernels of a Nonlinear System by Crosscorrelation*, Quarterly Progress Report 60, Research Laboratory of Electronics, M.I.T., Cambridge, Mass., 1961.

16. McFee, R., "Determining the Response of Nonlinear Systems to Arbitrary Inputs," *Trans. AIEE*, part II, *80*, 189 (1961).

17. Nieman, R. E., D. G. Fisher, and D. E. Seborg, "A Review of Process Identification and Parameter Estimation Techniques," *Intern. J. Control 13*, 309 (1971).

18. Roy, R. J., and P. M. DeRusso, "A Digital Orthogonal Model for Nonlinear Processes with Two-Level Inputs," *IEEE Trans. Auto. Cont. 7*, 93 (1962).

19. Roy, R. J., and J. Sherman, "A Learning Technique for Volterra Series Representation," *IEEE Trans. Auto. Cont. 12*, 761 (1967).

20. Sage, A. P., and J. L. Melsa, *System Identification*, Academic Press, New York, 1971.

21. Schetzen, M., *Measurement of the Kernels of a Non-Linear System of Finite Order*, Quarterly Progress Report 71, Research Laboratory of Electronics, M.I.T., Cambridge, Mass., 1963.

22. Schetzen, M., *Synthesis of a Class of Nonlinear Systems*, Quarterly Progress Report 68, Research Laboratory of Electronics, M. I. T., Cambridge, Mass., 1964, pp. 122–30.

23. Takaya, K., "The Use of Hermite Functions for System Identification," *IEEE Trans. Auto. Cont. 13*, 446 (1968).

24. Wiener, N., *Nonlinear Problems in Random Theory*, M.I.T. Press, Cambridge, Mass., 1942.

25. Wolf, A. A., "Some Recent Advances in the Analysis and Synthesis of Nonlinear Systems," *Trans. AIEE*, part II, *80*, 289 (1961).

26. Zadeh, L. A., "A Contribution to the Theory of Nonlinear Systems," *J. Franklin Inst. 255*, 387 (1953).

27. Zadeh, L. A., "On the Identification Problem," *IRE Trans. Circuit Theory 3*, 277 (1956).

PROBLEMS

10.1 Consider the linear filter system shown in Figure P10.1.

If $u(t)$ is a unit impulse, show that

$$r_0(t) = \sqrt{2}\, e^{-t}$$
$$r_1(t) = \sqrt{4}\,(-2e^{-t} + 3e^{-2t})$$
$$r_2(t) = \sqrt{6}\,(3e^{-t} - 12e^{-2t} + 10e^{-3t})$$

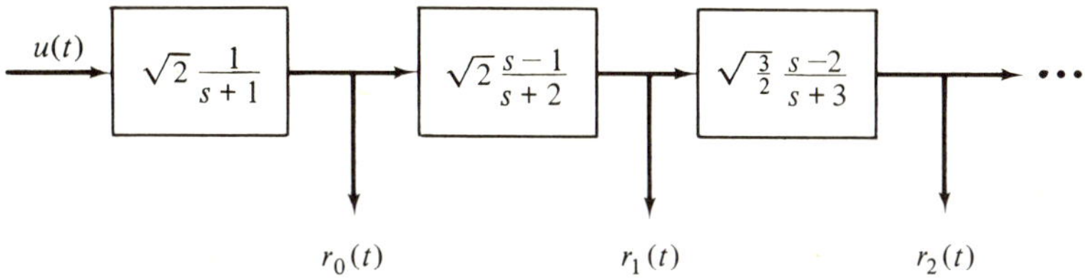

Figure P10.1. Linear filter system.

10.2 Comment on the use of the sequence filter method of generating $r_2(t)$ in Problem 10.1 as compared to generating each exponential component separately and then combining them in the proportion shown. *Hint:* Note that the maximum value of the combination occurs at $t = 0$ and is equal to $+1$. At any later time the sum has an absolute value less than 1.0, yet two of the exponential components are each an order of magnitude greater.

10.3 Show that the signals from Problem 10.1 form an orthogonal sequence over the interval $0 \leq t \leq \infty$. Note that this is equivalent to asking for the construction of orthonormal functions from the components $\{\exp[-nt]\}$, $n = 1, 2, \ldots$.

10.4 Using the model equations of Section 10.2.5 [14], apply the method of Section 10.3 to predict the system model. Use only linear and second-order terms in (10.3.5). How does the result compare with those in Figure 10.7?

Index

Index